Practical Electrodynamics with Advanced Applications

IOP Series in Emerging Technologies in Optics and Photonics

Series Editor

R Barry Johnson a Senior Research Professor at Alabama A&M University, has been involved for over 50 years in lens design, optical systems design, electro-optical systems engineering, and photonics. He has been a faculty member at three academic institutions engaged in optics education and research, employed by a number of companies, and provided consulting services.

Dr Johnson is an IOP Fellow, SPIE Fellow and Life Member, OSA Fellow, and was the 1987 President of SPIE. He serves on the editorial board of Infrared Physics & Technology and Advances in Optical Technologies. Dr Johnson has been awarded many patents, has published numerous papers and several books and book chapters, and was awarded the 2012 OSA/SPIE Joseph W Goodman Book Writing Award for Lens Design Fundamentals, Second Edition. He is a perennial co-chair of the annual SPIE Current Developments in Lens Design and Optical Engineering Conference.

Foreword

Until the 1960s, the field of optics was primarily concentrated in the classical areas of photography, cameras, binoculars, telescopes, spectrometers, colorimeters, radiometers, etc. In the late 1960s, optics began to blossom with the advent of new types of infrared detectors, liquid crystal displays (LCD), light emitting diodes (LED), charge coupled devices (CCD), lasers, holography, fiber optics, new optical materials, advances in optical and mechanical fabrication, new optical design programs, and many more technologies. With the development of the LED, LCD, CCD and other electo-optical devices, the term 'photonics' came into vogue in the 1980s to describe the science of using light in development of new technologies and the performance of a myriad of applications. Today, optics and photonics are truly pervasive throughout society and new technologies are continuing to emerge. The objective of this series is to provide students, researchers, and those who enjoy self-teaching with a wide-ranging collection of books that each focus on a relevant topic in technologies and application of optics and photonics. These books will provide knowledge to prepare the reader to be better able to participate in these exciting areas now and in the future. The title of this series is Emerging Technologies in Optics and Photonics where 'emerging' is taken to mean 'coming into existence,' 'coming into maturity,' and 'coming into prominence.' IOP Publishing and I hope that you find this Series of significant value to you and your career.

Practical Electrodynamics with Advanced Applications

Sergey Leble
Institute of Physics, Mathematics and Information Technology, Immanuel Kant Baltic Federal University, Kaliningrad, Russian Federation

IOP Publishing, Bristol, UK

ISBN 978-0-7503-2576-9 (ebook)
ISBN 978-0-7503-2574-5 (print)
ISBN 978-0-7503-2577-6 (myPrint)
ISBN 978-0-7503-2575-2 (mobi)

DOI 10.1088/978-0-7503-2576-9

Version: 20201201

IOP ebooks

British Library Cataloguing-in-Publication Data: A catalogue record for this book is available from the British Library.

Published by IOP Publishing, wholly owned by The Institute of Physics, London

IOP Publishing, Temple Circus, Temple Way, Bristol, BS1 6HG, UK

US Office: IOP Publishing, Inc., 190 North Independence Mall West, Suite 601, Philadelphia, PA 19106, USA

To my old and young students

Contents

Preface xix

Acknowledgement xxiii

Author biography xxiv

Epigraph xxv

1 Introduction **1-1**

1.1 General remarks: units 1-1
1.2 Inertial reference frames 1-2
1.2.1 Reference body: basic variables in mechanics 1-2
1.2.2 Observation of Galileo: relativity 1-3
1.2.3 Summarizing: Poincare group 1-4
1.3 Tensor fields 1-4
1.3.1 Definition of a tensor with respect to a group 1-4
1.3.2 Tensor field derivatives 1-5
1.3.3 Exercises 1-5
References 1-6

2 Basic notions and equations of electrodynamics **2-1**

2.1 Electrodynamics in vacuum 2-1
2.1.1 Definition of electric field vector $\vec{E}$ 2-1
2.1.2 Lorentz force: definition of magnetic induction field $\vec{B}$ 2-1
2.2 Maxwell's equations in integral form 2-5
2.2.1 I Maxwell's equation from Coulomb's law 2-5
2.2.2 II Maxwell equation: absence of magnetic charges 2-8
2.2.3 Faraday's law: III Maxwell equation 2-9
2.2.4 Ampère–Maxwell law: IV Maxwell equation 2-10
2.2.5 System of Maxwell's equations in differential form 2-11
2.3 Initial-boundary problem for Maxwell system in vacuum 2-12
2.4 Vector and scalar potentials 2-12
2.4.1 What possibilities give us the gauge choice?—Theoretic investigation 2-15
2.4.2 Hertz vector 2-15
2.5 Conservation principles: Poynting theorem 2-16
References 2-17

3 Electromagnetic waves in vacuum **3-1**

3.1 Wave equations 3-1

3.2 Harmonic plane wave in vacuum without charges 3-2

3.3 Wave packets 3-5

3.4 Cauchy Problem in 1 + 1 space–time 3-7

3.4.1 The initial (Cauchy) problem for plane wave with a fixed polarization 3-7

3.4.2 On projection method 3-7

3.5 Discussion and exercises 3-10

3.6 Inhomogeneous wave equation: wave generation 3-10

3.6.1 On the Green function method for wave equations 3-10

3.6.2 General formulas for electromagnetic field and wave emission 3-14

3.7 Emission of the isolated charged point particle 3-16

3.8 Emission of oscillating charged system of particles: multipole expansion 3-19

3.8.1 General remarks 3-19

3.8.2 Multipole expansion 3-20

3.8.3 The emission phenomenon in the dipole approximation 3-22

3.8.4 Next approximation: quadrupole momentum 3-23

3.8.5 Dipole magnetic moment definition and corresponding wave emission 3-25

3.8.6 Minidiscussion and exercises 3-27

References 3-29

4 Theory of relativity **4-1**

4.1 Lorentz transformation 4-1

4.2 Space–time geometry 4-4

4.3 Relativistic kinematics and four-vectors 4-6

4.4 Relativistic mechanics 4-8

4.5 Discussion 4-10

4.6 Exercises 4-11

4.7 A historical note: about a birth of new mechanics (theory of relativity) 4-11

References 4-12

5 Electromagnetic field in a matter **5-1**

5.1 Definition of vectors: polarization, electric induction, magnetization and magnetic field strength—Maxwell's equations for electromagnetic field in a matter 5-1

5.1.1 Implicit definition of the vectors: polarization, electric induction 5-1
5.1.2 On implicit definition of magnetization vector and magnetic field strength 5-3
5.1.3 Maxwell's equations for electromagnetic field in matter 5-4
5.1.4 Explicit definition of vectors: polarization, electric induction 5-5
5.1.5 On explicit empiric definition of vectors: magnetic field strength $\vec{H}$ and magnetic induction $\vec{B}$ 5-6
5.1.6 The fields measurements in experimental practice 5-7
5.1.7 Material equations as thermodynamic equations of state 5-8
5.2 Macroscopic Maxwell's equations, links to microscopic parameters 5-11
5.2.1 Resume, the empiric definition of the fields: macroscopic Maxwell's equations 5-11
5.2.2 A link between polarization vector and dipole momentum 5-12
5.2.3 A link between magnetization vector and magnetic dipole momentum 5-15
5.2.4 More comments on the definitions 5-16
5.2.5 Energy conservation/balance law derivation for matter 5-16
5.3 Classification of substances with respect to electric and magnetic properties 5-18
5.3.1 Classification of continuum matter on a base of material equations 5-18
5.3.2 Dielectrics 5-20
5.3.3 Magnetics 5-24
5.3.4 Ferromagnetism 5-27
5.3.5 Combined action: multiferroics 5-29
5.3.6 On dielectrics 5-30
5.3.7 Metamaterials 5-30
5.3.8 Exercises 5-31
References 5-31

6 Dispersion and transport 6-1

6.1 Dispersion account, operator material relations 6-1
6.1.1 Maxwell's equations: operators of dielectric permittivity and magnetic permeability 6-1
6.1.2 Energy density of wave packets in a dispersive medium 6-3
6.1.3 On Maxwell–Lorentz equations 6-7

6.2 Discussion 6-8
6.2.1 Polarization vector via a potential from coupled point charges 6-8
6.3 Dispersion in dielectrics, conductors and plasma 6-10
6.3.1 Lorentz model 6-10
6.3.2 Drude theory of metals: Ohm's law 6-12
6.4 Back to Ohm's law: Hall effect 6-14
6.4.1 On the DC Ohm law 6-14
6.4.2 Magnetic field account: Hall effect 6-16
6.4.3 Basic relations for 2D Hall resistance 6-17
6.4.4 On quantum electrodynamics manifestation of Hall effect 6-19
6.5 EM waves in isotropic conducting matter case 6-19
6.5.1 The linear equation and plane wave 6-19
6.5.2 The 1D nonlinear model outline 6-22
6.5.3 On dynamic projection method in linear problem 6-23
6.5.4 Dynamical projecting in a nonlinear problem 6-25
References 6-27

7 Plasma 7-1

7.1 Plasma types 7-1
7.1.1 General remarks: a matter as potential plasma 7-1
7.1.2 Atmosphere plasma 7-2
7.1.3 Solid state physics: electron–hole plasma 7-2
7.1.4 On stability of plasma 7-3
7.1.5 Tokamak plasma 7-6
7.2 Propagation of waves in a plasma: example of helicoidal waves 7-8
7.2.1 General remarks on plasma theoretical description 7-8
7.2.2 Simplifications and the linearized system: plasma waves 7-10
7.2.3 The initial problem formulation in matrix form: Fourier transform 7-13
7.2.4 The initial problem solution by dynamical projecting 7-15
7.3 The nonlinear case 7-17
References 7-21

8 Metamaterials 8-1

8.1 Research on metamaterials 8-1
8.1.1 Introduction: on the chapter content 8-1

8.2 Statement of problem: dispersion operator 8-3
8.2.1 Maxwell's equations: operators of dielectric permittivity and magnetic permeability 8-3
8.2.2 Boundary regime problem 8-5
8.3 Projecting operators 8-6
8.3.1 Projecting operators approach 8-6
8.3.2 Projecting operators construction 8-6
8.4 Separated equations and definition for left and right waves 8-9
8.5 Nonlinearity account 8-10
8.6 Wave propagation in a metamaterial within the lossless Drude dispersion and Kerr nonlinearity 8-12
8.6.1 Drude model for dispersion 8-12
8.6.2 Interaction of left and right waves with Kerr effect 8-13
8.6.3 Stationary problem solutions 8-15
8.7 Discussion and conclusion 8-19
References 8-19

9 Problems of electromagnetism in a piecewise continuous matter 9-1

9.1 Electro- and magneto-statics 9-1
9.2 Boundary conditions 9-2
9.2.1 Absence of surface charges and currents case 9-2
9.2.2 Conditions for magnetic and electric moments 9-4
9.2.3 Magnetic moment: spins contribution 9-6
9.2.4 Surface charges and currents 9-8
9.3 Demagnetization field 9-8
9.3.1 Instructive example 9-8
9.3.2 General demagnetization 9-9
9.4 Stray fields 9-9
9.4.1 On definition 9-9
9.4.2 Landau–Lifshits–Gilbert (LLG) equation and domain wall (DW) motion model 9-9
9.5 Microwire: DW and observations 9-11
9.5.1 Wire DW model 9-11
9.6 The stray field of the planar DW 9-12
9.6.1 Observation of domain walls in ferromagnetic microwires 9-14
References 9-15

10 Reflection and refraction of electromagnetic waves at a boundary 10-1

10.1 Reflection and transmission of a plane wave on a border 10-1

10.1.1 General textbook relations 10-1

10.1.2 Ampère–Maxwell equation for a wave with given frequency 10-2

10.1.3 Reflection and transmission of wave propagating orthogonally to a boundary 10-4

10.2 Problem of a plane wave with fixed frequency refraction 10-7

10.3 Boundary conditions impact 10-10

10.3.1 Snell law 10-10

10.3.2 To Fresnel formulas generalization 10-12

10.4 Energy density flux 10-13

10.4.1 Preparation: Pointing vector for a matter 10-13

10.4.2 Polarization choice 10-14

10.4.3 The generalized Snell law 10-15

10.5 Discussion 10-17

References 10-19

11 New dielectric guides techniques 11-1

11.1 Planar waveguides 11-1

11.1.1 Novel experiments in dielectric guides 11-1

11.1.2 Dielectric slab as a waveguide 11-3

11.2 Cylindrical dielectric waveguides 11-10

11.2.1 On the problem 11-10

11.2.2 Linear problem 11-11

11.2.3 Transformation to frequency domain 11-15

11.2.4 Projection operators in time domain 11-20

11.3 Including nonlinearity 11-21

11.3.1 Application of projection operators 11-29

References 11-30

12 Propagation of electromagnetic waves in exclusive dispersive media such as metamaterials 12-1

12.1 Electromagetic waves in metamaterial 12-1

12.1.1 On dispersion in 1D metamaterial 12-1

12.2 Directed modes in rectangular waveguides: polarization, dispersion, nonlinearity 12-3

12.2.1 Maxwell's equations for matter inside a waveguide 12-3

12.3 Boundary conditions: the transversal waveguide modes evolution 12-4

12.3.1 The boundary regime problem formulation for the transversal modes 12-7

12.3.2 Projecting operators 12-11

12.4 Rectangular waveguide filled with metamaterial: nonlinearity account 12-15

References 12-19

13 Plasma basic equations, waveguide formation 13-1

13.1 Maxwell-kinetic system 13-1

13.1.1 Joint electrodynamics—particles kinetics description 13-1

13.1.2 Kinetic equation: Vlasov plasma 13-3

13.2 Waves in homogeneous plasma 13-4

13.2.1 Cold plasma: general dispersion equation 13-4

13.2.2 Maxwell distribution background: Langmuir waves 13-6

13.2.3 More roots of the dispersion equation 13-8

13.3 Weakly inhomogeneous plasma 13-12

13.4 Plasma waveguides 13-14

13.4.1 On plasma confinement 13-14

13.4.2 The confined plasma perturbations 13-16

13.4.3 Hydrodynamic equations approach: flute instability 13-17

References 13-20

14 Helicoidal and other plasma wave phenomena 14-1

14.1 Helicoidal waves interactions 14-1

14.1.1 Basic equations 14-2

14.1.2 Introducing the projectors P_+ and P_- 14-4

14.1.3 Model with nonlinear term: the three-waves equation 14-5

14.1.4 The three-wave system in 1 + 1 case 14-9

14.2 Algebraic method of three-wave systems solution: solitons 14-9

14.2.1 Solutions derived using dressing by two-fold Darboux transformation (TfDT) 14-10

14.2.2 Solutions plots and discussion 14-11

14.3 Interaction of plasma waves 14-14

14.3.1 Interaction of Langmuir and ion waves 14-14

References 14-17

15 Diffraction in the presence of conductivity, x-rays manipulation and focusing 15-1

15.1 General remarks 15-1
15.2 Basic equations 15-3
15.3 Propagation of x-rays in vacuum 15-4
15.4 Approximation of electromagnetic field as a superposition of Gaussian beams 15-5
15.4.1 Paraxial equation for Kshevetskii–Wojda beam 15-5
15.4.2 Superposition of Gaussian beams 15-6
15.4.3 Quasi-exact solution of the Helmholtz equation 15-6
15.5 Oriented Gaussian beams method application to x-rays propagation through optical elements 15-8
15.6 Study of accuracy and efficiency of Gaussian beam methods 15-10
15.6.1 Estimation of convergence rate of solution obtained with superposition of oriented Gaussian beams to electric field described by boundary condition, which is a fast oscillating function 15-10
15.6.2 Propagation of x-rays through a lens and its aperture boundaries 15-12
15.7 Numerical calculations scheme 15-13
15.7.1 Implicit Runge–Kutta scheme 15-13
15.7.2 Algorithm: parameters of integration choice 15-14
15.8 The numerical simulations 15-15
15.8.1 General description 15-15
15.8.2 The first case, 33 aluminium lenses with 15 keV x-ray 15-15
15.8.3 The choice of space steps and errors 15-16
15.9 Results for ideal lenses and the bulk defects influence 15-18
15.9.1 Space steps choice and plots 15-18
15.9.2 Final remarks 15-19
References 15-20

16 Magnetic field dynamics, novel aspects of a theory based on Landau–Lifshitz–Gilbert equations 16-1

16.1 An exchange interaction concept 16-1
16.2 Heisenberg network dynamics 16-3
16.2.1 Heisenberg model: anisotropy 16-3
16.2.2 General continuum LLG equations 16-4

16.3 Walker theory 16-5
16.3.1 Walker solution of 1D LLG 16-5
16.3.2 Walker solution instability 16-6
16.3.3 Nanowires as Heisenberg chain 16-9
16.4 Propagation of domain wall in cylindrical amorphous ferromagnetic microwire 16-9
16.4.1 Introductory remarks 16-9
16.4.2 The LLG equation for cylindric microwires 16-10
16.4.3 LLG transforms: statement of problem 16-10
16.4.4 Basic equation in quadratic-linear approximation and its general solution 16-12
16.5 Average magnetization fields and DW dynamics 16-13
16.5.1 The third order nonlinearity account 16-13
16.5.2 Stationary background introduction 16-14
16.5.3 A linearization on a nonzero background 16-14
16.5.4 Averaging procedure and DW mobility 16-15
16.5.5 Velocity and acceleration 16-16
16.6 Exact particular solutions of LLG equation 16-16
16.6.1 Anisotropy coefficient coordinate dependence impact 16-17
16.6.2 The illustrations of DW form in 3D 16-18
16.6.3 Velocity of DW propagation: anisotropy constants determination 16-19
16.6.4 The field strength and induced magnetization by a coil 16-21
References 16-23

17 Condensed matter electrodynamics: equations of state by partition function 17-1

17.1 On derivation *ab initio* of an equation of state 17-1
17.1.1 The first law of thermodynamics forms by classical statistical physics 17-1
17.1.2 The first law of thermodynamics by quantum statistical physics 17-5
17.1.3 The scheme for two subsystems 17-5
17.2 Spin system and equations of state 17-5
17.2.1 Classical Langevin theory 17-5
17.2.2 Brillouin theory: space quantization 17-6
17.3 Heisenberg theory 17-8
17.3.1 Partition function 17-9

17.3.2 On Heisenberg equation solution 17-11
17.4 Para-, and ferro-magnetic matter 17-12
17.4.1 The magnetization curve for a paramagnetic 17-12
17.5 Problem of ferromagnetic state 17-12
17.5.1 Towards the Curie law 17-14
17.6 Multiferroics 17-19
17.6.1 Electric field action: Stark effect 17-19
17.6.2 Exchange integrals 17-20
17.6.3 Magneto-electric effect: material equation of state 17-20
17.6.4 On ferroelectricity 17-22
17.7 Fine particles case 17-22
17.7.1 Energy distribution 17-22
17.7.2 Back to statistical sum 17-24
17.7.3 Partition functions for a tiny particle 17-24
17.7.4 Numerical estimations and plot 17-25
References 17-26

18 More general material relations 18-1

18.1 A concept 18-1
18.1.1 E–D–B–H relations 18-1
18.1.2 Hydrodynamics–electrodynamics material relations 18-1
18.1.3 Continuum medium-electrodynamics material relations 18-2
18.1.4 Energy balance 18-4
18.2 Symmetry and groups 18-4
18.2.1 Crystallographic symmetry and groups 18-4
18.2.2 Tensorial symmetry with respect to indexes transpositions 18-5
18.2.3 Pauli symmetry with respect to electrons permutations 18-5
18.3 Euclidean and Lorentz symmetry 18-5
18.3.1 Euclidean group covariance 18-5
18.3.2 Lorentz group covariance 18-6
18.3.3 General tensors relations 18-7
18.4 Active dielectrics 18-8
18.4.1 Ferroelectrics 18-9
18.4.2 Piezoelectricity 18-9
18.4.3 Paraelectricity 18-10
18.4.4 Magnetoelasticity 18-10
18.5 Flexoelectricity 18-11

18.6 Ferroelasticity 18-11
References 18-12

19 On direct and inverse problems of electrodynamics 19-1

19.1 Direct problem of plane wave propagation in a layered medium 19-1
19.2 On inverse problem 19-3
19.2.1 Remarks 19-3
19.2.2 Some details of problem formulation 19-3
19.2.3 Equations of the inverse problem 19-5
19.3 Data collection methods: examples 19-6
19.3.1 Plasma Langmuir probe 19-6
19.3.2 Radar 19-9
19.3.3 Huygens' and Kirchhoff' formulas 19-10
19.3.4 Direct and inverse problems for a radar/lidar 19-12
19.4 Inverse problems as ill-posed one 19-13
19.4.1 The Tikhonov regularization 19-13
References 19-15

20 Advanced exercises 20-1

20.1 Short list of useful vector and tensor relations 20-1
20.2 A few definitions: curves, surfaces, integrals, etc 20-2
20.2.1 Curves 20-2
20.2.2 Surfaces 20-2
20.2.3 Integrals 20-3
20.2.4 Dirac delta-function 20-4
20.3 Projecting operators 20-6
20.4 Dressing method 20-6
20.5 Dielectric waveguides 20-7
20.5.1 Dielectric slab 20-7
20.5.2 Dielectric cylinder—optical fibers 20-7
20.5.3 Rectangular waveguide 20-8
20.6 Electromagnetic waves in metamaterials 20-8
20.6.1 Electromagnetic waves in metal rectangular waveguide, system derivation 20-8
20.7 Plasma confinement 20-10
20.8 Wave propagation at plasma 20-10
20.9 Refraction in presence of conductivity 20-11

20.10 Magnetism, a novel aspect 20-11

20.11 Condensed matter electrodynamics: equations of state by partition function 20-13

20.11.1 Paramagnetics and ferromagnetics 20-13

20.11.2 Multiferroics 20-13

20.12 General material relations 20-13

20.12.1 Piezoelectricity 20-14

20.13 Inverse problems of electrodynamics 20-14

References 20-14

Preface

There are rather many good textbooks on electrodynamics, different in style and approach.

One should be distinguished in a separate line—a splendid and comprehensive book of Jackson (Classical Electrodynamics. Third Edition 814 p, based on the University of California course [1]) and one—very comprehensive by Novozhilov and Yappa—my teachers—from St. Petersburg State University (352 p, St. Petersburg University Press [2]). Moreover the recent textbooks, e.g. of Bo Thidé Electromagnetic Field Theory (available http://www-f9.ijs.si/rok/sola/merjenja2/doc/EMFT_Book.pdf; Uppsala University) and of I Toptygin at which the electrodynamics is lifted up its quantum version [3]—with a great old exercise book; some papers-published, some other ones 'hanging' in the Internet. The exercises are presented with solutions! That is important, because the majority of problems are complicated.

As is typical for courses in theoretical physics, all accessible electrodynamics exercises are rather complicated, with extent solutions. So one of my important intentions is to suggest the simplest possible problems, which formulations containing answers for the main question: how to start to build a solution.

In Poland, where I have lectured in electrodynamics for 24 years, few original books on the field are issued. I would mention two.

Different aspects of phenomena related to interaction of charges via electromagnetic field are covered from the mentioned 'classic' ones (e.g. 'Elektodynamika klasyczna' [4]) with many details of charges collision theory up to the technical electrodynamics with its 'metal' content [5].

The general abundance of electrodynamic textbooks is quite visible via the Internet.

We however see a reason for writing the next one.

First, the significant parts in the abundance of textbooks are rather old (they may be very good!), in them, e.g., there no information on the topics in which important advances have been recently made:

1. New fiber optic techniques, including: possible applications in future computer techniques in the light use; nonlinear effects in dispersive media [6, 7].
2. Not much information about the propagation of electromagnetic waves in exclusive dispersive media, such as metamaterials with important modification of energy relations in conditions of negative dielectric permittivity and magnetic permeability [8].
3. Plasma confinement needs introduction, explanation and a discussion [9].
4. Diffraction in the presence of conductivity, x-rays manipulation and focusing.
5. Magnetism, such a novel aspect of a theory based on the Heisenberg theory, model and Llandau–Lifshitz–Gilbert equation [9].
6. More general material relations, e.g. for multiferroics [10].

7. Not enough material common with inverse problems, including information on the principles of the theory of radars (lidars) data processing.

Second, this textbook is maximally adapted to Polytechnic aims and scope, therefore, we could call it 'Applied Electrodynamics', but stopped on the more precise 'Practical Electrodynamics', by hint of the Editor (my thanks to Ashley Gasque!). 'It fills the gap between' classical 'and technical' descriptions; the word 'applied' means first of all applications in widely understanding physics, having in mind methods of measurements, data processing, monitoring, tending, for example, to such things as tomography. We give a detailed description of the electric and magnetic fields definitions paying attention to the practicality of the page to determine the basic sizes that lead to creating tables of fundamental physics, and, in measurements in geophysics, astrophysics, etc. So, besides the practical tools, we include both the theoretical basis of polarization and magnetization vectors introduction (definition!) in implicit ways as well as explicit one, that finally leads to direct determination via measurements.

The *third* original thing—we are going to rely only on the vectors themselves of electromagnetic field, avoiding the use of potentials. We want to show that it is possible and didactically 'economical'. We, however, show and compare it with the 'potentials' alternative, to demonstrate the preferences.

Further, each of the textbooks on electrodynamics contains a chapter on the theory of relativity, but there are shortages in the historical review and discrepancies in the very derivation of the Lorentz transformation. The deriving here, above all, follows the idea of Lorentz itself, operating on Maxwell's system. We follow the original Lorentz idea of the basic system covariance with respect to the space–time transformations, deriving the linearity of the transformations and its explicit form as a function of inertial reference frames relative to velocity.

We also want to pay attention to some mathematical aspects, necessary for modeling complicated phenomena, including—nonlinear, or when you cannot bring the Maxwell system into a linear (scalar) equation, say—differential dispersion relation. It is about such problems, for example as diffraction.

In the conventional student's manual, the technical university issue is subordinate formulas: energy conservation law, dipole field, point particles, etc. So we cannot avoid reproducing these patterns, but we try to serve in the spirit of the applied, again better to say, practical electrodynamics.

We also plan to not only introduce the basic notions within some of our own choice, but also to compare the definitions with other sources, discussing the relevant peculiarities.

For these definitions which there are in the literature, in the case of didactic discrepancies, we will quote the corresponding references. We refer to the list (literature), which is given at the end of each chapter as a guide. In the text, the discussion is separated as similar chapters (e.g., cf).

For example, take a definition of the magnetic field, which basic vector is introduced as:

1. A force acting on a magnetic point charge, which, with railway—physical—is treated as an effective dipole charge magnetic—magnetized thin wire (Vlasov [11]).
2. The natural definition via Lorentz force [2].

Some chapters, not included in the exam, but important for a perspective education, are essentially elevated to the second, advanced part of this book. At the end of each subsection we give a short review of the literature on basic definitions, steps in formulation of equations, derived patterns in illustrations. We also give simplest exercises with recommendation for algorithmic solution.

First of all there are a lot of differences in academic hours at universities. For example at Gdansk University—60, at Warsaw University—100, at Warsaw Politechnika—40, Gdansk University of technology—30. The same all over the world.

So we support a medium amount of academic hours, about 45 lectures/45 exercises and corresponding not very big text (about 200 pages).

We also plan to give a good reference list (a monograph style) for advanced students, that want to go deep into fundamentals (see e.g. [12]) or application [6, 7, 10, 13].

There are few books with a title that contains the word 'Applied':

1. Foundations of Applied Electrodynamics, 522 p, by Wen Geyi Waterloo, Canada.
2. Applied Electromagnetism by Gordon McKay School of Engineering and Applied Sciences, Harvard University.
3. AEP 5570—Applied Electrodynamics. Fall. 4 credits. Letter grades only.

Prerequisite: at the level of Jackson's Classical Electrodynamics. R Lovelace. Cornell University.

References

[1] Jackson J D 1998 *Classical Elektrodynamics* 3rd edn (New York: Wiley)

[2] Novozhilov Y V and Yappa Y A 1978 *Electrodynamics* (*Elektrodinamika*, in Russian) (New York: Academic)

[3] Batygin B B and Toptygin I N 2005 *Modern Electrodynamics (Sovremennaya elektrodinamika)* (Moskva, Izhevsk: IKI) ch 1

Batygin V and Toptygin I 1965 *Problems in Electrodynamics* (London: Academic)

[4] Ingarden R and Jamiolkowski A 1980 *Elektrodynamika klasyczna* (Warszawa: PWN)

[5] Turowski J 2014 *Elektrodynamika techniczna* (Warszawa: PWN)

[6] Leble S B 1991 *Nonlinear Waves in Waveguides* (Berlin: Springer)

[7] Leble S B 2002 *Nonlinear Waves in Optical Waveguides and Soliton Theory Applications* (Berlin: Springer)

[8] Leble S and Perelomova A 2018 *Dynamical Projectors Method in Hydro- and Electrodynamics* (Boca Raton, FL: CRC Press)

[9] Leble S 2019 *Waveguide Propagation of Nonlinear Waves Impact of Inhomogeneity and Accompanying Effects* (Berlin: Springer)

[10] Spaldin N A and Fiebig M 2005 The renaissance of magnetoelectric multiferroics *Science* **309** 391–2

[11] Vlasov A A 1955 *Makroskopicheskaya elektrodinamika* (Moskva: Gosudarstvennoe Izdatelstvo Tehniko-teoreticheskoj literatury)

[12] Fock V and Kemmer N 1964 *Theory of Space, Time and Gravitation* 2nd rev edn (Oxford: Pergamon)

[13] Fock V A and Armstrong J C 1967 Electromagnetic diffraction and propagation problems. International series of monographs in electromagnetic waves, vol 1 *Am. J. Phys.* **35** 362

Acknowledgement

A version of the base part of this manual was written (in Polish) by myself with the assistance of my student/PhD student Damian Rohraff. His notes of my lectures were used in formation of the preliminary version of the base part of this electrodynamics. He also partially made a set of illustrations.

I am using the patterns to some extent. I want to thank him for that and for important drums on the didactic topics of electrodynamics. He also contributed to the compilation of the historical origin of the relativity theory.

The text, however, underwent a complete transformation. From work and authorship Mr Damian Rohraff resigned because he wanted to focus on other business.

When writing the chapters on 'dielectric waveguides' we used the text of the PhD thesis by student/PhD Bartosz Reichel, discussions with him during the performance of the work (later—PhD dissertation) were very interesting and fruitful. The chapter on x-ray propagation and focusing was based on results of Sergey Ksheveskii and Pawel Wojda, whose assistance was helpful. The final sections of this chapter and numerical simulation were written on the base of the joint publication/master thesis of Mahmoud Elsawy, my cordial thanks for the collaboration. Similar thanks to N Chychkalo, our joint efforts with her master dissertation/publications were helpful for writing about '*ab initio*' theories, that result in material relations in magnetism.

I would also like to thank the students/co-workers of the faculty of physics of the Technical University of Gdansk and of Emmanuel Kant Baltic Federal University (kaliningrad) for comments on patterns and textures of the book.

Author biography

Sergey Leble

Prof Dr hab Sergey B Leble graduated from Leningrad State University in 1968 in theoretical physics, and in 1974, he recieved his PhD on the theory of elementary particles in spaces with curvature. He became senior lecturer in 1974, professor in 1988 and head of the theoretical physics department in 1989, all at Kaliningrad State University. In 1994 he was a visiting professor at Salamanca University (Spain). In 1995 he was professor of Gdansk Univ. of Technology (GUT, Poland), 2010 Head of differential equations and applied mathematics dep.,GUT, 2014 Professor at Atomic, Molecular and Optical Physics Department, and 2016 Professor of Emmanuel Kant Federal University (Russia, Kaliningrad), along with being a visiting professor (one month) at a few European Universities.

Prof Dr Leble has taught general and specialized courses in all areas of theoretical physics, assisted a number of graduate students to complete their PhD theses. He has also led several research groups, carrying out scientific research on topics in these broad areas, mostly on a contract basis with the Academy of Sciences of the USSR, the Russian Federal Property Fund grants, the Ministry of Education.

He has written chapters in collective monographs, authored numerous monographs, and contributed to review articles along with publishing approximately 200 scientific works in the field of theoretical and mathematical physics, as well as applied mathematics.

Epigraph

1. To base

   'Interpretation of light as electromagnetic phenomenon was so brave that it was most impressive for myself from all the things I met before.'

   —H Lorentz

2. To 'advanced'

   Высокий жар бесстрастных интегралов
   Один владел и сердцем, и умом.

   Алексей Лебедев, `Артиллерийская таблица.'

   High fever of impassive integrals
   Thou one possessed both heart and mind.
   Alexej Lebedev, Artillery table.

   (Google + myself translation)

   (И снова мысль боролась и искала,
   И в тишине, в безмолвии ночном)

Practical Electrodynamics with Advanced Applications

Sergey Leble

Chapter 1

Introduction

1.1 General remarks: units

Starting with the subject, under the name 'electrodynamics', we suppose that a reader passed exams from 'mechanics' and 'electricity and magnetism'. It would also be good to get through analytical mechanics and hydrodynamics to complete the set of basic notions of velocity and other fields. The division of physics and especially the electromagnetic science to such subjects is a tradition along which the basic variables and notions are introduced directly via experiment. In such 'first attack' the charge, current and fields are introduced in terms of scales (etalons), which construction and procedure of comparison is specified [4].

There are well-known, from primary school, SI units for mechanics: for a length 1 *meter* [m], the *second* [s] for time; the *kilogram* [kg] for mass and, finally the **newton** [N] for a force. Such quantities completely cover necessities of the second Newton law, but force formulas (Lorentz force) need electric charge and electromagnetic field units. Turning to thermodynamics naturally needs temperature.

Remark: quite recently a decision of the physicists' community https://www.bipm.org/en/the-si/ decided to determine principle units via universal constants, velocity of light in vacuum c for the 'meter', Plank constant h (mass) and electron electric charge e (current). Time is measured via hyperfine transition frequency of the cesium atom.

The charge is expressed in *coulomb*s [C], which conventionally is defined via a 'process of charging'—therefore it is directly linked to the *ampere* as a unit of a current. So, the unit electric charge is defined as the charge transported by a constant current of one ampere in one second: the fields $\vec{E}$ and $\vec{B}$ are defined via forces. The electric force per unit charge is $\vec{E}$, probing a point particle while the vector of magnetic induction $\vec{B}$ is defined via Lorentz force in more complicated scheme (see next section). Note a link to intensity of luminous radiation (candela).

doi:10.1088/978-0-7503-2576-9ch1

Remark: the characteristics of electromagnetic field in a matter need more vectors as polarization $\vec{P}$ and magnetization $\vec{M}$, see section 5.1.

There is another system of units, Lorentz–Heaviside, whose we use through this textbook.

Generally a choice of the unit system needs a complete formulation of electromagnetic and mechanical phenomena description see e.g. the fundamental book [2]. We will return to this question when a kind of formalism completeness would be achieved.

In the thermodynamic aspects of the theory the temperature unit Kelvin [K] is introduced (in the novel scale—Boltzmann constant), also mole [mol] as an 'amount of matter' (Avogadro constant) linked to mass unit.

1.2 Inertial reference frames

1.2.1 Reference body: basic variables in mechanics

The position and motion of a body is defined in relation to a reference frame (reference body + measurements tools). The fundamental scales of basic variables (MKS) are defined also with respect to such systems and include a comparison of the variables with the scales.

There is a special class of reference frames: inertial ones in which the mechanics is formulated in the simplest form.

Definition. Such a frame, in which a mass-point has a constant velocity if the sum of all forces acting at the point is zero, bore the name of inertial reference frame (IRF)

$$\vec{v}(t) = \frac{d\vec{r}}{dt} = \vec{\text{const}}. \tag{1.1}$$

We mean a class of such experiments in this frame: for any point mass for which $\vec{R} = \sum \vec{F}_i$ the velocity $\vec{v}$ of this body is constant.

A thesis that such reference frames exists (as experimental fact) in nature *is equivalent to* ***Newton's 1st law***, sometimes we name it the inertia law. It means that for many experiments described in the definition (with a controlled error) we have such a result.

Such RFs may be in a good approximation linked to stars (more exactly with a planet or star system mass center). Recent hypothesis about the motion of stars with acceleration needs very specific analysis from the definition and the law point of view.

In the strict sense and spirit of the definition the Earth is not a body with whom an inertial frame may be associated. The experimental proof of this fact one can find in analysis of Foucault pendulum (Rys. Figure 1.1).

Figure 1.1. 'Vous êtes invités à venir voir tourner la terre… ', 'Come and see as the Earth is rotating'—Jean Bernard Léon Foucault. (In the photo the Foucault pendulum in the main building of Gdan'sk University of technology (Politechnika Gdan'ska).)

We however often relate experiments, including electromagnetic, to the Earth as a reference body, considering it as an inertial one (neglecting 'Coriolis' forces).

Important note. The philosophy. From the 'metaphysical', philosophical point of view, it is impossible to speak about the only IRF, the scientific logic implies a comparison of results of experiments made at least in different places and times. Some equivalence of such reference frames is also desirable. It declares a possibility to recalculate results of measurements from one RF to another.

1.2.2 Observation of Galileo: relativity

A reference frame that moves with a constant velocity with respect to an inertial frame also is the inertial one.

Such observations are known under the title of the 'principle of relativity'. The short modern formulation of the principle looks as:

In every IRF all (physical) phenomena are described as equivalent.

It lies in a foundation of Maxwell theory [1] formulation, which is in a sense self-consistent. The experimental establishing (derivation) of the Faraday equation (one of the Maxwell system) introduces in **a IRF** a constant that is nothing but the velocity of light, while the system itself has the wave equation with the constant as velocity of propagation as the direct corollary.

An important enforcing of the observations related to IRFs in electrodynamics have been the Michelsona–Morley experiment [3]; which compares the velocity of electromagnetic waves propagation in places and velocities that correspond to the opposite phases of Earth rotation around the Sun. The identity of results confirms the hypothesis of independence of the light velocity on the reference frame moving one related to another.

1.2.3 Summarizing: Poincare group

All this means—invariance of a form of mathematical physics relations with respect to transformations from IRF S_1 to another IRF S_2, in other words, equations and relations of mathematical physics should have tensor form relatively the group of transformation [5]. The form of such transformations ($\vec{r}$, t − > $\vec{r}'$, t') was written by A Poincare, the most important ones for moving IRFs were established by Lorentz on the base of the complete set of Maxwell's equations covariance analysis (see section 4.1).

Finally, we declare an introduction of a physical variable with respect to a class of RFs, which is specified by a local Lorentz group of space–time (ST) transformations. The notion could be widen to a set of RFs linked by shifts in the ST, introducing already a corresponding field. Together the fundamental transformations form the Poincare group. The postulate of existence of such a set of equivalent IRF is considered as the modern **1st Newton Law formulation.**

An extension to a supersymmetry approach can be found in [6].

1.3 Tensor fields

1.3.1 Definition of a tensor with respect to a group

Definitions with respect to the group of rotations in 3D

It is important to stress that even the definition of physical fields via a force (related to a RF) needs the condition of inertiality. In fact, the absence of so-called inertial forces guarantee universality of such fields notion.

A group of transformations G that link different RFs is a key notion in definition of a tensor and tensor field. A good example is a rotation group $O^+(3)$ in R^3 as a group of 3 × 3 real matrices $A \in O^+(3)$ such that $A^T A = I$, $detA = 1$. It is easy to check the properties for a rotation matrix in the xy plane,

$$\begin{aligned} x' &= x \cos\alpha + y \sin\alpha, \\ y' &= -x \sin\alpha + y \cos\alpha, \\ z' &= z. \end{aligned} \tag{1.2}$$

It changes direction of Cartesian axis for angle α. A vector $\vec{V}$ components $V_x = V_1, \dots$ transform in accordance with

$$\begin{aligned} V_1' &= V_1 \cos\alpha + V_2 \sin\alpha, \\ V_2' &= -V_1 \sin\alpha + V_2 \cos\alpha, \\ V_3' &= V_3. \end{aligned} \tag{1.3}$$

Similar matrices transform coordinates and vector components with respect to arbitrary axis. There are two convenient forms for such transformations, the coordinate one

$$V_i' = A_{ik} V_k, \tag{1.4}$$

where summation by repeated index k is implied. The alternative matrix one

$$V' = AV, \tag{1.5}$$

where V, V' are columns with three elements and A is a square 3×3 matrix.

A tensor of the second rank with the components T_{ik} with respect to $O^+(3)$ is defined by the transformations

$$T_{ik}' = A_{il} A_{km} T_{lm}. \tag{1.6}$$

So, *if a set of quantities T_{ik} after transition to another RF linked with the original by* (1.6) *it represented a second rank tensor.*

Similar, a tensors of higher ranks are defined.

A generalization of the tensor notion to another group, say the Lorentz group L, is achieved by a simple change of the matrix A at equations (1.4) and (1.6) to a matrix $\Lambda \in L$. Some more details are given in section 4.2.

Lorentz group tensor in 3 + 1 space

Let us introduce the vector x components x_α, $\alpha = 0, 1, 2, 3.$ as notations in a ST, so, that $x_0 = ct$ for time variable. Then, for $\Lambda \in L$

$$V_\alpha' = \Lambda_{\alpha\beta} V_\beta \tag{1.7}$$

defines a four-vector in the ST. The tensors are defined by direct analog of equation (1.6).

1.3.2 Tensor field derivatives

Next, a vector $V_\alpha(x)$ as functions in the ST stand for a vector field in the ST. It is easy to prove that

$$\frac{\partial V_\alpha'}{\partial x_\beta'} = \frac{\partial V_\alpha'}{\partial x_\gamma}\frac{\partial x_\gamma}{\partial x_\beta'} = \Lambda_{\alpha\delta}\frac{\partial V_\delta}{\partial x_\gamma}\frac{\partial x_\gamma}{\partial x_\beta'} \tag{1.8}$$

form the second rank tensor in the ST. The derivatives $\frac{\partial x_\gamma}{\partial x_\beta'}$ are calculated by means of the same four-vector x_β' transformation (1.7). The higher rank tensor are differentiating similar.

More details are written in the sections 4.1 and 4.2 .

1.3.3 Exercises

(1) Write the transformation of the second rank tensor $\mathcal{T}$ in the space–time, with components $\mathcal{T}_{\alpha\beta}$ in the explicit form (see equation (1.6)).

(2) Prove that the set of derivatives (1.8) are components of the second rank tensor in the space–time.

References

[1] Maxwell J C 1865 A dynamical theory of the electromagnetic field *Philos. Trans. R. Soc. Lond.* **155** 459–512

[2] Jackson J D 1998 *Classical Electrodynamics* 3rd edn (New York: Wiley)

[3] Michelson A and Morley E 1887 *Am. J. Sci.* **XXXIV** 334–45

[4] Roller D E 1954 *The Development of the Concept of Electric Charge: Electricitry from the Greeks to Coulomb* (Cambridge: Harvard University Press)

[5] Fock V and Kemmer N 1955 *Theory of Space, Time and Gravitation* 2nd rev edn (Oxford: Pergamon)
Fok V A 1955 *Teoriya prostranstva, vremeni i tyagoteniya* (Russia: Moskva)

[6] Duplij S, Goldin G A and Shtelen V M 2007 Lagrangian and non-Lagrangian approaches to electrodynamics including supersymmetry *XXVI Workshop on the Geometrical Methods in Physics, (Bialystok), AIP Conf. Proc.* vol 956 ed. P Kielanowski, A Odzijewicz, M Schlichenmaier and T Voronov (College Park, MD: AIP) p 50

IOP Publishing

Practical Electrodynamics with Advanced Applications

Sergey Leble

Chapter 2

Basic notions and equations of electrodynamics

2.1 Electrodynamics in vacuum

It is known that the electromagnetic field appearance is a direct corollary of the charges presence, but the charges act across the domains without matter. Such domains we name 'vacuum' and study the electromagnetic fields description in them.

2.1.1 Definition of electric field vector $\vec{E}$

Let us start from a definition of electric field with respect to a given inertial reference frame (IRF) as a force acting for unit point charge, which conventionally is associated with the formula

$$\vec{E} = \frac{\vec{F}}{q}. \tag{2.1}$$

Such definition obviously implies a measurement of the force acting on a charged body whose dimension is much less than the field inhomogeneity scale. The force in non-relativistic theory forms a 3D (R^3) vector with respect to rotation group $O^+(3)$. The charge is considered as a scalar with respect to the same group [4]. The space–time nature of the fields is drastically different, see chapter 1 and section 20.1, and will be established in chapter 4 and section 20.1.

2.1.2 Lorentz force: definition of magnetic induction field $\vec{B}$

There are many modes in the magnetic field definition in textbooks, while the electric field is introduced in them in the same manner: equation (2.1). All of them are intimately connected with the Lorentz force formula

$$\vec{F}_B = \frac{q}{c}[\vec{v} \times \vec{B}]. \tag{2.2}$$

doi:10.1088/978-0-7503-2576-9ch2

One of the basic properties of the magnetic force is the fact that its work is always zero, because the scalar product $(\vec{F}_B, \vec{v}) = 0$, and the displacement $d\vec{r}$ is proportional to $\vec{v}$. A magnetic field does not change velocity module!

Let us mention here the following approaches;

i) *Via a force acting on 'a point magnetic charge' practically—a thin magnetized needle* [1].

In nature we have not found isolated magnetic charges (often named magnetic monopoles, see discussion in [7]) but there exist particles with natural magnetic momentum (e.g. electron (spin) magnetic momentum). As the magnetic momentum could be calibrated by means of a standard body with (scale, etalon) magnetic moment.

ii) *By means of Lorentz force expression which acts to a point particle. The (non-explicit!) formula (18.12) see, e.g. the textbook* [2].

iii) *In the book* [8] *the definition of the field $\vec{B}$ (to be precise—$\vec{H}$) is given* via *Lorentz force for a 'small interval of current' $d\vec{j}$, namely, to find the direction of the vector $\vec{H}$ we choose the direction of the vector $\vec{j}$ such that the force to be minimal.*

Next, we measure the force value with account of sign. Comment: such introduction allows to determine $\vec{H}$ in the theoretical manner, while a practical realization would need an isolation of the current element $d\vec{j}$. A difficulty of the field direction definition also looks obvious.

iv) *One may define the field $\vec{B}$ by means of a frame (loop) with current of known magnitude measuring torque ($\vec{N}$) of magnetic force. The forces acting upon the loop are equal and opposite, they both act to rotate the loop in the same direction. The method is used in practice in laboratories [9]. If the dipole momentum $\vec{\mu}$ is introduced (it is perpendicular to the current loop, its module is the product of current I and area S: IS), we have*

$$p = \vec{\mu} \times \vec{B}.$$

v) *The measurement of large magnetic fields on the order of a Tesla is often done by making use of the Hall effect. A thin film Hall probe is placed in the magnetic field and the transverse voltage (of the order of microvolts) is measured.* In such practice a thin copper film of thickness d of the order of 100 μm is used for a Hall probe. Taking the charge carrier density to be n (e m^{-3}), magnetic induction B (T), d (m) [10]. The Hall voltage (Volt) is

$$V_H = \frac{IB}{ned}.$$

The formula in fact is a direct corollary of the Lorentz force formula (see chapter 6, section 6.4).

Let us analyze definition (ii). To introduce a magnetic field definition, we choose a point particle with a charge q, which is moving in the field $\vec{B}$. It is under action of the magnetic force (Lorentz (2.2), also (18.12)) $\vec{F}_B$, which depends on the vector $\vec{v}$, so that its module value may change from 0 to maximum value, which we mark as $\vec{F}_{B\max}$. Let us note again that the magnetic force vector is always orthogonal to the particle velocity vector: $\vec{F}_B \perp \vec{v}$.

Let us take the vector product of the velocity vector and some vector which is specific for the magnetic field ($\vec{B}$) by the formula (2.2):

$$\vec{F}_B = \frac{q}{c}[\vec{v} \times \vec{B}]. \tag{2.3}$$

Let us mention that $\vec{B}$ was introduced by Lorentz [11].

This expression, (2.3), in fact defines the vector implicitly.

We use this expression (2.3) for the Lorentz force in Gauss units [12]. A choice of constants in this expression and in the Coulomb law defines the units system (see, e.g. [2]).

In accordance with equation (2.3), the direction of the magnetic induction vector $\vec{B}$ coincides or is opposite to the vector of a probing particle velocity if $\vec{F}_B = 0$. It allows to establish the line of the vector $\vec{B}$. One can choose the reference frame, such as the probing particle that would move along orthogonal to the vector $\vec{B}$ line whence the force module would have maximal value (along equation (2.3)). Formally, it could serve as the definition of magnetic field:

$$B = \frac{F_{B\max}}{qv}. \tag{2.4}$$

Because $\vec{B}$ enters the expression for (2.3) in implicit form, we should give explanations. The analysis looks as follows via equation (2.3) for a field $\vec{B}$ acting on a point charge.

In a case, when $\vec{v}$ is not parallel to $\vec{B}$, after projecting the vector equality (2.3), to Cartesian axes one has an algebraic system

$$\begin{aligned} F_{Bx} &= \frac{q}{c}[\vec{v} \times \vec{B}]_x = \frac{q}{c}(v_y B_z - v_z B_y), \\ F_{By} &= \frac{q}{c}(v_z B_x - v_x B_z), \\ F_{Bz} &= \frac{q}{c}(v_x B_y - v_y B_x). \end{aligned} \tag{2.5}$$

Suppose components of the force and velocity components are measured (known). The conventional way to solve equation (2.6) by means of Cramer's rule, fails!, because the system matrix determinant (denominator of the Cramers formula) is zero.

It means that the algorithm of the vector $\vec{B}$ determination should be different and consists of two stages:

1. Let the probing particle with the charge q fly along axis x, then

$$
\begin{aligned}
F_{Bx} &= \frac{q}{c}[\vec{v} \times \vec{B}]_x = \frac{q}{c}(v_y B_z - v_z B_y) = 0. \\
F_{By} &= \frac{q}{c}(v_z B_x - v_x B_z) = -\frac{q}{c} v_x B_z. \\
F_{Bz} &= \frac{q}{c}(v_x B_y - v_y B_x) = \frac{q}{c} v_x B_y.
\end{aligned} \tag{2.6}
$$

Those relations may give the components B_y and B_z in terms of F_{By} and F_{Bz} as follows

$$
\begin{aligned}
B_z &= -\frac{c}{qv_x} F_{By}. \\
B_y &= \frac{c}{qv_x} F_{Bz}.
\end{aligned} \tag{2.7}
$$

2. In the NEXT experiment, pushing the particle along y, with a velocity v' in a similar way one has B_x

$$
B_x = \frac{c}{qv'} F_{Bz}, \tag{2.8}
$$

via F_{Bx} measurement. It means that such measurement/definition may be used if field time dependence allows quick registration of both velocities in the vicinity of the same point.

The properties of magnetic field (e.g. ability to form a circumference trajectory without velocity module change) is often used in physics and techniques, see figures 2.1–2.3. Most known exemplary devices are the: Wilson chamber (1912) (Figure 2.1), cyclotron (1932) [13] (Figure 2.2), tokamak (1950) [14] (figure 2.3).

The full description of electromagnetic force on a point particle with a charge q may be presented as the sum

$$
\vec{F}_L = q\left(\vec{E} + \frac{1}{c}[\vec{v} \times \vec{B}]\right). \tag{2.9}
$$

The force $\vec{F}_L$ is named as Lorentz. The relation (2.9) is introduced by experiment analysis. As a classic example one can take the scholar example of two thin metal stripes (or cables) with current either unidirectional or moving in opposite directions [13].

After definition of the electromagnetic field, one may go to the equations for the fields. A system of equations for all field components is referred to as Maxwell's equations [15, 16], which expresses four basic laws of electromagnetism, namely:

- Coulomb's law (1785 r.),
- Biot–Savart's law (1820 r.), together with absence of magnetic charges!
- Faraday's law (1831 r.),
- Ampère's force law (1825),
- Ampère–Maxwell equation (1864 r.).

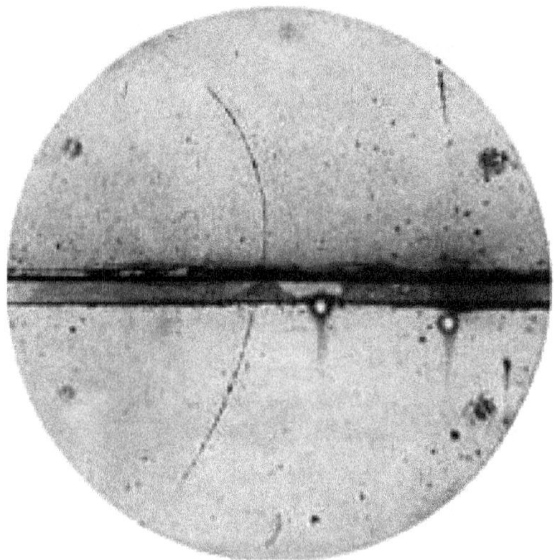

Figure 2.1. Cloud chamber photograph used to prove the existence of the positron. Observed by Carl D Anderson (Wiki, https://en.wikipedia.org/wiki/Cloud_chamber). Traces of charged particles in the Wilson chamber. One can see the curved trajectories under magnetic field action.

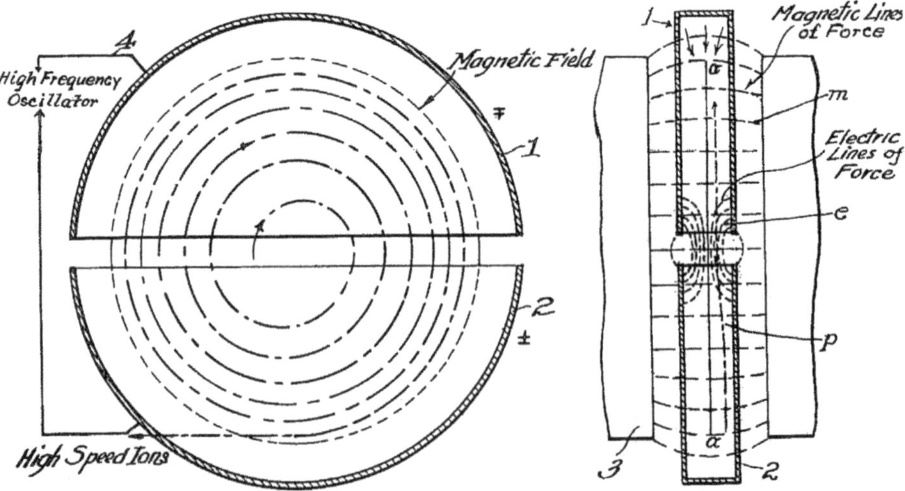

Figure 2.2. Image of the principles of a cyclotron. Cropped from US Patent 1 948 384—Ernest O Lawrence—Method and apparatus for the acceleration of ions.

2.2 Maxwell's equations in integral form

2.2.1 I Maxwell's equation from Coulomb's law

Consider an action of central field from a point charge q posed in (0,0,0)—Coulomb force $\frac{qQ}{r^2}\frac{\vec{r}}{r}$. Along the definition of electric field (2.1) taking into account the Coulomb force expression one writes:

Figure 2.3. Postal stamp. Tokamak (N Javlinskij chamber, 1954)—(**to**roidal'naya **ka**mera s **ma**gnitnymi **k**atushkami), toroidal chamber with magnetic coils. Credit: Mariluna, https://en.wikipedia.org/wiki/Tokamak.

$$\vec{E} = \frac{\vec{F}}{Q} = \frac{q}{r^2}\frac{\vec{r}}{r}. \tag{2.10}$$

Let the point charge q be surrounded by a sphere S_0 with the center in the coordinate origin (0,0,0) and radius r.

Along the definition of the flux of the field $\vec{E}$ through the surface S may be expressed as:

$$\Phi = \oint_S \vec{E} \cdot d\vec{S} = \oint_S (\vec{E}, \vec{n})dS. \tag{2.11}$$

Plugging the Coulomb force field (2.10) into (2.11), for the sphere S_0 we have:

$$\oint_{S_0} \vec{E} \cdot d\vec{S} = \oint_{S_0} \frac{q}{r^2}\left(\frac{\vec{r}}{r}, \vec{n}\right)dS = \frac{q}{r^2}\oint_{S_0} dS = 4\pi q. \tag{2.12}$$

Gauss noted that the formula (2.12) is valid for a system of point particles. He observed that surface deformation that preserves a convex geometry (see the figure 2.4) does not change the stream Φ. The statement

$$\oint_S \vec{E} \cdot d\vec{S} = \oint_{S_0} \vec{E} \cdot d\vec{S}$$

is based on the these:

1) the surface of the narrow cone cut from the surface S is almost the ellipse of semiaxes a and b and the a square πab, which lies at a distance $\approx R$, while from the unit sphere, the round of the square is equal πa_0^2,

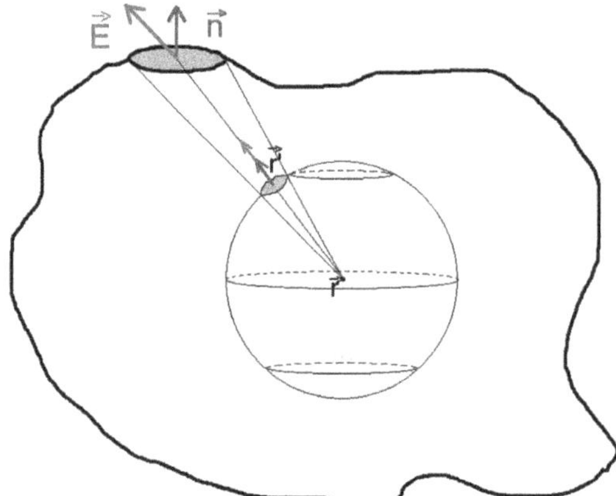

Figure 2.4. Point charge q is surrounded by a sphere with a radius r and a convex surface S.

2) the vector $\vec{E}$ projection to the normal unit $\vec{n}$ is proportional to $\cos\theta = a/b$,
3) the surface integral may be built via a Riemann sum by means of division of the surface to small pieces by a bundle of cones going from the origin until all the unit sphere is densely covered. We choose them so the relation $a/a_0 = r/R$ should hold.

For any number of charges q_i surrounded by the surface S, if accounting for the **superposition rule** $\vec{E} = \sum_i \vec{E}_i$, calculating E_i along equation (2.12) with $q \to q_i$, one has:

$$\oint_S \vec{E} \cdot d\vec{S} = 4\pi \sum_i q_i. \tag{2.13}$$

As a next step for the appropriate situation of plenty of charges we would introduce the density of electric charge, which could be defined as the following equality holds:

$$\int_V \rho(\vec{r})dV = Q, \tag{2.14}$$

where Q is the total charge in the volume V. The charge density is defined by $\rho(\vec{r}) = \lim_{\delta V \to 0} \frac{\delta Q}{\delta V}$, if a choice of the elementary volume δQ in a vicinity of $\vec{r}$ exists. Using a combination of equations (2.12) and (2.14) we arrive at

$$\oint_S \vec{E} \cdot d\vec{S} = 4\pi \int_V \rho(\vec{r})dV. \tag{2.15}$$

we shall name equation (2.13) or (2.15), **if there exist ρ, I Maxwell law in integral form.**

Next we apply the Gauss–Ostrogradski theorem

$$\oint_S \vec{K} \cdot d\vec{S} = \int_V (\nabla \cdot \vec{K})dV. \tag{2.16}$$

If the equality is valid for arbitrary subdomain of the whole V, there exists the transition of Gauss form of Coulomb's law to the differential form:

$$\nabla \cdot \vec{E} = 4\pi\rho. \tag{2.17}$$

We will name this as the first (I) Maxwell law in the differential form.

2.2.2 II Maxwell equation: absence of magnetic charges

Analyzing an equation similar to equation (2.17) but for a flux of the field $\vec{B}$, let us choose some closed surface, then calculate the flux through it (on the base of the normal component of the field vector measurements in the surface points S) or the integral along the Riemann definition:

$$\oint_S \vec{B} \cdot d\vec{S}.$$

Its value, as it may be numerically investigated gives (empiric!) zero. It means that the only source of magnetic field constitutes electric currents. Let us cite:

J Kovalik and J Kirschvink have done a search for magnetic monopoles trapped in earth rocks [17]. They studied a total of 643 kg of rock, including schists and manganese nodules, and 180 kg of seawater, using a superconducting detector, and found no [magnetic] *monopoles* (citation from [18]).

Repeat once more:

$$\oint_S \vec{B} \cdot d\vec{S} = 0.$$

Let us name this equality as the II Maxwell law in the integral form.

If a magnetic charge would exist, it could produce a magnetic field with force lines radially directed from a source. Such a field (similar to the electric one) would behave as $\frac{1}{r^2}$. The existence of the magnetic charge—solitary magnetic pole (about its properties and search of a magnetic monopole see also [2]), could be a reason, that from a domain that could contain such magnetic poles, an outgoing flux of the induction vector **B** could be observed. A symmetry between equations with charges and currents also should be taken into account.

Note, very recently it was reported a 'controlled creation of Dirac monopoles in the synthetic magnetic field produced by a spinor Bose–Einstein condensate' [7].

Application of the Ostrogradski–Gauss theorem (again as it was done for **E**), gives the equation

$$\nabla \cdot \vec{B} = 0. \tag{2.18}$$

We shall name it as the II Maxwell law in the differential form. In some textbooks this law (2.18) is named as 'Gauss's law' [19].

More about a search of magnetic monopole, its absence and corollaries, can be found in [20, 21] and [22].

In the conditions of the magnetic poles absence, magnetic field is produced only by moving particles with electric charge, that may be expressed by the empiric Biot–Savart law; let us evaluate divergence of the vector $\vec{B}$,

$$\nabla \cdot \vec{B} = \nabla \cdot \oint \frac{\vec{I} \times d\vec{r}}{r^2} = 0. \tag{2.19}$$

The result corresponds to equation (2.18) see also, e.g. [23].

2.2.3 Faraday's law: III Maxwell equation

Let us summarize generalizations of Faraday investigations (see, for example [23, 24]).

Let us consider a conductor in a magnetic field **B**. If the conductor is torn from the field, it is apparent that there is an electric current in the conductor. We say that a force is induced, its potential difference is called the *electromotive force* (EMF), by Faraday's terminology [25]. It also means, as could be experimentally verified, that in a varying magnetic field the EMF appears in a conductor. Under its action, a current is induced in this conductor.

In the modern terminology, EMF is a difference of potentials or voltage which characterize an electric field and that is the reason for the internal charges of the conductor movement (current). Then, the electromotive 'force' ε is expressed as

$$\varepsilon = \oint_L \vec{E}d\vec{l},$$

which, as Faraday observed is proportional to the speed of magnetic induction vector flux changing ($\frac{\partial \Phi}{\partial t}$, $\quad \Phi = \oint_S \vec{B}d\vec{S}$). The flux passes the surface restricted by a closed curve (F E Neumann 1845), as is formulated by Lenz:

Changes of magnetic flux through a surface closed by a contour leads to an induction current, which own magnetic field counteracts the stream changes of the external magnetic field.

When you enter the guide again to **B**, and rise again the EMF is that of the opposite sign.

Otherwise, counting the EMF as the integral of elementary work along the circuit L, surrounding the surface S, we have

$$\varepsilon = \oint_L \vec{E}d\vec{l} = -\frac{1}{c}\frac{\partial \Phi}{\partial t} = -\frac{1}{c}\frac{\partial}{\partial t}\oint_S \vec{B}d\vec{S}, \tag{2.20}$$

that gives the **III Maxwell law in the integral form.**

Important note: The coefficient of the equation is marked as $1/c$, where c enters the equations of propagation of electromagnetic waves in a vacuum, as the velocity of propagation (see chapter 3, section 3.1).

Applying Stokes theorem

$$\oint_L \vec{E}d\vec{l} = \oint_S rot\vec{E}d\vec{S}, \tag{2.21}$$

we arrive at the **III Maxwell law in the differential form.**

$$\frac{1}{c}\frac{\partial \vec{B}}{\partial t} = -rot\vec{E}. \tag{2.22}$$

2.2.4 Ampère–Maxwell law: IV Maxwell equation

The formulation of Ampère's law in the form

$$\oint_L \vec{B}d\vec{l} = 4\pi I/c, \tag{2.23}$$

very well interprets many phenomena where it is necessary to calculate the magnetic field (magnetic induction vector $\vec{B}$). Maxwell, however, noted the lack of symmetry between (2.23) and the Faraday equation (2.20) and added an element similar to the rate of flux vector magnetic induction changing, namely

$$\frac{1}{c}\frac{\partial \oint_S \vec{E}d\vec{S}}{\partial t}.$$

The decisive argument for Maxwell was a major effect of the new equation

$$\frac{1}{c}\oint_S \frac{\partial \vec{E}}{\partial t}d\vec{S} = \oint_L \vec{B}d\vec{l} - 4\pi I/c, \tag{2.24}$$

—the charge conservation law

$$\frac{dQ}{dt} = -\oint_S \vec{j}\,\vec{S}, \tag{2.25}$$

where the charge density ρ and the current density are introduced by the relation

$$Q = \int_V \rho(\vec{r})d\vec{r}, \quad I = \oint_S \vec{j}\,\vec{S}. \tag{2.26}$$

The Ampère–Maxwell equation puts the relationship between the magnetic field and electric field changes as well as the speed and velocity motion of electric charges.

Equation (2.24) will be to call a **IV Maxwell's law in the integral form.**

Transition to the differential equation is with the help of the Stokes formula (2.21) and (2.26):

$$\frac{1}{c}\frac{\partial \vec{E}}{\partial t} = rot\ \vec{B} - \frac{4\pi}{c}\vec{j}\,. \tag{2.27}$$

It is necessary to take into account the 'displacement current' $\frac{1}{c}\frac{\partial \vec{E}}{\partial t}$ in equation (2.27) in Maxwell times could not be verified by direct experiment, because typical speeds of the macroscopic bodies were small. We also understand the fact that only the term's account leads to the correct form of electromagnetic wave propagation equation (see chapter 3, section 3.1). In modern physics there are direct measurements that support the 'displacement current' presence in the equation (2.24) [26].

2.2.5 System of Maxwell's equations in differential form

Summarizing: The Maxwell's equations (in vacuum) in differential form, by means of the operator ∇, constitute the system: Coulomb's equation:

$$\nabla \cdot \vec{E} = 4\pi\rho, \tag{2.28}$$

Biot–Sawart's equation and magnetic charges absence:

$$\nabla \cdot \vec{B} = 0, \tag{2.29}$$

Faraday's law:

$$\frac{1}{c}\frac{\partial \vec{B}}{\partial t} = -\nabla \times \vec{E}, \tag{2.30}$$

Ampère–Maxwell's equation:

$$\frac{1}{c}\frac{\partial \vec{E}}{\partial t} = \nabla \times \vec{B} - \frac{4\pi}{c}\vec{j}, \tag{2.31}$$

where: $\rho(\vec{r}, t)$—electric charge density, $\vec{j}(\vec{r}, t)$—electric current density.

Using the properties of the Coulomb equation (2.28) and the Ampère–Maxwell one (2.31), we can derive the conservation of electric charge principle. Let's first apply the divergence operator to the equation (2.31):

$$\nabla \cdot [\nabla \times \vec{B}] = \frac{1}{c}\frac{\partial(\nabla \cdot \vec{E})}{\partial t} + \frac{4\pi}{c}(\nabla \cdot \vec{j}), \tag{2.32}$$

taking the identity

$$(\nabla, [\nabla \times \mathbf{B}]) = ([\nabla \times \nabla], \mathbf{B}) = 0. \tag{2.33}$$

into account, we have:

$$0 = \frac{1}{c}\frac{\partial(4\pi\rho)}{\partial t} + \frac{4\pi}{c}(\nabla \cdot \vec{j}). \tag{2.34}$$

Arranging the above equation and then dividing by the factor $\frac{4\pi}{c}$, we arrive at:

$$\frac{\partial \rho}{\partial t} + \nabla \cdot \vec{j} = 0. \tag{2.35}$$

The equation (2.35) represents the *conservation of electric charge law in the differential form* (= continuity equation). This law is not the assumption *a priori* but the direct corollary of Maxwell's equations so that they are indeed proved. It is satisfied (in the form (2.35)) in the model of continuous distribution of charges. Historically Maxwell, as mentioned, based on the law (2.35) in integral form (2.25) checking the necessity of adding the so-called 'displacement currents'—term $\frac{1}{c}\frac{\partial \vec{E}}{\partial t}$ to the Ampère equation. In a sense, the integral form is more general because it does not use the assumption about continuous distribution of charges and, moreover, the derivatives existence.

The introduction of electric and magnetic fields allows for independent examination of charges as field sources, which are also subject to the action of electromagnetic forces. For each (selected) charged body the presence of other charges affects its movement (e.g. accelerating it), which is described by the equation of motion. To close the system of Maxwell's equations (i.e. make a full description) it is necessary to add equations of motion of the Lorentz force on the right hand side. However, when it is possible to neglect the acceleration of the sources, the use of Maxwell's equations is obvious, considering the description of charges only based on given charge density and current density, as the data.

If a test charge will run the same strength, the distribution fields **E** and **B** are the same at a given space point. It is important to say that the electromagnetic field may exist in areas where there are no sources. An important fact is also that the fields, characterized by **E** and **B**, are carriers of energy, momentum and angular momentum.

2.3 Initial-boundary problem for Maxwell system in vacuum

Let us now analyze a mathematical basis for the description of fields **E** and **B**, namely the system of Maxwell's equations (2.28)–(2.31). The total number of Maxwell's equations is eight. Two of them (2.28) and (2.29) are equations that do not have a time derivative. This allows us at any time to express one of the components of both fields **E** and **B** in terms of the rest ones, then substitute them into the other dynamic (i.e. containing time derivatives) equations.

In other words, equations (2.28) and (2.29) introduce some links between basic variables (we mean **E** and **B**), for example:

$$B_z = -\int \left(\frac{\partial B_x}{\partial x} + \frac{\partial B_y}{\partial y}\right) dz. \tag{2.36}$$

These constraints are valid at any time, and therefore in the initial moment. This means that only four of the components are dynamically independent fields, so you can leave the four equations in obtaining functions of time.

It follows that the initial value problem formulation (i.e. Cauchy problem) for the Maxwell equations consists of four initial conditions. We will return to special problems formulation of applied mathematics topics in specific examples of appropriate chapters.

2.4 Vector and scalar potentials

Consider equations (2.29) and (2.31). Taking into account that $\nabla \cdot \vec{B} = 0$, the field $\vec{B}$ may be expressed by means of an auxiliary vector field usually named as the vector potential $\vec{A}$, that is defined by the implicit relation

$$\vec{B} = \nabla \times \vec{A}, \tag{2.37}$$

which automatically gives $\nabla \cdot [\nabla \times \vec{A}] = 0$. Let us substitute the vector potential $\vec{A}$ into the Faraday equation (2.30):

$$\nabla \times \vec{E} = -\frac{1}{c}\frac{\partial}{\partial t}[\nabla \times \vec{A}] \quad \Longrightarrow \quad \nabla \times \left(\vec{E} + \frac{1}{c}\frac{\partial \vec{A}}{\partial t}\right) = 0. \tag{2.38}$$

In electrostatics we defined a potential φ_e via:

$$\vec{E} = -\nabla \varphi_e. \tag{2.39}$$

On the basis of equations (2.38) and (2.39) we now introduce the general definition of scalar potential φ

$$\vec{E} + \frac{1}{c}\frac{\partial \vec{A}}{\partial t} = -\nabla \varphi, \tag{2.40}$$

that guarantees equation (2.38) identically, or:

$$\vec{E} = -\nabla \varphi - \frac{1}{c}\frac{\partial \vec{A}}{\partial t}. \tag{2.41}$$

In the case of potentials, independent of time, the formula (2.41) returns to electrostatics (2.39). While in electrodynamics the presence of the term $-\frac{1}{c}\frac{\partial \vec{A}}{\partial t}$ is very important (see the wave phenomena description: section 2.4). Substituting the definition of potentials (2.41) to the Coulomb equation (2.28), yields

$$\nabla \cdot \vec{E} = -\Delta \varphi - \frac{1}{c}\frac{\partial\left(\nabla \cdot \vec{A}\right)}{\partial t} = 4\pi\rho. \tag{2.42}$$

Similar action with equation (2.31) gives

$$\begin{aligned} \nabla \times \vec{B} = \nabla \times \left(\nabla \times \vec{A}\right) &= -\frac{1}{c}\nabla\frac{\partial \varphi}{\partial t} - \frac{1}{c^2}\frac{\partial^2 A}{\partial t^2} + \frac{4\pi}{c}\vec{j} \\ &= \nabla\left(\nabla \cdot \vec{A}\right) - \nabla^2 \vec{A}. \end{aligned} \tag{2.43}$$

In equation (2.43) we used the identity

$$\nabla \times \left(\nabla \times \vec{A}\right) = \nabla\left(\nabla \cdot \vec{A}\right) - \nabla^2 \vec{A}, \tag{2.44}$$

where:

$$\nabla \cdot \vec{A} = \frac{\partial A_x}{\partial x} + \frac{\partial A_y}{\partial y} + \frac{\partial A_z}{\partial z} = \operatorname{div} \vec{A},$$

that give for φ, $\vec{A}$ the equation

$$\nabla\left(\nabla \cdot \vec{A} + \frac{1}{c}\frac{\partial \varphi}{\partial t}\right) = \Delta\vec{A} - \frac{1}{c^2}\frac{\partial^2 \vec{A}}{\partial t^2} + \frac{4\pi}{c}\vec{j}\,. \tag{2.45}$$

The field $\vec{E}$, $\vec{B}$ express the potentials of using operators of differentiation, which means that

1) the following transformation

$$\vec{A} \to \vec{A}' + \nabla\Lambda(t, \vec{r}), \tag{2.46}$$

does not change the vector field $\vec{B}$ for arbitrary scalar $\Lambda(t, \vec{r})$! If relation (2.46) is substituted into equation (2.41), one has:

$$\vec{E} = \nabla\varphi - \frac{1}{c}\frac{\partial \vec{A}}{\partial t} - \frac{1}{c}\nabla\frac{\partial \Lambda}{\partial t} = -\nabla\left(\varphi + \frac{1}{c}\frac{\partial \Lambda}{\partial t}\right) - \frac{1}{c}\frac{\partial \vec{A}}{\partial t}. \tag{2.47}$$

2) Provided immutability of the field $\vec{E}$ so should replace the potential φ by potential φ':

$$\varphi' = -\frac{1}{c}\frac{\partial \Lambda(t, \vec{r})}{\partial t} + \varphi, \tag{2.48}$$

We can now describe the components of **E** of the electromagnetic field introduced by the potential $\vec{A}'$ and φ':

$$\vec{E} = -\nabla\varphi' - \frac{1}{c}\frac{\partial \vec{A}'}{\partial t}. \tag{2.49}$$

The transformation (2.46) and (2.48) is named the *gauge transformation*, or:

$$\begin{aligned} &\vec{A} \to \vec{A}' + \nabla\Lambda(t, \vec{r}) \\ &\varphi \to \varphi' + \frac{1}{c}\frac{\partial \Lambda(t, \vec{r})}{\partial t}. \end{aligned} \tag{2.50}$$

This transformation allows for some flexibility in the choice of vector potential. The function Λ is arbitrary, and it means that we can choose the functions φ and **A** to be able to simplify equation (2.45), imposing the following condition:

$$\frac{1}{c}\frac{\partial \varphi}{\partial t} + \nabla \cdot \vec{A} = 0. \tag{2.51}$$

The relation (2.51) is known under the name <u>*Lorentz gauge equation*</u>.

Note: A freedom of potentials choice remains after the introduction of the gauge condition. It can easily be seen that in the case of the Lorentz condition, formulas (2.46) and (2.48) keep their validity when Λ is a solution of the equation $\Box\Lambda = 0$, or

$$\Box\Lambda = -\frac{1}{c^2}\frac{\partial^2\Lambda}{\partial t^2} + \nabla \cdot (\nabla\Lambda) = 0. \tag{2.52}$$

In the case when

$$\nabla \cdot \vec{A} = 0, \tag{2.53}$$

we speak about the *Coulomb gauge condition*.

We use the representation of **E** and **B** by means of vector and scalar potentials when we want to simplify computations. Starting from Maxwell's equations we introduce unknown fields (E_x,E_y,E_z,B_x,B_y,B_z).

Going to vector and scalar potentials, we obtain only four variables (A_x, A_y, A_z, φ), or, as many as independent dynamic equations. The representation with nonzero vector potential **A** is used in evaluations of magnetic field (mainly magnetostatics), while the scalar potential φ is used in electrostatics.

2.4.1 What possibilities give us the gauge choice?—Theoretic investigation

The condition for Lorentz gauge is often used because of two very important properties. First of all we can get independent wave equations satisfied by $\vec{A}$ and φ: equation (2.42) and relation (2.45) are simplified

$$\Box\vec{A} = -\frac{4\pi}{c}\vec{j}, \tag{2.54a}$$

$$\Box\varphi = -4\pi\rho, \tag{2.54b}$$

where: $\Box = -\frac{1}{c^2}\frac{\partial^2}{\partial t^2} + \Delta$. It can always be chosen so that $\vec{A}$ and φ get this marking. Second property—an invariance of this condition (2.51) with respect to the Lorentz transformation (chapter 4)—conventionally used in the theory of relativity.

In the Coulomb gauge it is convenient to use when we can use the Poisson equation for the potential φ:

$$\frac{1}{c}\nabla\varphi = \Box\vec{A} + \frac{4\pi}{c}\vec{j}, \tag{2.55}$$

and

$$\Delta\varphi = -4\pi\rho. \tag{2.56}$$

After solving the above equations, we can calculate $\vec{A}$.

2.4.2 Hertz vector

Another popular example is the Hertz potential vector $\vec{Z}$, which is determined on the basis of the potentials φ, $\vec{A}$ via relationships:

$$\varphi = \operatorname{div}\vec{Z}, \qquad \vec{A} = \frac{\partial\vec{Z}}{\partial t}. \tag{2.57}$$

Its introduction shortens the description of electromagnetic field to one vector function, and therefore it is used in the theory of wave propagation, for example, in waveguides and resonators [27]. The choice of vector $\vec{Z}$ allows even greater freedom compared to the φ, $\vec{A}$ potentials.

2.5 Conservation principles: Poynting theorem

Conservation laws arise when integrating the dynamics equations as the 'first integrals', so after decreasing the order of these equations. On the other hand, in every field of physics, we have similar patterns. The reason is that they have a close relationship with the symmetry of space–time (relative to the Lorentz and Poincaré group), for the foundations of physics. They underlie all rights and phenomena occurring in nature.

These basic rights include the principle of conservation of energy, momentum, angular momentum—in accordance with generators named group [3]. Namely, from the viewpoint of electrodynamics, is the analysis which arose not only the mentioned principles, but also the very concept of space–time, the most we are interested in are the conservation of energy and momentum, which concern basic properties of the electromagnetic field and moving the charges [2, 23].

The electric charge conservation law stands apart from the space–time geometry and its group of motion [6], it is the so-called internal (gauge) symmetry, as we see, Maxwell's equations combine space–time symmetry and the internal one. In a sense Maxwell's electrodynamics can be 'derived' from the requirement of a local U (1) symmetry in the internal charge space [28].

We first consider the *principle of conservation of energy*, called also the *Poynting theorem*. For this purpose we use Maxwell's equations (2.30) and (2.31) to compute the following scalar products:

$$(\vec{B},\,\nabla\times\vec{E}) = -\frac{1}{c}\left(\vec{B},\,\frac{\partial\vec{B}}{\partial t}\right) = -\frac{1}{2c}\frac{\partial\vec{B}^2}{\partial t}, \tag{2.58a}$$

$$(\vec{E},\,\nabla\times\vec{B}) = \frac{1}{2c}\frac{\partial\vec{E}^2}{\partial t} + \frac{4\pi}{c}(\vec{E},\vec{j}\,). \tag{2.58b}$$

Subtracting by sides equations (2.58a) and (2.58b), we arrive at:

$$\begin{aligned}\frac{1}{2c}\frac{\partial\vec{E}^2}{\partial t} + \frac{4\pi}{c}\left(\vec{E},\vec{j}\,\right) + \frac{1}{2c}\frac{\partial\vec{B}^2}{\partial t} &= \frac{1}{2c}\frac{\partial\left(\vec{E}^2+\vec{B}^2\right)}{\partial t}\\ &+\frac{4\pi}{c}\left(\vec{E},\vec{j}\,\right) + \left(\vec{B},\,\nabla\times\vec{E}\right) - \left(\vec{E},\,\nabla\times\vec{B}\right) = 0.\end{aligned} \tag{2.59}$$

Then we use the vector identity:

$$\nabla\cdot[\vec{E}\times\vec{B}] = \left(\vec{B},\,\nabla\times\vec{E}\right) - \left(\vec{E},\,\nabla\times\vec{B}\right),$$

and, finally, we can write the law of conservation of energy in the form:

$$\frac{1}{8\pi}\frac{\partial}{\partial t}\left(\vec{E}^2 + \vec{B}^2\right) = -\frac{c}{4\pi}\left(\nabla \cdot [\vec{E} \times \vec{B}]\right) - \left(\vec{E}, \vec{j}\right). \tag{2.60}$$

The equation (2.60) has two terms: $\nabla \cdot \left(\frac{c}{4\pi}[\vec{E} \times \vec{B}]\right)$ and scalar product $(\vec{E}, \vec{j}\,)$ at the rhs. The first term $\left(\frac{c}{4\pi}[\vec{E} \times \vec{B}]\right)$ is called the *Poynting vector* and is denoted as follows

$$\vec{S} = \frac{c}{4\pi}[\vec{E} \times \vec{B}]. \tag{2.61}$$

It represents the energy flux density, i.e. the energy carried by the field per unit time per unit area. The ability to move the fields energy and storage is one of the the most important properties of the electromagnetic waves. Unit dimension of the Poynting vector is defined as $\frac{[\text{energy}]}{[\text{field surface}] \times [\text{time}]}$, which adopts the system of SI units form W m^{-2} [5].

The equation describing the law of conservation of energy, contains $\nabla \cdot \vec{S}$. This means that the Poynting vector is determined from equation (2.60) up to some constant. In our case it is arbitrary vector field $\vec{S}'$ which $\nabla \cdot \vec{S}' = 0$ (similar to the case of the vector and scalar potentials). The second term of $(\vec{E}, \vec{j}\,)$ determines power density, i.e. the work done per unit time per unit volume.

To summarize: *the principle of conservation of energy is the sum of electromagnetic energy changes per unit time in a volume and the electromagnetic energy flowing per unit of time from the volume through the bounding surface, and is equal to the work done per unit time by the sources in this volume.*

The energy density is defined as:

$$w_{\text{energy}} = \frac{\vec{E}^2 + \vec{B}^2}{8\pi}. \tag{2.62}$$

Integrating the relation (2.60) by volume and applying the Ostrogradski–Gauss identity one arrives to the energy conservation law in integral form.

Exercise 1. Please, do the integrations of equation (2.60) by volume in the explicit form.

References

[1] Vlasov A A 1955 *Makroskopicheskaya elektrodinamika (Macroscopic electrodynamics)* (Moskva: Gosudarstvennoe Izdatelstvo Tehniko-teoreticheskoi literatury)

[2] Jackson J D 1998 *Classical Electrodynamics* 3rd edn (New York: Wiley)

[3] Novozhilov Y V and Yappa Y A 1978 *Electrodynamics* (New York: Academic)

[4] Roller D E 1954 *The Development of the Concept of Electric Charge: Electricity from the Greeks to Coulomb* (Cambridge, MA: Harvard University Press)

[5] Smithonian Institute 2003 *Smithonian Physical Tables* 9th edn (Norwich: Knovel)

[6] Fock V and Kemmer N 1955 *Theory of Space, Time and Gravitation* 2nd rev edn (Oxford: Pergamon)
Fock V 1955 *Teoriya prostranstva, vremeni i tyagoteniya* (Russia: Moskva)
[7] Ray M W, Ruokokoski E, Kandel S, Mottonen M and Hall D S 2014 Observation of Dirac monopoles in a synthetic magnetic field *Nature* **505** 657–60
[8] Tamm I E 1989 *Osnovy teorii ellektrichestva* (Moscow: Izdatelstvo Nauka)
Tamm I E 1976 *Fundamentals of the Theory of Electricity* (Moscow: Mir)
[9] Dresselhaus M S *Solid State Physics, Part III, Magnetic Properties of Solids* http://web.mit.edu/course/6/6.732/www/6.732-pt1.pdf
[10] http://hyperphysics.phy-astr.gsu.edu/hbase/magnetic/hall.html
[11] Lorentz H A 1899 Simplified theory of electrical and optical phenomena in moving systems *KNAW, Proc., 1, 1898–1899 (Amsterdam)* pp 427–42
[12] Purcell E 1705 *Electricity and Magnetism by Edward Purcell*
[13] Halliday D and Resnick R 2019 *Fundamentals of Physics* 10th edn (New York: Wiley)
[14] Wesson J 2004 *Tokamaks* 3rd edn (Oxford: Oxford University Press)
[15] Maxwell J C 1865 A dynamical theory of the electromagnetic field *Philos. Trans. R. Soc. Lond.* **155** 459–512
[16] Maxwell J C 1865 *A Dynamical Theory of the Electromagnetic Field* vol CLV
[17] Kovalik J and Kirschvink J 1986 *Phys. Rev. A* **33** 1183
[18] Jeon H and Longo M 1995 Search for magnetic monopoles trapped in matter *Phys. Rev. Lett.* **75** 1443
[19] Chow T L 2006 *Electromagnetic Theory: A Modern Perspective* (Boston, MA: Jones and Bartlett) p 134
[20] Eberhard P H, Ross R R, Alvarez L W and Watt R D 1971 Search for magnetic monopoles in lunar material *Phys. Rev. D* **4** 3260–72
[21] Giacomelli G *et al* 2000 *Magnetic Monopole Bibliography* (arXiv: hep-ex 0005041)
[22] Kolm H H, Villa F and Odian A 1971 Search for magnetic monopoles *Phys. Rev. D* **4** 1285
[23] Griffiths D J 1989 *Introduction to Electrodynamics* 2nd edn (Englewood Cliffs, NJ: Prentice Hall)
[24] Feynman R P, Leighton R B and Sands M L 2006 *The Feynman Lectures on Physics* vol II (San Francisco, CA: Pearson/Addison-Wesley) pp 17–2
[25] Faraday M 1833 *Experimental Researches in Electricity* (London: Royal Society of London)
[26] Edwards B and Engheta N 2012 Experimental verification of displacement-current conduits in metamaterials-inspired optical circuitry *Phys. Rev. Lett.* **108** 193902
[27] Weinstein (Vaynshteyn) L A and Beckmann P 1969 *Open Resonators and Open Waveguides, (Golem Series in Electromagnetics)*
[28] Cheng T-P 2013 *Einstein's Physics: Atoms, Quanta, and Relativity—Derived, Explained, and Appraised* (Oxford: Oxford University Press)

IOP Publishing

Practical Electrodynamics with Advanced Applications

Sergey Leble

Chapter 3

Electromagnetic waves in vacuum

'Interpretation of light as electromagnetic phenomenon was so brave that it was most impressive for myself from all the things I met before.'

H Lorentz

3.1 Wave equations

Let us return to the Maxwell system in differential form (2.28)–(2.31). Dynamic equations (changes with time) link electric field with the magnetic one. There exists, however, a possibility of a separation of the fields that follows from the linear character of the equations with coefficients that do not depend on independent variables $\vec{r}$, t.

Consider III and IV Maxwell equations:

$$\frac{1}{c}\frac{\partial \vec{B}}{\partial t} = -\nabla \times \vec{E}, \tag{3.1a}$$

$$\frac{1}{c}\frac{\partial \vec{E}}{\partial t} = \nabla \times \vec{B} - \frac{4\pi}{c}\vec{j}\,. \tag{3.1b}$$

Differentiating equation (3.1b) with respect to time, dividing by c, we obtain

$$\frac{1}{c^2}\frac{\partial^2 \vec{E}}{\partial t^2} = \nabla \times \frac{1}{c}\frac{\partial \vec{B}}{\partial t} - \frac{4\pi}{c^2}\frac{\partial \vec{j}}{\partial t}. \tag{3.2}$$

Next, equation (3.2) is plugging into equation (3.1a), taking the identity $-\nabla \times (\nabla \times \vec{E}) == -\nabla(\nabla \cdot \vec{E}) + \nabla^2 \vec{E}$ and the first Maxwell equation (2.28) into account having

doi:10.1088/978-0-7503-2576-9ch3

$$\frac{1}{c^2}\frac{\partial^2 \vec{E}}{\partial t^2} = \Delta \vec{E} - \nabla(4\pi\rho) - \frac{4\pi}{c^2}\frac{\partial \vec{j}}{\partial t}. \tag{3.3}$$

The equality (3.3) has the form of inhomogeneous wave equation:

$$\Box \vec{E} = \vec{f}, \tag{3.4}$$

where $\Box$ is the d'Alembert operator (quabla), while

$$\vec{f} = -4\pi\left(\frac{1}{c^2}\frac{\partial \vec{j}}{\partial t} + \nabla\rho\right). \tag{3.5}$$

Solving equation (3.4) and taking into account equation (2.28) we should return to one from equations (3.1a) and (3.1b) to express the field $\vec{B}$ components, taking equation (2.29) into account.

Alternatively the system (3.1a, 3.1b) may be transformed to the wave equation for $\vec{B}$ and similar calculations of the components of the field $\vec{E}$.

3.2 Harmonic plane wave in vacuum without charges

Let us introduce some restrictions related to sources of electromagnetic field: the absence of charges in a domain. it means the zero values of $\rho = 0, \vec{j} = 0$ in Maxwell equations in the domain. The wave equation (3.4) for electric field simplifies because of $\vec{f} = 0$ in other words it takes the homogeneous form

$$\Box \vec{E} = 0. \tag{3.6}$$

This is the system of differential equations of the second order

$$\frac{1}{c^2}\frac{\partial^2 E_i}{\partial t^2} = \frac{\partial^2 E_i}{\partial x^2} + \frac{\partial^2 E_i}{\partial y^2} + \frac{\partial^2 E_i}{\partial z^2}, \quad i = 1, 2, 3. \tag{3.7}$$

Remember the condition (I Maxwell equation)

$$\operatorname{div} \vec{E} = 0, \tag{3.8}$$

that is valid in any time t.

Consider a solution of the system (3.6) by the Fourier method. The main idea of the method is a division of variables, i.e.

$$E_1(t, x, y, z) = T(t)X(x)Y(y)Z(z), \tag{3.9}$$

next, after integration by separation parameters, which act as the components of the wave vector $\vec{k}$: k_i, $i = 1, 2, 3$, one can built a general solution. The real exemplary particular solution of the wave equation is written as:

$$E_i = E_{i0}e^{i(\vec{k}\vec{r}-\omega t)} + c.\ c. = 2|E_{i0}|\cos(\vec{k}\vec{r} - \omega t + \psi_i), \tag{3.10}$$

while the scalar product $\vec{k}\vec{r} = k_x x + k_y y + k_z z$, contains the components $k_x = k_1$, $k_y = k_2$, $k_z = k_3$ (separation constants) and the complex amplitude $E_{i0} = |E_{i0}|\exp[i\psi_i]$. A similar algorithm of actions is delivered for magnetic field **B**, which results in expressions for **E** and **B** representing plane wave:

$$\vec{E} = \vec{E}_0 e^{i(\vec{k}\vec{r}-\omega t)} + c.\ c. \tag{3.11a}$$

$$\vec{B} = \vec{B}_0 e^{i(\vec{k}\vec{r}-\omega t)} + c.\ c, \tag{3.11b}$$

At formulas (3.11), the vector parameters $\vec{E}_0$ or $\vec{B}_0$ represent wave amplitude.

The next step is the account of equation (3.8), or

$$\operatorname{div} \vec{E}_0 e^{i(\vec{k}\vec{r}-\omega t)} = \vec{k} \cdot \vec{E}_0 e^{i(\vec{k}\vec{r}-\omega t)} = 0,$$

that means orthogonality of the electric field vector and the wave vector (show propagation wave direction)

$$\vec{k} \cdot \vec{E}_0 = 0. \tag{3.12}$$

To find a connection between the vectors $\vec{E}$ and $\vec{B}$, evaluate the rotation of the electric field in equation (3.11). Next we use the Faraday equation (3.1), see also equation (2.30)

$$\begin{gathered} \nabla \times \vec{E} = -\frac{1}{c}\frac{\partial \vec{B}}{\partial t} \\ -\frac{1}{c}(\vec{B}_0 e^{i\varphi}(-i\omega)) + c.\ c. = i\frac{\omega}{c}\vec{B}_0 e^{i\varphi} + c.\ c., \\ \nabla \times \vec{E} = [\nabla \times \vec{E}_0 e^{i\varphi}] + c.\ c. = \\ ie^{i\varphi}[\nabla\varphi \times \vec{E}_0] + c.\ c. = ie^{i\varphi}[\vec{k} \times \vec{E}_0] + c.\ c, \end{gathered} \tag{3.13}$$

while it is denoted $\varphi = \vec{k}\vec{r} - \omega t$. Next we express the amplitude of the magnetic field **B**:

$$\vec{B}_0 = \frac{c}{\omega}[\vec{k} \times \vec{E}_0]. \tag{3.14}$$

A similar algorithm we apply to evaluate rotation of the magnetic field. The definitions (3.11) and Ampère–Maxwell equation are used.

$$\nabla \times \vec{B} = -i\frac{\omega}{c}\vec{E}_0 e^{i\varphi} \tag{3.15a}$$

$$\nabla \times \vec{B} = ie^{i\varphi}[\vec{k} \times \vec{B}_0]. \tag{3.15b}$$

We should again equalize the sides of equation (3.15), while the amplitude of the field **B** extracted from equation (3.14)

$$\frac{c}{\omega}[\vec{k} \times [\vec{k} \times \vec{E}_0]] = -\frac{\omega}{c}\vec{E}_0. \tag{3.16}$$

The left side of relation (3.16) is simplified, if the vector identity *'bac-cab'* has been used:

$$[\vec{a} \times [\vec{b} \times \vec{c}]] = \vec{b}(\vec{a}, \vec{c}) - \vec{c}(\vec{a}, \vec{b}). \tag{3.17}$$

We arrive at:

$$\vec{k}(\vec{k}, \vec{E}_0) - \vec{E}_0\vec{k}^2 = -\frac{\omega^2}{c^2}\vec{E}_0. \tag{3.18}$$

The vectors **k** and **E** are orthogonal, expression (3.18) is reduced to the form

$$\left(\frac{\omega^2}{c^2} - \vec{k}^2\right)\vec{E}_0 = 0. \tag{3.19}$$

A solution of equation (3.19) and, hence of all the problem is nontrivial, or, $\vec{E}_0 \neq 0$ if:

$$\frac{\omega^2}{c^2} - \vec{k}^2 = 0, \tag{3.20}$$

the equation of the form (3.20) is known as the dispersion equation, that is the condition of resolvability of the system (3.11).

$$\omega = \pm c\sqrt{\vec{k}^2}. \tag{3.21}$$

We say that a wave is harmonic if, at zero time t_0, it is a function *sinus* or *cosinus*, depending on the choice of coefficients E_{0x}, E_{0y}, i.e.

$$E_x = E_{0x}e^{i(\vec{k}\vec{r}-\omega t)} + c.\ c. \tag{3.22a}$$

$$E_y = E_{0y}e^{i(\vec{k}\vec{r}-\omega t)} + c.\ c. \tag{3.22b}$$

$$E_z = E_{0z}e^{i(\vec{k}\vec{r}-\omega t)} + c.\ c. \tag{3.22c}$$

For such solutions the equation div $\vec{E} = 0$ gives the condition

$$k_xE_{0x} + k_yE_{0y} + k_zE_{0z} = 0. \tag{3.23}$$

The term $(\vec{k}, \vec{E}_0) = 0$ in equation (3.18) means that the field amplitude $\vec{E}_0$ (and, by this the magnetic field amplitude $\vec{B}_0$) is orthogonal to the wave vector $\vec{k}$. The vector $\vec{k}$ determines a propagation direction of the wave, so both vectors of electromagnetic field are perpendicular to the propagation direction which in turn coincide with the Poynting vector $\vec{S}$ direction (figure 3.2).

The case of interest may be illustrated by figure 3.1.

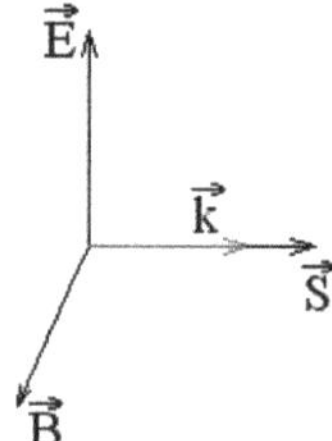

Figure 3.1. The vectors $\vec{E}$, $\vec{B}$, $\vec{k}$ of the harmonic plane wave are orthogonal in pairs, the Pointing vector $\vec{S}$ is parallel to the wave vector $\vec{k}$.

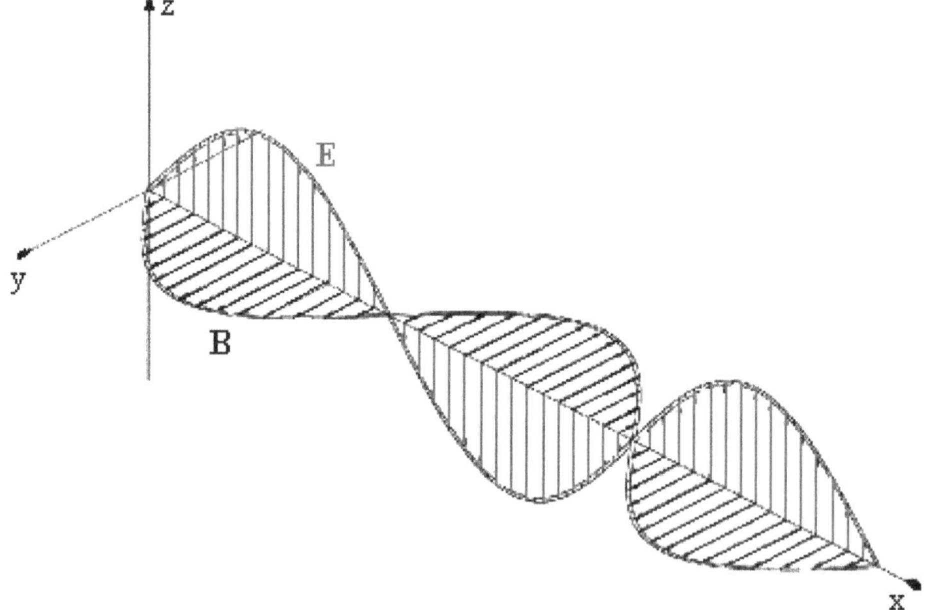

Figure 3.2. Plane monochromatic EM wave is characterized by electric field vector $\vec{E}$ and magnetic one $\vec{B}$, which are orthogonal to $\vec{k}$.

3.3 Wave packets

In real physical conditions a quasi-monochromatic wave with a frequency ν is represented by wave packets (wave train), such a wave may be generated by charges oscillating with frequency ν (see section 3.8). For example, if a generator is switched on in a moment $t = 0$ and switched off in a moment $\tau \gg \nu^{-1}$, the process of switching may be described by a modulating function.

Speaking again about waves in a vacuum we can start with harmonic waves (3.22) (nonzero in whole space R^3), but combining its Fourier superposition from k to $k + \delta k$.

Suppose that in the initial moment the x-component of the electric field is given by

$$E_x(\vec{r}, 0) = \phi_x(x, y, z) = \int \mathrm{d}k_x \mathrm{d}k_y \mathrm{d}k_z E_x(k_x, k_y, k_z), \tag{3.24}$$

—the Fourier integral for $E_x(k_x, k_y, k_z) = E_{0x}(k_x, k_y, k_z)e^{i(\vec{k}\vec{r})} + c.\ c.$ Recall that the dispersion relation (3.21) root with the sign '+' links $\omega = ck$. The amplitude function $E_{0x}(k_x, k_y, k_z)$ is calculated as the inverse Fourier transform.

The form of Fourier integral in three dimensions:

$$\vec{E}(\vec{r}, t) = \frac{1}{\sqrt{2\pi}} \int_{-\infty}^{+\infty} \mathrm{d}\vec{k}\vec{E}(\vec{k})e^{i(\vec{k}\vec{r}-ckt)}, \tag{3.25}$$

gives the general solution of a Cauchy problem for the homogeneous wave equation. The inverse transformation

$$E_j(\vec{k}) = \frac{1}{\sqrt{2\pi}} \int_{-\infty}^{+\infty} \mathrm{d}\vec{r}\phi_j(x, y, z)e^{-i(\vec{k}\vec{r})}, \tag{3.26}$$

gives the expression for all components $j = 1, 2, 3$ amplitudes $E_j(\vec{k})$ via the initial conditions $\phi_j(x, y, z)$. Remember the I Maxwell equation as a link between electric field components. For illustration we plot in figure 3.3 the electric field component $E = E_x$ with zero others, so that is

$$E = \exp\left[-\frac{(x - ct)^2}{3}\right]\cos(n(x - ct)).$$

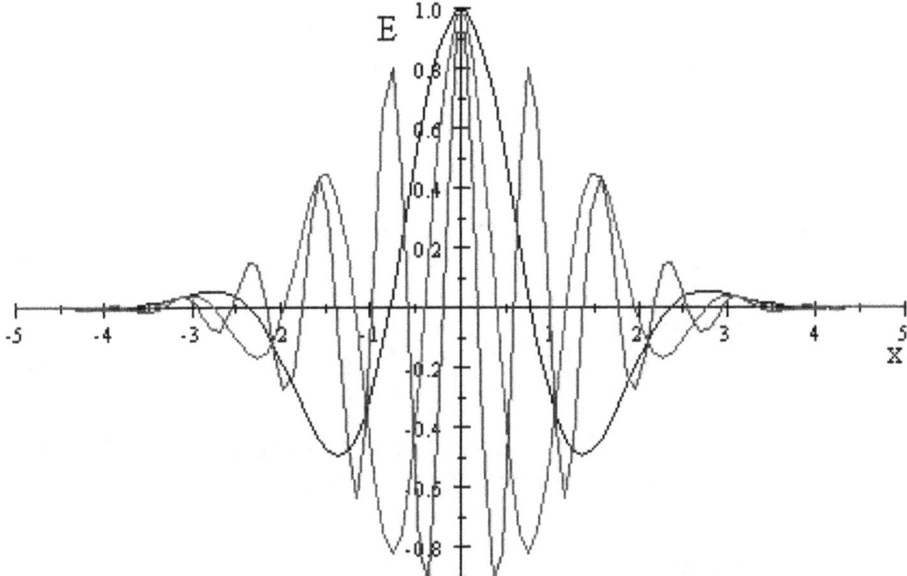

Figure 3.3. One-dimensional plot of a wave train (wave packet) of the electric field $x-$ component for $t = 0$, $n = 2,4,8$.

3.4 Cauchy Problem in 1 + 1 space–time

3.4.1 The initial (Cauchy) problem for plane wave with a fixed polarization

Our starting point is the Maxwell equations for vacuum with $\rho = 0, \vec{j} = 0$, in the Lorentz–Heaviside's unit system. For didactic reasons we start with the one-dimensional model as of [1], the x-axis chosen as the direction of a pulse propagation. The first Maxwell equation (Coulomb's law) (2.17) in this one-dimensional case means the simple form $\partial E_x/\partial x = 0$, hence, following the finite-energy field condition we assume that $E_x = 0$ and, similarly by (2.18) $B_x = 0$ taking into account the only polarization of electromagnetic wave, or $E_z = 0$, $B_y = 0$, denoting for brevity $E_y = E$, $B_z = B$. This allows us to rewrite the Maxwell equations as

$$\frac{1}{c}\frac{\partial E}{\partial t} = \frac{\partial B}{\partial x},$$
$$\frac{1}{c}\frac{\partial B}{\partial t} = \frac{\partial E}{\partial x}.$$

In this case we rewrite the system (3.27) as the matrix equation

$$\Psi_t = L\Psi,$$

where the field vector (3.27), denoting $\partial_x = \frac{\partial}{\partial x}$,

$$\Psi = \begin{pmatrix} E \\ B \end{pmatrix} \text{ and the matrix operator } L = \begin{pmatrix} 0 & c\partial_x \\ c\partial_x & 0 \end{pmatrix}, \tag{3.27}$$

that enter the matrix operator equation (3.27) as

$$\begin{pmatrix} \frac{\partial}{\partial t}E \\ \frac{\partial}{\partial t}B \end{pmatrix} = \begin{pmatrix} 0 & c\partial_x \\ c\partial_x & 0 \end{pmatrix}\begin{pmatrix} E \\ B \end{pmatrix} = \begin{pmatrix} c\frac{\partial}{\partial x}B \\ c\frac{\partial}{\partial x}E \end{pmatrix}. \tag{3.28}$$

Let us formulate now a Cauchy problem for the system equation (3.28) on $x \in (-\infty, \infty)$. Besides the equations (3.28) the problem contains two initial conditions

$$\Psi_0 = \begin{pmatrix} E(x, 0) \\ B(x, 0) \end{pmatrix} = \begin{pmatrix} \phi(x) \\ \chi(x) \end{pmatrix}. \tag{3.29}$$

It is easy to verify that the problem formulation has a unique solution, that is given and investigated in the next subsection.

3.4.2 On projection method

To solve the problem and highlight its vectorial nature, we would apply a special transformation to new dependent variables with simultaneous combination of equations. Such a transformation is very simple in the case of the example we investigate, but it will be more general and effective in forthcoming sections. As a first step to illustrate the method [3, 4], we define projection operators for the simple case of one-dimensional propagation in a vacuum we do consider. In this case we use equation (3.27) to construct the matrix equation

$$\Psi_t = L\Psi,$$

where the field vector (3.27) is used. Applying the Fourier transformation on x,

$$E = \frac{1}{\sqrt{2\pi}} \int \exp(ikx)\tilde{E}dk,$$

one goes to the ordinary differential equations (ODE) system, parameterized by k

$$\begin{pmatrix} \frac{d}{dt}\tilde{E} \\ \frac{d}{dt}\tilde{B} \end{pmatrix} = \begin{pmatrix} 0 & ick \\ ick & 0 \end{pmatrix} \begin{pmatrix} \tilde{E} \\ \tilde{B} \end{pmatrix}. \tag{3.30}$$

This is the ODE system which has an exponential $\exp(i\omega t)$ type solution, where $\omega = \pm ck$. We hence arrive at a simple 2×2 matrix eigenvalue problem which can be solved by applications of matrix projection operators.

Let us search for such matrices P_i, such that $P_1\Psi = \Psi_1$ and $P_2\Psi = \Psi_2$, Ψ_i are eigenvectors of the evolution matrix in equation (3.30) for arbitrary Ψ. Moreover, suppose the standard properties of orthogonal projecting operators

$$P_i * P_j = 0, \quad P_i^2 = P_i, \quad \sum_i P_i = 1 \tag{3.31}$$

are implied. For equation (3.30) the form of P_i may be derived from the mentioned conditions. Namely,

$$P_1 = \frac{1}{2}\begin{pmatrix} 1 & 1 \\ 1 & 1 \end{pmatrix}, \quad P_2 = \frac{1}{2}\begin{pmatrix} 1 & -1 \\ -1 & 1 \end{pmatrix}. \tag{3.32}$$

Technically we solve the problem in the k-space and next performing the inverse Fourier transform yields the x-representation of the operators [3]. In this simplest case the matrix elements of the projecting operators (3.32) in fact do not depend on k, (that is not the case if dispersion is accounted) hence its x-representation coincides with the k-representation equation (3.32).

Applying the projection operators to the vector Ψ (3.27), e.g. as

$$P_1\Psi = \frac{1}{2}\begin{pmatrix} E + B \\ E + B \end{pmatrix} = \begin{pmatrix} \Pi \\ \Pi \end{pmatrix}, \tag{3.33}$$

we can introduce new variables $\Lambda = \frac{1}{2}E + \frac{1}{2}B$ and $\Pi = \frac{1}{2}E - \frac{1}{2}B$ which correspond to the left and right direction of wave propagation [3]. Comparing new variables Λ and Π to ones presented in [5], under the name of 'hybrid field variables', ours have a similar form to their ones. Our form of equation (3.33) is exactly determined by dispersion relation $\omega = \pm ck$ from equation (3.28) and hence allows to account its arbitrary form. Moreover it allows us to present in an algorithmic way both the electric and magnetic field [3], in the simplest example we trace now,

$$E_y = \Lambda + \Pi, \quad B_z = (\Pi - \Lambda). \tag{3.34}$$

This correspondence equations (3.33) and (3.34) is a one-to-one local map and hence allows to determine initial conditions in the Cauchy problem for both the left and right wave variables (Λ, Π). It also gives a possibility to follow waves, extracting data in each time t by the corresponding projecting in a very general situation.

Applying the operators P_i equation (3.32) to equation (3.28) and respecting the projectors property $[P_i, (\frac{\partial}{\partial t} - L)] = 0$, yields

$$\left(\frac{\partial}{\partial t} - L\right)P_i\Psi = 0. \tag{3.35}$$

At this point we have to admit that Λ is strictly connected to the dispersion relation $\omega = -ck$ and analogously Π is related to $\omega = ck$. Hence this determines a sign, which stands as the parameter c in our equations. Finally, the result

$$\begin{pmatrix} \frac{\partial}{\partial t}\Pi \\ \frac{\partial}{\partial t}\Pi \end{pmatrix} - \begin{pmatrix} c\frac{\partial}{\partial x}\Pi \\ c\frac{\partial}{\partial x}\Pi \end{pmatrix} = 0, \tag{3.36}$$

presents (twice!) the evolution equation for right wave interacting with the left one.

Repeating our calculations from equation (3.35) to equation (3.36) with the use of the second projecting operator equation (3.32) we obtain a closed system of equations which describes the interaction between two waves propagating in opposite directions, see also chapter 20, section 20.3. The system of equations have the form

$$\begin{cases} \frac{\partial}{\partial t}\Pi - c\frac{\partial}{\partial x}\Pi = 0, \\ \frac{\partial}{\partial t}\Lambda + c\frac{\partial}{\partial x}\Lambda = 0. \end{cases} \tag{3.37}$$

The Cauchy problem includes initial conditions

$$\Lambda(x, 0) = \frac{1}{2}\phi + \frac{1}{2}\chi, \; \Pi(x, 0) = \frac{1}{2}\phi - \frac{1}{2}\chi. \tag{3.38}$$

The solutions of equations (6.148) with the initial conditions (3.39) we write

$$\Lambda(x, t) = \frac{1}{2}\phi(x - ct) + \frac{1}{2}\chi(x - ct), \; \Pi(x, t) = \frac{1}{2}\phi(x + ct) - \frac{1}{2}\chi(x + ct). \tag{3.39}$$

The electric and magnetic fields are expressed by equation (3.34).

In the conditions of the derivation, this system is equivalent to the vacuum Maxwell one-dimensional case and hence describes all linear polarized waves with arbitrary frequency. Let us discuss the results.

3.5 Discussion and exercises

The solution (3.39) is a *wave of translation*, typical for the dispersionless 1D wave phenomena. A textbook example for such a case is the conventional small amplitude string problem. It is quite visible that arbitrary initial pluse propagates with the velocity c, the constant first appeared in the Faraday experiment mathematical description.

A wave that propagates may be a carrier of information. In such a case it should be modulated, that means a change of amplitude (AM) or frequency (FM) imposed on a harmonic wave (3.22). Such a modified wave, e.g. short pulse, can inform us about whether it reached a body (np. antenna, receiver) or not. The velocity of a pulse propagation is defined by the derivative of the dispersion relation $\frac{d\omega}{dk}$ and is named as group velocity [3]. In free space (vacuum) such a derivative coincides with c, but, for example in a waveguide, it is less than c (see chapter 11, section 11.2).

Exercise 1.
From equations (3.1a) and (3.1b) derive the wave equation for $\vec{B}$ and calculate the components of the field $\vec{E}$. Check the validity of equation (2.28).

Exercise 2. Plot the one-dimension wavetrains

$$E(x, t) = \frac{1}{\sqrt{2\pi}} \int_{-\infty}^{+\infty} dk E(k) e^{ik(x-ct)}, \; E(-k) = E^{*}(k), \tag{3.40}$$

for

$$E(k) = \exp[-k^2], \tag{3.41}$$

evaluating the integral by a table of Fourier transform for $t = 0, 1, 2$. Take $c = 1$ (special units).

Exercise 3. What initial problem solves expression (3.40)?

3.6 Inhomogeneous wave equation: wave generation

3.6.1 On the Green function method for wave equations

In previous sections we have considered solutions of the 1D linear EM wave equation with zero rhs. In this section we study the 3D inhomogeneous wave equation, i.e. $f \neq 0$ case:

$$\Box u(t, \vec{r}) = f(t, \vec{r}). \tag{3.42}$$

We begin from the stationary equation for $u(\vec{r}, t) = u(\vec{r})$, $\vec{r} \in R^3$ (Poisson's equation):

$$\Delta u(\mathbf{r}) = f(\mathbf{r}), \tag{3.43}$$

in which $\vec{r} = (x, y, z) \in \mathbb{R}^3$. In electrostatics, in terms of the I Maxwell equation, f has the form: $f = 4\pi\rho$, while $\nabla\vec{E} = \nabla \cdot (\nabla\varphi) = \Delta\varphi = 4\pi\rho$, and $\rho(\vec{r})$ is the charge density that does not depend on time.

Next, we show that a solution of the Poisson equation (3.43) may be expressed by means of the Green function G. Along the definition

$$\Delta G(\vec{r}, \vec{r}') = \delta(\vec{r}, \vec{r}'), \quad \int \delta(\vec{r}, \vec{r}')f(\vec{r}')d\vec{r}' = f(\vec{r}). \tag{3.44}$$

It could be checked by a substitution to equation (3.43), that the function

$$u(\vec{r}) = \int G(\vec{r}, \vec{r}')f(\vec{r}')d\vec{r}', \tag{3.45}$$

where $d\vec{r}' = dx'dy'dz'$, is the solution of this equation.

As it follows from the definition, the Green function is a solution of the equation with point source, hence it has the spherical symmetry with the center in the point $\vec{r} = \vec{r}'$. Let us consider the equation for the function G in spherical coordinates r, θ, ϕ. In the forthcoming calculations we apply the following notations $|\vec{r} - \vec{r}'| = R$. Operator Δ has the form:

$$\Delta G = \frac{1}{R^2}\frac{\partial}{\partial R}\left(R^2\frac{\partial G}{\partial R}\right) + \frac{1}{R^2 \sin\theta}\frac{\partial}{\partial\theta}\left(\sin\theta\frac{\partial G}{\partial\theta}\right) + \frac{1}{R^2\sin^2\theta}\frac{\partial^2 G}{\partial\phi^2}. \tag{3.46}$$

The spherical symmetry of the function G means independence on the variables θ, ϕ. Developing the previous record:

$$\Delta_R G(R) = \delta(\vec{r}, \vec{r}'). \tag{3.47}$$

From the Green function properties along with the ones of the Dirac δ-function it is known that $\int d\vec{r}'\delta(\vec{r}, \vec{r}') = 1$. Integrating equation (3.47) by the sphere of radius R and center in the $R = 0$ domain yields:

$$\int_0^R \left(\int_0^{2\pi}\left(\int_0^{\pi} \Delta_R G(R)R^2 \sin\theta d\theta\right)d\phi\right)dR = 1, \tag{3.48}$$

rewriting as

$$R^2\Delta_R G = \frac{d}{dR}R^2\frac{dG}{dR}. \tag{3.49}$$

Equation (3.48) simplifies to the form:

$$4\pi\int_0^R (R^2\Delta_R G)dR = 4\pi R^2\frac{dG}{dR} = 1. \tag{3.50}$$

After trivial transformation and integration, one obtains the following differential equation:

$$\frac{dG}{dR} = \frac{1}{4\pi R^2}, \tag{3.51}$$

its solution has the form

$$G = -\frac{1}{4\pi R}. \tag{3.52}$$

Having the form of the Green function G, we can apply it to write a solution of the Poisson equation (3.43) in the form of integral (3.45)

$$u(\vec{r}) = -\int \frac{f(\vec{r}')}{4\pi|\vec{r} - \vec{r}'|} d\vec{r}', \tag{3.53}$$

recalling that $R = |\vec{r} - \vec{r}'|$ and

$$\Delta \frac{1}{4\pi R} = -\delta(\vec{r} - \vec{r}'). \tag{3.54}$$

Let us return to the wave equation (3.42). The short pulse electromagnetic wave is spherical and emitted from the point $\vec{r}'$. Supposing that an impulse is infinitely short, it represents the Green function of wave equation, which is the solution of

$$\Box G(\vec{r}, t; \vec{r}', t') = \delta(t, t'; \vec{r}, \vec{r}'). \tag{3.55}$$

The wave propagates in the whole space with the velocity c that has a spherical wave front at $R = |\vec{r} - \vec{r}'|$, while $\vec{r}$ is a point of observation.

The radius R may be described by the dependence $R = c(t - t')$, where $(t - t')$ is time, in which the wave reaches the distance $|\vec{r} - \vec{r}'| = R$. The scheme is illustrated by figure 3.4.

Equation (3.55) for the Green function is rewritten in the following way, for underlying time and space homogeneity

$$\Box G(\vec{r} - \vec{r}', t - t') = \delta(\vec{r} - \vec{r}')\delta(t - t'), \tag{3.56}$$

that is equivalent to writing in Cartesian coordinates

$$\frac{\partial^2 G}{\partial x^2} + \frac{\partial^2 G}{\partial y^2} + \frac{\partial^2 G}{\partial z^2} - \frac{1}{c^2}\frac{\partial^2 G}{\partial t^2} = \delta(x - x')\delta(y - y')\delta(z - z')\delta(t - t'). \tag{3.57}$$

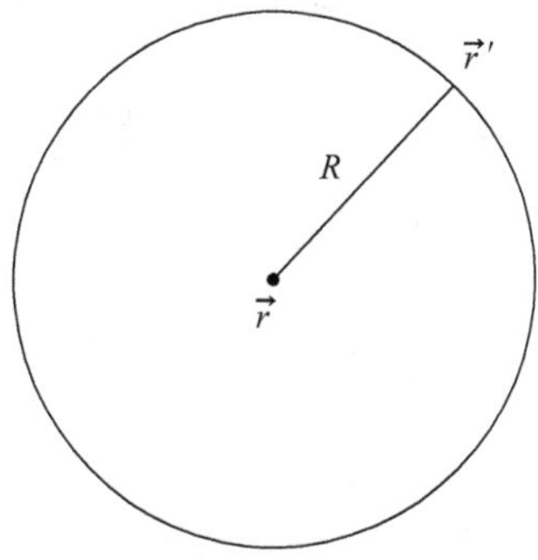

Figure 3.4. Sphere of radius R and the center at $\vec{r}$.

Summarizing, writing a solution of this equation at the whole space R^3 (3.57)

$$G(\vec{r},\, t;\, \vec{r}',\, t') = -\frac{\delta\left(t' - t + \dfrac{R}{c}\right)}{4\pi R}. \tag{3.58}$$

Decrease at infinity and define the divergent sphere wave by the construction.

The solution of the 3D wave equation (3.42) has the form

$$u(\vec{r}) = \int_V G(\vec{r},\, t;\, \vec{r}',\, t') f(\vec{r}',\, t') d\vec{r}' dt'. \tag{3.59}$$

Figure 3.5 represents the solution of the equation for the Green function (3.57) as a spherical wave.

The main property of the wave expressed by formula (3.58) is that it covers the distance $|\vec{r} - \vec{r}'|$ in definite time $t = t' + \frac{R}{c}$. The term $\frac{R}{c}$ relates to the retarding of the spherical wave.

Consider a wave, generated by a source with given frequency $\omega = ck$:

$$f(t,\, \vec{r}) = F(\vec{r})\exp[i\omega t] + c.\, c., \tag{3.60}$$

with a natural form of solution

$$f(t,\, \mathbf{r}) = F(\vec{r})\exp[i\omega t] + c.\, c., \qquad u(t,\, \vec{r}) = \exp[i\omega t]U(\vec{r}) + c.\, c. \tag{3.61}$$

Substitution to equation (3.42) gives the Helmholtz equation:

$$(k^2 + \nabla^2)U = -F(\vec{r}). \tag{3.62}$$

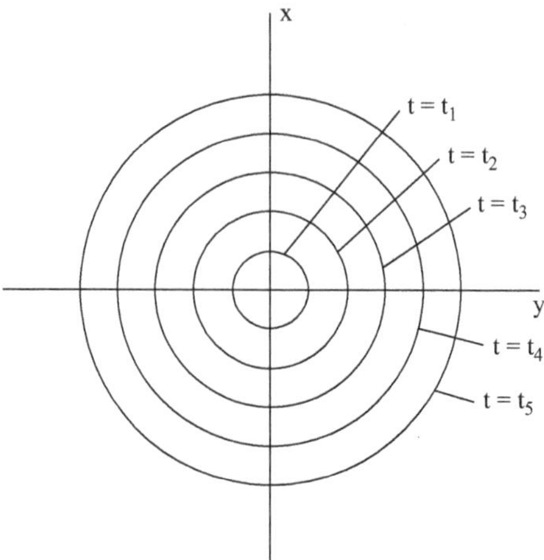

Figure 3.5. Solution of equation (3.57) for the Green function as a spherical wave. Time intervals are equal $t_i = i$, The circumferences are projections of the surfaces $G = \text{const}$ at the plane xy.

The Green function of Helmholtz equation (3.62) for a problem in whole space is given by

$$G = -\frac{e^{\pm ikR}}{4\pi R}, \tag{3.63}$$

where $R = |\vec{r} - \vec{r}'|$. It is verified by direct substitution in (3.62) and uses the definition of the Green function for Poisson equation.

3.6.2 General formulas for electromagnetic field and wave emission

The direct application of the relation (3.59), that constitutes the Green function method yields a solution of equation (3.4) as four-integral

$$\vec{E} = -4\pi \int G(\vec{r}, t; \vec{r}', t')\left(\frac{1}{c^2}\frac{\partial \vec{j}(\vec{r}', t')}{\partial t'} + \nabla\rho(\vec{r}', t')\right)dt' d\vec{r}'. \tag{3.64}$$

Integrating by parts produces a more compact expression with derivatives

$$\vec{E} = 4\pi \int \left(\frac{\vec{j}}{c^2}G_{t'} + \rho\nabla' G\right)dt' d\vec{r}'. \tag{3.65}$$

Performing differentiations, using the explicit form of the Green function (3.58) allows to proceed as

$$\begin{aligned} G_{t'} &= \frac{\partial}{\partial t'}\frac{\delta\left(t' - t + \frac{R}{c}\right)}{4\pi R} = \frac{\delta_{t'}(T)}{4\pi R} = \frac{1}{4\pi R}\frac{d\delta(T)}{dT}, \\ \nabla' G(\vec{r}, t; \vec{r}', t') &= \mathrm{grad}'\frac{\delta\left(t' - t + \frac{R}{c}\right)}{4\pi R} = \frac{\vec{n}}{4\pi R^2}\delta(T) - \frac{\vec{n}}{4\pi c R}\delta_{t'}(T), \end{aligned} \tag{3.66}$$

where $R = |\vec{r} - \vec{r}'|$, $T = t' - t + \frac{R}{c}$, $\vec{n} = \frac{\vec{r} - \vec{r}'}{R}$. Plugging it into equation (3.65) we derive the formula for the electric field

$$\vec{E} = 4\pi \int \left(\frac{\vec{j}}{c^2 4\pi R}\frac{d\delta(T)}{dT} + \rho\left[\frac{\vec{n}}{4\pi R^2}\delta(T) - \frac{\vec{n}}{4\pi c R}\frac{d\delta(T)}{dT}\right]\right)dt' d\vec{r}', \tag{3.67}$$

that, again integrating by parts, isolates the delta-function $\delta(T)$

$$\vec{E} = \int \left(\frac{-\vec{j}_{t'}(\vec{r}', t')}{c^2} + \rho(\vec{r}', t')\frac{\vec{n}}{R} + \rho_{t'}(\vec{r}', t')\frac{\vec{n}}{c}\right)\frac{\delta(T)}{R}dt' d\vec{r}'. \tag{3.68}$$

It allows to do the formal integration with respect to t' on a base of definition of the (one-dimensional) delta-function, which results in

$$\vec{E} = \int \left(\frac{\rho \vec{n}}{R^2} + \frac{\rho_{t'} \vec{n} - \vec{j}_{t'} / c}{cR} \right) d\vec{r}', \tag{3.69}$$

where $t' = t - R/c$. One recognizes the stationary Coulomb term and some dynamic ones, which are responsible for a phenomenon of electromagnetic wave generation. The important feature of such a term of the radiation phenomena is the more slow decrease of the field with $R \to \infty$, that means total energy losses for the emitting system of charges.

The similar equation for the magnetic field $\vec{B}$ may be derived by means of excluding the field $\vec{E}$ from equations (3.1a) and (3.1b), that yields

$$\frac{1}{c^2} \frac{\partial^2 \vec{B}}{\partial t^2} = \Delta \vec{B} + \frac{4\pi}{c} \text{rot} \vec{j} \, . \tag{3.70}$$

Applying the Green function method one arrives at

$$\vec{B} = \frac{4\pi}{c} \int G(\vec{r}, t; \vec{r}', t') \text{rot}' \vec{j} \, dt' d\vec{r}'. \tag{3.71}$$

An alternative version of the expression (3.71) is obtained integrating by parts of the rot operation components as

$$\int G(\vec{r}, t; \vec{r}', t') \text{rot}_x \vec{j} \, d\vec{r}' = \int G(\vec{r}, t; \vec{r}', t' \left(\frac{\partial j_{z'}}{\partial y'} - \frac{\partial j_{y'}}{\partial z'} \right) d\vec{r}' = - \int [\text{grad} G \times \vec{j} \,]_{x'} d\vec{r}'.$$

Using relation (3.66) and plugging the result into equation (3.71) allows to obtain a compact expression for the magnetic induction field

$$\begin{aligned} \vec{B} &= - \frac{4\pi}{c} \int [\text{grad} G \times \vec{j} \,] dt' d\vec{r}' \\ &= \frac{4\pi}{c} \int \left[\left(\frac{\vec{n}}{4\pi c R} \delta_{t'}(T) - \frac{\vec{n}}{4\pi R^2} \delta(T) \right) \times \vec{j} \, \right] dt' d\vec{r}' \\ &= - \frac{1}{c} \int \left(\frac{\vec{n} \times \vec{j}}{R^2} + \frac{\vec{n} \times \vec{j}_{t'}}{cR} \right) \delta(T) dt' d\vec{r}'. \end{aligned} \tag{3.72}$$

Or, after a little algebra

$$\vec{B} = - \frac{1}{c} \int_V \left(\frac{\vec{n} \times \vec{j}}{R^2} + \frac{\vec{n} \times \vec{j}_{t'}}{cR} \right) d\vec{r}', \tag{3.73}$$

where $t' = t - R/c$. The volume V contains all the sources as moving charged particles. We observe here the Biot–Sawart law and the term of emitted field, see also section 20.10.

Recalling the expression for the energy flux given by the Pointing vector (2.61), we would comment on the results from the point of energy conservation law. Suppose the volume V is finite, posed inside a sphere with radius R_0, so at $R \gg R_0$ the asymptotic behavior of both fields $\vec{E}$, $\vec{B}$ is defined by the terms $\sim 1/R$, hence the energy flux through a sphere of radius R is stationary.

3.7 Emission of the isolated charged point particle

A point charged particle with the position vector $\vec{s}(t)$ may be described by the Dirac $\delta-$ function so, that its charge density is represented by

$$\rho(\vec{r}, t) = q\delta(\vec{r} - \vec{s}(t)). \tag{3.74}$$

The particle velocity writes as the position time-derivative $\vec{v} = \dot{\vec{s}}(t)$, while the current density is defined similarly via

$$\vec{j}(\vec{r}, t) = q\delta(\vec{r} - \vec{s}(t))\dot{\vec{s}}(t). \tag{3.75}$$

The general expression (3.68) of the wave equation (3.4) defines a solution for the electric field via the Green function G; now, plugging in the densities (3.74) and (3.75) we get

$$\begin{aligned} \vec{E} = q\int_V \int_{-\infty}^{\infty} \Bigg(& \frac{\delta(\vec{r}\,' - \vec{s}(t'))\dot{\vec{s}}(t')}{c^2 R}\delta_{t'}(T) \\ & + \delta(\vec{r}\,' - \vec{s}(t'))\left[\frac{\vec{n}}{R^2}\delta(T) - \frac{\vec{n}}{cR}\delta_{t'}(T)\right]\Bigg) dt' d\vec{r}\,'. \end{aligned} \tag{3.76}$$

Again, as in section 3.6.2, the formal integrating with the delta-function $\delta(\vec{r}\,' - \vec{s}(t'))$ by space coordinates $\vec{r}\,'$ yields

$$\vec{E} = q\int_{-\infty}^{\infty}\left(\left[\frac{\dot{\vec{s}}(t')}{c^2 R} - \frac{\vec{n}}{cR}\right]\delta_T(T) + \frac{\vec{n}}{R^2}\delta(T)\right)dt', \tag{3.77}$$

where now $R = |\vec{r} - \vec{s}(t')|$, $\vec{n} = \frac{\vec{r} - \vec{s}(t')}{R}$. Integrating by t' gives

$$\vec{E} = \frac{q}{\kappa}\left(\left[\frac{\vec{n} - \vec{\beta}}{\kappa c R}\right]_{t'} + \frac{\vec{n}}{R^2}\right), \tag{3.78}$$

where the retarded time is $t' = t - R/c$, and, the acceleration is the velocity derivative $\vec{\beta} = \dot{\vec{s}}(t')/c$. The derivative by t'

$$T_{t'} = 1 - \vec{n}\cdot\vec{\beta} \equiv \kappa, \tag{3.79}$$

appears for the differentiation by T, for integration realization, should be changed to the differentiation by t'. Note that at the velocities of particles such that $|\vec{\beta}| \ll c$ the value of $\kappa \approx 1$. The following property of the delta-function (section 20.2.4)

$$\int_{-\infty}^{\infty} f(t)\delta(T(t))dt = \sum_{s}\int_{-\infty}^{\infty} f(t)\frac{\delta(t - t_s)}{|T_t(t_s)|}dt, \tag{3.80}$$

is used, where t_s are the simple roots of the equation $T(t) = 0$. In our case $T = t' - t + \frac{|\vec{r} - \vec{s}(t')|}{c}$ so the condition $T = 0$ defines the mentioned retarded time t'. The differentiation by t' gives the relations

$$\begin{aligned}
\left[\frac{\vec{n}}{cR} - \frac{\vec{s}(t')}{c^2R}\right]_{t'} &= \frac{1}{c}\frac{\vec{n}_{t'} - \vec{\beta}}{R} - \frac{\vec{n} - \vec{\beta}}{cR^2}R_{t'}, \\
\frac{1}{c}\vec{n}_{t'} &= -\frac{\vec{\beta} - \vec{n}(\vec{n}\cdot\vec{\beta})}{R} \\
\frac{1}{c}R_{t'} &= -(\vec{n}\cdot\vec{\beta}), \\
\kappa_{t'} &= -\vec{n}_{t'}\cdot\vec{\beta} - \vec{n}\cdot\vec{\beta}_{t'} \\
&= \frac{\vec{\beta}^2 - (\vec{n}\cdot\vec{\beta})^2}{R} - \vec{n}\cdot\dot{\vec{\beta}}.
\end{aligned} \tag{3.81}$$

Or, plugging the results in equation (3.78), finally writing

$$\vec{E} = \frac{q\vec{n}}{(\kappa R)^2} - \frac{q\vec{\beta}}{(\kappa R)^2} + \frac{q\vec{n}}{c\kappa}\left[\frac{1}{\kappa R}\right]_{t'} - \frac{q}{c\kappa}\left[\frac{\vec{\beta}}{\kappa R}\right]_{t'}. \tag{3.82}$$

For the field $\vec{B}$ we use the formula of similar structure, formula (3.72)

$$\begin{aligned}
\vec{B} &= q\frac{4\pi}{c}\int\left(\frac{\vec{n}}{4\pi cR}\delta_{t'}(T) - \frac{\vec{n}}{4\pi R^2}\delta(T)\right) \times \delta(\vec{r}' - \vec{s}(t'))\dot{\vec{s}}(t')dt'd\vec{r}' \\
&= q\int\left(\frac{\delta_T(T)}{cR} - \frac{\delta(T)}{R^2}\right)\vec{n}\times\vec{\beta}dt'.
\end{aligned} \tag{3.83}$$

Integrating by parts in t' in the first term (account equation (3.79)) and next in the rhs by $\vec{r}'$, it produces

$$\vec{B} = -q\int\left(\frac{\vec{n}\times\vec{\beta}}{R^2} + \left[\frac{\vec{n}\times\vec{\beta}}{\kappa cR}\right]_{t'}\right)\delta(T)dt'. \tag{3.84}$$

with $R = |\vec{r} - \vec{s}(t')|$. Finally, the formal integration with δ-function gives

$$\vec{B} = q\left(\frac{\vec{\beta}\times\vec{n}}{\kappa R^2} + \frac{1}{\kappa}\left[\frac{\vec{\beta}\times\vec{n}}{\kappa cR}\right]_{t'}\right), \tag{3.85}$$

with the retarded time $t' = t - R/c$. The first term in equation (3.85) reproduces the Biot–Sawart formula. As we see, it is derived directly from the Maxwell system. The terms in equations (3.69) and (3.85) with $\dot{\vec{s}}(t')$ relate to the emission because of the character of its behavior at infinity $\sim\frac{1}{R}$ (decreasing). Differentiating in equation (3.85), we arrive at

$$\vec{B} = q\left(\frac{\vec{\beta}\times\vec{n}}{\kappa R^2} + \frac{1}{c\kappa}\left[\frac{\vec{\beta}}{\kappa R}\right]_{t'}\times\vec{n} + \frac{\vec{\beta}}{\kappa^2 R^2}\times\vec{n}\right). \tag{3.86}$$

From equations (3.86) and (3.78) it is seen that

$$\vec{B} = \vec{n}\times\vec{E}, \tag{3.87}$$

that allows to evaluate easily the Pointing vector

$$\vec{S} = \frac{c}{4\pi}\vec{E}\times\vec{B} = \frac{c}{4\pi}\vec{E}\times[\vec{n}\times\vec{E}] = \frac{c}{4\pi}(\vec{n}E^2 - \vec{E}(\vec{n}\cdot\vec{E})), \tag{3.88}$$

where

$$\begin{aligned}\vec{E} &= \frac{q\vec{n}}{(\kappa R)^2} - \frac{q\vec{\beta}}{(\kappa R)^2} + q\frac{\vec{n}-\vec{\beta}}{c\kappa}\left[\frac{1}{\kappa R}\right]_{t'} - \frac{q}{c\kappa}\left[\frac{\vec{\beta}_{t'}}{\kappa R}\right] = \frac{q\vec{n}}{(\kappa R)^2} - \frac{q\vec{\beta}}{(\kappa R)^2}\\ &+ q\frac{\vec{n}-\vec{\beta}}{c\kappa^2}\left[\frac{c}{R^2}(\vec{n}\cdot\vec{\beta}) - \frac{1}{\kappa R}\left(\frac{\vec{\beta}^2 - (\vec{n}\cdot\vec{\beta})^2}{R} - \vec{n}\cdot\dot{\vec{\beta}}\right)\right] - \frac{q}{c\kappa}\left[\frac{\dot{\vec{\beta}}}{\kappa R}\right].\end{aligned} \tag{3.89}$$

As in the case of the magnetic field (3.86) one observes the terms that are proportional to the point acceleration $\dot{\vec{\beta}}$ and decrease as $1/R$ at infinity. Such terms define the emitted electromagnetic field at an arbitrary reference frame.

When one goes to the rest reference frame of the emitting particle, $\beta = 0$, $\kappa = 1$, the formulas significantly simplify.

$$\vec{E} = \frac{q\vec{n}}{R^2} - \frac{q}{cR}[\vec{n}(\vec{n}\cdot\dot{\vec{\beta}}) - \dot{\vec{\beta}}] = \frac{q\vec{n}}{R^2} + \frac{q}{cR}(\vec{n}\times[\vec{n}\times\dot{\vec{\beta}}]) = E_c + E_e, \tag{3.90}$$

$$\vec{B} = -\frac{q}{cR}[\vec{n}\times\dot{\vec{\beta}}]). \tag{3.91}$$

The energy conservation relation (Pointing theorem of section 2.5) characterizes the phenomenon of electromagnetic field generation in detail, important for applications. The Pointing vector (2.61), as of this contribution (second term in equation (3.90) only) may be evaluated either directly by

$$\vec{S} = \frac{c}{4\pi}[\vec{E}\times\vec{B}] = \frac{q^2}{c^3R^2}(\ddot{\vec{s}}^2\vec{n} - (\vec{n}\cdot\ddot{\vec{s}})\ddot{\vec{s}}) = \frac{q^2}{4\pi c^3 R^2}\ddot{\vec{s}}^2\sin^2\vartheta\vec{n}, \tag{3.92}$$

or by equation (3.88), taking into account that $(\vec{n}\cdot E_e) = \frac{q}{cR}(\vec{n}\cdot(\vec{n}\times[\vec{n}\times\dot{\vec{\beta}}]) = \frac{q}{cR}(\vec{n}\cdot(\vec{n}(\vec{n}\cdot\dot{\vec{\beta}}) - \dot{\vec{\beta}}) = 0$.

The angle between the vectors of acceleration $\ddot{\vec{s}}$ and normal unit one $\vec{n}$ is exactly the spherical angle ϑ if the z-axis is chosen along the $\ddot{\vec{s}}$. Integration of equation (3.92) over a sphere surface centered on $\vec{s}$ yields a known compact formula

$$\oint \vec{S} \cdot d\vec{\Sigma} = \frac{2q^2}{c^3}\ddot{\vec{s}}^2 \int_0^{\pi} \sin^3 \vartheta d\vartheta = \frac{2q^2}{3c^3}\ddot{\vec{s}}^2. \tag{3.93}$$

The expression (3.93) shows that the electromagnetic wave emission arises from moving charged particle only with nonzero acceleration.

Problems.

Exercise 1. Evaluate the intensity of wave emitted by a charge e rotating by a circumference of radius a.

Exercise 2. Average the result by the period of rotation.

3.8 Emission of oscillating charged system of particles: multipole expansion

The general formulas for the emitted electromagnetic field that have been derived in the previous sections are too complicated to use in applications. Moreover, the form in which the system of charges enter the integrands of the field components (3.82) or (3.73) need more details that can be supported by direct measurements. Therefore the standard notions of momenta descriptions such as multipole momenta are used in the context of the correspondent expansion. We start from an important case of waves with a fixed frequency.

3.8.1 General remarks

Let us consider a system of particles where charge and current densities change periodically with time.

$$\rho = \rho_0 e^{-i\omega t} + \rho_0^* e^{i\omega t}, \quad j_l = j_{0l} e^{-i\omega t} + j_{0l}^* e^{i\omega t}, \tag{3.94}$$

i.e. the rhs of equation (3.4) is also periodic in time function, i.e.

$$f_l = 4\pi\left(\frac{i\omega}{c^2}j_{0l} - \nabla\rho_{0l}\right)e^{-i\omega t} + c.\ c. = F_l e^{-i\omega t} + F_l^* e^{i\omega t}. \tag{3.95}$$

(index l marks specific components: $l = 1, 2, 3$); it is known that the solution of a boundary problem for equation (3.4) may be constructed as a superposition of the correspondent components:

$$E_l = U_l e^{-i\omega t} + U_l^* e^{i\omega t}, \tag{3.96a}$$

$$B_l = W_l e^{-i\omega t} + W_l^* e^{i\omega t}. \tag{3.96b}$$

If one substitutes equation (3.96a) to equation (3.4), taking into account equation (3.95), the inhomogeneous Helmholtz equation is obtained

$$(k^2 + \nabla^2)U_l = -F_l. \tag{3.97}$$

If one considers the problem of wave generation by a system of charged particles (from some closed volume V_g) at the physical space without boundaries, it is convenient to directly apply the Green function method (see section 3.6.2). The Green function for the whole 3D space for the Helmholtz equation (3.97) has the form (3.63):

$$G = \frac{e^{\pm ikR}}{4\pi R} \tag{3.98}$$

where $R = |\vec{r} - \vec{r}'|$. To apply the Green formula, we would determine amplitude of electric field components U_k as:

$$U_l = \int_V G(\vec{r}, \vec{r}')F_l(\vec{r}')d\vec{r}' + \int_S \left(U_l \frac{\partial G}{\partial n} - G \frac{\partial U_l}{\partial n} \right) dS, \tag{3.99}$$

where the second integral of the right-hand side goes to zero if the surface S by which we integrate tends to infinity. The first integral, that is calculated along the domain $V \to \infty$, is nonzero if $F_l \neq 0$, that is valid for $\vec{r}' \in V_g$. After that, from equality (3.99), it follows:

$$U_l = \int_{V_g} \frac{e^{ik|\vec{r}-\vec{r}'|}}{4\pi|\vec{r} - \vec{r}'|} F_l(\vec{r}')d\vec{r}'. \tag{3.100}$$

3.8.2 Multipole expansion

Suppose the charges (and, naturally, the currents) are absent outside some domain V_g, expanding the Green function in the Taylor series with respect to the vector $\vec{r}'$ components, gives:

$$\frac{e^{ik|\vec{r}-\vec{r}'|}}{|\vec{r} - \vec{r}'|} = \frac{e^{ikr}}{r}\left(1 - \left(ik - \frac{1}{r}\right)\frac{(\vec{r}, \vec{r}')}{r} + \cdots\right), \tag{3.101}$$

that is valid for $\vec{r} \in R^3 \backslash V_g$. A finite number of terms gives good approximation for the range in which $|\vec{r}| \gg |\vec{r}'|$. The function $F_l(\vec{r}')$ will be obtained if the function f_l definition (3.95) is used:

$$F_l(\vec{r}') = 4\pi\left(\frac{i\omega}{c^2}j_{0l} - \nabla_l \rho_0\right). \tag{3.102}$$

Hence

$$U_l = \frac{e^{ikr}}{r}\int_{V_g}\left(1 - \left(ik - \frac{1}{r}\right)\frac{(\vec{r}, \vec{r}')}{r} + \ldots\right)\left(\frac{i\omega}{c^2}j_{0l}(\vec{r}') - \nabla_l \rho_0(\vec{r}')\right)d\vec{r}'. \tag{3.103}$$

The first term of the *multipole expansion* for (3.103),

$$U_l^d = \frac{e^{ikr}}{r}\int_{V_g}\left(\frac{i\omega}{c^2}j_{0l} - \nabla_l\rho_0\right)d\vec{r}', \tag{3.104}$$

we name the **dipole approximation**. The next terms of the expansion in equation (3.103) yield the quadrupole approximation, magnetic dipole approximation and so on.

Let us make some transformations in equation (3.104); it is necessary to evaluate the integral, using the identity

$$\int_{V_g}\frac{\partial\rho_0}{\partial x_l}d\vec{r}' = 0, \tag{3.105}$$

which could be proven by representation of the integral in equation (3.104) as the triple one in Cartesian coordinates and direct use of integration by x_l with zero boundary conditions at the outer domain $\vec{r} \in R^3\backslash V_g$.

On the base of the continuity equation $-\omega\rho_0 + \text{div}\,\vec{j}_0 = 0$, we arrive at the following statement:

$$\int_{V_g} j_{0l}\,d\vec{r}' = -\int_{V_g} x_l'\nabla\cdot\vec{j}_0\,d\vec{r}' = -i\omega\int_{V_g} x_l'\rho_0(\vec{r}')d\vec{r}' = -i\omega p_l, \tag{3.106}$$

where

$$p_l = \int_{V_g} x_l'\rho_0(\vec{r}')d\vec{r}'. \tag{3.107}$$

This expression (3.107) defines the *dipole momentum*, the important notion that links the macroscopic polarization vector to appear within the matter electrodynamics theory (see chapter 5).

A proof of equation (3.106) relies upon a component consideration, e.g. $l = 1$

$$\int_{V_g} x_1'\nabla\cdot\vec{j}_0\,d\vec{r}' = \int_{V_g} x_1'\left(\frac{\partial j_{0x}}{\partial x_1'} + \frac{\partial j_{0y}}{\partial x_2'} + \frac{\partial j_{0z}}{\partial x_3'}\right)dx_1'dx_2'dx_3', \tag{3.108}$$

the derivative $\frac{\partial j_{0l}}{\partial x_l}$, similar to equation (3.105), give zero contribution for $l = 2, 3$. After integration by parts one has

$$= -\int_{V_g} j_{01}\,dx_1'dx_2'dx_3' = -\int_{V_g} j_{01}\,d\vec{r}'. \tag{3.109}$$

The rest components are similarly studied. Plugging equations (3.106) and (3.105) into equation (3.104) yields

$$U_l^d = \frac{\omega^2}{c^2}\frac{e^{ikr}}{r}p_l. \tag{3.110}$$

3.8.3 The emission phenomenon in the dipole approximation

We derived the explicit expression (3.110), that describes the dipole approximation (first approximation of the multipole expansion) in terms of the dipole momentum $\vec{p}$, (3.107):

$$U_l^d = k^2 \frac{e^{ikr}}{r} p_l, \tag{3.111}$$

the dispersion relation for vacuum $\omega = ck$ is used. The field $\vec{E}_l$ is evaluated along equation (3.96a).

$$E_l = k^2 \frac{p_l}{r} (e^{ikr - i\omega t} + e^{-ikr + i\omega t}). \tag{3.112}$$

To calculate the field $\vec{B}$ we use Faraday's law and take account of equation (3.96b):

$$\frac{-i\omega}{c} \vec{W}^d = -\nabla \times \vec{U}^d. \tag{3.113}$$

In tensor components notations (see chapter 20, section 20.1) we have

$$\frac{i\omega}{c} W_i^d = \varepsilon_{isl} \frac{\partial U_l^d}{\partial x_s} = k^2 \varepsilon_{isl} \frac{\partial}{\partial x_s} \left(\frac{e^{ikr}}{r} p_l \right). \tag{3.114}$$

Next, it results in the expression for the amplitude W_i^d in terms of the dipole momentum vector $\vec{p}$

$$W_i^d = -ik\varepsilon_{isl} \left(\left(ik - \frac{1}{r} \right) \frac{x_s}{r} e^{ikr} \right) p_l = -ik \left(\left(ik - \frac{1}{r} \right) \frac{e^{ikr}}{r} \right) [\vec{r} \times \vec{p}]_i. \tag{3.115}$$

Tensor notations link Cartesian components of the vectors and its direct product as follows $\varepsilon_{isl} x_s p_l = [\vec{r} \times \vec{p}]_i$.

If one supposes that the wavelength $\lambda = 2\pi/k$ is small compared to the distance between the source (V_g) ('antenna' domain) and the observation point $\vec{r}$, ($\lambda \ll r$), the term with r^{-1} in equation (3.115) will be small

$$W_k^d = k^2 \frac{e^{ikr}}{r} \left[\frac{\vec{r}}{r} \times \vec{p} \right]_k. \tag{3.116}$$

Summarizing, we arrive at the complete expression for the fields (3.96a) (taking into account the complex conjugate parts):

$$\vec{E}^d = 2k^2 \frac{\cos(kr - \omega t)}{r} \vec{p}, \tag{3.117a}$$

$$\vec{B}^d = 2k^2 \frac{\cos(kr - \omega t)}{r} \left[\frac{\vec{r}}{r} \times \vec{p} \right]. \tag{3.117b}$$

Let us now evaluate the energy flux vector (Pointing vector) by expression (2.61)

$$\vec{S}^d = \frac{c}{\pi}k^4\frac{\cos^2(kr - \omega t)}{r^2}\vec{p} \times \left[\frac{\vec{r}}{r} \times \vec{p}\right], \tag{3.118}$$

taking projecting of the Pointing vector (3.118) to the normal direction to a sphere which contains the observation point. The vector describes energy transport through the surface unit per unit time, therefore the value of $\vec{S}_i$ oscillates with time. Projecting equation (3.118) to the direction of propagation $\vec{n} = \vec{r}/r$ yields

$$S_n^d = \left(\vec{S}, \vec{n}\right) = \frac{\omega^4}{\pi r^2 c^3}\cos^2(kr - \omega t)\left(\vec{p}^2 - \left(\vec{p} \cdot \frac{\vec{r}}{r}\right)^2\right). \tag{3.119}$$

Finally: let similar the case of isolated particle emission (see the previous section), the vector $\vec{p}$ is directed along the z-axis and the angle between the axis and the vector $\vec{r}$ is equal ϑ, then

$$S_n^d = \frac{\omega^4}{\pi r^2 c^3}p^2\cos^2(kr - \omega t)\sin^2\vartheta. \tag{3.120}$$

Integrating over a sphere with radius r surface and with respect to t by a period $2\pi/\omega$, one arrives at

$$\frac{\omega}{2\pi}\int d\Sigma\int_0^{2\pi/\omega} dt S_n^d = \frac{4\omega^4}{3c^3}p^2, \tag{3.121}$$

the relation $\int_0^\pi \sin^3 x dx = \frac{4}{3}$ is used. See exercises 1 and 2 and alternative derivations in [6, 7].

There is a traditional *illustration*, which is known as the dipole system. It is a system of particles with oscillating magnitude of charges (see equation (3.94)) where charge density is modeled as a combination of point opposite charges

$$\rho_0(\vec{r}') = q(\delta(\vec{r}' - \vec{d}) - \delta(\vec{r}' + \vec{d})). \tag{3.122}$$

The system is presented in figure 3.6. Substituting equation (3.122) into the dipole momentum (3.107) definition yields

$$\vec{p} = \int \vec{r}' q(\delta(\vec{r}' - \vec{d}/2) - \delta(\vec{r}' + \vec{d}/2))d\vec{r}' = q\vec{d},$$

often used in the dipole system notations.

3.8.4 Next approximation: quadrupole momentum

To take into account the next approximation in the multipole expansion, let us return to the expression (3.103)

$$U_l = -\frac{e^{ikr}}{r}\int_{V_g}\left(1 - ik\frac{(\vec{r}, \vec{r}')}{r} + \dots\right)\left(\frac{i\omega}{c^2}j_{0l} + \nabla_l\rho_0\right)d\vec{r}'. \tag{3.123}$$

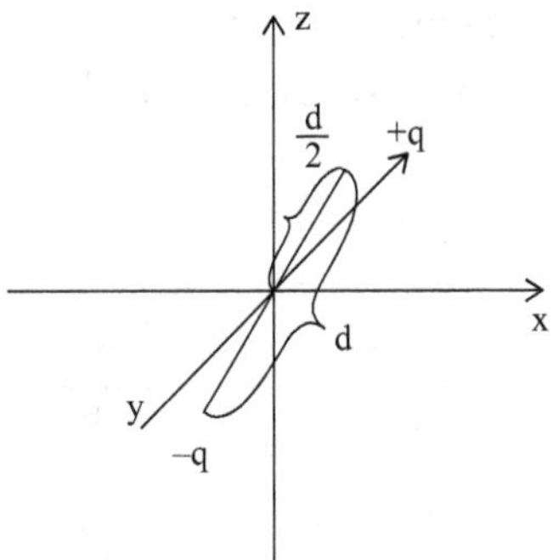

Figure 3.6. The dipole model charges position in space.

The integrals in equation (3.123) can be divided to one that is already expressed in terms of the dipole momentum, see equation (3.111). Hence we should transform the second integral

$$U_l^{(2)} = \frac{ik}{r^2}e^{ikr}\int_{V_g}(\vec{r}, \vec{r}')\left(\frac{i\omega}{c^2}j_{0l} + \nabla_l\rho_0\right)d\vec{r}', \tag{3.124}$$

in which we also take two terms

$$U_l^{(2,\,1)} = -\frac{k\omega}{c^2r^2}e^{ikr}\int_{V_g}(\vec{r}, \vec{r}')j_{0l}\,d\vec{r}', \tag{3.125a}$$

$$U_l^{(2,\,2)} = \frac{ik}{r^2}e^{ikr}\int_{V_g}(\vec{r}, \vec{r}')\nabla_l\rho_0 d\vec{r}'. \tag{3.125b}$$

The integral (3.125b) by the statement (3.105) is equal to zero because we suppose the system of particles under consideration (e.g. atom or antenna) to be neutral ($\int_{V_g}\rho(\vec{r})d\vec{r} = 0$). From the other side (3.125a) for the EM field component with $l = 1$ takes the form

$$U_1^{(2,\,1)} = -\frac{k\omega}{c^2r^2}e^{ikr}\int_{V_g}(x_1x_1' + x_2x_2' + x_3x_3')j_{0,\,1}\;dx_1'dx_2'dx_3'. \tag{3.126}$$

Similar to the case of the statement (3.106) a proof we want to express $j_{0,\,1}$ via the divergence of $\vec{j}_0$ under the integral. For this purpose the integral is transformed via by parts procedure as

$$\begin{aligned}\int_{V_g} x_1'j_{01}dx_1'dx_2'dx_3' &= \frac{x'^2_1j_{01}}{2}\Big|_{-\infty}^{+\infty} \\ &-\frac{1}{2}\int_{V_g} x'^2_1\left[\frac{\partial j_{01}}{\partial x_1} + \frac{\partial j_{02}}{\partial x_2} + \frac{\partial j_{03}}{\partial x_3}\right]dx_1'dx_2'dx_3',\end{aligned} \tag{3.127}$$

we have added two terms at square brackets that yield in zero contribution after integrating. Following the idea we add *zero* terms $\frac{\partial j_{0k}}{\partial x_k}$, $k = 2, 3$, so

$$\begin{aligned}\int_{V_g} x_l' j_{0l} dx_1' dx_2' dx_3' &= \frac{x_1'^2 j_{01}}{2}\Big|_{-\infty}^{+\infty} \\ &- \frac{1}{2}\int_{V_g} x_1'^2 \left[\frac{\partial j_{01}}{\partial x_1} + \frac{\partial j_{02}}{\partial x_2} + \frac{\partial j_{03}}{\partial x_3}\right] dx_1' dx_2' dx_3'.\end{aligned} \tag{3.128}$$

The divergence of the electric current allow us to take the electric charge conservation law (continuity equation):

$$\int_{V_g} x_1' j_{01} d\vec{r}' = \frac{i\omega}{2}\int_{V_g} x_1'^2 \rho_0(\vec{r}') d\vec{r}' = \frac{i\omega}{2} Q_{11}'. \tag{3.129}$$

Finally, the first component of the electric field is

$$U_1^{(2,1)} = \frac{ik\omega^2}{2c^2r^2} e^{ikr} x_l \int_{V_g} x_L'^2 \rho_0 \; d\vec{r}' = \frac{ik\omega^2}{c^2 r} e^{ikr} x_l Q_{l1}'. \tag{3.130}$$

Generally,

$$U_l^{(2)} = -\frac{k^2}{c^2r^2} e^{ikr} \left(\vec{r}, \int_{V_g} \vec{r}' j_{0l} d\vec{r}'\right), \tag{3.131}$$

$$U_s^Q = -\frac{ik}{c^2r^2} e^{ikr} x_p Q_{ps}', \tag{3.132}$$

where a tensor characteristic of the density distribution

$$Q_{ik}' = \int x_i' x_k' \rho_0(\vec{r}') d\vec{r}', \tag{3.133}$$

is introduced. Note, that quadrupole moment tensor is conventionally defined as

$$Q_{ik} = \int \rho_0(\vec{r}')(3x_i' x_k' - r^2 \delta_{ik}) d\vec{r}'. \tag{3.134}$$

The expression (3.131) represents the **Quadrupole approximation** for the emitted wave.

3.8.5 Dipole magnetic moment definition and corresponding wave emission

The magnetic dipole moment may be introduced applying the multipole expansion at the magnetic field components in a bit more direct way. Let us return to the basic representation (3.96a) and take the relation for the magnetic field, which gives the amplitude W_i, obtained quite similarly to the one for the electric field $\vec{E}$ in section 3.1.

Differentiating equation (3.1) with respect to time, we eliminate the electric field $\vec{E}$, similar to equation (3.3)

$$\frac{1}{c^2}\frac{\partial^2 \vec{B}}{\partial t^2} = \Delta \vec{B} + \frac{4\pi}{c}\mathrm{rot}\vec{j}\,. \tag{3.135}$$

Let us return to the basic representation (3.96b) and take the equation for the magnetic field that is obtained from equation (3.135) after substitution from equation (3.96b). For complex amplitudes one has the Helmholtz equation again. It has, however, the different right side:

$$(\Delta + k^2)W_j = F_{bj}, \tag{3.136}$$

where

$$\vec{F}_b = \frac{4\pi}{c}\mathrm{rot}\vec{j}_0\,.$$

Then, for a harmonic wave, the ω-Fourier component of the vector, $\vec{B}$, for each component of j, we have

$$W_j = \int_V G(\vec{r}, \vec{r}\,')F_{bj}(\vec{r}\,')d\vec{r}\,'. \tag{3.137}$$

Taking a part of the Green function expansion (3.101) that decreases up to $1/r^2$, supposing again $\lambda \ll r$

$$G \sim -\frac{ik}{4\pi r^2}e^{ikr}\left(\vec{r} - \frac{\vec{r}\,'}{2}\right)\cdot \vec{r}\,',$$

one, e.g. for the x-component, arrives at

$$W_x \;=\; -\frac{ik}{cr^2}e^{ikr}\left(\vec{r}\int \vec{r}\,'\left(\frac{\partial j_{0z}}{\partial y'} - \frac{\partial j_{0y}}{\partial z'}\right)d\vec{r}\,' - \frac{1}{2}\int \vec{r}\,'^2\left(\frac{\partial j_{0z}}{\partial y'} - \frac{\partial j_{0y}}{\partial z'}\right)d\vec{r}\,'\right). \tag{3.138}$$

The result transforms to

$$W_x = -\frac{ik}{cr^2}e^{ikr}\left[\left(y\int j_{0z}d\vec{r}\,' - z\int j_{0y}d\vec{r}\,'\right) - \frac{1}{2}\int \vec{r}\,'^2\left(\frac{\partial j_{0z}}{\partial y'} - \frac{\partial j_{0y}}{\partial z'}\right)d\vec{r}\,'\right], \tag{3.139}$$

by the identity

$$\vec{r}\int \vec{r}\,'\left(\frac{\partial j_{0z}}{\partial y'} - \frac{\partial j_{0y}}{\partial z'}\right)d\vec{r}\,' = \int\left(yj_{0z} - zj_{0y}\right)d\vec{r}\,' = y\int j_{0z}d\vec{r}\,' - z\int j_{0y}d\vec{r}\,',$$

that is obtained, integrating by parts.

The result is expressed via the dipole momentum (3.107) and account identities such as (see also the previous section)

$$\int \left(x' \operatorname{div} \vec{j}_0(\vec{r}\,')\right) d\vec{r}\,' = -\int j_{0x} d\vec{r}\,' = \iota\omega \int x' \rho(\vec{r}\,') d\vec{r}\,' = \iota\omega p_x,$$

which is a fundamental notion in the polarization phenomena description that constitutes in the *polarization vector* (see section 3.8.3).

The second term

$$\int \vec{r}\,'^2 \left(\frac{\partial j_{0z}}{\partial y'} - \frac{\partial j_{0y}}{\partial z'} \right) d\vec{r}\,' = \int [2y' j_{0z} - 2z' j_{0y}] d\vec{r}\,' = 2m_x, \tag{3.140}$$

may be transformed as follows and contains the *magnetic dipole vector* x-projection

$$m_x = \int \left(y' j_{0z}(\vec{r}\,') - z' j_{0y}(\vec{r}\,') \right) d\vec{r}\,'.$$

So, the magnetic dipole emission for equation (3.140) is defined by

$$W_x^m = \frac{ik}{cr^2} e^{ikr} m_x. \tag{3.141}$$

Generally, the whole three component object defines the ***magnetic dipole moment vector***

$$\vec{m} = \int [\vec{r}\,' \times \vec{j}_0] d\vec{r}\,', \tag{3.142}$$

and

$$\vec{W}^m = \frac{ik}{cr^2} e^{ikr} \vec{m}. \tag{3.143}$$

The full vectors of the electromagnetic fields and the Pointing one are composed as in section 3.8.3. Which in turn enters the magnetization vector of a matter (chapter 5, section 5.2.3) in a similar approach. The subsequent transition to electric field is realized on a base of the Faraday equation and the resulting expressions for the field B and E allows to evaluate the Pointing vector and next calculate the intensity of the emitted electromagnetic field as a function of the distance between the source and a detector, parameterized by electric and magnetic momenta.

3.8.6 Minidiscussion and exercises

The results of sections 3.7 and 3.8 in a sense reproduce the formulas of chapter 3 of [6]; (see also [7]); we, however, do not use the traditional Liénard–Wiechert potentials. The expressions for fields $\vec{E}$ (3.77) and for $\vec{B}$ (3.86) coincide with ones (14.1) from [6], where the detailed discussion of relativistic origin of the point particle emission formulas (3.88) and (3.89) is performed.

Problems

Exercise 1. Derive the total intensity of the electromagnetic field emitted by an electric dipole.

Plot the curves of a constant intensity in the xz plane, z-propagation axis in SI units for a given frequency and unit amplitude of oscillations.

Exercise 2. Derive the total intensity of the electromagnetic field emitted by a magnetic dipole.

Exercise 3. Derive the total intensity of the electromagnetic field emitted by an electrically charged point particle moving by round orbit with constant velocity.

Exercise 4. Derive the formulas of (Lorentz) transformations of electromagnetic field by means of the formulas for electrically charged point particle moving with constant velocity.

Exercise 5. Consider the transformation of an antisymmetric tensor component

$$A_{\alpha'\beta'} = \sum_{\alpha\beta} L_{\alpha'\alpha} L_{\beta'\beta} A_{\alpha\beta},$$

with respect to the Lorentz transformations L. Compare with the results of section 18.3, the matrix by equation (18.18), formulas for the fields $\vec{E}$, $\vec{B}$, identifying the tensor component with the vector fields components.

Exercise 6. Consider the quadrupole distribution of charges in the xy plane (double dipole), represented by

$$\rho_0(\vec{r}) = q\delta(\vec{r} - \vec{r}_1) + q\delta(\vec{r} - \vec{r}_2) - q\delta(\vec{r} - \vec{r}_3) + q\delta(\vec{r} - \vec{r}_4), \tag{3.144}$$

where $\vec{r}_1 = (d/2, d/2, 0)$, $\vec{r}_2 = (-d/2, -d/2, 0)$, $\vec{r}_3 = (d/2, -d/2, 0)$, $\vec{r}_4 = (-d/2, d/2, 0)$, see figure 3.7; and derive the field (3.124) by means of expression (3.132).

Exercise 7. Show that for a plane round current (I) with area S the magnetic dipole moment module is equal to $m = IS$ (see e.g. [2]).

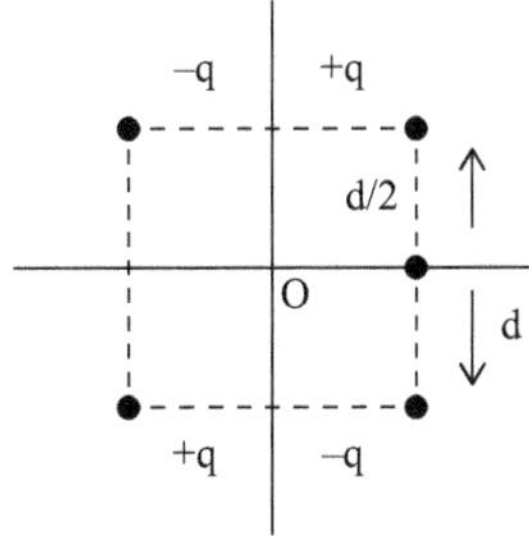

Figure 3.7. The quadrupole set of charges.

References

[1] Kuszner M and Leble S 2011 Directed electromagnetic pulse dynamics: projecting operators method *J. Phys. Soc. Jpn.* **80** 024002

[2] Griffiths D J 1989 *Introduction to Electrodynamics* 2nd edn (Englewood Cliffs, NJ: Prentice-Hall)

[3] Leble S 1990 *Nonlinear Waves in Waveguides* (Heidelberg: Springer)

[4] Leble S and Perelomova A 2018 *Dynamical Projectors Method in Hydro- and Electrodynamics* (Boca Raton, FL: CRC Press)

[5] Kinsler P, Radnor S B P and New G H C 2005 Theory of directional pulse propagation *Phys. Rev.* A **72** 063807

[6] Jackson J D 1998 *Classical Electrodynamics* 3rd edn (New York: Wiley)

[7] Stratton J A 1941 *Electromagnetic Theory* (New York: McGraw-Hill)

IOP Publishing

Practical Electrodynamics with Advanced Applications

Sergey Leble

Chapter 4

Theory of relativity

4.1 Lorentz transformation

The base, on which the so-called 'Theory of relativity' is built as a covariance of Maxwell's equations (2.28)–(2.31) with respect to transition from one inertial reference frame to another. Speaking more precisely we mean the *covariance- or form-invariance* of the system of differential equations (2.28)–(2.31) with respect to the group of special transformation of space–time [2]. The famous transformations, known under the name of *Lorentz transformations*, play an important role in modern physics. We shall use invariance of the important direct corollary of the Maxwell system—the wave equation for a component of electromagnetic field in vacuum (3.4) and (3.5) as a platform for the Lorentz transformations derivation.

So, the first aim of this chapter is a derivation of explicit expressions for the Lorentz transformations.

Invariance of the quabla operator ($\Box$) in this context one should understand as invariance of this operator form with respect to transition from one to another inertial reference frame (RF), L, L', that moves one relative to another with constant velocity. Such operator is the classical wave operator of the correspondent wave equation; as was established in section 3.1—the equation describes wave phenomena; propagation of the electromagnetic field from one space point to a different one. Thorough analysis of all physics allows to consider such wave phenomenon as the only possible carrier of information about position in space. Note the variables of position in space via $\vec{r}$, $\vec{r}'$ while the variables of time: t, t'. The condition of form-invariance (covariance) of the electromagnetic wave operator in vacuum for a transition of the frame L to L' we can write in a symbolic form:

$$\Box = \Box'. \tag{4.1}$$

It is important to stress that the electromagnetic wave velocity does not change while the transition between RFs is realized [3, 14].

doi:10.1088/978-0-7503-2576-9ch4

Simplifying the problem, suppose that the wave is plane and propagates along x. A generalization to the three-dimensional space is not complicated; may be realized by rotations in $\vec{r}$ space. Consider now a wave function U, of the electromagnetic wave, it may be a component of the electric (magnetic) field see once more (3.4) and (3.6); then the condition (4.1) reads

$$\frac{1}{c^2}\frac{\partial^2 U}{\partial t^2} - \frac{\partial^2 U}{\partial x^2} = \frac{1}{c^2}\frac{\partial^2 U}{\partial t'^2} - \frac{\partial^2 U}{\partial x'^2}. \tag{4.2}$$

Introduce functions, that translate variables x, t of the RF 'L' to 'primed' ones of L':

$$\begin{aligned} t' &= g(x, t), \\ x' &= f(x, t). \end{aligned} \tag{4.3}$$

Calculating derivatives of the implicit functions:

$$\frac{\partial U}{\partial t} = U_t = \frac{\partial U}{\partial x'}\frac{\partial x'}{\partial t} + \frac{\partial U}{\partial t'}\frac{\partial t'}{\partial t} = \frac{\partial U}{\partial x'}\frac{\partial f}{\partial t} + \frac{\partial U}{\partial t'}\frac{\partial g}{\partial t}, \tag{4.4a}$$

$$\frac{\partial U}{\partial x} = U_x = \frac{\partial U}{\partial x'}\frac{\partial x'}{\partial x} + \frac{\partial U}{\partial t'}\frac{\partial t'}{\partial x} = \frac{\partial U}{\partial x'}\frac{\partial f}{\partial x} + \frac{\partial U}{\partial t'}\frac{\partial g}{\partial x}. \tag{4.4b}$$

Differentiating once more, one have:

$$\begin{aligned} U_{tt} &= U_{t't'}g_t^2 + 2U_{t'x'}g_t f_t + U_{t'}g_{tt} + U_{x'x'}f_t^2 + U_{x'}f_{tt}, \\ U_{xx} &= U_{x'x'}f_x^2 + 2U_{t'x'}g_x f_x + U_{x'}f_{xx} + U_{t't'}g_x^2 + U_{t'}g_{xx}. \end{aligned} \tag{4.5}$$

Plugging to equation (4.2) yields

$$g_t f_t = c^2 g_x f_x, \tag{4.6a}$$

$$f_t^2 - c^2 f_x^2 = -c^2, \tag{4.6b}$$

$$g_t^2 - c^2 g_x^2 = 1, \tag{4.6c}$$

$$g_{tt} - c^2 g_{xx} = 0, \tag{4.6d}$$

$$f_{tt} - c^2 f_{xx} = 0. \tag{4.6e}$$

Consider the second of these equations

$$f_t^2 - c^2 f_x^2 = (f_t - cf_x)(f_t + cf_x) = -c^2. \tag{4.7}$$

Next, in characteristic variables $\xi = x - ct$, $\eta = x + ct$, we have

$$\frac{\partial f}{\partial \xi} = \frac{\partial f}{\partial t} - c\frac{\partial f}{\partial x}, \tag{4.8a}$$

$$\frac{\partial f}{\partial \eta} = \frac{\partial f}{\partial t} + c\frac{\partial f}{\partial x}. \tag{4.8b}$$

Or, denoting derivatives as indices it is written as

$$f_t - cf_x = f_\xi, \tag{4.9a}$$

$$f_t + cf_x = f_\eta, \tag{4.9b}$$

we derive for equation (4.7) the following,

$$f_\xi f_\eta = -c^2. \tag{4.10}$$

Differentiate the result by ξ, one obtains

$$f_{\xi\xi} f_\eta + f_\xi f_{\eta\xi} = 0. \tag{4.11}$$

Equation (4.6e) is equivalent to $f_{\eta\xi} = 0$, therefore

$$f_{\xi\xi} = 0.$$

In similar way we derive

$$g_{\xi\xi} = 0,$$

and, quite the same trick gives

$$f_{\eta\eta} = 0, \quad g_{\eta\eta} = 0.$$

Due to the fact that all second derivatives of the functions $f(\xi, \eta)$, $g(\xi, \eta)$ are equal to zero, the functions must be linear with respect to both variables ξ, η, or, due to its definitions, may be written in a form

$$x' = f(t, x) = at + bx, \tag{4.12a}$$

$$t' = g(t, x) = pt + qx. \tag{4.12b}$$

Substitute now equation (4.12) in the system (4.6). We obtain

$$p^2 - c^2q^2 = 1, \tag{4.13a}$$

$$a^2 - c^2b^2 = -c^2, \tag{4.13b}$$

$$pa = c^2qb. \tag{4.13c}$$

Recall now that, the 'primed' reference frame L' moves with a constant velocity $v < c$ with respect to the reference frame L. The motion is uniform, hence it is enough to choose the only point, for convenience $x' = 0$ (reference point of the frame L'); from the formula $f(t, x) = at + bx = 0$, it follows that $\frac{x}{t} = v = -\frac{a}{b}$. The following link of the parameters a, b, p, q with the motion parameter v appears;

$$a = -vb, \tag{4.14a}$$

$$p = -\frac{c^2}{v}q, \tag{4.14b}$$

that, after substitution to the equation (4.13b), allows to express them as

$$b = \frac{1}{\sqrt{1 - \frac{v^2}{c^2}}}. \tag{4.15}$$

From the relation $at + bx = x'$ one arrives at

$$x' = \frac{x - vt}{\sqrt{1 - \frac{v^2}{c^2}}}. \tag{4.16}$$

In the same order, from the identity (4.13a) it follows

$$q^2 = \frac{\frac{v^2}{c^2}}{c^2 - v^2}, \tag{4.17}$$

and

$$t' = \frac{t - \frac{v}{c^2}x}{\sqrt{1 - \frac{v^2}{c^2}}}. \tag{4.18}$$

The formulas (4.16) and (4.18) define Lorentz transformations, that connect the coordinates in frames L and L', moving one with respect to another with constant velocity v. The Lorentz transformation introduces a relation between space and time variables [14]. In the Galileo transformation the time variable figure out as absolute, time (clocks) goes in the same way and scales in all frames, in L and L'.

4.2 Space–time geometry

Considering equations (4.16) and (4.18) it is easy to check the important property of the Lorentz transformation

$$c^2t'^2 - x'^2 = c^2t^2 - x^2. \tag{4.19}$$

It may be thought of as invariance of the quadratic form

$$s^2 = c^2t^2 - x^2,$$

or, in direct analog, one for differentials,

$$ds^2 = c^2dt^2 - dx^2, \tag{4.20}$$

that may be names as (local) invariant for the differentials of x, t.

Natural generalization for space three dimensions (x, y, z), including time, named also as 3 + 1 dimensions, we write

$$ds^2 = c^2dt^2 - dx^2 - dy^2 - dz^2 = dx_0^2 - dx^2 - dy^2 - dz^2, \tag{4.21}$$

introducing the new variable $x_0 = ct$, that have the same physical units that a space coordinate (similarly—$dx_0 = cdt$) has.

Generally the form

$$ds^2 = \sum_{\nu,\mu=0}^{3} g_{\nu\mu} dx_\nu dx_\nu, \tag{4.22}$$

is used as the norm analog in the tangent space to the four-dimensional manifold (see, e.g. [2]), but it is obviously not Euclidean, because it does not satisfy a norm axiomatics, namely the positivity-definition [8]. The metric tensor $g_{\nu\mu} = \varepsilon_\nu \delta_{\nu\mu}$ numerated by Greek characters ($\mu, \nu = 0, 1, 2, 3$) is introduced. It is equation (4.21) diagonal with the diagonal elements $g_{00} = \varepsilon_0 = 1$, $g_{ii} = \varepsilon_i = -1$, $i = 1, 2, 3$. The relation corresponding to equation (13.53) gives a similar formula. As mentioned, such four-dimensional space is not a usual Euclidean one, it is named pseudo-Euclidean space or *Minkowski space*. We would accept it as a way to physical relativistic theory introducing also a scalar product analog of two vectors a, b with components a_μ, b_ν as the analog of a Euclidean-invariant *scalar product of two vectors*. It is following:

$$\{a, b\} = a_0b_0 - a_1b_1 - a_2b_2 - a_3b_3 = \sum_{\nu,\mu=0}^{3} g_{\nu\mu} a_\nu b_\mu = \sum_{\nu=0}^{3} \varepsilon_\nu a_\nu b_\nu. \tag{4.23}$$

In particular the time–space interval for a vector a is $\{a, a\} = a_0^2 - \vec{a}^2$.

Generally the Lorentz transformation is defined by a 4×4 matrix $L_{\nu\mu}$

$$x'_\mu = \sum_{\nu=0}^{3} L_{\mu\nu} x_\nu, \tag{4.24}$$

based on invariance of the (pseudo) scalar product (4.23). It is convenient to derive the general condition of invariance in terms of the matrix $L_{\nu\mu}$:

$$\{La, Lb\} = \sum_{\nu,\mu,\alpha,\beta=0}^{3} g_{\nu\mu} L_{\nu\alpha} L_{\mu\beta} a_\alpha b_\beta = \{a, b\}, \tag{4.25}$$

which yields

$$\sum_{\nu=0,\mu=0}^{3} g_{\nu\mu} L_{\nu\alpha} L_{\mu\beta} = g_{\alpha\beta}, \tag{4.26}$$

or, in matrix form

$$L^T G L = G. \tag{4.27}$$

This matrix condition defines the full Lorentz group.

Next we can introduce tensors in the Minkowski space (see also section 1.3 of chapter 1). The basic transformation of a vector A_ν component related to the 'first' reference frame gives values of the same vector in a 'second' reference frame

$$A'_\mu = \sum_{\nu=0}^{3} L_{\mu\nu} A_\nu, \tag{4.28}$$

along equation (4.24), that defines a vector as a tensor of the first rank.

For a second rank tensor with component $T_{\mu\nu}$ we have the following transformation rule

$$T'_{\mu\nu} = \sum_{\mu',\nu'=0}^{3} L_{\mu\mu'} L_{\nu\nu'} T_{\mu'\nu'}, \tag{4.29}$$

with a natural generalization to arbitrary rank. The difference between tensors in Euclid and Minkowski spaces is the contraction operation. For two 3 + 1 vectors with the components A_μ and B_ν, the invariant form is built as equation (4.23), also

$$S = \sum_{\mu=0}^{3} g_{\mu\mu} A_\mu B_\mu, \tag{4.30}$$

with a similar relation for a tensor of the second rank.

4.3 Relativistic kinematics and four-vectors

The general kinematic idea is arranging all physical variables into such groups, that each of them presents a tensor in Minkowski's space. Basic coordinates in space–time ($x_0 = ct$, $x_1 = x$, $x_2 = y$, $x_3 = z$) include the fourth (we call it zero) constituent ct and create the first order tensor—the vector with respect to Lorentz group. For the set of velocity components $v_i = \frac{dx_i}{dt}$ we naturally add the speed of light $c = \frac{dx_0}{dt}$. An important question arises: what is the temporal component p_0 of the momentum p_i? The sense of this additional component can be explained also by the invariant of a similar combination equation (4.19):

$$m_0^2 c^2 = p_0^2 - p_1^2 - p_2^2 - p_3^2, \tag{4.31}$$

marked on the right side with the inclusion, therefore the coefficient c^2 has been entered. In the rest system $p_i = 0$, so the zero-component of the quadrupole momentum p_0 is proportional to the invariant lhs $p_0 = m_0 c$. The parameter m_0 is named as invariant mass.

For further interpretation, consider the non-relativistic limit $v/c \ll 1$, which is equivalent up to $p \ll m_0 c$. For simplicity, let us look at the one-dimensional case

$p_2 = p_3 = 0$, $p_1 = p$, which gives the first order Taylor expansion with respect to the small parameter p/m_0c:

$$p_0 = \sqrt{m_0^2c^2 + p^2} \approx m_0c\left(1 + \frac{p^2}{2m_0^2c^2}\right) = m_0c + \frac{p^2}{2m_0c}, \tag{4.32}$$

what can be read as the sum m_0c and the well known formula for kinetic energy, divided by c. We come to the conclusion that the value p_0c is equivalent to the energy of the point body.

In the same way, we add the zero-component to the vector of the current density, $\vec{j}$, namely $j_0 = c\rho$. It's easy to check the validity of the charge conservation relation (2.35)

$$\frac{\partial c\rho}{\partial ct} = \frac{\partial j_0}{\partial x_0} = -\mathrm{div}\vec{j} = -\sum_{k=1}^{3}\frac{\partial j_k}{\partial x_k}, \tag{4.33}$$

or, in relativistic notations,

$$\sum_{\mu=0}^{3}\frac{\partial j_\mu}{\partial x_\mu} = 0,$$

is invariant with respect to the Lorentz transformation. To check this statement, let's apply the formula to the derivative of the implicit function (chain rule)

$$\sum_{\mu=0}^{3}\frac{\partial j'_\mu}{\partial x'_\mu} = \sum_{\nu,\alpha=0}^{3} L_{\mu\nu}\frac{\partial j_\nu}{\partial x_\alpha}\frac{\partial x_\alpha}{\partial x'_\mu}.$$

The Lorentz transformation formula (4.24), or its analog for the four-vector (4.28) applied to current density component yields

$$j'_\nu = \sum_{\mu=0}^{3} L_{\nu\mu} j_\mu. \tag{4.34}$$

The derivatives

$$\frac{\partial x_\alpha}{\partial x'_\mu} = [GL^TG]_{\alpha\mu},$$

are calculated by the inverse to equation (4.24) transform

$$x_\alpha = \sum_{\mu=0}^{3} L_{\alpha\mu}^{-1} x'_\mu, \tag{4.35}$$

with a simple expression obtained by equation (4.26), or (4.27)

$$L^{-1} = GL^TG, \tag{4.36}$$

when the $GG = I$ condition has been applied. So the statement under consideration may be written as four-dimensional divergence invariance

$$\sum_{\mu=0}^{3} \frac{\partial j'_\mu}{\partial x'_\mu} = \sum_{\mu=0}^{3} \frac{\partial j_\mu}{\partial x_\mu}.$$

4.4 Relativistic mechanics

A fundamental physical postulate is that a transition to a different reference frame should not break the sense of the Maxwell equations. Here, in electrodynamics, let us stress, that the complete set of Maxwell's equations includes **not only** field components but, as well, the charge and current density, that, for a charged particle is proportional to its velocity $\vec{j} = q\vec{v}$. It means that the components of velocity, that enters the mechanics, **should** transform as the vector components in the space–time [14]. As was mentioned in the previous section 4.3; the fourth component of the four-vector of velocity is the velocity of light. Next, it is important to note that the complete (closed) system of equations (Maxwell ones + Newtons) should include also the Lorentz force [7].

So, we arrive at the main principle of the modern theoretical physics: it is the *principle of general covariance* [2]. It reads as

The basic equations of physics should have the tensor form, each side of equation would have to be a some rank tensor with respect to the Lorentz group.

Consider the case of two reference systems U and U', which move one in relation to the other. Newton's equations of motion (Newton II law) in these systems are different as shown by the Lorentz transformations form (4.24) (see also the exercises). The road to build the Lorentz-covariant equations that would go into the Newtonian form in non-relativistic limit ($v \ll c$), is based on the introduction of invariant time

$$d\tau = ds/c = \sqrt{dt^2 - d\vec{r}^2/c^2} = dt\sqrt{1 - \frac{1}{c^2}\frac{d\vec{r}^2}{dt^2}} = dt\sqrt{1 - \frac{v^2}{c^2}}, \tag{4.37}$$

see a 1 + 1-dimensional analog pattern (4.20). Such a time can be interpreted as an '*own*' time of the point particle because it coincides with the time of the particle in its own reference frame ($v = 0$). This time allows us to enter the Lorentz *four-vector of velocity*

$$\frac{dx_\mu}{d\tau} = \left\{ \frac{c}{\sqrt{1 - \frac{v^2}{c^2}}}, \frac{\vec{v}}{\sqrt{1 - \frac{v^2}{c^2}}} \right\}, \tag{4.38}$$

and, after being multiplied by m_0, return to the four-dimensional Lorentz momentum:

$$p_\mu = m_0 \frac{dx_\mu}{d\tau}. \tag{4.39}$$

The formula (4.39) gives the expression for the energy via speed, see also equation (17.76) to get p_0.

$$p_0 = \sqrt{m_0^2c^2 + m_0^2\left(\frac{d\vec{r}}{d\tau}\right)^2} = m_0c\sqrt{1 + \frac{\frac{\vec{v}^2}{c^2}}{1 - \frac{v^2}{c^2}}} = \frac{m_0c}{\sqrt{1 - \frac{v^2}{c^2}}}. \tag{4.40}$$

Next, you'll enter *four-acceleration*

$$\frac{d^2x_\mu}{d\tau^2} = \frac{d}{d\tau}\left\{\frac{c}{\sqrt{1 - \frac{v^2}{c^2}}}, \frac{\vec{v}}{\sqrt{1 - \frac{v^2}{c^2}}}\right\}, \tag{4.41}$$

after differentiation we take into account equation (4.37), so

$$\frac{d^2x_\mu}{d\tau^2} = \frac{1}{\sqrt{1 - \frac{v^2}{c^2}}}\frac{d}{dt}\frac{1}{\sqrt{1 - \frac{v^2}{c^2}}}\{c, \vec{v}\}. \tag{4.42}$$

If you have multiplied it by the invariant m_0, the result can already be matched to the *Lorentz four-vector* F_μ. The most compact expression for the *equation of motion* of a point particle is the equality:

$$\frac{dp_\mu}{d\tau} = F_\mu. \tag{4.43}$$

The spatial part of this fundamental law of dynamics generalizes the Newtonian one in a way that results directly from equation (4.41)

$$m_0\frac{d}{d\tau}\frac{\vec{v}}{\sqrt{1 - \frac{v^2}{c^2}}} = \vec{F}. \tag{4.44}$$

We can derive the energy conservation law for a point particle, differentiated the mass square form (4.31)

$$\begin{aligned}\frac{dm_0^2c^2}{d\tau} &= \frac{d(p_0^2 - \vec{p}^2)}{d\tau} = 2\sum_{\mu=0}^{3}\varepsilon_\mu p_\mu\frac{dp_\mu}{d\tau} \\ &= 2\sum_{\mu=0}^{3}\varepsilon_\mu p_\mu F_\mu = 2p_0\frac{dp_0}{d\tau} - 2\vec{p}\vec{F} = 0,\end{aligned} \tag{4.45}$$

otherwise, substitution of expression for p_0 from equation (4.40), leads to the energy conservation law

$$\frac{dmc^2}{dt} = \vec{v}\vec{F}, \tag{4.46}$$

where the expression

$$E = mc^2 = \frac{m_0 c^2}{\sqrt{1 - \frac{v^2}{c^2}}},$$

we can interpret as energy, because the scalar product $\vec{v}\vec{F}$ is the power of the 3D force $\vec{F}$.

4.5 Discussion

Let us summarize the sequence of statements that constitutes the Lorentz-covariant mechanics, so-called relativity theory.

1. The Maxwell equations and the Lorentz force expression, entering covariant Newton equations, form the physical foundation of the theory, that unify the experimental facts of interaction between the electromagnetic field and charged particles.
2. The condition of symmetry (the covariance principle [2]) of the Maxwell equations defines Lorentz transformations of the variables of the equations. The transformations hence link position coordinates with time and, therefore, velocity of particles as well as field components between inertial reference frames.
3. The Lorentz transformations form the group that determines the space–time concept (Minkowski space) and its geometry properties via invariant pseudometrics as the group of the space motion [2].
4. The four-momentum definition that unify space momentum and energy forms relativistic kinematics of point particles. It is based on the invariant m_0^2, that m_0 is interpreted as the rest mass. The zero-component of the four momentum is hence interpreted as energy.
5. The Lorentz group-invariant fundamental pseudometrics ds of the space–time is used to determine four-vectors of velocity and acceleration that together with invariant mass definition allows to introduce covariant equations of motion that solves the problem formulation of a particle dynamics (together with the Maxwell equations and the expression for the Lorentz force). It, by construction, generalizes Newton's equation.

Let us remark also, that:

The energy–momentum four-vector and the corresponding conservation laws for its components have natural counterparts in the theory via E Noether theorem: it gives expressions for the energy and momentum in terms of a given Lagrangian, to be invariant with respect to translations at the space–time.

The extension of the Lorentz group, obtained by addition of translations of the space–time, that bear the name of Poincaré group have very deep significance not only in the space–time geometry and classical mechanics kinematics. Its transition to quantum mechanics, the quantum Poincaré group has the same algebra Lie, but different topology [18]. Its representations in Hilbert space of quantum states, yields the correct spin phenomena description. The spin notion in its deep intrinsic sense arises as a quantum vector operator (components in the rest reference frame s_i, $i = 1, 2, 3$) which square ($s_1^2 + s_2^2 + s_3^2$) eigenvalues characterizes (numerate) the irreducible representation of the quantum Poincaré group.

The natural (covariant) extension of the Newton dynamics second law (4.43) and its corollary (4.46)—energy conservation law is an important step towards an understanding of particles interaction features. It allows to analyze the particles scattering within the whole admitted velocity range $0 < v < c$. It is very important, for example, when the results of accelerator experiments are analyzed. One could arrive to important thesis that the rest (invariant) mass M of the fused particles with the rest mass m is bigger than $2m$. See, e.g. [17].

4.6 Exercises

Exercise 1. Consider the elastic scattering of two point particles on the base of energy–momentum (four-momentum) conservation laws.

Exercise 2. Consider the inelastic scattering of two identical point particles on the base of energy–momentum (four-momentum) conservation laws.

4.7 A historical note: about a birth of new mechanics (theory of relativity)

In the so-called relativity theory history the basic role belongs to Hendrik Lorentz *(1853–928)* who in 1895, published a paper in which he derived and interpret formulas that link a particle position and time coordinates in two inertial RFs, moving one with respect to another with velocity $\vec{v}$. Three years later, in 1898 Henri Poincarè *(1854–912)*, published the important work: 'La mesure du temps' [9], in which he explains a relativity of time coordinate and hence the sense of time transformation. 'Now the investigations published by Kaufmann [15] and Abraham in the past year have shown that the apparent mass is by no means to be discounted. It certainly forms a considerable part of the effective mass, and there is a possibility that in the end we shall have to ascribe apparent mass only and never true mass at all to electrons.'—from the Nobel lecture of Lorentz [13]. It is necessary to recall also papers of: 5 of June, 30 of June and 23 July in 1905, when the articles of Poincarè and Einstein were issued, that put the last points of the theory understanding.

The main issues of the theory development:

1864—J C Maxwell—expressed dependence between fields $\vec{E}$ and $\vec{B}$ as well as with densities of electric charge and current. The velocity of light propagation appears in the Faraday and Ampère–Maxwell laws as a parameter (c) of the theory in an inertial reference frame.

1879—Michelson and Morley—proved experimentally that the light velocity of propagation (of light, electromagnetic wave) in vacuum do not depend on a choice between moving uniformly inertial RFs; in a sense it is constant and coincides with the parameter of Maxwell's equations, i.e. c.

1892 (La théorie électromagntique de Maxwell et son application aux corps mouvants), 1895—H A Lorentz [4, 5], **1897**—Larmor [6], published the form of transformations that link time and space position in an inertial RF with the position and time coordinates in another inertial RF (named as Lorentz transformations, by Poincaré proposal).

1898—H Poincaré—'La mesure du temps' *(Time measurements)* [9].

1901—W Kaufmann—experimentally proved that the value of observable and $\frac{e}{m}$ depends on the electron velocity as prescribed by Lorentz transformations.

1898—905—Lorentz and Poincaré discussed eventual construction of space–time [8].

1900—H Poincaré—derived the formula for particle energy due to emission: $m = \frac{E}{c^2}$, that is equivalent to one $E = mc^2$ [10, 11].

1902—Nobel lecture of H A Lorentz [13].

1902—H Poincaré—'La Science et l'hypothèse' Science and hypothesis.

1904—H A Lorentz—'Electromagnetic phenomena in a system moving with any velocity less than that of light' [14].

1904—Conference in St. Louis during which Poincaré proposed that an updated relativity principle added up to five classical principles of physics:

'*...the principle of relativity, according to which the laws of physical phenomena should be the same, whether for an observer fixed, or for an observer carried along in a uniform movement of translation, so that we have not or could not have any means of discerning whether or not we are carried along in such a motion.*' (*H Poincaré*)

5 June 1905—H Poincaré—'Sur la dynamique de l'èlectron' (Paris) (on electron dynamics).

30 June 1905—A Einstein—'Zur Elektrodynamik der bewegten Körper'. To electrodynamics of moving body.

23 July 1905—H Poincaré—'Sur la dynamique de l'èlectron' (Palermo) (on electron dynamics).

5 May 1920—A Einstein—'L'èther et la teorie de la relativitè' conference in Leiden.

The review [16], related to Poincaré works, is very interesting.

References

[1] Korn G A and Korn T M 1968 *Mathematical Handbook for Scientists and Engineers* (New York: McGraw-Hill)

[2] Fock V and Kemmer N 1955 *Theory of Space, Time and Gravitation* 2nd rev (Oxford: Pergamon)
Fock V A 1955 *Teoriya prostranstva, vremeni i tyagoteniya* (Russia: Moskva)

[3] Michelson A and Morley E 1887 *Am. J. Sci.* **XXXIV** 334–45

[4] Lorentz H A 1892 La théorie électromagnétique de Maxwell et son application aux corps mouvants *Archives Néerlandaises* **25** 363–552

[5] Lorentz H A 1895 Het theorema van Poynting over de energie in het electromagnetisch veld en een paar algemeene stellingen over de voortplanting van het licht *Verslagen der Afdeeling Natuurkunde van de Koninklijke Akademie van Wetenschappen* **4** 176–87

[6] Larmor J 1897 On a dynamical theory of the electric and luminiferous medium, Part 3, relations with material media *Philos. Trans. R. Soc. A: Math. Phys. Eng. Sci.* **190** 205–300

[7] Lorentz H A 1899 Simplified theory of electrical and optical phenomena in moving systems *KNAW, Proc., 1, 1898–1899 (Amsterdam)* pp 427–42
[8] Poincaré H 1892a Non-Euclidian geometry *Nature* **45** 404–7
[9] Poincaré H 1898 La mesure du temps *Rev. Métaphys. Morale* **6** 1–13
[10] Poincaré H 1900 Les relations entre la physique expérimentale et la physique mathématique *Rev. Gén. Sci. Pures Appl.* **11** 1163–75
[11] Poincaré H 1900 Lorentz theory and principle of equality of action and reaction. Work of welcome offered by the authors to H.A. Lorentz, Professor of Physics at the University of Leiden, on the occasion of the 25th anniversary of his doctorate, the 11 Dec. 1900 *Arch. Nèerl. Sci. Exact. Nat. Ser. 2* **5** 252–78
[12] Poincaré H 1901 Sur les principes de la mécanique *Congrès Int. Philos. (1900-1903)* **3** 457–94
[13] Lorentz H A 1902 The theory of electrons and the propagation of light *Nobel Lecture*
[14] Lorentz H A 1904 Electromagnetic phenomena in a system moving with any velocity smaller than that of light *KNAW, Proc., 6, 1903–1904 (Amsterdam)* 6 pp 809–31
[15] Kaufmann W 1901 Die magnetische und elektrische Ablenkbarkeit der Bequerelstrahlen und die scheinbare Masse der Elektronen *Góttinger Nachrichten* **2** 143–68
Kangro H and Kaufmann W 1977 *Neue Deutsche Biographie (NDB)* Band 11 (Berlin: Duncker and Humblot) pp 352 f. (German).
[16] Logunov A A 2005 *Henri Poincaré and relativity theory* (arXiv:physics/0408077v4)
[17] Sharipov R A 1997 *Classic Electrodynamics and Relativity Theory* (Ufa: Bashkir State University)
[18] Novozhilov Y V 1975 Introduction to Elementary Particle Theory *Int. Series of Monographs in Natural Philosophy* 1st edn (Amsterdam: Elsevier)

IOP Publishing

Practical Electrodynamics with Advanced Applications

Sergey Leble

Chapter 5

Electromagnetic field in a matter

5.1 Definition of vectors: polarization, electric induction, magnetization and magnetic field strength—Maxwell's equations for electromagnetic field in a matter

5.1.1 Implicit definition of the vectors: polarization, electric induction

Up to this moment we have considered matter only as a set of charged point particles and electromagnetic field, implemented in vacuum. Now we intend to go to a description, typical for physics, of continuum: the charged particles and the field will be described on a base of densities: the density of charge and density of current together with the electromagnetic field defined for each point of space. We however suppose that in a vicinity of such point there is a domain for which one could neglect a change of number of points and correspondent inhomogeneities of electromagnetic field due to smallness of the points itself as sources of the field. Such logic mainly relies upon either an empiric introduction of the field in continuum or on the statistical physics approach. Statistical physics starts from the distribution function as probability density in phase space—state space of classical particles—and, in such way, also introduces fields of continuous media—by means of a procedure of averaging evaluation in velocity subspace eliminating so far too detailed a description. This description should take into account the quantum nature of microscopic particles and fields. We however, in standard conventional didactics (electrodynamics traditionally go ahead of quantum and statistical mechanics) cannot base neither on statistical physics nor quantum mechanics, hence we shall 'combine' possibilities of empiric approach. It is useful to understand the academic tradition.

Let us start with an introduction of basic notions of continuum on using so-called 'mind experiments'. We accept that these macroscopic properties of a matter firstly depend on properties of its microscopic constituents as the atoms or molecules. A given atom contains positively charged nucleus and a kind of negative charged electron clouds, that, by the way, arises from quantum description. In equilibrium an atom is neutral and has a symmetrical charge distribution.

In a moment when an atom is under the action of an external electric field, its charge distribution symmetry will be broken. The nuclei will be shifted in the direction of the electric field vector **E**, while the electron 'clouds' will be deformed in such a manner that their point of weight would be shifted in the opposite direction (figure 5.1). Considering such phenomenon we speak about induced polarization.

In a case when the field acting on a molecule induces partial separation of its charges, we say that it is polarized.

Let us make steps to a formal empiric (theoretic) definition of fields, describing polarization, consider now a medium, at the beginning, as a 'plasma', mixture of charges. While applying to such medium an electric field, a polarization appears, that, along the principal idea, relates to 'coupled' charges. These charges will be a source of a certain 'new', secondary field. Speaking precisely, a shift and even an induced movement of charged particles change the summary fields, e.g. a neutral atom obtains a dipole momentum, which produces a secondary field of correspondent structure.

Hence considering electric charge density in a material matter, we must divide it for 'coupled' ones (respected for polarization) and 'free' (all the rest that do not take part in polarization, which may leave atoms, making currents)

$$\rho = \rho_{zw} + \rho_{sw}. \tag{5.1}$$

By definition 'the coupled charges' will be understood as charged particles with restricted freedom of motion, e.g. electrons and nuclei that do not leave atoms in crystals or molecules in molecular dielectrics. A similar situation takes place in molecules (atoms) of gases. In solids it relates to atoms which glue crystal nets—electrons in full zones: dielectrics and metals, (quasi crystals, glasses). When one speaks about 'free charges', having in mind charges that can leave its localization (for example ions in gases or solutions, electrons in conductivity zone (valence) in metals or semiconductors).

Such division allows to rewrite the first Maxwell equation—the Coulomb–Gauss law (2.17) in the form

$$\operatorname{div} \vec{E} = 4\pi(\rho_{zw} + \rho_{sw}). \tag{5.2}$$

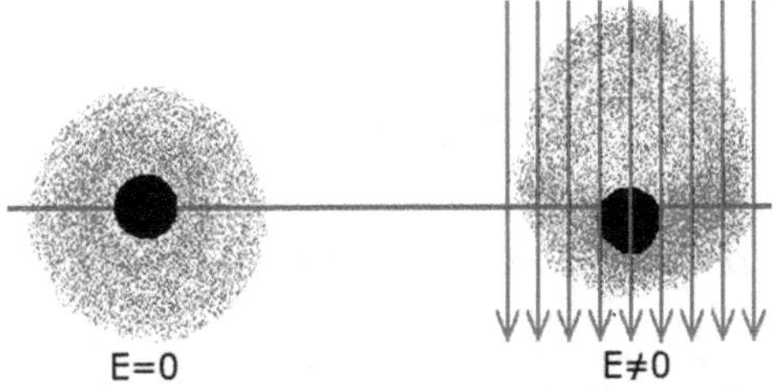

Figure 5.1. Schematic presentation of an atom polarization under external field $\vec{E}$ action.

Introducing (now implicit) *polarization vector* $\vec{P}$, that characterizes polarization via the equation

$$-4\pi\left(\operatorname{div}\vec{P}\right) = 4\pi\rho_{zw}, \tag{5.3}$$

having the Coulomb–Gauss law for matter in the form

$$\operatorname{div}\vec{D} = 4\pi\rho_{sw}. \tag{5.4}$$

The vector

$$\vec{D} = \vec{E} + 4\pi\vec{P}, \tag{5.5}$$

is named the *electric induction vector.*

Equation (5.4) is naturally referred to as the I Maxwell equation for matter; the index '*sw*' is usually omitted, having in mind a historically used continuous matter description with a 'liquid' electricity inside.

5.1.2 On implicit definition of magnetization vector and magnetic field strength

In an external electromagnetic field the electrons under action of the Lorentz force are moving both translationally and rotationally. hence each matter unit (either atom or molecule) represents a microscopic electric contour. Such contour have a magnetic moment (see equation (3.142)), which at the macroscopic level induces correspondent division of the current density, i.e. the vector $\vec{j}$.

So, in a polarizing matter, the electric current density $\vec{j}$ may be divided to three terms. As the first of them we take the density $\vec{j}_{zw}$ related to the matter magnetization, arisen from the electron rotational motion perturbation and spin (apropos, explained by relativistic quantum mechanics), linked to some extra magnetic property,

$$\vec{j}_{zw} = c\ \mathrm{rot}\vec{M}. \tag{5.6}$$

We will use relation (5.6) as the (implicit) definition of a *magnetization vector* $\vec{M}$.

Another group of terms is related to free current density j_{sw} which represents electrons/ions free motion. The third category is the density of polarization current $\vec{j}_p$, that arises due to translational motion of coupled charges as variation of electric polarization

$$\vec{j}_p = \frac{d\vec{P}}{dt}. \tag{5.7}$$

It means, that the full density of electric current is equal to

$$\vec{j} = \vec{j}_{zw} + \vec{j}_{sw} + \vec{j}_p = c\ \mathrm{rot}\vec{M} + \vec{j}_{sw} + \frac{d\vec{P}}{dt}. \tag{5.8}$$

When, in the Maxwell–Ampère equation (2.31), the terms (5.8) are taken into account, one has

$$\frac{1}{c}\frac{\partial \vec{D}}{\partial t} = \mathrm{rot}\vec{H} - \frac{4\pi}{c}\vec{j}_{sw}, \tag{5.9}$$

while the vector

$$\vec{H} = \vec{B} - 4\pi\vec{M} \tag{5.10}$$

is named as *magnetic field strength* or intensity of magnetic field.

Remark 5.1.1. The definition (5.10) of vector $\vec{H}$ is also implicit, as was the definition (5.5) of vector $\vec{D}$ by the similar reason: both vectors $\vec{P}$ and $\vec{M}$ cannot be expressed directly from equations (5.3) and (5.6) in the unique form.

5.1.3 Maxwell's equations for electromagnetic field in matter

Summarizing, insert relations (5.4) and (5.9) into the basic system of electromagnetic field equations for matter. The rest of the equations (see equations (2.28)–(2.31)) do not change because of the absence of isolated magnetic charges

$$\nabla \cdot \vec{B} = 0, \tag{5.11a}$$

$$\frac{1}{c}\frac{\partial \vec{B}}{\partial t} = -\nabla \times \vec{E}. \tag{5.11b}$$

As in section 2.2, it is possible to reproduce the integral form of Maxwell's equations for matter:

$$\int \vec{D}d\vec{S} = 4\pi q_{sw}, \tag{5.12a}$$

$$\int \vec{B}d\vec{S} = 0, \tag{5.12b}$$

$$+\frac{1}{c}\frac{\partial \int_S \vec{B}d\vec{S}}{\partial t} = -\int_L \vec{E}d\vec{l}, \tag{5.12c}$$

$$\frac{1}{c}\frac{\partial \int_S \vec{D}d\vec{S}}{\partial t} = \int_L \vec{H}d\vec{l} - \frac{4\pi}{c}\vec{I}_{sw}, \tag{5.12d}$$

where the total current through the surface S is determined as $I_{sw} = \int_S \vec{j}_{sw} d\vec{S}$. The Gauss and Stokes theorems within a choice of integration path L and the surface S have been applied (see details in section 2.2).

5.1.4 Explicit definition of vectors: polarization, electric induction

Equations (5.3) and (5.6) are conventionally used in textbooks for a definition of the vectors $\vec{P}$ and $\vec{M}$, however, as was mentioned, *such definition is not explicit*: to introduce the explicit one and for **practical purposes** we shall rely upon the following scheme, that should give *independent* definition of the vectors $\vec{E}$ and the vector $\vec{D}$. The same we should state about the pair $\vec{B}$ and the vector $\vec{H}$. It, naturally, may allow to define the constants ε, μ.

First, one applies the integral form of equation (5.12a) to a real physical configuration, choosing an appropriate domain volume, e.g. to a flat capacitor (see the figure 5.2). Second, one arrives at the definitions for this capacitor case.

`Definition of` D. *Absolute value of the vector* $\vec{D}$ *is determined by the formula*

$$D = 4\pi q_f/S \tag{5.13}$$

via *the charge* q_f, *that may be measured by ballistic galvanometer.*

`Definition of` E, `electric field in matter.` *The electric field* E *is defined by the procedure that is directly linked with its thermodynamic sense (see the next section) namely with definition of work that enter energy conservation law (see also section 17.1).*

The empiric (practical) definition of the vector $\vec{E}$ *module uses a simple measurement of potentials difference (work on unit charge transport)* V, *namely*

$$E = V/d. \tag{5.14}$$

Therefore the procedure of such experiment includes measurement of the potentials difference, taking into account the relation $U = Ed$ (in fact the experiment is based on the estimation of small current through voltmeter, using Ohm's rule). Recall the known from secondary school relation for the charge on a capacitor versus the potential difference $q = CV$, that is the direct corollary of the capacity C definition.

Next, the vector $\vec{P}$ may be expressed via equation (5.5).

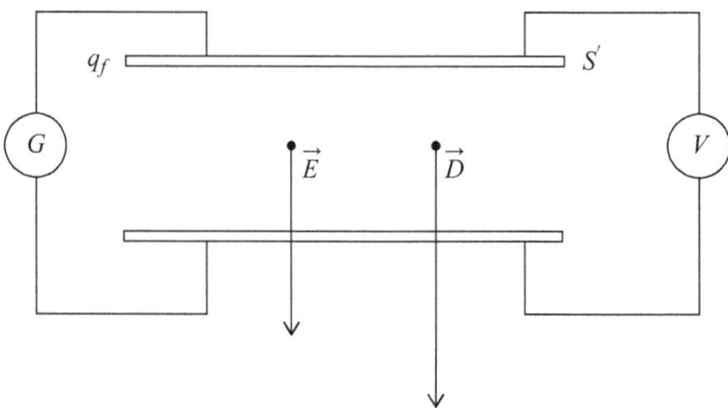

Figure 5.2. Schematic representation of the vectors: the electric field $\vec{E}$, via potential difference on the capacitor plates, as in equation (5.14) and the vector of electric induction $\vec{D}$ via a vector flux through the surface area S by equation (5.13), the charge q_f is measured by a ballistic galvanometer G.

Attention! There are theoretical proposals for a definition of the electric and magnetic fields in matter, used in the majority of textbooks [1] but such that, perhaps, are not used directly in practice. It is said about making cylindric holes of base diameter much more than its height. Definitions of such type have a difficulty: it implies a known direction of a field (either the vector $\vec{D}$ or the vector $\vec{H}$).

5.1.5 On explicit empiric definition of vectors: magnetic field strength $\vec{H}$ and magnetic induction $\vec{B}$

A practical way of a measurement—hence an *empiric definition* of magnetic part of the electromagnetic field in matter is based again on application of Maxwell's equations. More exact—it is use of Faraday's law (5.12c) for the magnetic induction vector $\vec{B}$ and the Ampère–Maxwell equation (9.11) in the case of the relatively small term: $\frac{\partial D}{\partial t} \ll 4\pi I$ (see also section 2.2.4). In fact such an approach is used for the definitions of the vectors $\vec{B}$ and $\vec{H}$ in the internal points of a long solenoid filled with an isotropic substance. In such geometry the vectors $\vec{B}$ and $\vec{H}$ are directed along the solenoid axis.

Let the integration path L_A be chosen as on figure 5.3, then the Ampère–Maxwell equation in an integral form looks as

$$\oint_{L_A} \vec{H}d\vec{l} = 4\pi I/c. \tag{5.15}$$

We choose the integration path so as the total current across the surface S, that has the path as the surface boundary L_A, is equal to $I = ni_t$, where n is the number of turns per unit length of the solenoid. The current in the solenoid wire is supposed to be i_t; the part of the integral along the path outside the solenoid (and small parts of the path of the integration across it) does not influence the result because the field outside is negligibly small.

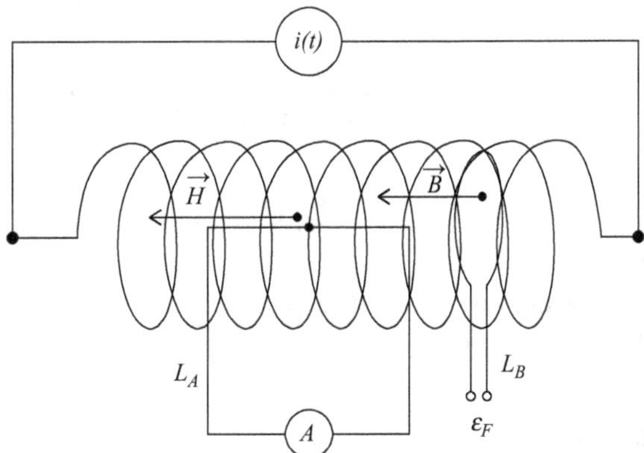

Figure 5.3. Schematic illustration of the measurements of the vectors of magnetic field strength $\vec{H}$ via current in the solenoid, multiplied by a number of turns inside the contour L_A as in (5.15), and the vector of magnetic induction $\vec{B}$ by the flux across a contour L_B (charge Q is obtained via the measurement by a ballistic galvanometer).

The value (module) of magnetic field strength H is expressed as

$$H = \frac{4\pi n i_t}{c}. \tag{5.16}$$

Similarly the (module) of the field B may be calculated from the Faraday equation, when other integration path L_B is used.

$$\oint_S \vec{B} d\vec{S} = -c \oint_{L_B} \vec{E} d\vec{l} = -c\varepsilon_F, \tag{5.17}$$

ε_F is Faraday's electromotive force. The form of the integration surface is chosen so as to cover the solenoid cross-section with the contour, parallel to the boundary of the solenoid.

If the surface area is denoted also by the character S then the field B module is calculated as:

$$B = -c\varepsilon_F / S. \tag{5.18}$$

In practice the module B of the vector $\vec{B}$ in a solenoid is evaluated and therefore *defined* by means of either the charge Q that passed through the contour till the magnetic field disappear, or, by given an alternate current $i(t) = I_0 \cos(\omega t)$, where the value $\varepsilon_F = U = iR$ is measured directly by a voltmeter.

In the first case the field is

$$B = -\int_0^\infty B_t dt = \frac{c}{S}\int_0^\infty \varepsilon_F dt = \frac{c}{S}\int_0^\infty i(t) R dt = \frac{cRQ}{S}. \tag{5.19}$$

In the second one:

$$B_0 = -\int_0^{\pi/2\omega} B_t dt = \frac{cR}{S}\int_0^{\pi/2\omega} I_0 cos(\omega t) dt = \frac{cI_0 R}{\omega S}. \tag{5.20}$$

Remark 5.1.2. Both empiric definitions of the magnetic and electric fields imply a kind of symmetry, that allows to fix the fields direction.

Such experiment setup allows to investigate magnetic properties of any material and phenomena such as hysteresis.

A theoretical (local) definitions of the fields $\vec{B}$, $\vec{H}$ uses some cylindric holes, similar to the electric fields case [1] (read also the discussion below).

5.1.6 The fields measurements in experimental practice

There are few approaches to vectors $\vec{E}$, $\vec{D}$ and $\vec{B}$, $\vec{H}$ measurements in the practical electrodynamics that is widely used in geophysics, medicine and experimental physics.

I. The fields $\vec{E}$, $\vec{D}$ measurements.

1) Atmospheric electric field measurements/monitoring. The local voltage in a space charge cloud (for example of charged mist or dust) can be measured with a voltage follower probe or with an electrostatic field meter acting as a potential probe.
   The electric field at the sensing.
2) In medicine, an example of an E-field probe being used in the feed back loop of an immunity test system use of near-field, quasi-electrostatic measurements ('electric field sensing') for measuring the configuration of human bodies [2].

II. The fields $\vec{B}$, $\vec{H}$ measurements.

1) **Magnetometers**. An example of such device is one constructed for the Earth magnetic field monitoring: measurement variations of the field in time. As an example mention 'Kvartz-4', a recent IZMIRAN [Institute of Terrestrial Magnetism, Ionosphere and Radio Wave Propagation named after Nikolay Pushkov of the Russian Academy of Sciences or IZMIRAN for short] construction [3], using quartz magnetic sensors comprises magneto resistive elements and permanent magnet films, which are combined together. It measures three components of a planet magnetic field with errors of about 2000 nT [4].
2) **Fluxmeters**. A device is constructed for various magnets field measurements, the meter that measures magnetic flux by the current it generates in a coil [5].

5.1.7 Material equations as thermodynamic equations of state

If one calculates the number of fields $\vec{E}$, $\vec{B}$, $\vec{D}$ and $\vec{H}$ (three × four = 12) and compare it with the actual number of equations (8), which had not changed compared with vacuum—see also the section 2.2.5, one understands that it is necessary to close (complete) the systems (5.4) and (5.9). The closure means addition of relations that have the transparent sense from the point of view of general thermodynamics, because the variable $\vec{D}$ is dual to the variable $\vec{E}$, while $\vec{H}$ is dual to $\vec{B}$. One of the basic principles claims the equations of state existence, which introduces a dependence of the *internal* parameters (the components D_x, D_y, D_z here) on the *external* ones (E_x, E_y, E_z) see [6], which contains very instructive and interesting discussion as well as precise definitions. It is easy to see from the energy balance for a matter (5.51), written in SI units, that the elementary work of the electric field per unit volume is equal to

$$\delta A_e = \vec{E} \cdot \delta \vec{D}, \tag{5.21}$$

while the elementary work of the magnetic field

$$\delta A_m = \vec{H} \cdot \delta \vec{B}, \tag{5.22}$$

that points out also on the relation between $\vec{B}$, $\vec{D}$ like the external parameter and $\vec{E}$, $\vec{H}$ like the internal one. Unfortunately there are no known possibility to introduce the relations like equations (5.21) and (5.22) on a solid base of energy conservation law in the general case; only in the case of the local tensor link as equation (5.28) between internal and external fields, or for wave packet (see the comments in section 6.1.1).

Therefore the equation of state of the dielectric matter is defined by the relation

$$\vec{D} = \vec{D}(E_x, E_y, E_z), \tag{5.23}$$

that in a very general approach may be nonlocal one (see the discussion section 6.1) and in yet more general case D_i may depend on B_i, see section 5.3.1, chapter 18.

For the magnetic fields it looks similarly as

$$\vec{H} = \vec{H}(B_x, B_y, B_z), \tag{5.24}$$

the generalizations are in section 6.1 and more general section 5.3.1, chapter 18. Both equations (5.23) and (5.24) are referred to as *material equations.*

The simplest case of such (dielectric) material relations is very popular in textbooks, the *local* one is written as:

$$\vec{D} = \varepsilon\vec{E}, \tag{5.25}$$

which is claimed to be applied easily in the case of isotropic linear media, but not very *practical one*, see section 5.3.1. This supposition relies upon the same direction of the field $\vec{D}$ as the vector $\vec{E}$, it is a kind of *a priori* one: it means a symmetry with respect to rotations existence. In this case it is enough to find a relation between modules. The coefficient ε is named as the '*dielectric constant*' or '*dielectric permittivity*'. For a given isotropic (linear dielectric) material it is easy to establish such relation experimentally if applying the empiric definitions of the vector $\vec{D}$ via equation (5.13), if putting a plate of such dielectric between the capacitor plates, see figure 5.4 and the vector $\vec{E}$ by equation (5.14):

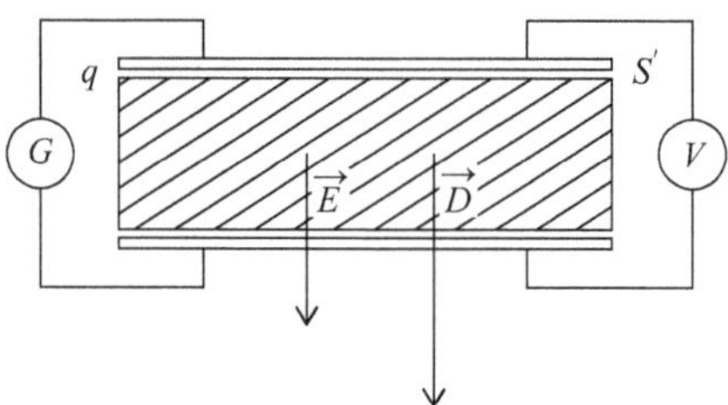

Figure 5.4. If the space between the capacitor plates is filled with a dielectric, we can determine (measure) the dielectric constant by the relation $\varepsilon = D/E$. Schematic representation of the vectors: the electric field $\vec{E}$ and the vector of electric induction $\vec{D}$ for the dielectric permittivity ε measurement (the charge q is measured by a ballistic galvanometer).

$$\varepsilon = D/E = \frac{4\pi dq}{SV}. \tag{5.26}$$

Introducing an empiric link between the charge and voltage on a capacitor (see figure 5.4) yields

$$q = CV = CEd. \tag{5.27}$$

One also obtains the formulas for the coefficient C if expressing the charge via D from equation (5.13) and substituting it into equation (5.27), that yields $C = \frac{\varepsilon S}{4\pi d}$ for the flat capacitor.

One of the practical definitions of the dielectric constant of isotropic matter ε is $\varepsilon = C/C_0$, or the ratio the capacities of the matter and the vacuum (matter absence).

The more complicated case relates to the anisotropic medium (e.g. a crystal)

$$D_i = \sum_{k=1}^{3} \varepsilon_{ik} E_k, \tag{5.28}$$

which is also local but exhibits the tensor link between the fields. The experimental definition of the tensor needs three measurements in this case. Most convenient to do it by a flat capacitor again, using three specimen made of the crystal cut in three principal directions, in which the tensor ε is diagonal with the values ε_{ii}.

In practice, one establishes the relation between the vectors $\vec{D}$ or $\vec{P}$ and the vector $\vec{E}$ for a kind of the matter by the measurements at capacitors described above. After that it is used in a theoretical description of some experimental situation of interest. For example, one can derive the formula for a flat capacitor with the isotropic dielectric inside $C = \frac{\varepsilon S}{4\pi d}$ on the basis of the matter equation (5.25).

It the case of magnetic fields the formulas (5.16) with constant current $i(t) = i_0$ and (5.19) allow to express the coefficient

$$\mu = \frac{c^2 RQ}{4\pi n i_0 S}$$

directly from the measurement, similar to one described above (see figure 5.3). A tradition introduces the magnetic permittivity μ for a isotropic medium by the link

$$\vec{B} = \mu \vec{H}, \tag{5.29}$$

in which the external parameter, the field $\vec{B}$ and the internal one $\vec{H}$ are interchanged in comparison with equation (5.25). For anisotropic matter we write

$$B_i = \sum_{k=1}^{3} \mu_{ik} H_k. \tag{5.30}$$

The (macroscopic) magnetization vector is extracted from the relation (5.10). The figure 5.5 shows a scheme that is used for a measurement of the magnetic permittivity μ of an isotropic medium. A specimen made of the given material in

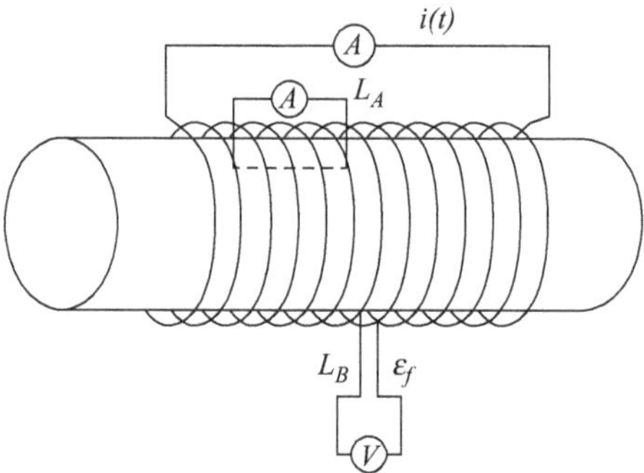

Figure 5.5. Schematic illustration of the measurements of the vectors of magnetic field strength $\vec{H}$, and the vector of magnetic induction $\vec{B}$ to obtain the magnetic permittivity μ. The contour L_A is inside a narrow slit of a magnetic (rod). If the magnetic is a conductor, the wire of L_A should not touch the rod matter. The continuity of the tangential component of the magnetic field strength guarantee the measurement quality. The plane of the coil of the contour L_B is orthogonal to the solenoid axis and also lies inside a slit in the rod. Its area is marked as S. The continuity of the normal component of the magnetic induction vector $\vec{B}$ is important to be sure that the measurement in the slit gives the correct module of $\vec{B}$ value inside the rod. The mentioned continuity conditions are derived in section 9.2.1.

a form of the cylindric rod is placed inside the solenoid with n coils which is connected to a source (battery), look at figure 5.5.

We would repeat that the medium is supposed to be isotropic. In the case of anisotropic matter a permittivity tensor μ_{ik} is introduced, a similar technique to electric fields may be used. The specimen (cylindric rod) may be cut from a crystal along three principal directions.

5.2 Macroscopic Maxwell's equations, links to microscopic parameters

5.2.1 Resume, the empiric definition of the fields: macroscopic Maxwell's equations

The formulation of definitions in previous sections, besides the preferences have some obvious necessity to complete the description in terms of the growing number of variables from $\vec{E}$ and $\vec{B}$ to $\vec{E}$, $\vec{B}$, $\vec{D}$ and $\vec{H}$. It means that, for the mathematically closing of the system (5.11a) and (5.11b) one needs more relations between the fields components. In the case of the continuum theory those are the material relations, to define a dependence between $\vec{E}$ and $\vec{D}$, similarly between $\vec{B}$ and $\vec{H}$ as well as a relation between current density and the fields, known under the name of Ohm: chapter 6, section 6.3.2, equation (6.79) or (6.82).

A thermodynamic consider the relation between the vector $\vec{D}$ with the vector $\vec{E}$ ($\vec{B}$ and $\vec{H}$—in the same manner) as an equation of state.

More generally, taking into account dispersion, we go to the operator link between polarization vector and electric field, that yields

$$\vec{D} = \hat{\varepsilon}\vec{E}.$$

Similarly we have

$$\vec{B} = \hat{\mu}\vec{H},$$

see for details section 8.2. Further generalization includes a dependence of polarization vectors on magnetic field and magnetization vector on electric field, see section 5.3. Such a case implies such phenomena that takes place in multiferroics and attracts great attention in modern investigations [7], see also section 5.3.1.

5.2.2 A link between polarization vector and dipole momentum

We would like to turn back to the *implicit theoretical definition of the field* $\vec{D}$ in terms of the polarization vector $\vec{P}$, that is determined by formula (5.3) taking into account equation (5.7). We first must solve equation (5.3) with respect to $\vec{P}$ so condition (5.7) is fulfilled automatically. In the Lorentz approach (from 'micro' to 'macro') the field is 'produced' by *microscopic* charges, but we measure it by means of macroscopic 'probing' (small) charged particle. Besides that in reality the elementary charges in atoms are moving extremely quick and along quantum mechanics rules. A transition to continuum is performed by means of the procedure of averaging ('smearing') over 'control domain' V_e, similar to that we had introduced when introducing the notion of charge density and of current density. If one does not apply such a procedure, the very inhomogeneous description occurs (with eventual singularities in the case of point charges models used) and complicated expression for the vector $\vec{P}$ (see further section 6.1.3 (6.58)). Such 'averaging' is introduced at a basis of superposition principle and we would denote it as

$$\langle A \rangle = \frac{1}{V_a}\frac{1}{\delta t}\int_{\delta t}\sum_{V_e}\langle A(\vec{r}_a)\rangle_{kw}, \tag{5.31}$$

integrating by time and summing over positions of particles of the medium that bear charges inside (attached to) (atoms or molecules). The sum in this definition represents the sum of electric fields of coupled charges in the domain V_e. Namely in such a manner it would look like the classic approach (of Lorentz), which does not take quantum peculiarities into account (see discussion and references to literature at the end of this section). The quantum description is taken into account by additional mean value operation $\langle\ \rangle_{kw}$. Such averaging has natural restriction by the field change scale (large wavelength compared with atomic dimensions and frequency compared with time of atomic transitions).

For example, the vector of polarization used in continuum description is defined by the formula:

$$\vec{P} = \langle \vec{P}' \rangle. \tag{5.32}$$

Next, returning to the relation (5.3) we write

$$-\langle \rho_{zw} \rangle = \operatorname{div}\langle \vec{P}' \rangle, \tag{5.33}$$

supposing the operations of div and averaging commute.

Denote the electric field $\vec{E}$ averaged in the same area by the same character

$$\langle \text{div}\, \vec{E} \rangle = \text{div}\langle \vec{E} \rangle = 4\pi(\langle \rho_{sw} \rangle + \langle \rho_{zw} \rangle); \tag{5.34}$$

and start with the definition in 'micro' scale.

The charge density is then determined as (the mean value by time and quantum state) the proton number minus electron number at the (macroscopic) volume unit V_a—atomic cell. In the case of charges (protons and electrons) coupled in atoms (molecules) supposing the atom is neutral electrically, taking an index 'a' as denoting an atom as a whole. The position of the atom 'a' (its mass center) is marked by the vector $\vec{r}_a$.

Consider (as the simplest case) a medium built from atoms of the only sort (chemical kind).

Now integrate equation (5.33), multiplied by the vector $\vec{r}' = \vec{r} - \vec{r}_a$ components along a domain that is defined by the same variable as $|x'| = |x - x_a| \leqslant \varepsilon/2$, $|y - y_a| \leqslant \varepsilon/2$, $|z - z_a| \leqslant \varepsilon/2$. Suppose that the introduced cube contains the whole atom 'a' with some vicinity, let also the concentration of such atoms is equal to $n_a = \varepsilon^{-3}$.

For example, using the relation $\frac{\partial}{\partial x} = \frac{\partial}{\partial x'}$ and integrating by parts, one has

$$\begin{aligned} &-\int_{-\varepsilon/2}^{\varepsilon/2}\int_{-\varepsilon/2}^{\varepsilon/2}\int_{-\varepsilon/2}^{\varepsilon/2} x'\, \text{div}\,'\vec{P}'dx'dy'dz' \\ &= -x'P_x'|_{-\varepsilon/2}^{\varepsilon/2}\varepsilon^2 + \int_{-\varepsilon/2}^{\varepsilon/2}\int_{-\varepsilon/2}^{\varepsilon/2}\int_{-\varepsilon/2}^{\varepsilon/2} P_x'dx'dy'dz' \\ &- \int_{-\varepsilon/2}^{\varepsilon/2}\int_{-\varepsilon/2}^{\varepsilon/2}\int_{-\varepsilon/2}^{\varepsilon/2} x'\frac{\partial P_y'}{\partial y}dx'dy'dz' - \int_{-\varepsilon/2}^{\varepsilon/2}\int_{-\varepsilon/2}^{\varepsilon/2}\int_{-\varepsilon/2}^{\varepsilon/2} x'\frac{\partial P_z'}{\partial z}dx'dy'dz' \\ &= P_x'\varepsilon^3 - x'P_x'|_{x'=-\varepsilon/2}^{x'=\varepsilon/2}\varepsilon^2 - \frac{\varepsilon^2}{2}P_y'|_{y'=-\varepsilon/2}^{y'=\varepsilon/2} - P_z'|_{z'=-\varepsilon/2}^{z'=\varepsilon/2}. \end{aligned} \tag{5.35}$$

The first term (after averaging) gives the value of polarization vector, the rest, stream across surfaces of the cube. An important, already mentioned condition is that inhomogeneity of the fields ($\vec{P}$, $\vec{E}$) are much more less that the dimension of the one introduced domain of integration. In the language of wave phenomena it means that a wavelength of the wave λ is much larger than atoms a ($\lambda \gg a$). Such suppose it is important for all points excluding ones close to physical boundaries, where the stream in equation (5.35) gives a density of surface charges.

Hence for internal points we have:

$$P_x' = \varepsilon^{-3}\int_{-\varepsilon/2}^{\varepsilon/2}\int_{-\varepsilon/2}^{\varepsilon/2}\int_{-\varepsilon/2}^{\varepsilon/2} x'\rho_{zw}dx'dy'dz' = n_a p_{ax}, \tag{5.36}$$

where

$$\vec{p}_a = \int_{-\varepsilon/2}^{\varepsilon/2}\int_{-\varepsilon/2}^{\varepsilon/2}\int_{-\varepsilon/2}^{\varepsilon/2} \vec{r}\rho_{zw;a}(\vec{r})dxdydz, \tag{5.37}$$

is the dipole momentum of atom (see equation (20.16) at section 3.8.2). Introducing the density of electrons (with account of nuclei)

$$\rho_{zw,a} = \sum_i \rho_{zw,ai}. \tag{5.38}$$

After averaging, we have

$$\langle \vec{p}_a \rangle = \sum_i \int_{-\varepsilon/2}^{\varepsilon/2} \int_{-\varepsilon/2}^{\varepsilon/2} \int_{-\varepsilon/2}^{\varepsilon/2} \langle x' \rho_{zw;ai} \rangle dx' dy' dz'. \tag{5.39}$$

In general such averaging relates to account of the fact that the atoms (its electrons and nuclei) are quantum particles, the quantum mean value in example of the only electron means

$$\langle x\rho_{zw;ai} \rangle_{kw} = e \int_{\infty}^{\infty} \int_{\infty}^{\infty} \int_{\infty}^{\infty} x |\psi_i(x, y, z)|^2 dxdydz, \tag{5.40}$$

where ψ_i is wave function of the electron i, that represents the atom electrons cloud (in equation (5.40); within the choice $\vec{r}_a = 0$), exited by external field $\vec{E}$. Namely in such a way the electric field parameterizes dipole momentum, that in the theory introduces the so-called material equation ('thermal equation of state'—in thermodynamics) for polarization vector (see previous section (section 5.1.7)).

A majority of textbooks introduces the link between polarization vector and the dipole electric momentum on the base of potentials (see discussion in section 5.2.4, where we reproduce the approach, e.g. of [8]).

Consider now the case, in which the medium is a mixture of atoms (molecules) of the following concentrations $n_a = \varepsilon_a^{-3}$. The electric charge density is the sum of the densities $\rho_{zw} = \sum_a \rho_{zw;a}$, then the polarization vector is expressed as

$$\vec{P} = \sum_a n_a \int_{V_a} \langle \vec{r}' \rho_{zw;a} \rangle d\vec{r}' = \sum_a \vec{p}_a, \tag{5.41}$$

for each atom (or molecule) the averaging is provided in a specific manner (e.g. along equation (5.40)), the shorthand $\int_{V_a}$ stands for the integral of equation (5.37).

So we have found a link between the polarization vector $\vec{P}$ and dipole momentum vector $\vec{p}_a$ of some medium unit. In the case of isotropic one-component medium the relation (5.36) is the simplest:

$$\vec{P} = N\vec{p}, \tag{5.42}$$

here it is denoted $n_a = N$.

In respect to the complete mean value evaluation (5.31), as it already was mentioned, in theoretical physics the quantum-statistical averaging is taken into account, as in it—quantum properties of the field itself (photons content) and integration by time and space, that accounts effects of a macroscopic measurement device (see the discussion in section 5.2.4).

Historically first 'closed' description, based on yet classic atom model was the Lorentz theory, in which each electron is modeled as harmonic oscillator in a field $\vec{E}$. The Lorentz theory is reproduced in section 6.3.

5.2.3 A link between magnetization vector and magnetic dipole momentum

In a direct analogy with the previous section we start from the relation (5.6) as a definition of the magnetization vector $\vec{M}$

$$c\text{rot}\vec{M} = \vec{j}_{zw}, \tag{5.43}$$

where $j_{zw} = \sum_a j_{zw;a}$, a superposition of atomic currents. Vector-multiplying it by $\vec{r}' = \vec{r} - \vec{r}_a$ and integrating the result by the atom vicinity (some matter unit) volume (V_a), we do it similar to equation (5.35). The left side x-component reads

$$\begin{aligned} &-\int_{-\varepsilon/2}^{\varepsilon 2}\int_{-\varepsilon/2}^{\varepsilon/2}\int_{-\varepsilon/2}^{\varepsilon/2} c[\vec{r}' \times \text{rot}'\vec{M}]_x dx'dy'dz' \\ &= -c\int_{-\varepsilon/2}^{\varepsilon/2}\int_{-\varepsilon/2}^{\varepsilon/2}\int_{-\varepsilon/2}^{\varepsilon/2}\left[y'\left(\frac{\partial M_y}{\partial x'} - \frac{\partial M_x}{\partial y'}\right) - z'\left(\frac{\partial M_x}{\partial z'} - \frac{\partial M_z}{\partial x'}\right)\right]dx'dy'dz' \\ &= 2c\int_{-\varepsilon/2}^{\varepsilon 2}\int_{-\varepsilon/2}^{\varepsilon/2}\int_{-\varepsilon/2}^{\varepsilon/2} M_x dx'dy'dz' + \cdots \end{aligned} \tag{5.44}$$

up to surface terms noted by '…'. If the terms are neglected and withdraw the slow-varying x-component of the magnetization vector M_x out of the integral, the complete equality yields

$$M_x = \frac{1}{2c\varepsilon_a^3}\int_{V_a}\left[\vec{r}' \times \vec{j}_{zw;a}\right]_x dxdydz, \tag{5.45}$$

within the conditions about space scale of the M_x variations and the coordinates choice. The relation (5.45) and the similar ones for the 'y' and 'z' components link the magnetization vector $\vec{M}$ with the atomic magnetic moment $\vec{\mu}_a = \frac{1}{2c}\int_{V_a}\sum_a\left[\vec{r}' \times \vec{j}_{zw;a}\right]dxdydz$ (see equation (3.142) at section 3.8.5). It looks like

$$\vec{M} = \sum_a n_a\vec{\mu}_a, \tag{5.46}$$

where $n_a = \frac{1}{\varepsilon_a^3}$.

The density ρ_{sw} represents free charges whose value may change (in dependence of the free charges proportion—negative contra positive). It could result in a positive or negative (or neutral) of the matter units (e.g. atoms) The density of charges coupled to atoms ρ_{zw} relates to charged particles with restricted freedom, they are in a domain close to atomic nuclei. Their position is characterized by small deviation of an equilibrium one. This phenomenon is named as polarization and described by the polarization vector $\vec{P}$. At the beginning of this section there is an illustration of the phenomenon representing an atom, as basic unit of matter. For a isotropic substance built from identical atoms, the expression (5.59) is reduced to the form:

$$\vec{M} = N\vec{\mu}, \tag{5.47}$$

where N is a number of magnetic dipoles in unit volume.

5.2.4 More comments on the definitions

On the field $\vec{B}$ definition

Let us return to the field definition B in a solenoid filled with a substance. In practice the charge is measured by means of ballistic galvanometer, when the electromotive force $\varepsilon = |\Phi_t|$ exists in the short time. Then, for a measurement coil, the diameter of which is (almost) equal to diameter of specimen

$$\Delta B = \frac{\Delta \Phi}{S},$$

where the variation of the magnetic flux $\Delta\Phi$ is proportional of the maximum deviation of the ballistic galvanometer, see e.g. http://lab2.phys.spbu.ru/.

A sensitive galvanometer was designed by William Thomson (later Lord Kelvin) in 1858 to detect the current through the Atlantic cable. A small mirror is suspended by a thread between two coils, and on the back of the mirror are glued several short and light magnets.

Some particular discussions on the phenomena in solids described in this subsection are presented for example in [9] and [10].

Let us discuss the introduction to fields in matter in more details.

On definition of the field $\vec{E}$ at matter

The only relation within the set (5.12a)–(5.12d), that contains the field $\vec{E}$ is the Faraday equation (5.12c), which we now choose as convenient for the empiric definition. If the contour (L) for integration in this equation will be chosen parallel to the field $\vec{E}$, figure 5.2, the part inside the capacitor gives Ed, the rest one is approximately zero, while the second term (relating to a voltmeter contour) is proportional to $\frac{\partial \Phi}{\partial t}/c$, so the module of electric field yields E extracted from this measurement.

5.2.5 Energy conservation/balance law derivation for matter

Let us multiply the Ampère–Maxwell equation (5.12d) by the vector $\vec{E}$

$$\vec{E} \cdot \mathrm{rot}\vec{H} = \frac{1}{c}\vec{E} \cdot \frac{\partial \vec{D}}{\partial t} + \frac{4\pi}{c}\vec{E} \cdot \vec{j}_{sw}, \tag{5.48}$$

and the Faraday equation by the vector $\vec{H}$,

$$\vec{H} \cdot \mathrm{rot}\vec{E} = -\frac{1}{c}\vec{H} \cdot \frac{\partial \vec{B}}{\partial t}. \tag{5.49}$$

The identity

$$\mathrm{div}[\vec{E} \times \vec{H}] = \nabla \cdot [\vec{E} \times \vec{H}] = -\vec{E} \cdot [\nabla \times \vec{H}] + \vec{H} \cdot [\nabla \times \vec{E}], \tag{5.50}$$

allows to join the equations (5.48) and (5.49) as

$$\mathrm{div}[\vec{E} \times \vec{H}] + \frac{1}{c}\vec{H} \cdot \frac{\partial \vec{B}}{\partial t} + \frac{1}{c}\vec{E} \cdot \frac{\partial \vec{D}}{\partial t} + \frac{4\pi}{c}\vec{E} \cdot \vec{j}_{sw} = 0. \tag{5.51}$$

In the case of the rather general tensor equations of state for anisotropic matter (5.28) and (5.86) one easily proves

$$\vec{E} \cdot \frac{\partial \vec{D}}{\partial t} = \frac{1}{2}\frac{\partial \vec{E} \cdot \vec{D}}{\partial t},$$

and for the magnetic fields

$$\vec{H} \cdot \frac{\partial \vec{B}}{\partial t} = \frac{1}{2}\frac{\partial \vec{H} \cdot \vec{B}}{\partial t}.$$

Substituting these relations into equation (5.52) leads to the differential law of energy conservation for a medium with the current (density $\vec{j}_{sw}$) as a source.

$$\frac{c}{4\pi}\mathrm{div}[\vec{E} \times \vec{H}] + \frac{1}{8\pi}\frac{\partial(\vec{E} \cdot \vec{D} + \vec{H} \cdot \vec{B})}{\partial t} + \vec{E} \cdot \vec{j}_{sw} = 0. \tag{5.52}$$

The form of relation (5.52) allows to introduce the expressions for the *energy density*

$$W_m = \frac{1}{8\pi}(\vec{E} \cdot \vec{D} + \vec{H} \cdot \vec{B}), \tag{5.53}$$

that after plugging in equations (5.28) and (5.30) it reads

$$W_m = \frac{1}{8\pi}(\varepsilon_{ik}E_iE_k + \mu_{ik}H_iH_k), \tag{5.54}$$

summation by i,k is implied and the *Pointing vector*

$$\vec{S}_m = \frac{c}{4\pi}[\vec{E} \times \vec{H}], \tag{5.55}$$

for matter. Note, that the expression for the *power density*

$$P_m = \vec{E} \cdot \vec{j}_{sw},$$

rest the same as for the vacuum approach description, but for the free charges current density.

Generally the energy integral of the elementary work does not exist because the Pfaff form

$$\frac{1}{4\pi}(\vec{E}\cdot\delta\vec{D}+\vec{H}\cdot\delta\vec{B}), \tag{5.56}$$

that determines the field work, may be not exact, see e.g. [6].

5.3 Classification of substances with respect to electric and magnetic properties

5.3.1 Classification of continuum matter on a base of material equations

Electric properties

Some natural classification of matter relates to its phase state that, in turn, determines surrounding of microscopic 'units' (atoms, molecules, etc) and, hence, a state of charges eventually belonging to it. Speaking about electric and magnetic properties one should remember about such 'units' type. The modern physics points out the gas, liquid, solid and plasma matters. Each case exploits a different vision of free charges—ions that could dynamically appear and disappear (recombine). Gases and liquids consider electrons and ions as isolated particles with essentially translation movement. At solids an electronic state belongs to the body as a whole, so its position may be not localized. The same one can say about 'holes'—regions in which lack of electrons is interpreted as a positive charge. The energy of the states form so-called 'bands' of quasi-continuous (discrete, but very close) levels which are divided by 'gaps' of prohibited energy values (bands). One should note that a de-localization is typical for crystal solids but changes in amorphous matter (e.g. glasses) where only a local order is saved. Some kind of a 'soft matter' forms at solid surfaces, liquids and even in gases when so-called 'clusters' or 'nanostructures' are formed. The solids are subdivided to dielectrics, semiconductors and metals; the classes differ by the 'gap' width, which is biggest in the case of dielectric. It means that the electrons in dielectric solids cannot be accelerated because for a velocity (hence-energy) change there needs a place at some higher level; and, therefore cannot normally be free (contribute in ρ_{sw}).

Such structural properties intersect with macroscopic properties related to equations of state. As was mentioned, a dependence of the fields $\vec{D}$ and $\vec{H}$ on $\vec{E}$ and $\vec{B}$ correspondingly defines the elementary work and, hence the energetic relations between magnetic and electric parts.

Speaking about gases and liquids we first of all could divide them for two classes of magnetics and dielectrics. As was mentioned the gases and liquids electromagnetic properties are determined mainly by the properties of their units so as the polarization and magnetization per unit are represented as a sum of corresponding moments of the units.

Let us rewrite the link between the polarization vector per unit volume and dipole momenta (14.47) as follows

$$\vec{P}=\sum_a \vec{p}_a, \tag{5.57}$$

where dipole moment, as it was said in section 5.2.2, is the sum of the own atomic moment and induced one

$$\vec{p}_a = \vec{p}_{aw} + \vec{p}_{ai}. \tag{5.58}$$

Molecules of a dielectric may be divided into two groups: one that is made of molecules with constant electric dipole moment (e.g. water), while second ones, such that are polarized, the polarization appears *only* in the presence of external electric field (molecules have so-called induced dipole moment, see equation (5.58)). The induced dipole moment exists only in the presence of electric field and, for small enough values of E, it is proportional to the electric field, as in equation (5.28). In the previous section the dipole moment appears in the atom only in the presence of external field **E** (formulas (5.40)), it was the case $\vec{p}_{aw} = 0$. In nature we meet also molecules with nonzero dipole moment in the absence of external field.

The well known representative of polar molecules is the molecule of water. On a magnitude of the p_{aw} decides first of all the molecule form of the character 'V' with the oxygen atom in the middle. Such built as well as the molecule electron's states characteristics H_2O holds electrons close to the oxygen 'screening' positive charge of the hydrogen nuclei. Such separation of charges in the molecule explains the permanent dipole moment, that for the water is equal to 1.84 D (Debye).

Magnetic properties

An atom consists of an atomic nuclei and electrons, that surround the nuclei. Due to the electrons motion (rotations and translations), each can be considered as a closed electric circuit. Such an atomic system is represented by magnetic dipole moment, that we define, integrating by a small macroscopic sphere

$$\vec{M} = \sum_a \vec{\mu}_a, \tag{5.59}$$

where $\vec{M}$ describes magnetization (total dipole magnetic moment) related to unit domain, while $\vec{\mu}_a$ expresses magnetic properties of the unit subsystem (atom, molecule). Magnetic dipole moment, as was already mentioned is the sum of its proper (eigen) moment and induced one

$$\vec{\mu}_a = \vec{\mu}_{aw} + \vec{\mu}_{ai}.$$

Very important contributions to magnetization give the eigen magnetic moment of elementary particles. The proper (eigen) magnetic moment arises in relativistic quantum theory from spins (intrinsic, proper angular moment, see e.g. [11] of the elementary particles of a molecule (atom)). Note, that the spin moment in quantum mechanics, as all dynamic variables, observables are represented by operators, so we use here the averaged values of it. The basic relation between the proper angular (mechanic) momentum (spin) for the electron looks as

$$\vec{\mu}_s = \frac{e}{mc}\vec{s}.$$

So, finally

$$\vec{\mu}_a = \vec{\mu}_{aw} + \vec{\mu}_{ai} + \vec{\mu}_{as}. \tag{5.60}$$

The last thing that enters the matter description in macroscopic electrodynamics relates to free charges; it is a link between the charge density, current density and the fields. The thermodynamic level implies simply a dependence

$$\begin{aligned} \rho &= \rho(\vec{E}, \vec{B}) \\ \vec{j} &= \vec{j}(\vec{E}, \vec{B}). \end{aligned} \tag{5.61}$$

The most known relation of such a type is Ohms' law for isotropic medium:

$$\vec{j} = \sigma\vec{E}, \tag{5.62}$$

such closure we use in the sections 6.4 and 6.5, when studying waves propagation in a conducting medium. The charge density changing in the electric and magnetic fields is observed in such phenomena as electro- and magneto-striction.

The concrete examples of the equations we discuss below.

5.3.2 Dielectrics

As was already mentioned in section 5.3.1, the main property of a dielectric (insulator) is the absence of free charges. It means that the current density $j_{sw} = 0$ (see section 5.3.6 'discussion'). The electric field can penetrate the dielectric and induce a polarization. The next step in the classification needs to fix the equations of state (5.23) and (5.24). Dielectric materials can be solids, liquids, or gases.

Solid dielectrics are perhaps the most commonly used dielectrics in electrical engineering, and many solids are very good insulators. Some examples include glass, porcelain, and most plastics.

Liquid dielectrics like water were described at the previous section, some are used in transformers (transformer oil). Also in [12]

Air, nitrogen and sulfur hexafluoride are the three most commonly used *gaseous dielectrics.*

Linear susceptibility, isotropic case

Formulating the relations in terms of the multipole expansion one says that the matter units may have dipole momentum and higher electric momenta.

For expressing the polarization vector that determine a polarization degree of a dielectric, in the first approximation we study the sum of all molecule dipole moments.

For example, in an isotropic homogeneous medium, i.e. in such matter which physical properties do not depend on direction, relation (5.23) and (5.25) for the vectors $\vec{D}$ and $\vec{P}$ are following:

$$\vec{D} = \varepsilon\vec{E} \tag{5.63a}$$

$$\vec{P} = \chi\vec{E}, \tag{5.63b}$$

by the way (Gauss unite) $\varepsilon = 1 + 4\pi\chi$, where ε is electric permittivity of a dielectric (sometimes named as dielectric constant, see sections about waves), while the parameter χ defines its electric susceptibility (dimensionless parameter). For example, the constant for dielectric on normal condition for hydrogen is $\varepsilon_H = 1.000\,270$, next, for water $\varepsilon_{H_20} = 78.3$ is much bigger, that is explained by constant dipole moment of its molecule.

Optic properties (refraction, reflection) will be described in section 10.1.

Anisotropic dielectrics

More generally, as was mentioned in section 5.1.7, the link is a tensor one, see equation (5.28), that is unrolled as

$$D_i = \varepsilon_{i1}E_1 + \varepsilon_{i2}E_2 + \varepsilon_{i3}E_3. \tag{5.64}$$

The second rank tensor ε is symmetric, i.e. $\varepsilon_{ik} = \varepsilon_{ki}$.

Let us take this example to analyze the tensor character of this equation (5.64) link, based on a tensor definition (see chapter 1.3.1) of the entries. If one goes to the reference frame that is obtained from the initial by a rotation R, $R^T R = I$ i.e.

$$\begin{aligned} D_i &= \sum_{s=1}^{3} R_{is}D_s{}' \\ E_k &= \sum_{p=1}^{3} R_{kp}E_p{}' \\ \varepsilon_{ik} &= \sum_{s=1,p=1}^{3} R_{is}^T R_{kp}\varepsilon_{sp}{}', \end{aligned} \tag{5.65}$$

or, in matrix form, for columns E, E', D, D' and square matrices ε, ε'

$$D = RD', \quad E = RE', \quad \varepsilon = R\varepsilon' R^T. \tag{5.66}$$

To verify it we plug equation (5.65) in equation (5.64), having

$$D_i = \sum_{s=1}^{3} R_{is}D_s' = \sum_{p} \varepsilon_{ip}E_p = \sum_{p=1}^{3} \varepsilon_{ip} \sum_{s=1}^{3} R_{ps}E_s'. \tag{5.67}$$

Hence, for the rotation matrix property ($R^T R = I$, or $\sum_{j=1}^{3} R_{rj}^T R_{js} = \delta_{rs}$)

$$\sum_{j=1}^{3} R_{rj}^T \sum_{s=1}^{3} R_{js}D_s' = D_r' = \sum_{s=1}^{3} \varepsilon_{rs}'E_s'. = \sum_{j=1}^{3} R_{rj}^T \sum_{p=1}^{3} \varepsilon_{jp} \sum_{s=1}^{3} R_{ps}E_s'. \tag{5.68}$$

Finally,

$$\varepsilon'_{rs}. = \sum_{j=1}^{3} R^T_{rj} \sum_{p=1}^{3} \varepsilon_{jp} R_{ps} = \sum_{j,p=1}^{3} R^T_{rj} \varepsilon_{jp} R_{ps}. \tag{5.69}$$

The last relation in equation (5.66) means that there is an orientation of coordinate axis in which the dielectric permittivity tensor transforms to the diagonal form

$$\varepsilon' = \begin{pmatrix} \varepsilon'_1 & 0 & 0 \\ 0 & \varepsilon'_2 & 0 \\ 0 & 0 & \varepsilon'_3 \end{pmatrix}. \tag{5.70}$$

In applications the transition to such coordinates is used for experimental determination of the tensor ε. The measurement of ε'_i gives the tensor in an arbitrary coordinate system (reference frame) by the transform

$$\varepsilon_{ik} = \sum_{s=1,p=1}^{3} R^T_{is} R_{kp} \varepsilon'_{sp} = \sum_{s=1,p=1}^{3} R^T_{is} R_{kp} \varepsilon'_s \delta_{sp} = \sum_{s=1}^{3} R^T_{is} R_{ks} \varepsilon'_s. \tag{5.71}$$

This property is general for the second rank tensor.

In optics there is a phenomenon that is explained by anisotropy of a crystals such as a famous Iceland spar. The Iceland spar is the traditional name of a mineral that is chemically clear, transparent crystallized calcite ($CaCO_3$), of romboedric type. The phenomenon is the birefringence, or double refraction, two different values of ε, hence refraction indices, for two polarizations (the field components), therefore the decomposition of a ray of light into two rays (the ordinary ray and the extraordinary ray) see section 10.1. Namely such material is used for the fabrication of the famous Nicola biprizm (William Nicola 1828) used as a polarizer.

Nonlinear susceptibility: Kerr effect

More generally the polarization vector can be written (in SI) as

$$\mathbf{P} = \varepsilon_0(\chi^{(1)}\mathbf{E} + \chi^{(2)} \vdots \mathbf{EE} + \chi^{(3)} \vdots \mathbf{EEE} + \cdots), \tag{5.72}$$

where $\chi^{(1)}$ is the linear dielectric susceptibility, $\chi^{(2)}$ is the second order dielectric susceptibility, and $\chi^{(3)}$ is the third order one.

In this expression (5.72) the notation

$$\vdots \mathbf{EE},$$

denotes the tensor product, so that its term in equation (5.72) reads:

$$P_i^{(2)} = \varepsilon_0 \, \chi_{ikl}^{(2)} E_k E_l, \tag{5.73}$$

the third term is written similar.

The second order dielectric susceptibility $\chi^{(2)}$ is comparably small in materials with symmetrical molecule. From higher order dielectric susceptibilities we left the third order one $\chi^{(3)}$.

In optics, the third order susceptibility is responsible for nonlinear refraction of light, self phase modulation (SPM) and cross phase modulation (XPM).

Considering wavetrains (section 3.3), such that for impulses longer than 0.1 ps one can treat a response of a medium as instantaneous, we write

$$\mathbf{P}_{NL}(t) = \varepsilon_0 \chi^{(3)}(t,\, t,\, t) \vdots \mathbf{E}(t)\mathbf{E}(t)\mathbf{E}(t). \tag{5.74}$$

The third order dielectric susceptibility $\chi^{(3)}$ for isotropic media is discussed, e.g. in the paper [13], and the book [14]. Based on it we arrive at

$$\chi_{ijkl} = \chi_{xxxx}\delta_{ij}\delta_{kl} + \chi_{xyxy}\delta_{ik}\delta_{jl} + \chi_{xyyx}\delta_{il}\delta_{kj}, \tag{5.75}$$

$$\chi_{xxxx} = \chi_{yyyy} = \chi_{zzzz} = \chi_{xxyy} + \chi_{xyxy} + \chi_{xyyx}, \tag{5.76}$$

$$\chi_{xxyy} \simeq \chi_{xyxy} \simeq \chi_{xyyx}, \tag{5.77}$$

In such approximation, plugging the relation equations (5.72), (5.74)–(5.77) into the Maxwell equations in SI units, we write the wave equation for nonlinear case

$$\nabla^2\mathbf{E} - \mu_0\varepsilon_0\varepsilon\frac{\partial^2\mathbf{E}}{\partial t^2} = \mu_0\varepsilon_0\chi_{xxxx}\frac{\partial^2}{\partial t^2}|\mathbf{E}|^2\mathbf{E}. \tag{5.78}$$

All components of the electric field are in standard form for a wavetrain:

$$E_m = \frac{1}{2}\tilde{E}_m e^{i\omega t} + c.c., \tag{5.79}$$

inserting this relation into equation (5.74), we get nonlinear polarization as

$$P_m = \frac{\varepsilon_0}{8}\chi_{xxxx}\sum_j\left(2\tilde{E}_m|\tilde{E}_j|^2 + \tilde{E}_j^2\tilde{E}_m^*\right)e^{i\omega t} + c.c., \tag{5.80}$$

(non-resonant terms are removed) and the indexes are marked as $m, j = x, y, z$. The amplitude $\tilde{E}^*$ is a complex conjugate of $\tilde{E}$.

For example, for the z component we have

$$\begin{aligned} P_z = {} & \frac{3}{8}\chi_{xxxx}\varepsilon_0\left\{\left[|\tilde{E}_z|^2 + \frac{2}{3}\left(|\tilde{E}_x|^2 + |\tilde{E}_y|^2\right)\right]\tilde{E}_z\right\}e^{i\omega t} \\ & + \frac{\varepsilon_0}{8}\chi_{xxxx}\left\{\tilde{E}_z^*\left(\tilde{E}_x^2 + \tilde{E}_y^2\right)\right\}e^{i\omega t} + c.c. \end{aligned} \tag{5.81}$$

We have the same equation as in the texts of [14, 15] with applications in fiber optics. Next we can put it into the Maxwell equations. (We introduce nonlinearity into the Maxwell equations (5.78) in the form of the Kerr effect (5.72), with the assumption of small nonlinearity [14].) The result allows to study propagation of nonlinear waves in a medium with Kerr effect and different kinds off dispersion and polarization, for advanced studies see chapter 11.

5.3.3 Magnetics

Bodies with magnetic properties that are such where the magnetization vector (5.59) can be nonzero in some conditions ($\mu \neq 1$, or $\chi \neq 0$), are named magnetics. Diamagnets were first discovered when Sebald Justinus Brugmans observed in 1778 that bismuth and antimony were repelled by magnetic fields. The term diamagnetism was coined by Michael Faraday in September 1845, when he realized that every material responded (in either a diamagnetic or paramagnetic way) to an applied magnetic field [16].

Its origin is an important problem of modern physics which is considered from a quantum point of view [17]. One of the most important issues on the topic, *IEEE Transactions on Magnetics* publishes scholarly articles of archival value as well as tutorial expositions and critical reviews of classical subjects and topics of current interest. These are science and technology related to the basic physics and engineering of magnetism, magnetic materials, applied magnetics, magnetic devices, and magnetic data storage.

Classification of matter whose magnetic properties are discussed in connection with problems considered includes wide variety of magnetization vector formulas (5.10) and (5.24), starting from the simplest (5.29)

$$\vec{M} = \chi_m \vec{B}, \tag{5.82}$$

In this case the vector $\vec{M}$ is one, which direction is along the magnetic field acting on a dielectric, where the coefficient χ_m, named the *magnetic permeability* is obviously linked to μ via (5.24),

$$\mu = 1 + 4\pi\chi_m. \tag{5.83}$$

Let us repeat, that a magnetic is named as an isotropic one if its magnetization vector is parallel to the vector of magnetic induction $\vec{B}$.

Analyzing behavior of isotropic bodies (see again equation (5.25)) in a magnetic field one can specialize the following groups:

$$\text{Diamagnetics: } \mu < 1$$
$$\text{Paramagnetics: } \mu > 1$$
$$\text{Ferromagnetics: } \mu \gg 1.$$

Note, that in the last case the μ as a function of H is not unique. Let us characterize them in more details.

Diamagnetics

In a sense, all substances are diamagnetic: the strong external magnetic field speeds up or slows down the electrons orbiting in atoms in such a way as to oppose the action of the external field in accordance with Lenz law. The theory of diamagnetism was successfully explained due to the Lorentz force $\vec{F} = e\vec{E} + \frac{e}{c}[\vec{v} \times \vec{H}]$, acting on an orbiting electron when a magnetic field H is applied. Here $\vec{v}$ is the velocity of the electron and 'e' is the electronic charge (Langevin 1905, Pauli 1920) [16]. This changes the electric current in the electron orbit, in a way that gives an induced magnetization. The diamagnetism of some materials, however, is masked either by a weak magnetic attraction (paramagnetism) or a very strong attraction (ferromagnetism) [16]. The magnetic moment of a current loop is equal to the current times the area of the loop (see section 3.8.5, exercise 7). It is therefore

$$M_d = -NZe^2H < r^2 > /4\,m, \tag{5.84}$$

where $<r^2>^{1/2}$ is the quantum average radius of the electron orbit in the atom, with the negative sign implying that the induced effect opposes the external applied one, 'N' is the number of atoms per unit volume and 'Z' is the atomic number, i.e. the number of electrons per atom and 'm' is the electronic mass.

Diamagnets are substances built from particles that obtain magnetic moment only in the presence of external magnetic field. From an empiric point of view the diamagnetics are such bodies that to be placed in a magnetic field are expelled from it (diamagnetic levitation, see e.g. Wiki) (figure 5.6). In the term of magnetic magnetic permeability $\chi_m < 0$ is typically near negative one-millionth. It looks so as the field inside a magnetic is less than outside.

The examples of typical diamagnetics: water, quartz, gold, silver, copper, diamond and noble gases, so such kind of matter which molecules (or atoms) have electron shells (by hydrogen classification) totally filled.

E.g. usually water is a diamagnetic with the susceptibility—0.79 [18].

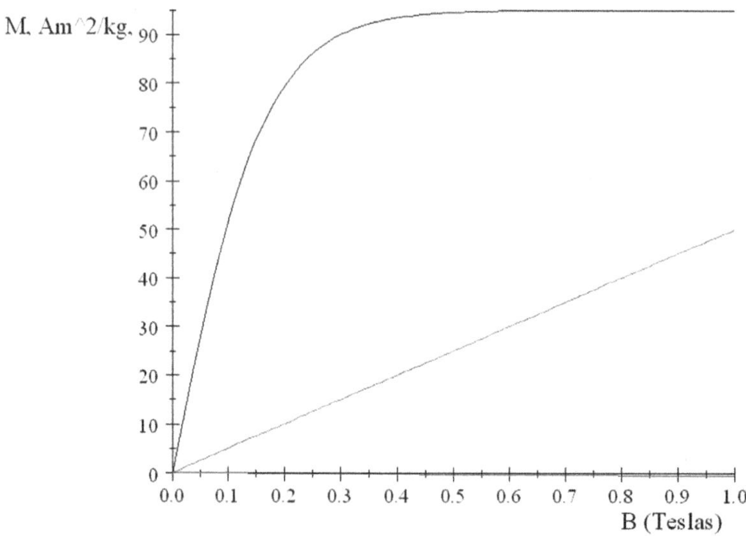

Figure 5.6. A comparison of diamagnetics (red, $M = -0.62$ B, as for quartz, lies below x-axis, very close), paramagnetics (green) and ferromagnetics (navy) in homogeneous magnetic field. In the vicinity of $B = 0$, it is approximated by equation (5.82).

The microscopic origin of the diamagnetism arises from electron states of atoms in magnetic field. Diamagnetism is believed to be due to quantum mechanics. In the case of localized electrons it is understood in terms of the Lorentz force action to the electron at atomic orbit as in equation (5.84).

For free electrons of crystals, or, more generally for delocalized electrons of a solid, so-called Landau levels [19] represent its state. To be precise, a theory of such levels appeared in an earlier but advanced theory by Vladimir Fock [20]. Let us cite it. Introducing the Larmor frequency $\omega_l = eH/2mc$, and ω_0—the frequency of the oscillator, he derives energy levels as

$$E = (n_1 - n_2)\,\hbar\omega_l + (n_1 + n_2 + 1)\,\hbar\sqrt{\omega_0^2 + \omega_l^2}, \tag{5.85}$$

and wave functions expresses via Laguerre polynomials. The Landau formula is obtained as a limit from equation (5.85). The systematic evaluation of the vector $\vec{M}$ needs application of quantum statistical physics. The expression is finally obtained by the thermodynamic potential differentiation, which in turn is derived from a statistical sum built from the Hamiltonian matrix elements in wave functions mentioned.

The Bohr–van Leuwen theorem [21] proves that there cannot be any diamagnetism or paramagnetism in a purely classical system. Yet the classical theory for Langevin diamagnetism, estimated by equation (5.84) gives the same prediction as the quantum theory.

Anisotropic magnetics

A generalization of the relation (5.82) is the tensor analog of equation (5.28). Its use accounts for anisotropy, its simplest version is realized via surface effects (such anisotropy is mentioned in section 5.3.4), for example small ferromagnets [22]. A static magnetic memory includes a layer having a plurality of vertically oriented and shape-anisotropic elongated ferromagnetic particles. The general linear relation between the fields is

$$B_i = \sum_{k=1}^{3} \mu_{ik} H_k, \tag{5.86}$$

Its tensor properties are studied in section 5.3.2 on the example of relation between vectors $\vec{D}$ and $\vec{E}$.

What is spin?

To prepare the transition to para- and ferromagnetism, let us repeat (mentioned in sections 5.1.2 and 5.3.1), that this a particle property, spin, has quantum origin, hence the corresponding magnetic moment does not enter the classical electrodynamics. It could be taken into account as an extra term in magnetic dipole moment as in section 5.3.1, but no classical rotation can explain its value. Consequently we should study a phenomenon of para-ferromagnetism within quantum approach, its correct interpretation was started from the seminal paper of Heisenberg [23] where not only was

Weiss electric interaction explained as quantum exchange effect, but *ab initio* theory, based on statistical distribution/sum was built.

Paramagnets

Atoms, that in spite of absence of external magnetic field, have nonzero magnetic moment are named paramagnetics. One of the main reasons of such evidence may be the presence of electrons' spin moments and the proportion to the magnetic ones.

Therefore, if the external magnetic field is switched on, a tendency appears for electrons that their own mechanical and magnetic moments would be along the direction of magnetic field $\vec{B}$. In real matter, usually solid, this tendency is restricted by heat motion of particles. Paramagnetic properties conventionally demonstrate metals (alkalic and rare earths). Some exclusive behavior in a gaseous state, in specific conditions, demonstrates an example below: 'In 2009, a team of MIT physicists demonstrated that a lithium gas cooled to less than one kelvin can exhibit ferromagnetism. The team cooled fermionic lithium-6 to less than 150 billionths of one kelvin above absolute zero using infrared laser cooling. This demonstration is the first time that ferromagnetism has been demonstrated in a gas' (wiki; lithium gas https://news.mit.edu/2009/magnetic-gas-0918). The ferromagnetic property demonstrates also some organic materials.

5.3.4 Ferromagnetism

The phenomenon

Let us start with a definition of the ferromagnetic in terms of the equation of state ($\vec{B}(\vec{H})$ dependence). Ferromagnets are conventionally solids (see the previous subsection) in which a dependence between the vectors: magnetic moment $\vec{M}$ and magnetic field $\vec{H}$ is nonlinear and not unique. It means that a growth of the magnetic field stress yields changes in magnetization velocity of a ferromagnetic. So happens till some extremal level is reached, that is named magnetic saturation, $\vec{M} \to \vec{M}_s$.

The next feature is typical for a ferromagnetic in some temperature range, restricted by Curie temperature: it is a phenomenon of hysteresis. It depends on *not only* external magnetic field, but also on a level of magnetization of a specimen at

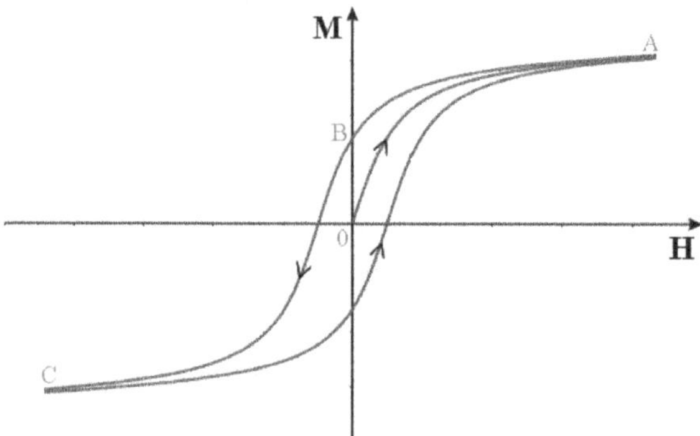

Figure 5.7. The typical hysteresis loop for the bulk ferromagnetic matter [24].

$t = 0$. Such situation one can present by the plot, so-called hysteresis loop, figure 5.7. Switching on the magnetic field at ($t = 0$), corresponds to the time of magnetization start. In the time marked $t = 1$, the hysteresis curve reaches its maximum, saturation point A. Next, switching off the magnetic field, the magnetization of the body diminishes down to the value, corresponding to point B. One can state experimentally that the specimen will have some magnetization, while the external magnetic field is absent. To eliminate the magnetization of the ferromagnetic body, it is necessary to do some actions, applying a magnetic field of the opposite direction, repeating this a few times. The work of such action magnitude is equal to the area of the hysteresis curve marked at the plot.

On the theory

From the microscopic point of view, the main reason for the ferromagnetism is the presence of the spin part of the magnetic moment of electrons (see equation (5.60)) in the context of Pauli's principle, but more importantly that it is a collective effect which is expressed in the ability to have non-compensate (unpaired) total spin at some area (domain). In other words the ferromagnetic materials have some unpaired electrons so their atoms have a net magnetic moment. They get their magnetic properties due to the presence of so-called magnetic domains, see figure 5.8. In these domains, large numbers of unit's (electron) moments of order from 10^{12} to 10^{15} are aligned parallel so that the magnetic force within the domain is strong [24]. A characteristic feature of a ferromagnet is also that their ferromagnetic properties conventional take place in the solid state, typical examples of which may be iron, nickel or cobalt. The condition of this could be the domains existence. The ferromagnetic properties can exhibit the crystalline structure of the matter and amorphous as well [25]. According to Kondorsky classification [26] there are three main causes of hysteresis.

1. Hysteresis due to a delay in the displacement of boundaries between domains.

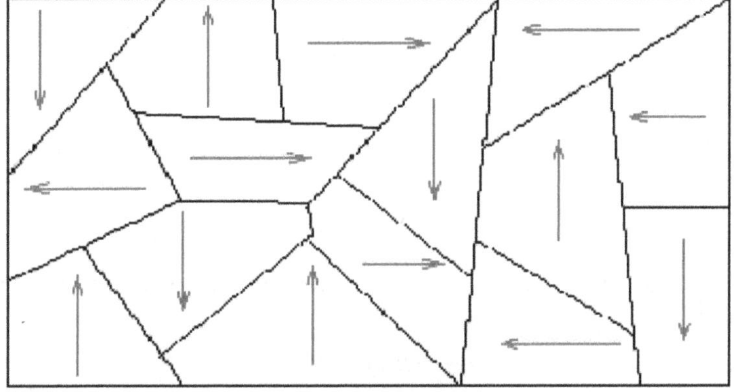

Figure 5.8. Schematic presentation of magnetic domains distribution in a ferromagnetic.

2. Hysteresis due to growth retardation of the magnetization reversal of nucleation.
3. Hysteresis due to irreversible rotation.

Domains properties take place only within a temperature interval. The temperature increase leads to kinetic energy growth in the crystal atomic net. So at high temperatures, an ordering of a structure, including domains disappear. Each ferromagnetic has some critical temperature, under the name of Curie temperature, above which ferromagnetic properties transform to paramagnetic ones. In other words, one says that below a certain temperature, the Curie point, a body can possess magnetization in the absence of an external magnetic field; when a ferromagnetic material is in the demagnetized state, the domains are nearly randomly oriented and the magnetic field from the whole body is about zero. When a magnetizing force is applied, the domains become aligned to produce a strong magnetic field within the part. Components with these materials are commonly inspected using the magnetic particle method [27].

It is interesting and instructive to touch the theory of this phenomenon, that go up to the pioneering work of Heisenberg [23].

A plurality of writing conductors are adjacent to the layer, and the conductors selectively apply magnetic fields to selected regions of the layer by directing electrical current to magnetize the particles in an up or down direction. Static reading means detecting the direction of magnetization. The particles may include a soft magnet portion and a hard magnet portion. In another preferred embodiment, a material and a method of making same includes providing a matrix full of elongated holes, depositing a first magnetic material having a first coercivity into the holes, and then depositing a second magnetic material having a second coercivity into the holes to form a composite elongated particle in each hole.

5.3.5 Combined action: multiferroics

The linear version of the equations of state (5.23) and (5.24) of section 5.1.7 are written as

$$D_i = \varepsilon_{ik} E_k + \varepsilon'_{ik} B_k, \tag{5.87}$$

$$H_i = \tilde{\mu}_{ik} B_k + \tilde{\mu}'_{ik} E_k, \tag{5.88}$$

More general equations we use in chapter 18, section 18.1. Let us take the particular 1D cases:

$$D = \varepsilon E + \varepsilon' B, \tag{5.89}$$

that show an eventual impact of magnetic field on the polarization. Opposite, the electric field may change the magnetization:

$$H = \tilde{\mu} B_k + \tilde{\mu}' E. \tag{5.90}$$

Such dependence allows to manipulate by polarization and magnetization of matter, because, the relations $D = E + 4\pi P$, $B = H + 4\pi M$ are still valid. We also can write

$$P = \chi E + \chi' B, \tag{5.91}$$

$$M = \tilde{\chi}_m B_k + \tilde{\chi}'_m E. \tag{5.92}$$

Both phenomena are known under the name of *magnetoelectric effect.* It was shown, theoretically and after that experimentally, that such a magnetic linear magneto-electric effect is possible. The starting point on a real crystal was marked by Dzyaloshinskii [28] who predicted the effect in the antiferromagnetic Cr_2O_3. After this prediction, Astrov [29] reported the experimental observation of the phenomenon. Later, the linear magnetoelectric effect has been observed in many magnetic compounds, including for example, yttrium iron garnet. The concept within the electrodynamics is properly discussed in [30], see also more recent [31].

5.3.6 On dielectrics

In section 5.3.2: an example of a method of dielectric anisotropy measurements is given in [32].

The matter division we considered is rather relative. For example each dielectric has (weak) magnetic properties. So, a marble, being a good insulator, is a diamagnetic with the susceptibility—0.7 [18]. Ionization transforms dielectrics to a conductor. Let us mention two books in the titles of which the word 'dielectric' is presented, but in fact the content relates more to plasma physics. The first one is devoted to ionization and of liquid dielectrics (I Adamczewski [12]); the second is the named 'Physics of dielectrics' (G Skanavi [33]) and devoted to a leakage (punch?) phenomena inside dielectrics in strong electric fields.

If the external field may be strong enough, the condition $j_{sw}(\vec{E}) = 0$ holds only in the vicinity of the point $\vec{E} = 0$.

5.3.7 Metamaterials

Very recent applications are connected with metamaterials [34].

All transparent or translucent materials possess positive refractive index—a refractive index that is greater than zero. However, is there any fundamental reason that there should not be materials with negative refractive index? This question was asked by Victor Veselago. In 1968, Veselago published a theoretical analysis of the electromagnetic properties of materials with negative permittivity and negative permeability [35], see chapter 8, section 6.1.2. The electric permittivity and the magnetic permeability are commonly used material parameters that describe how materials polarize in the presence of electric and magnetic fields. Maxwell's equations relate the permittivity and the permeability to the refractive index as follows (i initial angle, r angle of refraction):

$$n = \frac{c}{v} = \frac{\sin i}{\sin r} = \sqrt{\varepsilon\mu}. \tag{5.93}$$

For the origin of the first equality see section 6.5.1, relation (6.110) for zero conductivity. The second equality is derived in chapter 10, section 10.1.

The sign of the index is usually taken as positive. However, Veselago showed that if a medium has both negative permittivity and negative permeability, this convention must be reversed: we must choose the negative sign of the square root! As metamaterial example we mention the paper [36], where the authors study the effective dielectric tensor for a composite of spheres with anisotropic dielectric tensor embedded in an isotropic uniform medium.

5.3.8 Exercises

Exercise 1: Following the method of section 5.3.2 for the tensor ε_{ik}, express the components of the tensor of magnetic permeability in main axes $\mu'_{ik} = \delta_{ik}\mu'_l$.

References

[1] Jackson J D 1998 *Classical Electrodynamics* 3rd edn (New York: Wiley)

[2] *Electric Field Measurements* http://www.ets-lindgren.com/pdf/Better_Field_Measurements_king.pdf

[3] Bobrov V N and Kulikov N D 1969 *Quartz Magnetic Sensor. Geomagnetism and Aeronomy* vol 9 p 307

[4] Magnetometer Sites http://www.magnetometer.org/ http://www.windows.ucar.edu/tour/link=/teacher_resources/magnetometer_edu.html

[5] Fluxmeters site http://www.trifield.com/DCModel1.htm

[6] Guggenheim E A 1967 *Thermodynamics: An Advanced Treatment for Chemists and Physicists* (Amsterdam: Elsevier)

[7] Spaldin N A and Fiebig M 2005 The renaissance of magnetoelectric multiferroics *Science* **309** 391–2

[8] Thide B 2004 *Electromagnetic Field Theory* (Sweden: Upsilon)

[9] Ashcroft N W and David Mermin N 1976 *Solid State Physics* 1st edn (Boston, MA: Cengage Learning)

[10] Azaroff L V 1984 *Introduction to Solids* (New York: McGraw-Hill) (first published 1960)

[11] Fock V A 1978 (1982, 1986) *Fundamentals of Quantum Mechanics* (Moscow: Mir)

[12] Adamczewski I 1965 *Ionization and Conductivity of Liquid Dielectrics* (Warsaw: PWN) (Polish)

[13] Maker P D and Terhune R W 1966 Study of optical effects due to an induced polarization third order in the electric field strength *Phys. Rev. A* **137** 801
Maker P D and Terhune R W 1966 *Phys. Rev. A* **148** 990 (Erratum)

[14] Agrawal G P 2000 *Nonlinear fiber optics Nonlinear Science at the Dawn of the 21st Century* (Berlin: Springer) pp 195–211

[15] Menyuk C R 1987 Stability of solitons in birefringent optical fibers. I: equal propagation amplitudes *Opt. Lett.* **12** 628–30

[16] Vonsovsky S V 1971 *Magnetism. Magnetic properties of dia-, para-, ferro-, antiferro- and ferrimagnets* (Russian)
Vonsovsky S V and Katsnelson M I 1989 *Quantum Solid State Physics* (New York: Springer)

[17] Dattagupta S, Jayannavar A M and Kumar N 2001 *Landau Diamagnetism Revisited (arXiv: cond-mat/0106646)*

[18] Smithonian Institute 2003 *Smithonian Physical Tables* 9th edn (Norwich: Knovel)
[19] Landau L 1930 Diamagnetismus der Metalle *Z. Phys.* **64** 629
[20] Fock V 1928 A comment on quantization of the harmonic oscillator in a magnetic field *Z. Phys.* **47** 446
[21] Bohr N 1911 *Dissertation Copenhagen*
Van Leuwen J H 1921 *J. Phys.* **2** 361
[22] Sattler-Ed K D 2011 *Handbook of Nanophysics: Nanoparticles and Quantum Dots* (Boca Raton, FL: CRC Press)
[23] Heisenberg W 1928 Zur Theorie des Ferromagnetismus *Z. Phys.* **49** 619–36
[24] Coey J M D 2009 *Magnetism and Magnetic Materials* (Cambridge: Cambridge University Press)
[25] Väzquez M 2015 *Magnetic Nano- and Microwires: Design, Synthesis, Properties and Applications. Woodhead Publishing Series in Electronic and Optical Materials* (Amsterdam: Elsevier)
[26] Borovik E S, Eremenko V V and Milner A S 2005 *Lectures on Magnetism* 3rd edn (Moscow: FIZMATLIT) (Russian)
[27] IEAE 2000 *Liquid Penetrant and Magnetic Particle Testing at Level 2* (Vienna: International Atomic Energy Agency)
[28] Dzyaloshinski I E 1959 *Zh. Eksp. Teor. Fiz.* **37** 881
[29] Astrov D N 1960 *Zh. Eksp. Teor. Fir.* **38** 984
[30] O'Dell T H 1970 *The Electrodynamics of Magneto-electric Media* (London: North-Holland)
[31] Siratori K, Kohn K and Kita E 1992 Magnetoelectric effect in magnetic materials *ACTA Phys. Polonica A* **81** 413–66
[32] Zvyagintsev A A, Strizhachenko A V and Chizhov V V 2002 Measurement of dielectric permittivity tensor, main directions and optical axises in single-axis and bi-axis crystals *Microwave and Telecommunication Technology* pp 544–5
[33] Skanavi G I 1958 *Physics of Dielectrics (Strong-Field Region)* (Moscow: Fizmatgiz)
Skanavi G I 1958 *Physics of Dielectrics (Region of Strong Fields)* (Moscow: State Physical and Mathematical Press)
Skanavi G I 1958 *Dielectric Physics* translated by ed Chen Y H (Beijing: High Educational)
[34] http://www.ee.duke.edu/drsmith/
[35] Veselago V G 1968 The electrodynamics of substances with simultaneously negative values of ε and μ *Sov. Phys. Uspekhi* **10** 509–14
[36] Hinsen K, Bratz A and Felderhof B U 1992 Anisotropic dielectric tensor and the Hall effect in a suspension of spheres *Phys. Rev. B* **46** 14995–5003

Chapter 6

Dispersion and transport

6.1 Dispersion account, operator material relations

6.1.1 Maxwell's equations: operators of dielectric permittivity and magnetic permeability

Our starting point is the Maxwell equations for a simple case of linear isotropic but dispersive dielectric media, in the SI unit system:

$$\mathrm{div}\vec{D}(\vec{r}, t) = 0, \tag{6.1}$$

$$\mathrm{div}\vec{B}(\vec{r}, t) = 0, \tag{6.2}$$

$$\mathrm{rot}\vec{E}(\vec{r}, t) = -\frac{\partial\vec{B}(\vec{r}, t)}{\partial t}, \tag{6.3}$$

$$\mathrm{rot}\vec{H}(\vec{r}, t) = \frac{\partial\vec{D}(\vec{r}, t)}{\partial t}. \tag{6.4}$$

The basic functions $\vec{E}(\vec{r}, t), \vec{B}(\vec{r}, t), \vec{D}(\vec{r}, t), \vec{H}(\vec{r}, t)$ of the electromagnetic field are named t-representation or a time domain. The domain of Fourier images of the vectors $\mathcal{E}(\vec{r}, \omega)$, $\mathcal{B}(\vec{r}, \omega)$, $\mathcal{D}(\vec{r}, \omega)$, $\mathcal{B}(\vec{r}, \omega)$ we will call $\omega-$representation or a frequency domain. The Fourier images of the fields components E_i, B_i,D_i and H_i are defined in a similar manner:

$$E_i(\vec{r}, t) = \frac{1}{\sqrt{2\pi}}\int_{-\infty}^{\infty}\mathcal{E}_i(\vec{r}, \omega)\exp(i\omega t)d\omega. \tag{6.5}$$

We begin with linear material equations in $\omega-$representation as:

$$\mathcal{D} = \varepsilon_0\varepsilon(\omega)\mathcal{E}, \tag{6.6}$$

doi:10.1088/978-0-7503-2576-9ch6

$$\mathcal{B} = \mu_0\mu(\omega)\mathcal{H}. \tag{6.7}$$

Here: $\varepsilon(\omega)$—dielectric permittivity of medium, ε_0—dielectric permittivity of vacuum. $\mu(\omega)$—magnetic permeability of medium and μ_0—magnetic permeability of vacuum. The relations are conventionally found as empiric or from the Lorenz model, enforced by quantum theory (section 6.3.1). For practical uses and further steps in the theory development we need to go to t-representation. In this representation ε and μ become integral operators. Then:

$$\begin{aligned}\vec{D}(\vec{r}, t) &= \frac{1}{\sqrt{2\pi}}\int_{-\infty}^{\infty} \mathcal{D}(\vec{r}, \omega)\exp(i\omega t)d\omega \\ &= \frac{\varepsilon_0}{\sqrt{2\pi}}\int_{-\infty}^{\infty} \varepsilon(\omega)\mathcal{E}(\vec{r}, \omega)\exp(i\omega t)d\omega,\end{aligned} \tag{6.8}$$

$$\begin{aligned}\vec{B}(\vec{r}, t) &= \frac{1}{\sqrt{2\pi}}\int_{-\infty}^{\infty} \mathcal{B}(\vec{r}, \omega)\exp(i\omega t)d\omega \\ &= \frac{\mu_0}{\sqrt{2\pi}}\int_{-\infty}^{\infty} \mu(\omega)\mathcal{H}(\vec{r}, \omega)\exp(i\omega t)d\omega.\end{aligned} \tag{6.9}$$

Plugging

$$\mathcal{E}(\vec{r}, \omega) = \frac{1}{\sqrt{2\pi}}\int_{-\infty}^{\infty} E(\vec{r}, s)\exp(-i\omega s)ds \tag{6.10}$$

into equation (6.8) we obtain the expression that contains a double integral:

$$\vec{D}(x, t) = \frac{\varepsilon_0}{2\pi}\int_{-\infty}^{\infty} \varepsilon(\omega)\int_{-\infty}^{\infty} \vec{E}(\vec{r}, s)\exp(-i\omega s)ds\,\exp(i\omega t)d\omega. \tag{6.11}$$

If the electric field $\vec{E}(\vec{r}, s)\exp(-i\omega s)$ components satisfy the basic conditions of Fubini's theorem, we can change the order of integrations:

$$\begin{aligned}\vec{D}(\vec{r}, t) &= \frac{\varepsilon_0}{2\pi}\int_{-\infty}^{\infty}\int_{-\infty}^{\infty} \varepsilon(\omega)\exp[i\omega(t-s)]d\omega\vec{E}(\vec{r}, s)ds \\ &= \int_{-\infty}^{\infty} \tilde{\varepsilon}(t-s)\vec{E}(\vec{r}, s)ds,\end{aligned} \tag{6.12}$$

where the kernel

$$\tilde{\varepsilon}(t-s) = \frac{\varepsilon_0}{2\pi}\int_{-\infty}^{\infty} \varepsilon(\omega)\exp[i\omega(t-s)]d\omega, \tag{6.13}$$

defines the integral operator of convolution type

$$\hat{\varepsilon}\psi(x, t) = \int_{-\infty}^{\infty} \tilde{\varepsilon}(t-s)\psi(x, s)ds. \tag{6.14}$$

Finally, the relation (6.12) reads:

$$\vec{D}(\vec{r}, t) = \hat{\varepsilon}\vec{E}(\vec{r}, t), \tag{6.15}$$

that expresses the material relation for the definition (6.6) in the operator form.

In a similar way, for the components of the magnetic fields, we introduce the integral operator:

$$\vec{B}(\vec{r}, t) = \int_{-\infty}^{\infty} \tilde{\mu}(t - s)\vec{H}(\vec{r}, s)ds = \hat{\mu}\vec{H}(\vec{r}, t), \tag{6.16}$$

with the kernel

$$\tilde{\mu}(t - s) = \frac{\mu_0}{2\pi}\int_{-\infty}^{\infty} \mu(\omega)\exp[i\omega(t - s)]d\omega. \tag{6.17}$$

The inverse links follow from equations (6.12), (6.16) are

$$\begin{aligned} \vec{E}(\vec{r}, t) &= \int_{-\infty}^{\infty} \tilde{e}(t - s)\vec{D}(\vec{r}, s)ds, \ \vec{H}(\vec{r}, t) \\ &= \int_{-\infty}^{\infty} \tilde{m}(t - s)\mathcal{B}(\vec{r}, s)ds = \hat{\mu}^{-1}\vec{B}(\vec{r}, t), \end{aligned} \tag{6.18}$$

with kernels

$$\begin{aligned} \tilde{e}(t - s) &= \frac{1}{2\pi}\int_{-\infty}^{\infty} \varepsilon^{-1}(\omega)\exp(i\omega(t - s))d\omega, \\ \tilde{m}(t - s) &= \frac{1}{2\pi\mu_0}\int_{-\infty}^{\infty} \mu^{-1}(\omega)\exp(i\omega(t - s))d\omega. \end{aligned} \tag{6.19}$$

The transforms define the fields in time domain, using the conventional continuation of the fields to the half space $t < 0$ and causality condition [3].

6.1.2 Energy density of wave packets in a dispersive medium

The derivation originates from [1], see also [2], but this presentation is more close to one proposed by Millonni in 2005 [3]. In the equality (5.52) the expression

$$\frac{1}{c}\vec{H} \cdot \frac{\partial\vec{B}}{\partial t} + \frac{1}{c}\vec{E} \cdot \frac{\partial\vec{D}}{\partial t}, \tag{6.20}$$

gives, in cases, specified in the previous section, the speed of energy density change. Generally, in analogy with Pointing's theorem it is equal to the sum of energy flux plus the heat transfer from current, see again equation (5.52).

The case we consider, as in the precedent section, the electric displacement field $\vec{D}$ is expressed via convolution-type integral operator $\hat{\varepsilon}$ action on the electric field $\vec{E}$. We start from

$$\vec{D}(x, t) = \frac{\varepsilon_0}{2\pi} \int_{-\infty}^{\infty} \int_{-\infty}^{\infty} \varepsilon(\omega)\exp(i\omega(t - s))d\omega \vec{E}(x, s)ds \\ = \int_{-\infty}^{\infty} \tilde{\varepsilon}(t - s)\vec{E}(x, s)ds, \tag{6.21}$$

where the kernel of the operator is expressed by

$$\tilde{\varepsilon}(t - s) = \frac{\varepsilon_0}{2\pi} \int_{-\infty}^{\infty} \varepsilon(\omega)\exp(i\omega(t - s))d\omega. \tag{6.22}$$

Let the electric and magnetic field be described by the relation

$$\vec{E} = \vec{\mathbb{E}}(x, t)\exp(-i\omega t) + c.c., \tag{6.23}$$

as the wavetrain (section 3.3), where the function $\vec{\mathbb{E}}(x, t)$ represents a slow varying amplitude. After Fourier transformation of the amplitude:

$$\vec{\mathbb{E}} = \frac{1}{\sqrt{2\pi}} \int_{-\infty}^{\infty} \vec{e}(x, \omega')\exp(i\omega' t)d\omega', \tag{6.24}$$

we have the electric field in the form:

$$\vec{E} = \frac{1}{\sqrt{2\pi}} \exp(-i\omega t) \int_{-\infty}^{\infty} \vec{e}(x, \omega')\exp(i\omega' t)d\omega' + c.c., \tag{6.25}$$

Substitution of equation (6.25) into equation (6.21) gives:

$$\vec{D}(x, t) = \frac{\varepsilon_0}{(2\pi)^{3/2}} \int_{-\infty}^{\infty} \int_{-\infty}^{\infty} \varepsilon(w)\exp(iw(t - s))dw \exp(-i\omega s) \\ \times \int_{-\infty}^{\infty} \vec{e}(x, \omega')\exp(i\omega' s)dsd\omega' + c.c. \tag{6.26}$$

Changing the order of integration we continue as:

$$\vec{D}(x, t) = \frac{\varepsilon_0}{(2\pi)^{3/2}} \int_{-\infty}^{\infty} \varepsilon(w)\exp(iwt) \\ \int_{-\infty}^{\infty} \int_{-\infty}^{\infty} \exp(-i\omega s)\exp(iws)\exp(i\omega' s)dsdw\vec{e}(x, \omega')d\omega' + c.c. \tag{6.27}$$

Integrating by ds yields

$$\vec{D}(x, t) = \frac{\varepsilon_0}{(2\pi)^{1/2}} \int_{-\infty}^{\infty} \varepsilon(w)\exp(iwt) \int_{-\infty}^{\infty} \delta(w + \omega' - \omega)dw\vec{e}(x, \omega')d\omega' + c.c. \tag{6.28}$$

The use of the $\delta-$ function definition gives:

$$\vec{D}(x, t) = \frac{\varepsilon_0}{(2\pi)^{1/2}} \int_{-\infty}^{\infty} \varepsilon(w)\exp(iwt)\vec{e}(x, \omega - w)dw + c.c. \tag{6.29}$$

We introduce the shift Δ from maximum value of the wavetrain envelope in $\omega-$ domain $\vec{e}(x, \omega')$ at $\omega' = 0$:

$$\Delta = \omega - w,$$

having

$$w = \omega - \Delta,$$
$$dw = d(\omega - \Delta).$$

The variable ω is the parameter, carrier frequency, induced by the source, then $dw = -d\Delta$. Hence

$$\vec{D}(x, t) = -\exp(i\omega t)\frac{\varepsilon_0}{(2\pi)^{1/2}}\int_{-\infty}^{\infty}\vec{e}(x, \Delta)\varepsilon(\omega + \Delta)\exp(i\Delta t)d\Delta + c.c. \tag{6.30}$$

Assuming that the spectrum of the wavetrain is localized around the carrier frequency ω, we expand $\varepsilon(\omega + \Delta)$ in the Taylor series, leaving only the first two terms:

$$\varepsilon(\omega - \Delta) \approx \varepsilon(\omega) - \Delta\varepsilon_\omega, \tag{6.31}$$

where

$$\varepsilon_\omega \equiv \frac{d\varepsilon}{d\omega},$$

or, plugging equation (6.31) into equation (6.30) yields

$$-\vec{D}(x, t) \approx \exp(i\omega t)\frac{\varepsilon_0}{(2\pi)^{1/2}}\int_{-\infty}^{\infty}\vec{e}(x, \Delta)(\varepsilon(\omega) - \Delta\varepsilon_\omega)\exp(i\Delta t)d\Delta + c.c., \tag{6.32}$$

next,

$$\begin{aligned}\vec{D}(x, t) \approx &- \exp(i\omega t)\frac{\varepsilon_0}{(2\pi)^{1/2}}\{\varepsilon(\omega)\int_{-\infty}^{\infty}\vec{e}(x, \Delta)\exp(i\Delta t)d\Delta \\ &+ i\varepsilon_\omega\frac{\partial}{\partial t}\int_{-\infty}^{\infty}\vec{e}(x, \Delta)\exp(i\Delta t)d\Delta\} + c.c.\end{aligned} \tag{6.33}$$

Fourier transformation by Δ yields:

$$\vec{D}(x, t) \approx -\varepsilon_0\exp(i\omega t)\{\vec{\mathbb{E}}(x, t)\varepsilon(\omega) + i\mathbb{E}_t\varepsilon_\omega\} + c.c., \tag{6.34}$$

with the derivative

$$\vec{\mathbb{E}}_t \equiv \frac{\partial\vec{\mathbb{E}}}{\partial t},$$

then

$$\begin{aligned}\vec{D}_t(x, t) \approx &-\varepsilon_0 i\omega\exp(i\omega t)\{\vec{\mathbb{E}}(x, t)\varepsilon(\omega) \\ &+ i\vec{\mathbb{E}}_t\varepsilon_\omega\} - \varepsilon_0\exp(i\omega t)\{\vec{\mathbb{E}}_t(x, t)\varepsilon(\omega) + i\vec{\mathbb{E}}_{tt}\varepsilon_\omega\} + c.c.\end{aligned} \tag{6.35}$$

Or, in the first order, in the more compact form:

$$\vec{D}_t(x, t) \approx \varepsilon_0 \left[\vec{\mathbb{E}}_t \frac{d\omega\varepsilon(\omega)}{d\omega} - i\omega\varepsilon(\omega)\vec{\mathbb{E}}(x, t) \right] \exp(i\omega t) + c.c. \tag{6.36}$$

The energy density arises from equation (6.20):

$$\vec{E}\vec{D}_t + \vec{H}\vec{B}_t.$$

The scalar product of $\vec{E}$ from equation (6.23) times $\vec{D}_t$ from (6.36),

$$\vec{E}\vec{D}_t = \varepsilon_0(\vec{\mathbb{E}} \exp(-i\omega t) + c.c.) \cdot \left[\left[\vec{\mathbb{E}}_t \frac{d\omega\varepsilon(\omega)}{d\omega} - i\omega\varepsilon(\omega)\vec{\mathbb{E}}(x, t) \right] \exp(i\omega t) + c.c. \right], \tag{6.37}$$

contributes as:

$$\vec{E}\vec{D}_t \approx \varepsilon_0[\vec{\mathbb{E}} \cdot \vec{\mathbb{E}}_t^* + \vec{\mathbb{E}}^* \cdot \vec{\mathbb{E}}_t] \frac{d\omega\varepsilon(\omega)}{d\omega} + \cdots, \tag{6.38}$$

we do not show terms with double carrier frequency, neglecting the period-averaged energy. Finally,

$$\vec{E}\vec{D}_t \approx \varepsilon_0[\vec{\mathbb{E}} \cdot \vec{\mathbb{E}}^*]_t \frac{d\omega\varepsilon(\omega)}{d\omega} + \cdots, \tag{6.39}$$

that defines the electric part of energy density in the approximations we have outlined:

$$W_E \approx \varepsilon_0 \vec{\mathbb{E}} \cdot \vec{\mathbb{E}}^* \frac{d\omega\varepsilon(\omega)}{d\omega} + \cdots, \tag{6.40}$$

Note, that

$$\vec{E} \cdot \vec{E}^* = \mathbb{E} \cdot \mathbb{E} \exp(-2i\omega t) + 2\vec{\mathbb{E}} \cdot \vec{\mathbb{E}}^* + \mathbb{E}^* \cdot \mathbb{E}^* \exp(2i\omega t) \sim 2\vec{\mathbb{E}} \cdot \vec{\mathbb{E}}^* \tag{6.41}$$

in the same paradigm.

In a similar manner we transform the wavetrains for magnetic fields $\vec{B}$ and $\vec{H}$, that gives

$$\vec{H}\vec{B}_t \approx \mu_0[\vec{\mathbb{H}} \cdot \vec{\mathbb{H}}^*]_t \frac{d\omega\mu(\omega)}{d\omega} + \cdots, \tag{6.42}$$

the complete energy density is written as

$$W_E + W_H \approx \varepsilon_0 \vec{\mathbb{E}} \cdot \vec{\mathbb{E}}^* \frac{d\omega\varepsilon(\omega)}{d\omega} + \mu_0 \vec{\mathbb{H}} \cdot \vec{\mathbb{H}}^* \frac{d\omega\mu(\omega)}{d\omega}, \tag{6.43}$$

cf [2]. This result is important in the theory of metamaterials, when both ε and μ are negative, see chapter 8, section 8.6.1.

6.1.3 On Maxwell–Lorentz equations

Two different methods were found for formulating the electrodynamics equations in continuum in the basis of vacuum Maxwell's equations. The first one goes up to Maxwell himself [4] that lies upon declaration of the equation in the form of the equations (6.1)–(6.4) with the further addition of the (empiric) equations of state (5.23) and (5.24); read the textbooks of Reichl, [5], including the statistical description; interestingly, Novozhilov and Yappa [6] introduce both vectors $\vec{E}$ and $\vec{D}$ at the very beginning that allows universal description (including matter), tracing the thermodynamic energy–work relations. See also the important viewpoint of [7].

The second method is based on a pure theoretic description, namely on the microscopic equation in vacuum with moving point charged particles. This system is now named Maxwell–Lorentz [8]. The equations unify the proper Maxwell equations (2.28)–(2.31) and the relativistic Newton equation (4.43) with the Lorentz force (2.9); the closure of the system is achieved by the link of the density current with particle velocity $\vec{j} = e\vec{v}$. The density of charge in such theory is represented by the use of the Dirac delta function

$$\rho(\vec{r}, t) = \sum_i q_i \delta(\vec{r} - \vec{r_i}(t)), \tag{6.44}$$

where $\vec{r_i}$ is a position of a point charge number i, which velocity is $\vec{\dot{r}_i}$, hence the current density is represented by

$$\vec{j}(\vec{r}, t) = \sum_i q_i \vec{\dot{r}_i}(t) \delta(\vec{r} - \vec{r_i}(t)). \tag{6.45}$$

The most advanced form of a point particle motion is written by space–time tensors of relativistic theory for each particle (see section 4.4):

$$\frac{dp_\mu}{d\tau} = F_\mu, \tag{6.46}$$

$\mu = 0, 1, 2, 3$, that close the description, because the rhs is chosen as a four-Lorentz force. Such an approach is effectively used in the modern theoretical investigations, for example the *free-electron laser*, or *cyclotron resonance maser* can be described by solving the Lorentz–Maxwell equations.

The simplest link between micro and macro descriptions is built via relations (5.10) and (5.5) in which the polarization and magnetization vectors are introduced via the sum of dipole moments of atoms in the text above equation (5.29) (e.g. see [9]) taking into account the further averaging procedure. The classic procedure is described in the textbook of Jackson [10] (generally it should be quantum, see e.g. [11], with second quantization account).

6.2 Discussion

Most textbooks (let us compare with e.g. [12], pp 84–85, [13]) establish the link between polarization vector and dipole (multipole) moment via the potential notion. We reproduce the guidelines of such theory with comments discussing the 'weak points' of such construction as follows.

6.2.1 Polarization vector via a potential from coupled point charges

Let us recall equation (5.3) that defines the vector $\vec{P}$ as a field from microscopic charges, represented by the coupled charge density

$$\operatorname{div} \vec{P} = -\rho_{zw}. \tag{6.47}$$

Next, consider the known equation for a potential from the coupled charges

$$\Delta \varphi_{zw} = -4\pi \rho_{zw}. \tag{6.48}$$

Such implicit link may be thought as a representation of electric field, polarization vector $\vec{P}$ in Coulomb gauge (see section 2.4).

Let us write the potential from coupled charges inside the volume V expressed by means of the Green function

$$\varphi_{zw} = -4\pi \int_V G(\vec{r}, \vec{r}\,')\rho_{zw}(\vec{r}\,')d\vec{r}\,'. \tag{6.49}$$

The Green function at the whole space looks as

$$G = \frac{1}{4\pi|\vec{r} - \vec{r}\,'|}, \tag{6.50}$$

to be substituted in equation (6.49) yields

$$\varphi_{zw} = -\int_V \frac{\rho_{zw}(\vec{r}\,')}{|\vec{r} - \vec{r}\,'|} d\vec{r}\,'. \tag{6.51}$$

Let us divide the volume V to subvolumes of basic elements of a matter (atoms or molecules) as

$$V = \sum_a V_a. \tag{6.52}$$

After substitution of equation (6.52) in equation (6.51) we obtain

$$\varphi_{zw} = -\sum_a \int_{V_a} \frac{\rho_{zw_a}(\vec{r}\,')}{|\vec{r} - \vec{r}\,'|} d\vec{r}\,'. \tag{6.53}$$

Let us apply now the method of multipole expansion (section 3.8.2). Expanding the Green function in the expression for φ_{zw} in the Taylor series in the vicinity of the central point of an atom $\vec{r}_a$ (being near its nuclei), supposing $(\vec{r}\,' = \vec{r}_a + \vec{r}\,'', |\vec{r}\,''| \ll |\vec{r} - \vec{r}_a|)$, we obtain, in the first (dipole) approximation

$$\frac{1}{|\vec{r} - \vec{r}_a - \vec{r}\,''|} = \frac{1}{|\vec{r} - \vec{r}_a|} + \left[\nabla_{\vec{r}\,''}\frac{1}{|\vec{r} - \vec{r}_a - \vec{r}\,''|}\right]_{\vec{r}\,''=0} \cdot \vec{r}\,''. \tag{6.54}$$

Next, substituting equation (6.54) into equation (6.51) yields

$$\varphi_{zw} = -\sum_a \frac{\int_{V_a} \rho_{zw;a} d\vec{r}\,''}{|\vec{r} - \vec{r}_a|} - \sum_a \left[\nabla_{\vec{r}\,''}\frac{1}{|\vec{r} - \vec{r}_a - \vec{r}\,''|}\right]_{\vec{r}\,''=0} \cdot \int_{V_a} \rho_{zw;a}(\vec{r}\,'')\vec{r}\,'' d\vec{r}\,''. \tag{6.55}$$

The first part is zero because of the zero atomic charge

$$q_{zw,a} = \int_V \rho_{zw,a} d\vec{r}\,'. \tag{6.56}$$

On the other hand in the second term the dipole moment appears $\vec{p}_a = \int_V \rho_{zw_a}(\vec{r}\,'')\vec{r}\,'' d\vec{r}\,''$.

Now it is the time to recall the macroscopic fields definition in matter. The standard approach introduces the macroscopic volume $V \gg V_a$ in such a way that the variations of the fields inside it are small and the fields relate to some central point $\vec{r}_0$ of the volume V. Hence we pick up the fields from the subvolumes of atoms but relate it to this central point. Going to integration over the volume V by $\vec{r}_a$, considering the sum as the Riemann one, we can introduce the dipole moment density $\vec{\Pi}(\vec{r}_a)$ (some value for each V_a) and stress that, generally, the moment depends on the external field $\vec{E}$. The dependence is introduced by the atom/molecule model.

So, changing $\vec{r}_a \to \vec{r}\,'$, we arrive at the representation

$$\varphi_{zw} = -\int_V \left[\nabla_{\vec{r}\,''}\frac{1}{|\vec{r} - \vec{r}_a - \vec{r}\,''|}\right]_{\vec{r}\,''=0} \cdot \vec{\Pi}(\vec{r}\,')d\vec{r}\,'. \tag{6.57}$$

This expression can be interpreted as the superposition of potentials of the dipoles, representing each electron.

Let us note

$$\left[\nabla_{\vec{r}\,''}\frac{1}{|\vec{r} - \vec{r}_a - \vec{r}\,''|}\right]_{\vec{r}\,''=0} = \left[\nabla_{\vec{r}\,'}\frac{1}{|\vec{r} - \vec{r}\,'|}\right],$$

$(\vec{r}_a = \vec{r}\,'$ is supposed) that yields in

$$\varphi_{zw} = -\int_V \left[\nabla_{\vec{r}'} \frac{1}{|\vec{r} - \vec{r}'|} \right] \cdot \vec{\Pi}(\vec{r}')$$
$$d\vec{r}' = -\int_V \left[\text{div}\left(\frac{\vec{\Pi}}{|\vec{r} - \vec{r}'|} \right) - \left(\frac{\text{div}\, \vec{\Pi}}{|\vec{r} - \vec{r}'|} \right) \right] d\vec{r}', \tag{6.58}$$

where the identity

$$\text{div}\left(\frac{\vec{\Pi}}{|\vec{r} - \vec{r}'|} \right) = \left(\frac{\text{div}\, \vec{\Pi}}{|\vec{r} - \vec{r}'|} \right) + grad\left(\frac{1}{|\vec{r} - \vec{r}'|} \right) \cdot \vec{\Pi}, \tag{6.59}$$

was used, and all the derivatives are evaluated with respect to $\vec{r}'$.

The first term, by means of the Gauss theorem transforms to the surface integral, which is neglected if the volume V is inside the matter type to be considered, hence

$$\varphi_{zw} = \int_{V_a} \frac{\text{div}\, \vec{\Pi}(\vec{r}')}{|\vec{r} - \vec{r}'|} d\vec{r}'. \tag{6.60}$$

The next step is to average the vector $\vec{P}'$ so that the vector $\vec{\Pi}$ may be identified as $<\vec{P}'> = \vec{P}$ (see equation (6.47)).

The link between the vector of magnetization $\vec{M}$ and the dipole moment (5.59) may be obtained in a similar way.

What we would take into account:

1. Such 'derivation' of the link between the polarization vector and dipole moments of a medium units is based on assumptions typical for electrostatics: the formula for the potential (6.51) arises from equation (3.43) that is generally valid only for Coulomb gauge (2.53) which, however, influences the link of the potential with the field $\vec{E}$ (2.41).
2. The formulas give no hint how the dipole moment is introduced in the classical (e.g. via Lorentz model (see section 6.3.1)) or quantum model of matter. Moreover, if one considers matter as a set of dipoles and try to evaluate a field by direct averaging, some divergences appear.
3. In some texts the term 'polarization vector' is changed to 'dipole moment' (of unit volume).

6.3 Dispersion in dielectrics, conductors and plasma

6.3.1 Lorentz model

When plane electromagnetic wave with fixed frequency propagates in a 'cloud' of atoms it exits some extra radiation which interfere with the incident field. This radiation is a superposition of the same and 'combined' frequencies (Raman scattering or effect, may be described by nonlinear polarization section 5.3.2) of the incident and eigen (atomic) ones. A realistic description of such rather complicated phenomena gives quantum mechanics [14, 15], its application allows to derive expressions for extra dipole momentum and hence built a polarization vector of individual atom without damping (relaxation) account. The damping may be introduced either by phenomenology or by next level theory—quantum electrodynamics. Some historically interesting and still valuable description is based on Lorentz theory which is outlined below.

In the previous section we introduced the polarization vector which for an isotropic homogeneous medium of identical molecules may be presented by the expression (5.42), namely as the sum of all molecules polarization vectors

$$\vec{P} = N\vec{p}, \tag{6.61}$$

(the number of molecules N at unit volume).

Let us consider a model of the molecule as a system of independent oscillators (see the relation in section 3.8.5), one for each 'optic' electron, that contribute in the emission process. Let the position of one of them coincide with the origin so as

$$\vec{p} = e\vec{r}, \tag{6.62}$$

where e means value of electron charge.

For electrons, under the action of external electric field $E_j = E_{0j}e^{-i\omega t} + c.c.$, acts as the driving force with components

$$F_k = eE_k. \tag{6.63}$$

In this model a dependence of the field on $\vec{r}$ is not taken into account because of the general approximation about scale of the field inhomogeneity $\lambda \gg a$, compared with the molecule dimension a.

Next, the direction of the x-axis is chosen along the field $\vec{E}$.

The oscillator equation of motion is

$$m\frac{d^2x}{dt^2} + m\gamma\frac{dx}{dt} + m\omega_0^2 x = eE_0e^{-i\omega t} + c.c., \tag{6.64}$$

where m—reduced electron mass, γ—relaxation parameter, ω_0 and ω as correspondent frequencies: Eugen (one of spectral frequencies of the atom) and the frequency of the external wave. The expression

$$x = x_0e^{-i\omega t} + c.c., \tag{6.65}$$

describes a stationary state, in which the charge system oscillates with the frequency ω. When one substitutes the expression (6.65) into (6.64), an expression for the complex amplitude as a function of the external electric field is obtained:

$$x_0 = \frac{eE_{0x}}{m(\omega_0^2 - \omega^2 - i\gamma\omega)}. \tag{6.66}$$

The electric dipole moment of an atom is expressed by means of equation (6.62) for real $\vec{E}_0$ as

$$\vec{p} = Re\left[\frac{e}{m(\omega_0^2 - \omega^2 - i\gamma\omega)}\right]\vec{E}. \tag{6.67}$$

If the description is widened to the N-electron system of a molecule, the vector $\vec{p}$ will be a superposition with weights f_j (see also equation (17.33)). It gives for the vector of polarization:

$$\vec{P} = N\sum_{j=1}^{n} Re\left[\frac{e^2 f_j}{m\left(\omega_{0j}^2 - \omega^2 - i\gamma_j\omega\right)}\right]\vec{E}, \tag{6.68}$$

where N is the number of optical electrons while f_j is named as oscillator forces, it is introduced in the Lorentz model as a set of phenomenological parameters, while $\sum_j f_j = 1$. The term 'oscillator force' have now rather historical meaning, introduced in terms of classic physics.

The parameters f_j can be evaluated in quantum mechanics on the base of relations (6.68). In accordance with equation (5.63b), the electric susceptibility has the form:

$$\chi = \frac{Ne^2}{m}Re\left[\sum_j \frac{f_j}{\omega_{0j}^2 - \omega^2 - i\gamma_j\omega}\right]. \tag{6.69}$$

The dielectric permeability is obtained from the relation (5.63) which joins (6.61) with

$$\varepsilon = 1 + 4\pi\chi. \tag{6.70}$$

See the plot depicting (6.69) (figure 6.1)

6.3.2 Drude theory of metals: Ohm's law

The Lorentz model for dielectrics may be applied to conductors (metals or, generally for plasma), considering this case as a limit $\omega_0 = 0$. The result is achieved because of relaxation mechanism is taken into account. The limit may be interpreted as a transition to free electrons of a metal (generally free charges, ions or holes in a solid).

It is considered as a version of the Drude model [12] as a response of a conducting matter to a time-dependent electric field with an angular frequency ω (cf the DC case in the next section).

An opposite limit $\omega_0 \ll \omega$ may be studied for high frequencies.

The complex amplitude of oscillations in a case of a conductor is given by the formula

$$x_0 = \frac{-eE_{0x}}{m(\omega^2 + i\gamma\omega)}. \tag{6.71}$$

We will consider relations for the complex amplitudes for the exponential form of solutions.

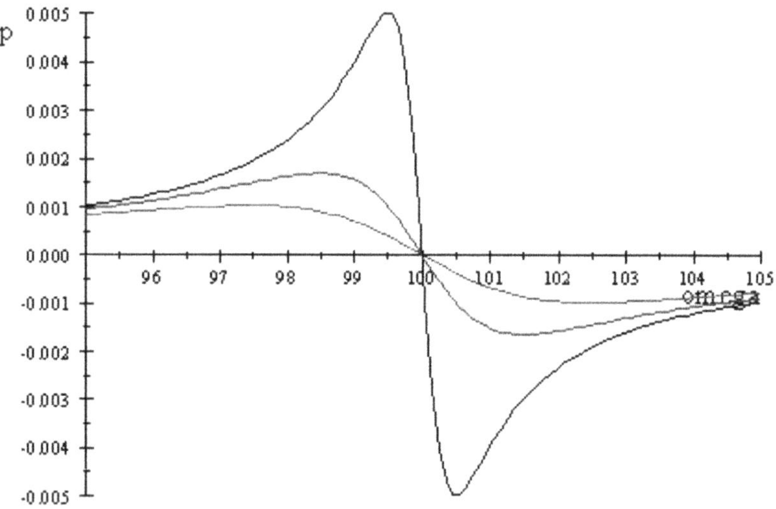

Figure 6.1. Schematic illustration of the dependence the leading term of $\frac{m}{Ne^2}\chi$ on the frequency ω with $\omega_0 = 100$ and $\gamma_1 = 1.0$—black, $\gamma_1 = 3.0$—red, $\gamma_1 = 5.0$—green, $f_1 = 1$.

The polarization vector will hence have the form

$$\vec{P} = \frac{e^2N}{m}Re\left[\sum_k \frac{-1}{\omega^2 + i\gamma_k\omega}\right]\vec{E}, \tag{6.72}$$

while the dielectric permeability we represent as

$$\chi = -\frac{Ne^2}{m}\sum_k Re\left[\frac{1}{\omega^2 + i\gamma_k\omega}\right], \tag{6.73a}$$

$$\varepsilon = 1 - \frac{4\pi Ne^2}{m}\sum_k Re\left[\frac{1}{\omega^2 + i\gamma_k\omega}\right]. \tag{6.73b}$$

Consider now the Maxwell–Ampère equation for matter (6.4). If one writes $\frac{1}{c}\frac{\partial\vec{D}}{\partial t}$, and next plug the derived expression for the dielectric permeability (6.73b), we find the following equality

$$\frac{1}{c}\frac{\partial\vec{D}}{\partial t} = \frac{i\omega}{c}\left(1 - \frac{4\pi Nq^2}{m}\sum_j \frac{1}{\omega^2 + i\gamma_j\omega}\right)\vec{E}_0e^{i\omega t} + c.c.= \varepsilon\frac{1}{c}\frac{\partial\vec{E}}{\partial t} + \frac{4\pi}{c}\sigma\vec{E}, \tag{6.74}$$

in which the expression under summation is written as

$$\frac{1}{\omega^2 + i\gamma_j\omega} = \frac{1}{\omega^2 + \gamma_j^2} - \frac{i\gamma_j}{\omega^3 + \gamma_j^2\omega}. \tag{6.75}$$

Plugging equation (6.75) into (6.76) yields

$$\frac{1}{c}\frac{\partial \vec{D}}{\partial t} = \frac{i\omega}{c}\left(1 - \frac{4\pi Nq^2}{m}\sum_j \frac{1}{\omega^2 + \gamma_j^2} - \frac{4\pi Nq^2}{m}\sum_j \frac{i\gamma_j}{\omega^3 + \gamma_j^2\omega}\right)\vec{E}_0 e^{i\omega t} + c.c. \tag{6.76}$$
$$= \varepsilon\frac{1}{c}\frac{\partial \vec{E}}{\partial t} + \frac{4\pi}{c}\sigma\vec{E}.$$

Finally we obtain

$$\sigma = \frac{Nq^2}{m}\sum_j \frac{\gamma_j}{\omega^2 + \gamma_j^2}, \tag{6.77a}$$

while the Maxwell–Ampère equation would have the form

$$\frac{1}{c}\frac{\partial \vec{D}}{\partial t} = \mathrm{rot}\vec{H} - \frac{4\pi}{c}\sigma\vec{E}, \tag{6.78}$$

that finally allows to write Ohm's law as

$$\vec{j} = \sigma\vec{E}. \tag{6.79}$$

For a simple model, made from the hydrogen-like atoms (taking the complex conjugate part of equation (6.75)) put for electron number $N = n$, $q = e$ as the charge carrier, yields

$$\frac{1}{c}\frac{\partial \vec{D}}{\partial t} = \frac{\varepsilon}{c}\frac{\partial \vec{E}}{\partial t} - \frac{4\pi}{c}\frac{ne^2\gamma}{m(\omega^2 + \gamma^2)}\vec{E}. \tag{6.80}$$

When the obtained relation is compared with equation (6.78), the dependence:

$$\frac{ne^2\gamma}{m(\omega^2 + \gamma^2)} = \sigma, \tag{6.81}$$

is derived and applied in some specific conditions of relaxation time and frequency [16]. For the DC $\omega \to 0$ case it simplifies as:

$$\frac{ne^2}{m\gamma} = \sigma. \tag{6.82}$$

6.4 Back to Ohm's law: Hall effect

6.4.1 On the DC Ohm law

The charge transport in the DC case is a nonequilibrium process hence in general theory such a relation as equation (6.79) is derived via the kinetic equation solution with averaging procedure application [5]. In textbooks there often used a simplified theory, that in my imagination can be understood as follows. The quasiparticle which represents the charge transport (say, electron, more precise—polaron) is moving in the phonon (quasiparticle of atom net oscillations) gas of a solid as a drop falling in air. After the start it is accelerated by constant electric field, but soon the acceleration decreases because the electric force is compensated by friction that is proportional to the electron velocity. The equation of motion (of a small sphere) hence is modeled as

$$m\frac{d\vec{v}}{dt} = e\vec{E} - 6\pi\mu R\vec{v}, \tag{6.83}$$

where the second, Stokes drug term is proportional to the radius R and dynamical viscosity μ. The viscosity coefficient in turn is determined by properties of the gas, for example kinetic (Chapman–Enskog) theory for Maxwellian particles gives

$$\mu = n_p k_B T/\lambda, \tag{6.84}$$

where n_p is the gas concentration, k_B—Boltzmann constant, T—temperature and the constant λ—collision integral eigenvalue.

In our case it is the phonon gas, the simplest Einstein model with the delta-function density of phonon states [5] gives the mean phonon number

$$n_p = \frac{1}{\exp[h\nu_e/kT] - 1}, \tag{6.85}$$

where ν_e related to the 'Einstein temperature'. Note, that the more correct description gives the Debye model.

The solution of equation (6.83) with zero initial condition is given by the formula

$$\vec{v} = \frac{e}{m}\tau\vec{E}(1 - \exp[-t/\tau]), \tag{6.86}$$

where $\tau = m(6\pi\mu R)^{-1}$.

The stationary state after some relaxation time τ, corresponds to zero acceleration, or

$$0 = \frac{e}{m}\vec{E} - \frac{\vec{v}_s}{\tau}, \tag{6.87}$$

that yields for the stationary velocity the expression:

$$\vec{v}_s = \frac{e\tau}{m}\vec{E} = \frac{e}{6\pi\mu R}\vec{E}, \tag{6.88}$$

or, multiplying the velocity (6.88) by number of current carriers n and electron charge we obtain

$$\vec{j} = en\vec{v}_s = \frac{e^2 n\tau}{m}\vec{E}. \tag{6.89}$$

It is instructive to link the dynamical viscosity $\mu = \nu/n_p$ with the kinematic viscosity ν and the phonon number n_p as density of the medium in which the electron moves.

Consider an intimate relation with Ohm's law derived within the Drude theory for AC current (6.82) that gives an expression for

$$\sigma = \frac{e^2 n\tau}{m}, \tag{6.90}$$

because $\gamma = \tau^{-1}$. As a conclusion, note, that the Drude model provides a good explanation of DC and AC conductivity in metals, the Hall effect, and the magnetoresistance in metals near room temperature.

6.4.2 Magnetic field account: Hall effect

Generalizing Ohm's law (6.79) similar to the material relations would has the tensor form

$$j_i = \sigma_{ik} E_k, \tag{6.91}$$

where σ_{ik} is the conductivity tensor. Its physics, e.g. arises from magnetic field presence.

So, next, let us account for the magnetic field in the Newton vector equation of motion (6.83)

$$m\frac{d\vec{v}}{dt} = e\vec{E} + \frac{e}{c}[\vec{v}, \vec{B}] - \frac{m\vec{v}}{\tau}, \tag{6.92}$$

The correspondent system of differential equations yields the generalization of the Lorentz theory, including Hall effect.

The stationary case of equation (6.92)

$$0 = e\vec{E} + \frac{e}{c}[\vec{v}, \vec{B}] - \frac{m\vec{v}}{\tau}, \tag{6.93}$$

is equivalent to

$$6\pi\mu Ren\vec{v} - \frac{e}{c}[en\vec{v}, \vec{B}] = e^2 n\vec{E}, \tag{6.94}$$

or

$$\frac{\vec{j}}{\tau} - \frac{e}{c}[\vec{j}, \vec{B}] = e^2 n\vec{E}, \tag{6.95}$$

in components

$$\begin{aligned} j_x - \frac{e\tau}{c}(j_y B_z - j_z B_y) &= e^2 n\tau E_x, \\ j_y - \frac{e\tau}{c}(j_z B_x - j_x B_z) &= e^2 n\tau E_y, \\ j_z - \frac{e\tau}{c}(j_x B_y - j_y B_x) &= e^2 n\tau E_z. \end{aligned} \tag{6.96}$$

Solving equation (6.96) with respect to j_x, j_y, j_z yields

$$j_x = ne^2\tau\frac{c^2 + e^2\tau^2B_x^2}{c^2 + e^2\tau^2B_x^2 + e^2\tau^2B_y^2 + e^2\tau^2B_z^2}E_x + ne^3\tau^2\frac{cB_z + e\tau B_xB_y}{c^2 + e^2\tau^2B_x^2 + e^2\tau^2B_y^2 + e^2\tau^2B_z^2}E_y$$
$$- ne^3\tau^2\frac{cB_y - e\tau B_xB_z}{c^2 + e^2\tau^2B_x^2 + e^2\tau^2B_y^2 + e^2\tau^2B_z^2}E_z,$$

$$j_y = - ne^3\tau^2\frac{cB_z - e\tau B_xB_y}{c^2 + e^2\tau^2B_x^2 + e^2\tau^2B_y^2 + e^2\tau^2B_z^2}E_x + ne^2\tau\frac{c^2 + e^2\tau^2B_y^2}{c^2 + e^2\tau^2B_x^2 + e^2\tau^2B_y^2 + e^2\tau^2B_z^2}E_y$$
$$+ ne^3\tau^2\frac{cB_x + e\tau B_yB_z}{c^2 + e^2\tau^2B_x^2 + e^2\tau^2B_y^2 + e^2\tau^2B_z^2}E_z,$$

$$j_z = ne^3\tau^2\frac{cB_y + e\tau B_xB_z}{c^2 + e^2\tau^2B_x^2 + e^2\tau^2B_y^2 + e^2\tau^2B_z^2}E_x - ne^3\tau^2\frac{cB_x - e\tau B_yB_z}{c^2 + e^2\tau^2B_x^2 + e^2\tau^2B_y^2 + e^2\tau^2B_z^2}E_y$$
$$+ ne^2\tau\frac{c^2 + e^2\tau^2B_z^2}{c^2 + e^2\tau^2B_x^2 + e^2\tau^2B_y^2 + e^2\tau^2B_z^2}E_z,$$

that gives the tensor components σ_{ik} from equation (6.91).

6.4.3 Basic relations for 2D Hall resistance

It is convenient from a didactic point of view to start with two-dimensional problem. The geometry for the Hall measurements is shown in figure 6.2, see also [17]. The two voltages V_x (driving voltage) and V_H (Hall voltage) are measured. The usual convention that $\vec{B}$ points along the z-axis holds, then, for $E_z = 0, j_z = 0$, $B_x = 0$, $B_y = 0$, directly from equation (6.96) one obtains

$$\begin{aligned} j_x &= c^2n\tau\frac{e^2}{c^2 + e^2\tau^2B_z^2}E_x + cn\tau^2B_z\frac{e^3}{c^2 + e^2\tau^2B_z^2}E_y, \\ j_y &= - ne^3\tau^2\frac{cB_z}{c^2 + e^2\tau^2B_z^2}E_x + ne^2\tau\frac{c^2}{c^2 + e^2\tau^2B_z^2}E_y. \end{aligned} \tag{6.97}$$

In general, the conductivity tensor (6.91) relates the current density ($\vec{j}$) and the electric field ($\vec{E}$) vectors, and the tensor relations in the matrix form in 2D are written as:

$$j = \begin{pmatrix} j_x \\ j_y \end{pmatrix} = \begin{pmatrix} \sigma_{xx} & \sigma_{xy} \\ \sigma_{yx} & \sigma_{yy} \end{pmatrix}\begin{pmatrix} E_x \\ E_y \end{pmatrix}. \tag{6.98}$$

The expressions for j_x, j_y from equation (6.97) give the matrix elements σ_{ik}.

The geometry of the Hall voltage experiment is shown in figure 6.2.

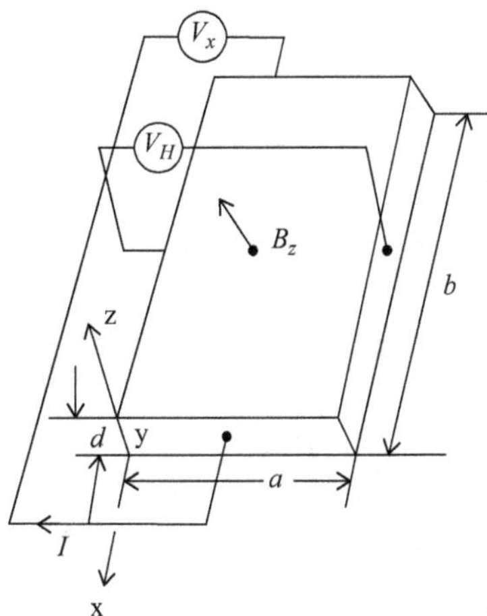

Figure 6.2. Schematic illustration of the experiment of the Hall voltage and current measurements [17].

The simple 'school' relations give

$$\begin{aligned} I_x &= adj_x = nad(-v_x)(-e) \\ I_y &= bdj_y = nbd(-v_y)(-e) \\ E_y &= V_H/a \\ E_x &= V_x/b, \end{aligned} \tag{6.99}$$

Multiplying equations (6.98) by ad (first) and by bd (second), after some algebra we have

$$\begin{aligned} I_x &= adj_x = \frac{ad}{b}\sigma_{xx}bE_x + d\sigma_{xy}aE_y, \\ I_y &= dbj_y = d\sigma_{yx}bE_x + b\sigma_{yy}dE_y, \end{aligned} \tag{6.100}$$

where $\rho_{xx} = \dfrac{ad}{b}\sigma_{xx}$, $\rho_{xy} = d\sigma_{xy}$, $\rho_{yx} = d\sigma_{yx}$, $\rho_{yy} = b\sigma_{yy}$, while $bE_x = V_x$, $aE_y = V_y$. The voltage $V_y = V_H$ is called the Hall voltage. Then, equation (6.100) may be rewritten as

$$V = \begin{pmatrix} V_x \\ V_y \end{pmatrix} = \begin{pmatrix} R_{xx} & R_{xy} \\ R_{yx} & R_{yy} \end{pmatrix} \begin{pmatrix} I_x \\ I_y \end{pmatrix} = RI, \tag{6.101}$$

the resistivity tensor R is introduced as inverse to ρ, so $R = \rho^{-1}$.

The longitudinal (R_x) and Hall (R_H) resistances are defined in terms of the current I_x as:

$$\begin{aligned} V_x &= R_x I_x \\ V_H &= R_H I_x. \end{aligned} \tag{6.102}$$

An especially interesting implication of these formula is that in a 2D system, when $R_{xx} = 0$ but $R_{xy} \neq 0$, then R_{xx} is also zero (and vice versa). This means that (as long

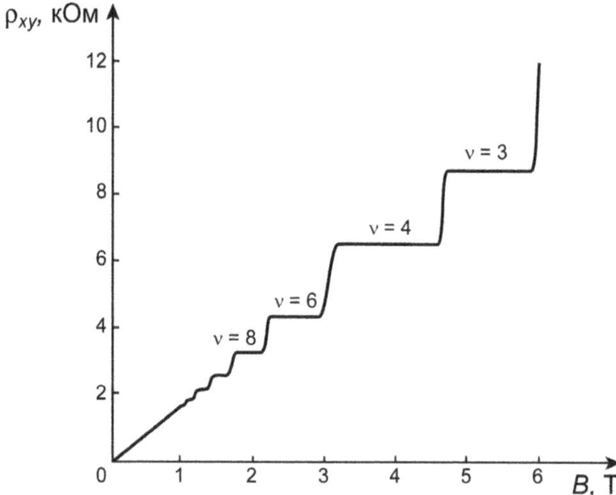

Figure 6.3. Longitudinal and transverse (Hall) resistivity, ρ_{xx} and ρ_{xy}, of a two-dimensional electron gas as a function of magnetic field. The inset shows $1/\rho_{xx}$ divided by the quantum unit of conductance e^2/h as a function of the filling factor ν [18, 19]. Author: Alexander Mayorov / Source: Cool Draw.

as R_{xy} is finite), the vanishing of the longitudinal conductivity implies that the longitudinal resistivity also vanishes. This is precisely the situation that occurs in the quantum Hall effect, and is fundamental to this phenomenon. Namely, the Hall resistivity exhibits quantized plateaus at values of $\rho_{xy} = h/nc^2$. The longitudinal resistivity vanishes at the plateaus, $\rho_{xx} = 0$, as shown as in figure 6.3.

6.4.4 On quantum electrodynamics manifestation of Hall effect

It is interesting to note that in special conditions of strong magnetic field and low temperature the dependence of the Hall resistance/conductance $n\frac{e^2}{h}$ on magnetic field (B) exhibit quantum behavior: plateaus at quantized integer values R_H in the Hall resistance and, at the same range, the disappearance of the sample's electrical resistance. The possibility of its precise measurement is important to be used as a scale in the resistivity definition as well as in quantum electrodynamics to be very precise determination of the fine-structure constant. We only mention the original publication of the Von Klitzing *et al* [20] and recent review [18] (there is access via Wiki).

6.5 EM waves in isotropic conducting matter case

6.5.1 The linear equation and plane wave

Let us differentiate the macroscopic Ampère–Maxwell equation (5.9), with respect to time

$$\mathrm{rot}\frac{\partial \vec{H}}{\partial t} = \frac{1}{c}\frac{\partial^2 \vec{D}}{\partial t^2} + \frac{4\pi}{c}\frac{\partial \vec{j}}{\partial t}, \tag{6.103}$$

considering $\vec{j}$ as the density of free electrons and apply the relevant material equations for isotropic medium (5.29) in the form,

$$\vec{D} = \varepsilon\vec{E}, \vec{H} = \frac{1}{\mu}\vec{B}, \tag{6.104}$$

that yields

$$\frac{1}{\mu}\text{rot}\frac{\partial\vec{B}}{\partial t} = \frac{\varepsilon}{c}\frac{\partial^2\vec{E}}{\partial t^2} + \frac{4\pi}{c}\frac{\partial\vec{j}}{\partial t}, \tag{6.105}$$

while the Faraday equation reads

$$\frac{\partial\vec{B}}{\partial t} = -c\text{rot}\vec{E}. \tag{6.106}$$

Hence

$$-\frac{1}{\mu}\text{rotrot}\vec{E} = \frac{\varepsilon}{c^2}\frac{\partial^2 E}{\partial t^2} + \frac{4\pi}{c^2}\frac{\partial\vec{j}}{\partial t}, \tag{6.107}$$

Finally, taking into account equation (3.17), suppose the expression for a free current density is given by Ohm's law $\vec{j} = \sigma\vec{E}$ (the case of isotropic medium), arriving at

$$\frac{1}{\mu}\Delta\vec{E} = \frac{\varepsilon}{c^2}\frac{\partial^2\vec{E}}{\partial t^2} + \frac{4\pi\sigma}{c^2}\frac{\partial\vec{E}}{\partial t}. \tag{6.108}$$

The result produces wave equation

$$v^2\Delta\vec{E} = \frac{\partial^2\vec{E}}{\partial t^2} + a\frac{\partial\vec{E}}{\partial t}, \tag{6.109}$$

in which the velocity of wave propagation in isotropic medium is defined as

$$v = \frac{c}{\sqrt{\varepsilon\mu}}, \tag{6.110}$$

while the absorption connected with losses action on current exiting is proportional to

$$a = \frac{4\pi\sigma}{\varepsilon}. \tag{6.111}$$

Similar transformations may be delivered for the magnetic induction vector $\vec{B}$. Starting with the equation (8.4), let us differentiate it by t

$$\frac{\partial^2 \vec{B}}{\partial t^2} = -c\text{rot}\frac{\partial \vec{E}}{\partial t}, \tag{6.112}$$

and take the Ampère–Maxwell equation again

$$\frac{\varepsilon}{c}\frac{\partial \vec{E}}{\partial t} = \text{rot}\vec{H} - \frac{4\pi\sigma}{c}\vec{E}, \tag{6.113}$$

with equation (6.104) account,

$$\frac{\partial \vec{E}}{\partial t} = \frac{c}{\mu\varepsilon}\text{rot}\vec{B} - \frac{4\pi\sigma}{\varepsilon}E, \tag{6.114}$$

into account, substituting it in equation (6.112), obtaining

$$\frac{\partial^2 \vec{B}}{\partial t^2} = -c\text{rot}\left[\frac{c}{\mu\varepsilon}\text{rot}\vec{B} - \frac{4\pi\sigma}{\varepsilon}E\right]. \tag{6.115}$$

Ordering results, one arrives at the equation

$$\frac{\partial^2 \vec{B}}{\partial t^2} - v^2\Delta\vec{B} - \frac{4\pi\sigma}{\varepsilon}\frac{\partial \vec{B}}{\partial t} = 0, \tag{6.116}$$

which describes propagation of wave under conditions of equation (6.104). The form of the operator in equation (6.109) coincides with the one of equation (6.116).

The plane wave solution of equation (6.109) is based on

$$\vec{E} = \vec{E}_0\exp[i\vec{k}\vec{r} - i\omega t] + c.c., \tag{6.117}$$

that leads to the complex dispersion relation

$$v^2k^2 = \omega^2 + ia\omega. \tag{6.118}$$

There are two possibilities of direct application of the plane wave solution.

First one corresponds to a boundary regime with fixed frequency ω, that, for the 1D case means

$$\vec{E} = \vec{E}_0\exp[-i\omega t] + c.c., \tag{6.119}$$

at $z = 0$.

If ω is real, $k = k_1 + ik_2$ should be complex, or

$$(k_1 + ik_2)^2 = k_1^2 + 2ik_1\cdot k_2 - k_2^2 = v^{-2}(\omega^2 + ia\omega). \tag{6.120}$$

Equalizing real and imaginary parts gives

$$k_1^2 - k_2^2 = v^{-2}\omega^2,\ 2k_1\cdot k_2 = v^{-2}a\omega. \tag{6.121}$$

The solution of it is given by the roots of

$$k_{1,\pm}^2 = \frac{\omega^2}{2v^2}\left[1 \pm \sqrt{1 + \frac{a^2}{\omega^2}}\right] \approx \frac{\omega^2}{2v^2}\left[1 \pm \left(1 + \frac{a^2}{2\omega^2}\right)\right], \tag{6.122}$$

for small a/ω and the parameter that defines the scale of attenuation $\lambda_2 = k_2^{-1}$ is

$$\lambda_2 = \frac{2k_1}{v^{-2}a\omega}. \tag{6.123}$$

This approximate solution corresponds the case of small $\frac{a^2}{\omega^2} \ll 1$. One observes two propagating modes and two decaying ones.

More details are given by

$$k_{1,+}^2 = \frac{\omega^2}{2v^2}\left[1 + \sqrt{1 + \frac{a^2}{\omega^2}}\right] \approx \frac{\omega^2}{v^2} + \frac{a^2}{4v^2} \tag{6.124}$$

$$k_{1,-}^2 = \frac{\omega^2}{2v^2}\left[1 - \sqrt{1 + \frac{a^2}{\omega^2}}\right] \approx -\frac{a^2}{4v^2}. \tag{6.125}$$

An important question arises: how the waves (four modes) contribute a boundary regime?

In one-dimensional case it yields a biquadratic equation for the real part of k, which allows to obtain k_2.

$$k_1^4 - k_1^2\frac{\omega^2}{v^2} - \left(\frac{a\omega}{2v^2}\right)^2 = 0,\ k_2 = \frac{a\omega}{2k_1v^2}. \tag{6.126}$$

In the 1D case with the condition (6.119) for $k_{1,2}$ given by (6.126)

$$\vec{E} = \vec{E}_0 \exp[ik_1z - k_2z - i\omega t] + c.c. \tag{6.127}$$

The second possibility implies a bit artificial situation (from physical viewpoint) of Cauchy problem. The initial condition should be created in a whole space, with a possible localization, if necessary.

More advanced applications of the results is based on Fourier transformation applications, adjusted to boundary or initial problems.

6.5.2 The 1D nonlinear model outline

We restrict ourselves to a one-dimensional model as in [21], the x-axis is chosen as the direction of a pulse propagation. For the 1D case we put $D_x = 0$ and $B_x = 0$ taking into account the only polarization of electromagnetic wave. The first pair of Maxwell's equations for a matter coincide with equations (6.1) and (6.2), while the second pair (6.3) and (6.4) for the case $\vec{j} = 0$ we rewrite in the Lorentz–Heaviside's unit system to fit notations of the source [22]

$$\nabla \times \mathbf{E} = -\frac{1}{c}\frac{\partial \mathbf{B}}{\partial t}, \tag{6.128}$$

$$\nabla \times \mathbf{H} = \frac{1}{c}\frac{\partial \mathbf{D}}{\partial t}. \tag{6.129}$$

The nonlinearity is included in the polarization vector $\mathbf{P}$ dividing it in the sum of linear $\mathbf{P}_L = \chi^{(1)}E_y$ and nonlinear terms

$$\mathbf{P} = \mathbf{P}_L + \mathbf{P}_{NL}. \tag{6.130}$$

The material relations have the standard form for this units choice

$$\mathbf{D} = \mathbf{E} + 4\pi\mathbf{P}, \quad \mathbf{H} = \mathbf{B}, \tag{6.131}$$

we do not take magnetization into account.

Therefore its 1D form may be expanded with the susceptibility $\chi^{(1)}$ introduced as

$$D_y = E_y + 4\pi\chi^{(1)}E_y + (P_{NL})_y, \tag{6.132}$$

while the Maxwell equations are simplified till

$$\begin{gathered} \frac{\partial D_y}{\partial y} + \frac{\partial D_z}{\partial z} = 0, \quad \frac{\partial B_y}{\partial y} + \frac{\partial B_z}{\partial z} = 0, \\ \frac{1}{c}\frac{\partial D_y}{\partial t} = \frac{\partial B_z}{\partial x}, \quad \frac{1}{c}\frac{\partial B_z}{\partial t} = \frac{\partial E_y}{\partial x}. \end{gathered} \tag{6.133}$$

We conventionally choose the concrete case of isotropic optically non-active media [23], cutting the nonlinear part of polarization amplitude expansion (5.72) at the third order, supposed to have the simplest form:

$$(P_{NL})_y = 4\pi\chi^{(3)}E_y^3. \tag{6.134}$$

The nonlinearity certainly introduce interaction between unidirectional waves of section 3.4.2, see the modes definitions (3.34). Its realization in linear isotropic matter is quite similar and gives first steps towards the short pulse equation (SPE) through application of the projection operators.

6.5.3 On dynamic projection method in linear problem

Following the idea of the Cauchy problem solution in section 3.4.2, we would develop the method of [24, 9]. Let us define projection operators for the case of linear isotropic dielectric media for the material relations $D_y = \varepsilon E_y$ for $\varepsilon = 1 + 4\pi\chi^{(1)}$ as the linear part of equation (6.132). We as in section 3.4.2 reformulate equation (6.133) as the matrix evolution equation

$$\Psi_t = L\Psi,$$

for the field vector

$$\Psi = \begin{pmatrix} E_y \\ B_z \end{pmatrix}, \tag{6.135}$$

while the evolution operator reads

$$L = \begin{pmatrix} 0 & \frac{c}{\varepsilon}\partial_x \\ c\partial_x & 0 \end{pmatrix}. \tag{6.136}$$

In such notations the matrix operator equation is written as

$$\frac{\partial}{\partial t}\begin{pmatrix} E_y \\ B_z \end{pmatrix} = \begin{pmatrix} 0 & \frac{c}{\varepsilon}\partial_x \\ c\partial_x & 0 \end{pmatrix}\begin{pmatrix} E_y \\ B_z \end{pmatrix} = \begin{pmatrix} \frac{c}{\varepsilon}\frac{\partial B_z}{\partial x} \\ c\frac{\partial E_y}{\partial x} \end{pmatrix}. \tag{6.137}$$

The Cauchy problem for the system (6.137) includes an initial condition.

$$\Psi = \begin{pmatrix} E_y \\ B_z \end{pmatrix} = \begin{pmatrix} \phi(x) \\ \psi(x) \end{pmatrix}, \tag{6.138}$$

with arbitrary field components ϕ, ψ.

Going to the space of Fourier transforms by x, e.g. as

$$E_y = \frac{1}{\sqrt{\pi}} \int \exp(ikx)\tilde{E}_y dx$$

we get the system of ordinary differential equations,

$$\begin{pmatrix} \frac{\partial}{\partial t}\tilde{E}_y \\ \frac{\partial}{\partial t}\tilde{B}_z \end{pmatrix} = \begin{pmatrix} \frac{ck}{\varepsilon}\tilde{B}_z \\ kc\tilde{E}_y \end{pmatrix}, \tag{6.139}$$

with k as the parameter.

Such a system of differential equations with t-independent coefficients has an exponential $\exp(i\omega t)$ solution, where $\omega = \pm\frac{ck}{\sqrt{\varepsilon}}$. We hence arrive at the 2×2-matrix eigenvalue problem.

Namely, let us search projection operators as matrices P_i that act as $P_1\Psi = \Psi_1$ and $P_2\Psi = \Psi_2$ to be eigenvectors of the matrix of evolution in equation (6.139). We use of the standard properties of orthogonal projecting operators

$$P_i P_j = 0, \quad P_i^2 = P_i, \quad \sum_i P_i = 1, \tag{6.140}$$

as algebraic conditions.

For the evolution operator of equation (6.139), the matrices P_i get the form

$$P_1 = \frac{1}{2}\begin{pmatrix} 1 & \frac{1}{\sqrt{\varepsilon}} \\ \sqrt{\varepsilon} & 1 \end{pmatrix}, \quad P_2 = \frac{1}{2}\begin{pmatrix} 1 & -\frac{1}{\sqrt{\varepsilon}} \\ -\sqrt{\varepsilon} & 1 \end{pmatrix}, \tag{6.141}$$

that is slight generalization of ones from section 3.4.2.

Thus we have solved the problem in the k-space and after that, performing the inverse Fourier transform, we reproduce the x-space representation of the operators [24]. In this dispersionless case we obtain matrix elements of the projecting operators (6.141) that do not depend on k (quite different is the case if dispersion is taken into account). It means that its inverse Fourier transform yields x-space representation, that is identical to the k-representation, see again equation (6.141).

Along the conventional projecting procedure we apply the operators to the vector Ψ equation (6.135) and introduce new variables $\Lambda = \frac{1}{2}E_y + \frac{1}{2\sqrt{\varepsilon}}B_z$ and $\Pi = \frac{1}{2}E_y - \frac{1}{2\sqrt{\varepsilon}}B_z$, for example as

$$P_1\Psi = \begin{pmatrix} \frac{1}{2}E_y + \frac{1}{2\sqrt{\varepsilon}}B_z \\ \frac{1}{2}\sqrt{\varepsilon}E_y + \frac{1}{2}B_z \end{pmatrix} = \begin{pmatrix} \Pi \\ \sqrt{\varepsilon}\Pi \end{pmatrix}, \tag{6.142}$$

which correspond to left (Λ) and right (Π) direction of the wave propagation [24]. Comparing new variables Λ and Π to ones presented by Kinsler *et al* [5] ours have a similar form to their ones. Our form of equation (6.142) is exactly determined by dispersion relation $\omega = \pm\frac{ck}{\sqrt{\varepsilon}}$ from equation (6.139). Then it allows us to present in algorithmic way both electric and magnetic field [24], in the simplest example we trace now, it is

$$E_y = \Lambda + \Pi \quad B_z = \sqrt{\varepsilon}(\Pi - \Lambda). \tag{6.143}$$

This map equations (6.142) and (6.143) is time-independent one-to-one local correspondence and hence allows to determine initial conditions in the Cauchy problem for both left and right wave variables (Λ, Π) by means of equation (6.138).

It also allows to follow waves, extracting data in each time t by the corresponding projecting in a very general situation besides the example we study.

6.5.4 Dynamical projecting in a nonlinear problem

The nonlinearity (section 6.5.2), see equation (6.134) account adds the polarization term, then equation (6.137) that gives

$$\frac{\partial}{\partial t}\Psi - L\Psi = N(\Psi). \tag{6.144}$$

Such an operator form of nonlinear evolution equation is general [9], so we demonstrate the dynamical projecting method application to the problems with (weak) nonlinearity account. Plugging the nonlinear term of equation (6.134) to equation (6.144) yields

$$\frac{\partial}{\partial t}\begin{pmatrix} E_y \\ B_z \end{pmatrix} - \begin{pmatrix} 0 & \frac{c}{\varepsilon}\partial_x \\ c\partial_x & 0 \end{pmatrix}\begin{pmatrix} E_y \\ B_z \end{pmatrix} = \frac{4\pi}{\varepsilon}\frac{\partial}{\partial t}\begin{pmatrix} \chi^{(3)}E_y^3 \\ 0 \end{pmatrix}. \tag{6.145}$$

Applying the operators P_i equation (6.141) to equation (6.145) and use the basic projectors property $[P_i, \left(\frac{\partial}{\partial t} - L\right)] = 0$, lead us to

$$\left(\frac{\partial}{\partial t} - L\right)P_i\Psi = P_iN(\Psi). \tag{6.146}$$

We have to admit that Λ is still aligned to the dispersion equation root $\omega = \frac{-ck}{\sqrt{\varepsilon}}$ and analogously Π is related to $\omega = \frac{ck}{\sqrt{\varepsilon}}$. This option determines a sign, which stands at the parameter c in our equations. So we arrive at the result

$$\begin{pmatrix} \frac{\partial}{\partial t}\Pi \\ \frac{\partial}{\partial t}\sqrt{\varepsilon}\Pi \end{pmatrix} - \begin{pmatrix} \frac{c}{\varepsilon}\frac{\partial}{\partial x}\sqrt{\varepsilon}\Pi \\ c\frac{\partial}{\partial x}\Pi \end{pmatrix} = 4\pi\frac{\partial}{\partial t}\begin{pmatrix} \frac{1}{\varepsilon}\chi^{(3)}(\Lambda + \Pi)^3 \\ \frac{1}{\sqrt{\varepsilon}}\chi^{(3)}(\Lambda + \Pi)^3 \end{pmatrix}, \tag{6.147}$$

that presents in two lines the evolution equation for right wave interacting with the left one.

Applying the second operator (6.141) to equation (6.147) we obtain a system of equations which describes the interaction between two waves propagating in opposite directions. The system of equations have the form

$$\begin{cases} \frac{\partial}{\partial t}\Pi - \frac{c}{\sqrt{\varepsilon}}\frac{\partial}{\partial x}\Pi = \frac{2\pi}{\varepsilon}\frac{\partial}{\partial t}\chi^{(3)}(\Lambda + \Pi)^3 \\ \frac{\partial}{\partial t}\Lambda + \frac{c}{\sqrt{\varepsilon}}\frac{\partial}{\partial x}\Lambda = \frac{2\pi}{\varepsilon}\frac{\partial}{\partial t}\chi^{(3)}(\Lambda + \Pi)^3 \end{cases}. \tag{6.148}$$

In the conditions of the derivation this system is equivalent to the original Maxwell system (6.133).

Problems.

Exercise 1. Write down the table of the conductivity tensor components σ_{ik} in 3D from the relations (6.96), marking $j_x = j_1, \ldots$

Exercise 2. Study 2D (6.92) in the case of AC, taking $\vec{E} = \vec{E}_0 \exp[i\omega t] + c.c.$. Derive the conductivity tensor components σ_{ik}.

Exercise 3. Calculate the components of the resistivity tensor in 2D, as prescribed by equations (6.97), (6.100) and (6.101).

Exercise 4. Consider equations (6.133) with arbitrary constant μ and reduce it to the system for directed waves.

Exercise 5. Derive the separate evolution equations for the directed waves from equation (6.142).

Exercise 6. Solve and plot the solution of equation (6.148) for the case $\Lambda = 0$ numerically for $\chi^{(3)} = 1$.

References

[1] Brillouin L (ed) 1960 *Wave Propagation and Group Velocity* (New York: Academic)

[2] Veselago V G 1966 *Fiz. Tverd. Tela* **8** 3571
Veselago V G 1967 *Sov. Phys. Solid State* **8** 2853
Veselago V G 1968 The electrodynamics of substances with simultaneously negative values of ε and μ *Sov. Phys. Usp.* **10** 509–14

[3] Millonni P W 2005 *Fast Light, Slow Light and Left-Handed Light* (Bristol: IOP Publishing)

[4] Maxwell J C 1865 *A Dynamical Theory of the Electromagnetic Field* vol CLV

[5] Reichl L E 1998 *A Modern Course in Statistical Physics* 2 edn (New York: Wiley)

[6] Novozhilov Y V and Yappa Y A 1978 *Elektrodinsmika* (Moscow: Mir)

[7] Guggenheim E A 1967 *Thermodynamics An Advanced Treatment for Chemists and Physicists* (Amsterdam: Elsevier)

[8] Lorentz H A 1899 Simplified theory of electrical and optical phenomena in moving systems *KNAW, Proc., 1, 1898–1899 (Amsterdam)* pp 427–42

[9] Leble S and Perelomova A 2018 *Dynamical Projectors Method in Hydro- and Electrodynamics* (Boca Raton, FL: CRC Press)

[10] Jackson J D 1998 *Classical Electrodynamics* 3rd edn (New York: Wiley)

[11] Tamm I E 1989 *Fundamentals Of The Theory Of Electricity* (Moscow: Nauka) (Russian)
Tamm I E 1989 *Fundamentals Of The Theory Of Electricity* (Moscow: Mir) (English)

[12] Drude P 1900 Zur Elektronentheorie der Metalle *Ann. Phys.* **306** 566–613

[13] Kondratjev E 1998 *Lectures in Electromagnetism. Short Course. Parts 1, 2* (Kaliningrad: Kaliningrad State University)

[14] Fock V A 1978 (1982 and 1986) *Fundamentals of Quantum Mechanics* (Moscow: Mir)

[15] Fock V A and Armstrong J C 1967 Electromagnetic diffraction and propagation problems. International series of monographs in electromagnetic waves, vol 1 *Am. J. Phys.* **35** 362

[16] Scheffler M, Dressel M, Jourdan M and Adrian H 2005 Extremely slow Drude relaxation of correlated electrons *Nature* **438** 1135–7

[17] Dresselhaus M S 2001 *Solid State Physics, PART III, Magnetic Properties of Solids* http://web.mit.edu/course/6/6.732/www/6.732-pt1.pdf

[18] von Klitzing K 2004 *25 years of Quantum Hall Effect, Poincaré Seminar* (Basel: Birkhäuser)

[19] https://en.wikipedia.org/wiki/Quantum_Hall_effect

[20] Klitzing K V, Dorda G and Pepper M 1980 New method for high-accuracy determination of the fine-structure constant based on quantized Hall resistance *Phys. Rev. Lett.* **45** 494–7

[21] Chung Y, Jones C K R T, Schäfer T and Wayne C E 2005 Ultra-short pulses in linear and nonlinear media *Nonlinearity* **18** 1351–74

[22] Kuszner M and Leble S 2011 Directed electromagnetic pulse dynamics: projecting operators method *J. Phys. Soc. Jpn.* **80** 024002

[23] Boyd R W 1992 *Nonlinear Optics* (Boston, MA: Academic)

[24] Leble S 1990 *Nonlinear Waves in Waveguides* (Heidelberg: Springer)

Chapter 7

Plasma

7.1 Plasma types

7.1.1 General remarks: a matter as potential plasma

One of the important applications of electrodynamics is plasma physics. From the point of view of the basic relations (5.61) the presence of the movable, free, in the sense of the section 5.1.1, positive and negative charges (ions) make matter 'a plasma'. In this sense any kind of matter in an ionized (at least—partially) state is or may be a plasma. Even a good insulator becomes a conductor when it is subjected to ionization by some particles impact or under action of a strong electric field.

The plasma notion formal description is based on a closure of the Maxwell equations via formulas such as (5.61), which expresses the charge density and current density in terms of electric and magnetic fields.

More technically namely a classification of materials by means of matter equations forms a base of the practical (applied) electrodynamics.

So, the essential in a plasma description is the closed matter+field consideration.

Next it is important to recall that the motion of ions particles needs to include the equations of mechanics. In section 6.1.3 we discuss the Lorentz model of matter that accounts for oscillations of the charged particles, including the limiting case of the zero frequency that corresponds to eventual free motion (Drude theory of metals).

A popular and convenient model for plasma is the magnetic hydrodynamics in which the continuum approach allows to introduce mean particles motion as the hydrodynamical parameters (mean velocity, mass and charge densities). In such case a closure is performed by continuity, energy and momentum conservation equations of fluid mechanics.

More advanced plasma description is based on kinetic theory [1], that starts either from Boltzmann or from its collision-less version of Vlasov [2]. Such approach in problems of electrodynamics combines kinetic description formulated in terms of distribution function and one for electromagnetic field that rather has hydrodynamic form. Therefore such description should have an additional kind of projecting from

doi:10.1088/978-0-7503-2576-9ch7

space of distribution functions space formed as functions of momentum (velocity) and position variables to the space of function only position coordinates [3].

The more complicated problems, such as plasma confinement, are considered in chapter 13.

7.1.2 Atmosphere plasma

Any gas becomes a plasma after a process of ionization-recombination is initiated by an external source of radiation, high energy interplanetary particles or collisions between molecules. We speak about electrically charged gases, or 'ionized' gases, that are found in nature. For this case I Langmuir, a researcher working to understand electric discharges, was the first to use the term 'plasma' [4]. In literature plasma is named as the fourth state of matter. Such matter is electrically conductive, and its electrical resistance, unlike metals, decreases with increasing temperature.

Ionized atoms, molecules as well as free electrons, are present in the environment of neutral molecules, but the whole volume occupied by the gaseous plasma in a large scale is electrically neutral. In planetary or stellar atmospheres a field due to steady-state plasma polarization exists because the upwards accelerated ions are partially lost at greater heights through recombination.

The atmosphere plasma is intensely studied [5], its concentration height dependence has a strong impact on radio-waves propagation [6] and serves as an excellent tool for monitoring of waves and other phenomena in geophysics, see, e.g. [10].

7.1.3 Solid state physics: electron–hole plasma

As we have written in the previous chapters, any conductor should contain free charges, as in Ampère–Maxwell equation (5.9). In solids such presence is modeled by notions of quasifree electrons and holes, whose quantum description put them into so-called admissible energy bands. In such a model that goes up to the charges that are referred to as electron–hole plasma [8], mainly used in the theory of semiconductors, including its thermodynamic properties [11]. The dynamics of such charges should account for interaction with electromagnetic field in a form of other quasiparticles, see, as an example, [7].

Such important phenomenon as plasmons is found in high speed, high power semiconductor, such as GaAs oscillators and switching processes [9].

The properties of the electron–hole plasma are used to describe, for example, the operation of high-gain photoconductive semiconductor switches, resonant tunneling diodes, impact ionization avalanche transit time devices, and Gunn oscillators. Models for electronic polarizability, plasma resonances, transit time effects, and plasma cooling in diodes, transistors, two-dimensional field effect transistors, and quantum wires must reach beyond single particle interactions and include collective, many-body effects and hydrodynamic equations to obtain reasonable agreement with measurements, reviewed in [12].

7.1.4 On stability of plasma

A zoo of plasma instabilities

There is a huge list of kinds of plasma instabilities [13]. There are a few reasons for such behavior:

1. The charged particles have a tendency to recombination, forming neutral atoms or molecules.
2. The next important physical reason is electromagnetic waves radiation, as energy losses.
3. The Lorenz magnetic field force is orthogonal to the magnetic field, so to stabilize the magnetic plasma we need its inhomogeneity.

A recombination of ions prevails in the absence of ionization factors as external radiation. Electromagnetic radiation of frequency ω can ionize an atom at energy level n if the corresponding ionization energy is large enough $\hbar\omega > E_{in}$. Such case may be important in a planet atmosphere plasma and ion fluences its ions concentration variations dependent on a star activity. An electromagnetic waves radiation is explained by inelastic collisions between ions and neutral atoms or molecules, dependent on the gas temperature.

The phenomenon of stability/instability and the term itself has its origin in the description of mechanical systems, such as a ball on a surface of a given profile [13]. In hydrodynamics as fluid mechanics there are own instabilities, such as listed in [14]. Mathematically the instabilities in the linear regime are studied by searching of solutions in the Fourier component form with the factor $\exp[i\omega t]$, that gives the dispersion equation. The nonzero negative imaginary part of the frequency $\omega = \omega_1 + i\omega_2$, $\omega_2 < 0$ may point to an instability existence. A nonlinearity account may lead either to stabilization of such a case or to destabilize a system, stable in the linear approximation by such a phenomenon as wave breaking [19].

Magnetic plasma confinement

In a uniform magnetic field a charged particle moves along a helical path, see figure 7.1, with a radius equal to the Larmor radius

$$r_L = \frac{mv_t}{qB}$$

(by the name of J Larmor, who discovered the rotation of the charge particles in a magnetic field). Here m is mass, q—charge, v_t—tangent component of velocity of the charged particle, B—induction of magnetic field. Same parameters define the frequency of rotation in the helix, named as cyclotron frequency

$$\omega_c = \frac{qB}{m},$$

see also exercise 1 of this chapter.

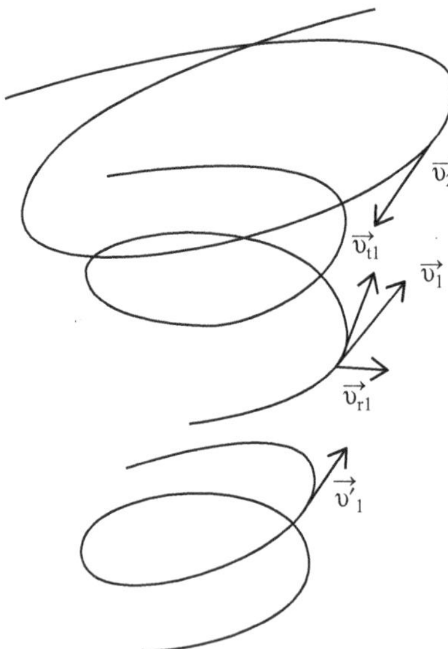

Figure 7.1. The charged particles paths with initial velocities $\vec{v}_1 = v_{t1}\vec{t} + v_{r1}\vec{n}$, $\vec{v}_2$, and final one $\vec{v}_1'$, after collision, in magnetic field with induction $\vec{B}$.

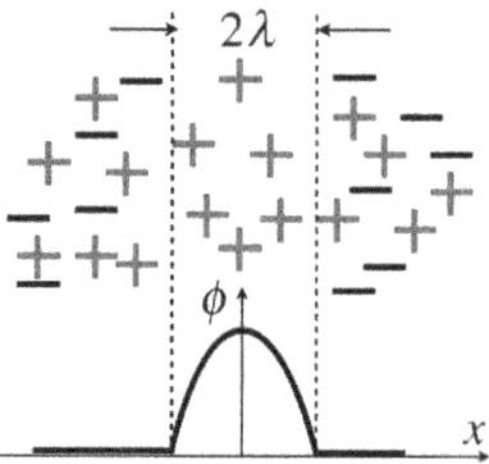

Figure 7.2. The charged particles are displaced by external field as shown. Such charge distribution produces the displacement field as in equation (7.4). Courtesy of Paulo C Lozano [17].

It, in a very simple scope, constitutes the idea of magnetic retention (confinement) or magnetic thermo-isolation of plasma. Thus, in the absence of collisions, a charged particle is connected to the magnetic field and does not go to the wall of a camera.

The localization of a charged particle is broken by collisions with other particles, look also the same figure 7.1. Hence, when an electron collides with an ion, an electron momentum jumps, just as in a collision with a heavy wall. After the collision, the electron will continue to revolve around the new axis, relative to the initial value in the order of the electron Larmor radius. So, the magnetic trap of such a simple construction that confines the particles needs to take the collisions and other factors into account.

Another important (collective) parameter of plasma is plasma frequency: if the electrons are displaced by a distance from the ions in a finite volume of plasma with density of electrons n_e and ions n_i, see figure 7.2, [17], where 1D picture is studied as follows. Given the density of charges creates the electric field potential that satisfies the Poisson equation, chapter 2, section 2.4

$$\Delta\phi = -\frac{e}{\varepsilon_0}(n_i - n_e). \tag{7.1}$$

The 1D model we use is the following. Having zero ion density $n_i = 0$ inside the interval $x \in [-\lambda, \lambda]$ and constant n_e we integrate the equation $\phi_{xx} = \frac{en_e}{\varepsilon_0}$ as $\phi = \frac{en_e}{2\varepsilon_0}x^2 + cx + d$. Outside the interval $x \in [-\lambda, \lambda]$, the charge density $n_i - n_e \approx 0$, equation (7.1) simplifies to the Laplace one. For the 1D problem its solution is given by the linear function $\phi = ax + b$, using the gauge freedom (section 2.4) we choose the potential outside as zero. It allows to formulate the zero boundary conditions at $x = \pm\lambda$ for the continuous potential on the whole axis, that yields

$$\phi = \frac{en_e\lambda^2}{2\varepsilon_0}\left[1 - \frac{x^2}{\lambda^2}\right], \tag{7.2}$$

this gives rise to a displacement field

$$E = -\frac{d\phi}{dx} = \frac{en_e x}{\varepsilon_0}, \tag{7.3}$$

where e is the elementary charge. The parameter λ is evaluated from the energy conservation relation. Energy of electron transition from origin to the point with coordinate $x = \lambda$ is evaluated as the work

$$W = \int_0^\lambda eEdx = \int_0^\lambda \frac{en_e x}{\varepsilon_0}dx = \frac{en_e\lambda^2}{2\varepsilon_0}. \tag{7.4}$$

If such energy is supported by the thermal one k_BT, equalizing them:

$$W = k_BT = \frac{en_e\lambda^2}{\varepsilon_0}, \tag{7.5}$$

we obtain the Debye length λ_D

$$\lambda_D = \sqrt{\frac{\varepsilon_0 k_B T}{en_e^2}}, \tag{7.6}$$

The Debye length provides an estimate for the size of non-neutral regions in a plasma and is critical to understanding the action of plasmas in contact with other bodies. For instance, a charged body immersed in a plasma becomes shielded by the production of a Debye layer. A speed of changes at plasma is characterized by the equation of motion of an electron displacement, that have the oscillator form

$$m\frac{d^2x}{dt^2} = -eE = -\frac{e^2n_e}{\varepsilon_0}x. \tag{7.7}$$

In a collision-less plasma, this is equivalent to having an oscillator with frequency

$$\omega_p = \sqrt{\frac{e^2 n_e}{m\varepsilon_0}}, \tag{7.8}$$

which describes the dynamics of Debye layer formation. With m the electron mass (in principle it should be the reduced mass $Mm/(M + m)$ for the corresponding ion and electron mass should be used here, but as $m \ll M$ this is practically identical with m; in any case, the usually assumed ion plasma frequency (where M replaces m in the equation above) does not occur here unless one is dealing with an ion–ion plasma). Then, if the plasma is subjected to a well defined external perturbation like for instance electromagnetic waves, a precisely defined driven oscillation can be maintained in principle indefinitely (see Plasma Oscillations, section 13.2, [3]).

It is however obvious that classically a steady-state field cannot be maintained as it would accelerate the ions and consequently make the plasma volume unstable (further processes can of course re-stabilize the situation; see below). The only two solutions classically possible: either the electrons lose most of their kinetic energy transfer to the ions (so that both diffuse with the same velocity), or that the electrons oscillate with regard to the ion background (see Plasma Oscillations) (a further possibility is of course that the charges cannot return in the original place because they are already confined locally by a magnetic field). See the website 'plasma physics': [4].

7.1.5 Tokamak plasma

Main ideas

One can create a magnetic field with a solenoid wound around a cylindrical tube, and then roll the tube into a torus to prevent departure of particles along the magnetic field. A D Sakharov and I E Tamm in 1951 proposed a modified scheme formulating a theoretical basis for a thermonuclear reactor where the plasma would have a torus shape and be held by a magnetic field [15].

Such a trap was named a Tokamak, **to**roidal'naja **ka**mera s **ma**gnitnymi **k**atushkami (in Russian) [18]. Its essential element is a torus-mounted formatter that creates a vortex electric field causing a current in the plasma, which heats the plasma.

The plasma description via averaged (hydrodynamic) values of its parameters such as velocity is called magneto-hydro-dynamics (MHD). This complex name simply covers the notion of fluid (hydro) in movement (dynamics) in a magnetic field (magneto), which applies very precisely to what happens inside a tokamak. To confine a hot plasma in an immaterial container formed by the magnetic field lines is a bit like wanting to contain a gas under pressure in a tyre inner tube. See for more details http://physci.llnl.gov/Research/Tokamak/, the spherical alternative is presented in figure 7.3. The general view of the largest international project JET tokamak European Torus, built in England, is given in figure 7.4. The following records certain plasma parameters: temperature 320 million K, concentration 4×10^{20} m^{-3}, energy retention time (see below) 1.8 s. Magnetic field in this setup

Figure 7.3. Thermonuclear system. Spherical Tokamak Globus-M of the A F Ioffe Institute of Physics and Technology in St. Petersburg. RIA 'Novosti' © Photo: Gleb Kurskiev, 03/18/2020. https://ria.ru/20200318/1568798938.html. Author: Mariluna, https://en.wikipedia.org/wiki/Tokamak.

Figure 7.4. The inside of the largest JET tokamak. Credit © ITER Organization [23].

is almost 4 T, so at a temperature of 100 million K, ions Larmor, the measuring radius is 3.5 mm.

Today (2020) the tokamak-ITER ('The Way' in Latin) project reaches its engineering height, being one of the most ambitious projects of energetics in the world to date. It intends to be the heart of the electric plant at the Cadarache facility [16], the ITER Tokamak will be the largest and most powerful fusion device in the world. It should achieve a deuterium–tritium plasma in which the reaction is sustained through internal heating.

Stability of confinement

It is important to note, a plasma due to electromagnetic field momentum, such as atomic gas, is characterized by magnetic pressure. The configuration of a tokamak implies specific conditions to confine the plasma ions by compensating the pressure of the plasma. An equilibrium, if easy to achieve, may become unstable, i.e. a small disturbance may grow with time and, in certain circumstances, lead to a loss of confinement: such phenomenon is referred to as disruption.

Disruptions

Any sudden decrease in either current or magnetic field induces mirror currents or magnetic fields inside the tokamak torus. The higher the initial plasma huge current (mega amperes in a modern tokamak) the shorter time it takes to disappear; hence it leads to the quick current variations, and more dangerous disruption can take place. A base of fundamental understanding of electromagnetic field behavior in plasma needs its mode content definition and its interaction investigation, see section 7.2 [27]. The most important mode, the entropy one, determines the behavior of the plasma in the nonlinear regime via so-called heating phenomenon [22, 24]. Disruptions at the limiting plasma density in these experiments are associated with nonlinear engagement of the internal and external screw modes. An analysis of abundant studies indicates that the sequence of development of perturbations in the process of disruption depends on a combination of physical phenomena, determined both by the local plasma parameters and by the external conditions of the experiment. Despite the significant progress achieved at present in experimental and theoretical studies (see review papers [20, 21]), a number of features of the development of a breakdown and, firstly, the interaction of various types of relaxation vibrations, remains not fully understood. Under these conditions, the study of the physical mechanisms of relaxation processes and plasma breakdowns, as well as the development of reliable methods for their stabilization, is one of the most important areas of scientific research on the program for the construction of a cost-effective fusion reactor [25].

7.2 Propagation of waves in a plasma: example of helicoidal waves

7.2.1 General remarks on plasma theoretical description

Generally plasma is a rich medium for waves initiation and propagation [26, 27]. As an example of plasma dynamics we choose a description of propagation of helicoidal waves in a model plasma with immobilized positive ions and movable electrons [27]. It is a good example to demonstrate some normal for plasma theory model approximations. The main difference with the previous section is an account of ions dynamics (via momentum balance). Namely such option is a natural step towards a closed matter-field description. We also restrict ourselves by the case of hydrodynamic approximation of the plasma interacting with electromagnetic field. Such description in a frame of small Knudsen numbers $Kn = \frac{\nu_W}{\nu} \ll 1$ (ν is collisions frequency, ν_w is a wave frequency), collisions between particles play a basic role, hence plasma may be considered as a conducting fluid. The electrodynamic equations are then in a correspondence with the hydrodynamic level description of particles. Starting with the first Maxwell's equation ('Coulomb law')

$$\operatorname{div} \vec{E} = 4\pi \sum_a e_a n_a, \tag{7.9}$$

where $e_1 = e_-$ is electron charge and $n_1 = n_-$ denotes electron concentration, the rest indices values mark ions. Here the density of charge in equation (11.16) is written as the sum of ion densities with the charges factors account. Adding the absence of magnetic charges statement

$$\operatorname{div} \vec{B} = 0, \tag{7.10}$$

we reproduce Faraday's law of section 6.1.1.

$$\operatorname{rot} \vec{E} = -\frac{1}{c}\frac{\partial \vec{B}}{\partial t}, \tag{7.11}$$

and close the Maxwell system by the Maxwell–Ampère one

$$\operatorname{rot} \vec{B} = \frac{1}{c}\frac{\partial \vec{E}}{\partial t} + \frac{4\pi}{c}\sum_a e_a n_a \vec{v}_a. \tag{7.12}$$

The density of current (see again section 5.2.2) is written as the sum of ion currents densities modeled as hydrodynamic velocities with the charges factors account. Similar to that we write a continuity equation for concentrations n_a:

$$\frac{\partial n_a}{\partial t} + \operatorname{div}(n_a \vec{v}_a) = 0. \tag{7.13}$$

The ion momentum balance (Newton law) without ions collisions (friction) account is described by the equation

$$\frac{\partial \vec{v}_a}{\partial t} + (\vec{v}_a \cdot \vec{\nabla})\vec{v}_a = \frac{e_a}{m_a}\left\{\vec{E} + \frac{1}{c}[\vec{v}_a \times \vec{B}]\right\}. \tag{7.14}$$

We would follow a model, in which the nuclei and neutral particles are considered as immobile surrounded by electron gas whose motion we take into account, that corresponds to frequency range chosen in the introduction. Equations in such approximation form still rather the extent system for the electron density $n_- = n$, mass $m_- = m$ and electron mean velocity $\vec{v}_- = \vec{v}$ and electromagnetic field. They are:

1. The continuity one (charge conservation law)

$$\frac{\partial n_-}{\partial t} + \operatorname{div}(n_- \vec{v}_-) = 0. \tag{7.15}$$

2. The momentum balance for electrons (Newton law)

$$\frac{\partial \vec{v}_-}{\partial t} + (\vec{v}_- \cdot \vec{\nabla})\vec{v}_- = \frac{e_-}{m_-}\left\{\vec{E} + \frac{1}{c}[\vec{v}_- \times \vec{B}]\right\}. \tag{7.16}$$

The last equations are

3. The Maxwell electrodynamics in which charged particles densities are placed on the rhs

$$\operatorname{div}\vec{E} = 4\pi\sum_a e_a n_a \approx 4\pi e_- n_-, \tag{7.17}$$

$$\operatorname{div}\vec{B} = 0, \tag{7.18}$$

4. Faraday's law

$$\operatorname{rot}\vec{E} = -\frac{1}{c}\frac{\partial\vec{B}}{\partial t}, \tag{7.19}$$

and in the,

5. Ampère law we neglect the ion's terms (zero velocity of positive ions approximation)

$$\operatorname{rot}\vec{B} = \frac{1}{c}\frac{\partial\vec{E}}{\partial t} + \frac{4\pi}{c}\sum_a e_a n_a \vec{v}_a \approx \frac{1}{c}\frac{\partial\vec{E}}{\partial t} + \frac{4\pi}{c}e_- n_- \vec{v}_-. \tag{7.20}$$

In the next sections we omit the index '–'.

7.2.2 Simplifications and the linearized system: plasma waves

Note, that a more advanced approach and model for plasma waves is studied in chapter 14. In accordance with the earlier assumptions, our consideration relates mainly to electrons, hence in further relations we will omit the index (–). Let us simplify equations (7.16) and (7.20) taking into account the following model assumption: for small variations of electromagnetic field, characterized by $\frac{1}{c}\frac{\partial\vec{E}}{\partial t}$ and, hence small acceleration $\frac{\partial\vec{v}}{\partial t} \to 0$ of electrons compared to the rest terms of equation [27]. Equation (7.16) is shortened up to

$$\vec{E} + \frac{1}{c}[\vec{v} \times \vec{B}] = 0, \tag{7.21}$$

the second term in equation (7.16) is also neglected due to the general assumption about linearity of the first approximation. Equation (7.20) at low frequencies simplifies as

$$\operatorname{rot}\vec{B} = \frac{4\pi}{c}en\vec{v}. \tag{7.22}$$

The linearization relates also to the division of the magnetic fields in two parts

$$\vec{B} = \vec{B}_0 + \vec{B}', \tag{7.23}$$

the second part $\vec{B}'$ describes the wave disturbance, which module is considered as a small one compared to a basic field $\vec{B}_0$.

$$B' \ll B_0.$$

Evaluating div from equation (7.20) and applying to equation (7.22) yields:

$$\operatorname{div}(\operatorname{rot} \vec{B}) = \operatorname{div}\left(\frac{4\pi}{c} en\vec{v}\right). \tag{7.24}$$

From equation (7.24), recalling that $\operatorname{div}(\operatorname{rot} \vec{B}) = 0$, we derive:

$$0 = \frac{4\pi e}{c} \operatorname{div}(n\vec{v}). \tag{7.25}$$

Plugging relation (7.25) into equation (7.15) gives:

$$\frac{\partial}{\partial t}(en) = 0. \tag{7.26}$$

The initial system of seven equations is reduced now to six (without continuity equation). From equality (7.21) it follows:

$$\vec{E} = -\frac{1}{c}[\vec{v} \times \vec{B}]. \tag{7.27}$$

Let us multiply vectorially both sides by vectors $\vec{B}$ as:

$$[\vec{B} \times \vec{E}] = -\frac{1}{c}[\vec{B} \times [\vec{v} \times \vec{B}]], \tag{7.28}$$

and, taking into account the identity:

$$\vec{A} \times [\vec{B} \times \vec{C}] = \vec{B} \cdot (\vec{A} \cdot \vec{C}) - \vec{C} \cdot (\vec{A} \cdot \vec{B}), \tag{7.29}$$

equation (7.28) simplifies as:

$$[\vec{B} \times \vec{E}] = -\frac{1}{c}[\vec{v} \cdot (\vec{B} \cdot \vec{B}) - \vec{B} \cdot (\vec{B} \cdot \vec{v})].$$

We remember, that $(\vec{B} \cdot \vec{v}) = 0$, therefore

$$c[\vec{B} \times \vec{E}] = -\vec{v} \cdot B^2.$$

Or, neglecting $\vec{B}'$ in the denominator it results in:

$$\vec{v} = -\frac{c[\vec{B} \times \vec{E}]}{B_0^2}. \tag{7.30}$$

Taking into account the result (7.30), equation (7.25) takes the form

$$\mathrm{div}\left(n\frac{c}{B_0^2}[\vec{B}_0 \times \vec{E}]\right) = 0, \tag{7.31}$$

and equation (7.22) is written as

$$\mathrm{rot}\,\vec{B} = -4\pi\frac{en}{B_0^2}[\vec{B}_0 \times \vec{E}]. \tag{7.32}$$

Write equation (7.18) as:

$$\frac{\partial B_y}{\partial y} = -\left(\frac{\partial B_x}{\partial x} + \frac{\partial B_z}{\partial z}\right). \tag{7.33}$$

From equation (7.17) it follows:

$$\frac{\partial E_z}{\partial z} = 4\pi en - \frac{\partial E_x}{\partial x} - \frac{\partial E_y}{\partial y}. \tag{7.34}$$

In the model we consider it is assumed that the number of electrons is equal to the number of protons. It means that the plasma is quasineutral, or $n = 0$, then:

$$\mathrm{div}\,\vec{E} = 0. \tag{7.35}$$

From equation (7.35) we have

$$\frac{\partial E_z}{\partial z} = -\frac{\partial E_x}{\partial x} - \frac{\partial E_y}{\partial y},$$

deviation from the $n = 0$ condition may be taken into account when E_z is evaluated.

Differentiating equation (7.32) by time yields:

$$(\vec{\nabla} \times \vec{B}_t) = -4\pi\frac{en}{B_0^2}[\vec{B}_0 \times \vec{E}_t]. \tag{7.36}$$

Using the Faraday equation (7.19) and the celebrated identity (7.29) we obtain

$$[\vec{\nabla} \times \vec{B}_t] = \vec{\nabla} \times [-c[\vec{\nabla} \times \vec{E}]] = -c\vec{\nabla} \times [\vec{\nabla} \times \vec{E}]$$
$$= -c[\vec{\nabla}(\vec{\nabla}\cdot\vec{E}) - \Delta \cdot \vec{E}].$$

We remember from equation (7.35) that $(\vec{\nabla}\cdot\vec{E}) = \mathrm{div}\,\vec{E} = 0$, then equation (7.36) reads:

$$\Delta\vec{E} = \Delta(E_x\vec{i} + E_y\vec{j}) = \frac{4\pi en}{cB_0}\left[-\frac{\partial E_y}{\partial t}\vec{i} + \frac{\partial E_x}{\partial t}\vec{j}\right]. \tag{7.37}$$

This results in components

$$\begin{cases} \Delta E_x = -\dfrac{4\pi en}{cB_0}\dfrac{\partial E_y}{\partial t}, \\ \Delta E_y = \dfrac{4\pi en}{cB_0}\dfrac{\partial E_x}{\partial t}. \end{cases} \tag{7.38}$$

Such a system may be considered as a base of Cauchy problem with the following initial conditions

$$\begin{cases} E_x(x, y, z, 0) = \phi(x, y, z), \\ E_y(x, y, z, 0) = \psi(x, y, z). \end{cases} \tag{7.39}$$

7.2.3 The initial problem formulation in matrix form: Fourier transform

In this section we apply the dynamic projecting operator method as in section 6.5.3, developing it for the three-dimensional case. The system of equations (7.38) may be written in matrix form as the evolution system:

$$\begin{cases} \dfrac{\partial E_x}{\partial t} = \dfrac{cB_0}{4\pi en}\Delta E_y, \\ \dfrac{\partial E_y}{\partial t} = -\dfrac{cB_0}{4\pi en}\Delta E_x. \end{cases} \tag{7.40}$$

The system in matrix operator form is written as

$$\frac{\partial \psi}{\partial t} + L\psi = 0, \tag{7.41}$$

where:

$$\psi = \begin{pmatrix} E_x \\ E_y \end{pmatrix};\ L = \begin{pmatrix} 0 & \dfrac{cB_0}{4\pi en}\Delta \\ -\dfrac{cB_0}{4\pi en}\Delta & 0 \end{pmatrix}. \tag{7.42}$$

The Fourier transforms of the basic variables in equation (7.40) are written as:

$$E_x(t, \vec{r}) = \frac{1}{\sqrt{2\pi}} \int \tilde{E}_x(t, \vec{k}) e^{-i(\vec{k}\vec{r})}, \tag{7.43}$$

$$E_y(t, \vec{r}) = \frac{1}{\sqrt{2\pi}} \int \tilde{E}_y(t, \vec{k}) e^{-i(\vec{k}\vec{r})}. \tag{7.44}$$

Therefore:

$$\Delta \to -\left(k_x^2 + k_y^2 + k_z^2\right) = -\vec{k}^2. \tag{7.45}$$

The system (7.40) after Fourier transformation takes the form:

$$\begin{cases} \dfrac{\partial \tilde{E}_x}{dt} + \dfrac{cB_0}{4\pi en}k^2\tilde{E}_y = 0, \\ \dfrac{\partial \tilde{E}_y}{dt} - \dfrac{cB_0}{4\pi en}k^2\tilde{E}_x = 0. \end{cases} \tag{7.46}$$

The system (7.46) is conventionally written as a matrix one:

$$\frac{\partial \tilde{\psi}}{\partial t} + \tilde{L}\tilde{\psi} = 0, \tag{7.47}$$

where:

$$\tilde{\psi} = \begin{pmatrix} \tilde{E}_x \\ \tilde{E}_y \end{pmatrix};\ \tilde{L} = \begin{pmatrix} 0 & -\dfrac{cB_0k^2}{4\pi en} \\ \dfrac{cB_0k^2}{4\pi en} & 0 \end{pmatrix}. \tag{7.48}$$

To apply the projecting operator method, we solve the eigen problem for the matrix (7.48), parametrized by $\vec{k}$:

$$\tilde{L}\varphi = \lambda\varphi \tag{7.49}$$

We continue to work in the Fourier ($\vec{k}-$) representation

$$\frac{cB_0k^2}{4\pi en}\begin{pmatrix} 0 & -1 \\ 1 & 0 \end{pmatrix}\begin{pmatrix} \phi_1 \\ \phi_2 \end{pmatrix} = \lambda\begin{pmatrix} \phi_1 \\ \phi_2 \end{pmatrix}. \tag{7.50}$$

The condition of nonzero solution

$$det\begin{pmatrix} -\lambda & -\dfrac{cB_0k^2}{4\pi en} \\ \dfrac{cB_0k^2}{4\pi en} & -\lambda \end{pmatrix} = 0,$$

defines the eigenvalues

$$\lambda_{\pm} = \pm i\frac{cB_0k^2}{4\pi en}. \tag{7.51}$$

Eigenvectors of the operator in equation (7.48) are marked by $\pm$ so that the link between components is the following:

For $\lambda_+ = i\frac{cB_0k^2}{4\pi en}$;.

$$\phi_{2+} = -i\phi_{1+};\ \phi_{+} = \begin{pmatrix} 1 \\ -i \end{pmatrix}\phi_{1+}, \tag{7.52}$$

while for $\lambda_{-} = -i\frac{cB_0k^2}{4\pi en}$;

$$\phi_{2-} = -i\phi_{1-};\ \phi_{-} = \begin{pmatrix} 1 \\ i \end{pmatrix}\phi_{1-}. \tag{7.53}$$

7.2.4 The initial problem solution by dynamical projecting

Let us introduce the projecting operators P_+ and P_- with the standard properties (as in section 6.5):

$$(1)\ P_+P_- = 0, \tag{7.54}$$

$$(2)\ P_{\pm}^2 = P_{\pm}, \tag{7.55}$$

$$(3)\ P_+ + P_- = I. \tag{7.56}$$

One could repeat the useful exercise using the general property for $P_{\pm}$ and equations (7.52) and (7.53):

$$P_{\pm}\begin{pmatrix} e_{1\pm} \\ e_{2\pm} \end{pmatrix} = \phi_{1\pm}\begin{pmatrix} 1 \\ \mp i \end{pmatrix}. \tag{7.57}$$

Solving the system with respect to matrix elements of $P_{\pm}$ taking into account equations (7.55) and (7.54) one arrives at

$$P_+ = \frac{1}{2}\begin{pmatrix} 1 & i \\ -i & 1 \end{pmatrix}, \tag{7.58}$$

and

$$P_- = \frac{1}{2}\begin{pmatrix} 1 & -i \\ i & 1 \end{pmatrix}. \tag{7.59}$$

We as in the case of section 6.5 have the happy case of projectors whose matrix elements do not depend on $\vec{k}$. It means that its inverse transform ($\vec{r}$-representation) coincides with the $\vec{k}$-representation.

Hence we may act directly by the operator P_+ and next—by P_- on the basic equation (7.41) (in the coordinate representation '$\vec{r}$'):

$$\begin{aligned} P_+\psi_t + P_+L\psi &= 0, \\ P_-\psi_t + P_-L\psi &= 0, \end{aligned} \tag{7.60}$$

using the commutation property, we write the first of equation (7.60) as

$$\begin{pmatrix} E_1 + iE_2 \\ -iE_1 + E_2 \end{pmatrix}_t + \frac{cB_0}{4\pi en}\Delta\begin{pmatrix} iE_1 - E_2 \\ E_1 + iE_2 \end{pmatrix} = 0, \tag{7.61}$$

or, reading the first line,

$$(E_1 + iE_2)_t + i\frac{cB_0}{4\pi en}\Delta(E_1 + iE_2) = 0. \tag{7.62}$$

The second one of equation (7.60) gives

$$(E_1 - iE_2)_t - i\frac{cB_0}{4\pi en}\Delta(E_1 - iE_2) = 0. \tag{7.63}$$

Let us recall the main property (7.56), of the projectors (completeness), acting as:

$$(P_+ + P_-)\Psi = \Psi, \tag{7.64}$$

where:

$$P_+\Psi = \begin{pmatrix} \frac{1}{2} & \frac{1}{2}i \\ -\frac{1}{2}i & \frac{1}{2} \end{pmatrix}\begin{pmatrix} E_1 \\ E_2 \end{pmatrix} = \begin{pmatrix} E_+ \\ -iE_+ \end{pmatrix},$$

and

$$P_-\Psi = \begin{pmatrix} \frac{1}{2} & -\frac{1}{2}i \\ \frac{1}{2}i & \frac{1}{2} \end{pmatrix}\begin{pmatrix} E_1 \\ E_2 \end{pmatrix} = \begin{pmatrix} E_- \\ -iE_- \end{pmatrix}.$$

Hence it is convenient to introduce the new variables:

$$E_\pm = \frac{1}{2}(E_1 \pm iE_2), \tag{7.65}$$

writing equations (7.62) and (7.63) as:

$$(E_\pm)_t \pm i\aleph\Delta(E_\pm) = 0, \tag{7.66}$$

where

$$\aleph = \frac{cB_0}{4\pi en}.$$

Equations (7.66) are of the Schrödinger evolution type. Its solution is well described in textbooks on mathematical physics, e.g. in [28]. The fields $E_{1,2}$ are recovered via the inverse of equation (7.65), that for given and recalculated for equation (7.66) initial conditions solve the Cauchy problem, see exercises 4 and 5. Let us repeat, that in the case we do consider, the nonlinear term is not taken into account, it will be derived and added in the next section.

7.3 The nonlinear case

Let us return to the basic system (7.15)–(7.20). It contains nonlinear terms, for example the 'convective term':

$$(\vec{v}\nabla)\vec{v}. \tag{7.67}$$

Reproduce again equations (7.15)–(7.20), that account a few ions contributions:

$$\frac{\partial n_a}{\partial t} + \operatorname{div}(n_a v_a) = 0,$$

$$\frac{\partial \vec{v}_a}{\partial t} + (\vec{v}_a \cdot \vec{\nabla})\vec{v}_a = \frac{e_a}{m_a}\left\{\vec{E} + \frac{1}{c}[\vec{v}_a \times \vec{B}]\right\},$$

$$\operatorname{div}\vec{E} = 4\pi \sum_a e_a n_a,$$

$$\operatorname{rot}\vec{B} = \frac{1}{c}\frac{\partial \vec{E}}{\partial t} + \frac{4\pi}{c}\sum_a e_a n_a \vec{v}_a,$$

$$\operatorname{div}\vec{B} = 0,$$

$$\operatorname{rot}\vec{E} = -\frac{1}{c}\frac{\partial \vec{B}}{\partial t},$$

$$\frac{d\rho_a}{dt} + \rho_a \operatorname{div}(\vec{v}_a) = 0.$$

In the way of simplification, for only the same electron subsystem, omitting the index –, but taking into account the nonlinear convective term $(\vec{v}\nabla)\vec{v}$ in:

$$\frac{\partial n}{\partial t} + \operatorname{div}(n\vec{v}) = 0, \tag{7.68}$$

neglecting again the acceleration term for the low frequency range, in the momentum balance equation:

$$(\vec{v}\cdot\vec{\nabla})\vec{v} = \frac{e}{m}\left\{\vec{E} + \frac{1}{c}[\vec{v} \times \vec{B}]\right\}, \tag{7.69}$$

simplifying as

$$\operatorname{div}\vec{E} = 4\pi \sum en = 4\pi\rho, \tag{7.70}$$

and, similar

$$\operatorname{rot}\vec{B} = \frac{1}{c}\frac{\partial \vec{E}}{\partial t} + \frac{4\pi}{c}en\vec{v}, \tag{7.71}$$

omitting as well indices in the continuity equation for electrons, we have

$$\frac{d\rho}{dt} + \rho \cdot \operatorname{div}(\vec{v}) = 0. \tag{7.72}$$

Calculating (div) of the Ampère equation (7.71) and, next, plugging the result in equation (7.72),

$$0 = \frac{4\pi}{c} \sum \operatorname{div}(en\vec{v}),$$

one states:

$$\frac{d}{dt}(ne) = 0.$$

Transformation of the equation (7.69) approximation is done by iterations in electron's velocity:

$$\vec{v} = \vec{v}_{(1)} + \vec{v}_{(2)} = -\frac{c}{B_0^2}[\vec{B}_0 \times \vec{E}\,] + \vec{v}_{(2)},$$

taking into account only the first and second order. Plugging the first order velocity into the second order terms, yields

$$\begin{aligned}(\vec{v},\, \nabla)\vec{v} &= \frac{c^2}{B_0^4}([\vec{B}_0 \times \vec{E}\,],\, \nabla) \cdot [\vec{B}_0 \times \vec{E}\,] \\ &= \frac{c^2}{B_0^4}(\vec{B}_0 \times ([\vec{B}_0 \times \vec{E}\,],\, \nabla)) \cdot \vec{E},\end{aligned} \tag{7.73}$$

or, including the Lorentz force,

$$\frac{c^2}{B_0^4}[\vec{B}_0 \times ([\vec{B}_0 \times \vec{E}\,],\, \nabla)\vec{E}\,] = \frac{e}{m}\left(\vec{E} + \frac{1}{c}[\vec{v} \times \vec{B}_0]\right), \tag{7.74}$$

by next rearranging we have

$$\frac{mc^3}{eB_0^4}[\vec{B}_0 \times ([\vec{B}_0 \times \vec{E}\,],\, \nabla)\vec{E}\,] - c\vec{E} = \vec{v} \times \vec{B}_0.$$

Plugging the expansion (7.27) we express the velocity as:

$$\vec{v} = \frac{mc^3}{eB_0^6}[\vec{B}_0 \times [\vec{B}_0 \times ([\vec{B}_0 \times \vec{E}\,],\, \nabla)\vec{E}\,]] - \frac{c}{B_0^2}[\vec{B}_0 \times \vec{E}\,]. \tag{7.75}$$

This resulting expression for velocity we substitute in the rhs of the basic Ampère equation (7.71), neglecting the time derivative of the field $\vec{E}$ at the low frequency range:

$$\operatorname{rot} \vec{B} = \frac{4\pi}{c} en\vec{v},$$

arriving at

$$\begin{aligned}\operatorname{rot}\vec{B} = &-\frac{8\pi en}{B_0^2}[\vec{B}_0 \times \vec{E}\,] \\ &+\frac{4\pi nmc^2}{B_0^6}[\vec{B}_0 \times [\vec{B}_0 \times ([\vec{B}_0 \times \vec{E}\,],\, \nabla)\vec{E}\,]].\end{aligned} \tag{7.76}$$

Applying the rot operator to the Faraday equation

$$\operatorname{rot}\vec{E} = -\frac{1}{c}\frac{\partial\vec{B}}{\partial t},$$

and using the BAC–CAB identity we derive

$$\Delta\vec{E} = -\frac{1}{c}[\nabla \times \vec{B}_t].$$

Differentiating equation (7.76) by time (marked as index by $\vec{E}$) and using the last equation we arrive at

$$\begin{aligned}\Delta\vec{E} \;=\; &-\frac{8\pi en}{cB_0^2}[\vec{B}_0 \times \vec{E}_t] \\ &-\frac{4\pi nmc}{B_0^6}[\vec{B}_0 \times [\vec{B}_0 \times ([\vec{B}_0 \times E_t],\, \nabla)\vec{E} + \vec{B}_0 \times ([\vec{B}_0 \times \vec{E}\,],\, \nabla)\vec{E}_t]].\end{aligned} \tag{7.77}$$

Going to components, let us recall that the vector $\vec{B}_0$ is chosen along the 'z'-axis, getting finally:

$$\begin{aligned}\frac{8\pi en}{cB_0}\frac{\partial E_y}{\partial t} = &-\Delta E_x + \frac{4\pi nmc}{B_0^3}\left(\frac{\partial E_x}{\partial t}\frac{\partial}{\partial y} - \frac{\partial E_y}{\partial t}\frac{\partial}{\partial x}\right)E_y \\ &+\frac{4\pi nmc}{B_0^3}\left(-E_y\frac{\partial}{\partial x} + E_x\frac{\partial}{\partial y}\right)\frac{\partial E_y}{\partial t}\end{aligned} \tag{7.78}$$

$$\begin{aligned}\frac{8\pi en}{cB_0}\frac{\partial E_x}{\partial t} = &-\Delta E_y + \frac{4\pi nmc}{B_0^3}\left(\frac{\partial E_x}{\partial t}\frac{\partial}{\partial y} - \frac{\partial E_y}{\partial t}\frac{\partial}{\partial x}\right)E_x \\ &+\frac{4\pi nmc}{B_0^3}\left(-E_y\frac{\partial}{\partial x} + E_x\frac{\partial}{\partial y}\right)\frac{\partial E_x}{\partial t}.\end{aligned} \tag{7.79}$$

Now we would go to the complex components by means of projecting operator P_+ and P_- by equations (7.58) and (7.59):

$$P_+ = \frac{1}{2}\begin{pmatrix} 1 & i \\ -i & 1\end{pmatrix}, \quad \wedge \quad P_- = \frac{1}{2}\begin{pmatrix} 1 & -i \\ i & 1\end{pmatrix}. \tag{7.80}$$

Acting by the projectors on the system (13.25) and (7.79) and going to the variables

$$\Pi = E_x + iE_y \qquad \longleftrightarrow \qquad E_x = \frac{1}{2}(\Pi + \Lambda), \tag{7.81}$$

and

$$\Pi = E_x - iE_y \qquad \longleftrightarrow \qquad \Lambda = -\frac{1}{2}i(\Pi - \Lambda), \tag{7.82}$$

its plugging into equation (13.25) yields the final result

$$\begin{aligned}&\frac{8\pi en}{cB_0}\frac{\partial}{\partial t}\Lambda + \frac{2\pi nmc}{B_0^3}\\ &\left(i\frac{\partial}{\partial t}(\Pi - \Lambda)\frac{\partial}{\partial x}\Pi + \frac{\partial}{\partial t}(\Pi + \Lambda)\frac{\partial}{\partial y}\Pi + i(\Pi - \Lambda)\frac{\partial^2}{\partial x \partial t}\Pi + (\Pi + \Lambda)\frac{\partial^2}{\partial y \partial t}\Pi\right)\\ &-i\Delta\Lambda = 0,\end{aligned} \tag{7.83}$$

$$\begin{aligned}&\frac{8\pi en}{cB_0}\frac{\partial}{\partial t}\Pi + \frac{2\pi nmc}{B_0^3}\\ &\left(i\frac{\partial}{\partial t}(\Pi - \Lambda)\frac{\partial}{\partial x}\Lambda + \frac{\partial}{\partial t}(\Pi + \Lambda)\frac{\partial}{\partial y}\Lambda + i(\Pi - \Lambda)\frac{\partial^2}{\partial x \partial t}\Lambda + (\Pi + \Lambda)\frac{\partial^2}{\partial y \partial t}\Lambda\right)\\ &-i\Delta\Pi = 0.\end{aligned} \tag{7.84}$$

This system describes interaction between Λ and Π hybrid waves. In chapter 14, section 14.1.3 we present the electric field as a superposition of wavepackets with E_i amplitudes, group speeds v_i and the carrier frequencies ω_i. Assume the first of the waves will be an ion-acoustic wave, while the second and third are then helicoid waves.

The presented type of force can be used in the description of ionospheres and for some approximation. In solid state physics (e.g. in textbooks [11, 26]), calculations were carried out for the system in which $T \ll \frac{m^* v_\phi^2}{k_B}$, m^*—effective mass, v_ϕ—phase velocity of wave.

Exercise 1. Derive the expression for Larmor radius and cyclotron frequency from mechanics of a charged particle in the magnetic field whose action is expressed as the Lorenz force.

Exercise 2. Plugging the solution of the 1D equation for potential into boundary conditions $\phi(\pm\lambda) = 0$, derive the formula (7.2).

Exercise 3. Derive the explicit form for the projecting operators (7.58) and (7.59).

Exercise 4. Solve the Scrödinger evolution equation (7.66) by Fourier method for the given initial condition as a Gaussian function.

Exercise 5. Recalculate the initial conditions for equation (7.66) by the given initial conditions for $E_{1,2}$ and solve the Scrödinger evolution equation by Fourier method. Find the fields $E_{1,2}$ for $t > 0$.

References

[1] Reichl L E 1998 *A Modern Course in Statistical Physics* 2nd edn (New York: Wiley)

[2] Colonna G and D'Angola A (ed) 2018 *Plasma Modeling Methods and Applications* (Bristol: IOP Publishing)

[3] Leble S 1990 *Nonlinear Waves in Waveguides* (Heidelberg: Springer)

[4] http://www.plasmaphysics.org.uk/

[5] Rycroft M J and Harrison R G 2012 Electromagnetic atmosphere-plasma coupling: the global atmospheric electric circuit *Space Sci. Rev.* **168** 363–84

[6] Seybold J S 2005 *Introduction to RF Propagation* (New York: Wiley) pp 3–10

[7] Leitenstorfer A, Huber R, Tauser F and Brodschelm A 2003 How fast do charged particles get dressed? *Phys. Status Solidi* B **238** 455–61

[8] Vladimirov V V 1975 Helical instability in electron–hole plasma in semiconductors *Am. Inst. Phys. Sov. Phys. Usp.* **18** 37–50

[9] Moradia A 2016 Surface and bulk plasmons of electron–hole plasma in semiconductor nanowires *Phys. Plasmas* **23** 114503

[10] De Santis A, Marchetti D and Haagmans R 2019 Geophysical hazards, precursory worldwide signatures of earthquake occurrences on Swarm satellite data *Sci. Rep.* **9** 20287

[11] Klingshirn C F 2012 The electron–hole plasma *Semiconductor Optics. Graduate Texts in Physics* (Berlin: Springer)

[12] Zutavern F J, Baca A G, Chow W, Hafich M J, Hjalmarson H, Loubriel G M, Mar A, O'Malley M W, Roose L D and Vawter G A 2002 Electron–hole plasmas in semiconductors *Conference: Pulsed Power Plasma Science, 2001. PPPS-2001. Digest of Technical Papers*

[13] https://en.wikipedia.org/wiki/Plasma_stability

[14] https://en.wikipedia.org/wiki/List_of_hydrodynamic_instabilities_named_after_people

[15] Wesson J 2004 *Tokamaks* 3rd edn (Oxford: Oxford University Press)

[16] https://www.iter.org/

[17] https://ocw.mit.edu/courses/aeronautics-and-astronautics/16-522-space-propulsion-spring-2015/lecture-notes/MIT16_522S15_Lecture8.pdf

[18] Arcimovich L A *et al* 1969 *Experiments in Tokamak Devices* (Vienna: International Atomic Energy Agency (IAEA))

[19] Scott B D 2005 Drift wave versus interchange turbulence in tokamak geometry: linear versus nonlinear mode structure *Phys. Plasmas* **12** 062314

[20] Savrukhin P, Lyadina E, Martynov D, Kislov D and Poznyak V 1994 Coupling of internal m = 1 and m = 2 modes at density limit disruptions in the T-10 tokamak *Nucl. Fusion* **34** 317

[21] Morris A A V 1992 MHD instability control, disruptions, and error fields in tokamaks *Plasma Phys. Control. Fusion* **34** 1871–80

[22] Perelomova A 2020 On description of periodic magnetosonic perturbations in a quasi-isentropic plasma with mechanical and thermal losses and electrical resistivity *Phys. Plasmas* **27** 032110

[23] https://www.iter.org/mach

[24] Perelomova A 2018 Magnetoacoustic heating in a quasi-isentropic magnetic gas *Phys. Plasmas* **25** 042116

[25] Savrukhin P V and Klimanov I V 2001 Tangential x-ray imaging system for analysis of the small-scale modes in the T-10 tokamak *Rev. Sci. Instrum.* **72** 1668–71

[26] Bellan P M 2009 *Fundamentals of Plasma Physics* (Cambridge: Cambridge University Press)
[27] Leble S and Rohraff D W 2006 Nonlinear evolution of components of an electromagnetic field of helicoidal waves in plasma *Phys. Scr.* **123** 140–4
[28] Tikhonov A N and Samarskii A A 2011 *Equations of Mathematical Physics. Dover Books on Physics* (New York: Dover)

IOP Publishing

Practical Electrodynamics with Advanced Applications

Sergey Leble

Chapter 8

Metamaterials

8.1 Research on metamaterials

8.1.1 Introduction: on the chapter content

The specific properties of the artificial materials (called *metamaterials*) as a medium for electromagnetic waves propagation force us to include it into the basic part of this textbook, and continue it in the advanced part of chapter 12. Having in mind the material relations specific advance, we state the problem and show, how the material relations change while accounting for dispersion. In correspondence with the practical technique construction, we show, how projecting operators look in ω and t representations (domains), with nontrivial integral operator matrix elements. It allows to simplify the main task of the solutions space separation in the universal scheme. Next, it allows to incorporate a nonlinearity account and realize the important example of the Kerr one, deriving the system of the directed waves interaction. Going to the Drude dispersion, the technique includes realization of the program along with the Kerr nonlinearity model and hence finalize the main result of the directed waves interaction system for this model, important in metamaterials investigations. An example of a class of stationary solutions includes also the case that shows a difference between conventional and Veselago materials.

Description of metamaterials is intimately linked to plasma physics because of the Drude formula application for both media, applied for dielectric permittivity and magnetic permeability.

The first research on metamaterials was provided by Bengali polymath, physicist J C Bose in 1898 [1]. He studied the rotation of the plane of polarization of electric waves by a twisted structure he created. In 1946 W E Kock suggested to use a mixture of metal spheres as a refractive material [2]. He coined the term artificial dielectric, which has been later used in the microwave literature [3].

One such interesting problem for researchers is metamaterials with simultaneously negative dielectric permittivity and magnetic permeability. In 1968, Victor Veselago [4] wrote about the general electrodynamic properties of metamaterials,

doi:10.1088/978-0-7503-2576-9ch8

but only in 2000 did David Smith and his group create such structures [7]. An interaction of ultra short pulses with ordinary materials is well understood in nonlinear optics [31]. Structures with simultaneously negative dielectric permittivity and magnetic permeability has been called by many names: Veselago media, negative-index media, negative-refraction media, etc [6]. Since the discovery of materials with negative refractive index, it has been possible to built new devices that use metamaterials ability to control the path of electromagnetic energy, manipulating by Pointing vector direction [5]. The applications for metamaterials are broad and varied from the celebrated electromagnetic cloaking [8], to new imaging capabilities [9].

To achieve negative values of the constitutive parameters ε and μ, metamaterials must be dispersive, i.e. their permittivity and permeability must be frequency dependent, otherwise they would not be causal [10]. As is shown [11] if we have frequency dispersion, the full energy density of wavepacket electromagnetic field will be (see the derivation in section 6.1.2):

$$W = \varepsilon_0 \frac{\partial(\omega\varepsilon(\omega))}{\partial\omega} E^2 + \mu_0 \frac{\partial(\omega\mu(\omega))}{\partial\omega} H^2, \tag{8.1}$$

$W > 0$ if:

$$\frac{\partial(\omega\varepsilon(\omega))}{\partial\omega} > 0, \ \frac{\partial(\omega\mu(\omega))}{\partial\omega} > 0.$$

That's not contradicted with simultaneously negative $\varepsilon < 0$ and $\mu < 0$ [4].

The materials with typical plasma dispersion (Drude formula) for both $\varepsilon(\omega)$ and $\mu(\omega)$ (so-called two-time derivative Lorentz material (2TDLM) model) are often discussed [14]:

$$\chi = \frac{\omega_p^2\chi_a + i\omega_p\chi_\beta\omega - \chi_\gamma\omega^2}{\omega_0^2 + i\omega\Gamma - \omega^2}. \tag{8.2}$$

Γ is plasma collision frequency, used as a parameter for a metamaterial. There would be independent models for the permittivity and permeability

$$\varepsilon(\omega) = (1 + \chi_e), \ \mu(\omega) = (1 + \chi_m). \tag{8.3}$$

The real part of this permittivity is negative for all $\omega < \sqrt{\omega_p^2 + \Gamma^2}$. The Drude model is obtained with the fakir's bed medium [13].

In most cases, metamaterials are fabricated using planar lithography techniques which cannot produce large, volumetric metamaterials. This is particularly true at optical frequencies, but even at THz and microwave frequencies achieving volumetric rather than two-dimensional samples typically requires manual assembly of two-dimensional metamaterial boards or sheets. The simplest example of drawn metamaterial is the fiber array, with an effective permittivity along the fibers having a plasmonic response that can be adjusted through the diameter and separation of the wires [15].

8.2 Statement of problem: dispersion operator

8.2.1 Maxwell's equations: operators of dielectric permittivity and magnetic permeability

Our starting point is the Maxwell equations for a simple case of linear isotropic but general dispersive dielectric media, in the SI unit system. We reproduce here dynamic, Faraday and Ampère equations, that have coefficients, different from the ones in Gauss units

$$\frac{\partial \vec{B}(\vec{r}, t)}{\partial t} = -\mathrm{rot}\vec{E}(\vec{r}, t), \tag{8.4}$$

$$\frac{\partial \vec{D}(\vec{r}, t)}{\partial t} = \mathrm{rot}\vec{H}(\vec{r}, t). \tag{8.5}$$

The principle features of general dispersion may be demonstrated within the 1D model as in Schäfer, Wayne [24] and Kuszner, Leble [21, 25] works, where the x-axis is chosen as the direction of an electromagnetic wavepacket propagation. As with the mentioned authors, we assume $D_x = 0$ and $B_x = 0$, that means the minimal version of the theory with only the polarization of electromagnetic waves, in the following sections we rely upon [27]. This allows us to write the Maxwell equations as:

$$\begin{aligned} \frac{\partial D_y}{\partial t} &= -\frac{\partial H_z}{\partial x}, \\ \frac{\partial B_z}{\partial t} &= -\frac{\partial E_y}{\partial x}. \end{aligned} \tag{8.6}$$

Let us denote the Fourier transforms $\mathcal{E}, \mathcal{B}, \mathcal{D}, \mathcal{H}$ of the fields E, B, D and H that, by definition are obtained by

$$E(x, t) = \frac{\varepsilon_0}{\sqrt{2\pi}} \int_{-\infty}^{\infty} \mathcal{E}(x, \omega)\exp(i\omega t)d\omega, \tag{8.7}$$

$$B(x, t) = \frac{1}{\sqrt{2\pi}} \int_{-\infty}^{\infty} \mathcal{B}(x, \omega)\exp(i\omega t)d\omega, \tag{8.8}$$

$$D(x, t) = \frac{1}{\sqrt{2\pi}} \int_{-\infty}^{\infty} \mathcal{D}(x, \omega)\exp(i\omega t)d\omega, \tag{8.9}$$

$$H(x, t) = \frac{\mu_0}{\sqrt{2\pi}} \int_{-\infty}^{\infty} \mathcal{H}(x, \omega)\exp(i\omega t)d\omega. \tag{8.10}$$

The indexes for this 1D case are omitted. The Fourier images we'll call ω—representation or a frequency domain. The fields E, B, D, H are named as t-representation or marked as variables from time domain. Let the material relations in ω—representation be linear:

$$\mathcal{D} = \varepsilon_0\varepsilon(\omega)\mathcal{E}, \tag{8.11}$$

$$\mathcal{B} = \mu_0\mu(\omega)\mathcal{H}. \tag{8.12}$$

The notations are the following:

$\varepsilon(\omega)$—dielectric permittivity of the matter of wave propagation, $\mu(\omega)$—magnetic permeability of the medium, while ε_0, μ_0—dielectric permittivity and magnetic permeability of vacuum. For the final theory formulation we need to use time domain representation, where the material relations become integral operators:

$$\begin{aligned} D(x, t) &= \frac{1}{\sqrt{2\pi}}\int_{-\infty}^{\infty} \mathcal{D}(x, \omega)\exp(i\omega t)d\omega \\ &= \frac{\varepsilon_0}{\sqrt{2\pi}}\int_{-\infty}^{\infty} \varepsilon(\omega)\mathcal{E}(x, \omega)\exp(i\omega t)d\omega, \end{aligned} \tag{8.13}$$

$$\begin{aligned} B(x, t) &= \frac{1}{\sqrt{2\pi}}\int_{-\infty}^{\infty} \mathcal{B}(x, \omega)\exp(i\omega t)d\omega \\ &= \frac{\mu_0}{\sqrt{2\pi}}\int_{-\infty}^{\infty} \mu(\omega)\mathcal{H}(x, \omega)\exp(i\omega t)d\omega. \end{aligned} \tag{8.14}$$

If substitute the inverse transform

$$\mathcal{E}(x, \omega) = \frac{1}{\sqrt{2\pi}}\int_{-\infty}^{\infty} E(x, s)\exp(-i\omega s)ds, \tag{8.15}$$

into equation (8.13) we arrive at the material relation

$$D(x, t) = \frac{\varepsilon_0}{2\pi}\int_{-\infty}^{\infty} \varepsilon(\omega)\int_{-\infty}^{\infty} E(x, s)\exp(-i\omega s)ds\,\exp(i\omega t)d\omega, \tag{8.16}$$

in the form of double integral. If $E(x, s)\exp(-i\omega s)$ satisfies the basic conditions of Fubini's theorem, we can change the order of integration as:

$$\begin{aligned} D(x, t) &= \frac{\varepsilon_0}{2\pi}\int_{-\infty}^{\infty}\int_{-\infty}^{\infty} \varepsilon(\omega)\exp(i\omega(t - s))d\omega E(x, s)ds \\ &= \int_{-\infty}^{\infty} \tilde{\varepsilon}(t - s)E(x, s)ds, \end{aligned} \tag{8.17}$$

having the integral operator with the kernel

$$\tilde{\varepsilon}(t - s) = \frac{\varepsilon_0}{2\pi}\int_{-\infty}^{\infty} \varepsilon(\omega)\exp(i\omega(t - s))d\omega, \tag{8.18}$$

of convolution type. The operator link then is

$$\widehat{\varepsilon}\psi(x, t) = \int_{-\infty}^{\infty} \tilde{\varepsilon}(t - s)\psi(x, s)ds, \tag{8.19}$$

or, as a shorthand

$$D(x, t) = \hat{\varepsilon} E(x, t), \tag{8.20}$$

that corresponds the material relation for the origin (8.11). Manipulating similarly with E and magnetic fields, we write the direct and inverse integral operators:

$$\begin{aligned} E(x, t) &= \int_{-\infty}^{\infty} \tilde{e}(t - s)D(x, s)ds, \\ B(x, t) &= \int_{-\infty}^{\infty} \tilde{\mu}(t - s)H(x, s)ds = \hat{\mu} H(x, t), \end{aligned} \tag{8.21}$$

$$H(x, t) = \int_{-\infty}^{\infty} \tilde{m}(t - s)\mathcal{B}(x, s)ds = \hat{\mu}^{-1}B(x, t), \tag{8.22}$$

with corresponding kernels

$$\begin{aligned} \tilde{e}(t - s) &= \frac{1}{2\pi}\int_{-\infty}^{\infty} \varepsilon^{-1}(\omega)\exp(i\omega(t - s))d\omega, \\ \tilde{\mu}(t - s) &= \frac{\mu_0}{2\pi}\int_{-\infty}^{\infty} \mu(\omega)\exp(i\omega(t - s))d\omega, \\ \tilde{m}(t - s) &= \frac{1}{2\pi\mu_0}\int_{-\infty}^{\infty} \mu^{-1}(\omega)\exp(i\omega(t - s))d\omega. \end{aligned} \tag{8.23}$$

The formulas as (8.22) define the fields in the time domain, using the conventional continuation of the fields to the half space $t < 0$, for which the causality condition holds [28].

8.2.2 Boundary regime problem

The original system (8.6), closed by the operator material relations (8.20) and (8.21) is formulated as:

$$\begin{aligned} \frac{\partial B}{\partial t} &= -\frac{\partial E}{\partial x}, \\ \frac{\partial \hat{\varepsilon} E}{\partial t} &= -\frac{\partial \hat{\mu}^{-1} B}{\partial x}. \end{aligned} \tag{8.24}$$

The action of the operators $\hat{\varepsilon}$ and $\hat{\mu}^{-1}$ is determined by equations (8.19) and (8.22). The full statement of boundary problem includes the boundary regime conditions:

$$E(0, t) = j(t),\ B(0, t) = k(t), \tag{8.25}$$

that guarantees the uniqueness of a solution, for arbitrary functions $j(t)$ and $k(t)$, continued to the half space $t < 0$ as antisymmetric:

$$j(-t) = -j(t),\ k(-t) = -k(t). \tag{8.26}$$

8.3 Projecting operators

8.3.1 Projecting operators approach

In our book we use the projecting operators approach, originating from [22]. That's a general tool of theoretical physics to split evolution system to a set of equations of the first order in time that naturally includes equations for unidirectional waves or non-wave perturbations corresponding to elementary roots of dispersion equation [23]. It is based on a complete set of projecting operators, each for a dispersion relation root that fixes the corresponding subspace of a linearized fundamental system such as Maxwell's equations. The method, compared to the one used in [18, 19, 26], allows us to combine *equations* of the complex basic system in an algorithmic way with dispersion, dissipation and, after some development, a nonlinearity taken into account and also, introduces combined (hybrid) fields as basic modes. It therefore allows us to formulate effectively a corresponding mathematical problem: initial or boundary conditions in an appropriate physical language in a mathematically correct form.

As part of this method we have a transition to new variables, e.g. of the form

$$\psi^{\pm} = \varepsilon\frac{1}{2}E_i \pm \mu\frac{1}{2}H_j,$$

as did Fleck [18], Kinsler [19, 26] and Amiranashvili [20] in their works. This part is in a sense similar to the projection operator method, use of which we demonstrate here.

In this section we'll apply the mentioned method of projecting operators to the problem of wave propagation in 1D-metamaterial with dispersion of both material relations, determined by $\varepsilon(\omega)$ and $\mu(\omega)$. The main aim of this section is very similar to the recent [26] and [21], in that we do want to derive an evolution equation for the mentioned conditions with the minimal simplifications. The methodical differences and results are highlighted and discussed.

8.3.2 Projecting operators construction

We begin with the system of equations (8.24), using the transformations (8.13) and (8.14)

$$\begin{aligned}&\frac{\varepsilon_0}{\sqrt{2\pi}}\frac{\partial}{\partial t}\left(\int_{-\infty}^{\infty}\varepsilon(\omega)\mathcal{E}(x,\,\omega)\exp(i\omega t)d\omega\right)\\&=-\frac{1}{\mu_0\sqrt{2\pi}}\frac{\partial}{\partial x}\left(\int_{-\infty}^{\infty}\frac{\mathcal{B}(x,\,\omega)}{\mu(\omega)}\exp(i\omega t)d\omega\right).\end{aligned} \tag{8.27}$$

Its inverse Fourier transformation origin of the first equation of (8.24) looks as

$$\frac{\partial\mathcal{B}}{\partial x} = -i\omega\mu_0\varepsilon_0\mu(\omega)\varepsilon(\omega)\mathcal{E}. \tag{8.28}$$

Similarly, Fourier transform of the Faraday equation (8.4) gives:

$$\frac{\partial \mathcal{E}}{\partial x} = -i\omega \mathcal{B}. \tag{8.29}$$

By definition $c^2 = \frac{1}{\varepsilon_0 \mu_0}$, where c is the velocity of light in vacuum. Put the notation:

$$\mu_0 \varepsilon_0 \varepsilon(\omega)\mu(\omega) \equiv c^{-2}\varepsilon(\omega)\mu(\omega) \equiv a^2(\omega). \tag{8.30}$$

The system (8.28) and (8.29) simplifies as:

$$\frac{\partial \mathcal{B}}{\partial x} = -i\omega a^2(\omega)\mathcal{E}, \tag{8.31}$$

$$\frac{\partial \mathcal{E}}{\partial x} = -i\omega \mathcal{B}. \tag{8.32}$$

Conventionally, glance at [23], we write the system (8.24) in the matrix form, introducing the matrices: column $\tilde{\Psi}$:

$$\tilde{\Psi} = \begin{pmatrix} \mathcal{B} \\ \mathcal{E} \end{pmatrix}, \tag{8.33}$$

and 2×2 matrix $\mathcal{L}$ as evolution operator along x:

$$\mathcal{L} = \begin{pmatrix} 0 & -i\omega a^2(\omega) \\ -i\omega & 0 \end{pmatrix}. \tag{8.34}$$

The matrix form of the evolution equations (8.28) and (8.29) in ω-representation is:

$$\frac{\partial \tilde{\Psi}}{\partial x} = \mathcal{L}\tilde{\Psi}. \tag{8.35}$$

Equation (8.35) is a system of ordinary differential equations with constant coefficients that have exponential-type solutions. Following the technique described in [23], we arrive at a 2×2 eigenvalue problem. Let us look for such matrices $P^{(i)}$, $i = \overline{1,2}$ that $P^{(i)}\Psi = \Psi_i$ that are eigenvectors of the evolution matrix $\mathcal{L}$, defined by equation (8.34). We also use the standard properties of orthogonal projecting operators:

$$\begin{aligned} P^{(1)}P^{(2)} &= 0,\ i \neq j, \\ P^{(i)} \cdot P^{(i)} &= P^{(i)}, \\ P^{(1)} + P^{(2)} &= I. \end{aligned} \tag{8.36}$$

By the algorithm we use, (see exercises 1 and 3 of section 20.3) the matrices $P^{(i)}$ in the frequency domain have the form:

$$P^{(1)}(\omega) = \frac{1}{2}\begin{pmatrix} 1 & -a(\omega) \\ -\dfrac{1}{a(\omega)} & 1 \end{pmatrix}, \tag{8.37}$$

$$P^{(2)}(\omega) = \frac{1}{2}\begin{pmatrix} 1 & a(\omega) \\ \dfrac{1}{a(\omega)} & 1 \end{pmatrix}. \tag{8.38}$$

Using the inverse Fourier transformation for a matrix $\widehat{\mathbf{P}}^{(i)} = \mathcal{F}P^{(i)}\mathcal{F}^{-1}$, where $\mathcal{F}$ is the operator of the Fourier transformation, leads to projectors in t-representation:

$$\widehat{\mathbf{P}}^{(1,2)}(t) = \frac{1}{2}\begin{pmatrix} 1 & \widehat{P}_{12}^{(1,2)} \\ \widehat{P}_{21}^{(1,2)} & 1 \end{pmatrix}, \tag{8.39}$$

The diagonal elements of projectors have δ-function kernel, i.e. they are proportional to identity:

$$\begin{aligned} \widehat{P}_{11}^{(1,2)}\xi(x, t) &= \frac{1}{4\pi}\int_{-\infty}^{\infty} \exp(i\omega t - i\omega\tau)\xi(x, \tau)d\omega \\ &= \frac{1}{2}\int_{-\infty}^{\infty} \delta(t - \tau)\xi(x, \tau)d\tau = \frac{1}{2}\xi(x, t), \\ \widehat{P}_{22}^{(1,2)}\eta(t) &= \frac{1}{4\pi}\int_{-\infty}^{\infty} \exp(i\omega t - i\omega\tau)\eta(x, \tau)d\omega \\ &= \frac{1}{2}\int_{-\infty}^{\infty} \delta(t - \tau)\eta(x, \tau)d\tau = \frac{1}{2}\eta(x, t). \end{aligned} \tag{8.40}$$

Non-diagonal elements of the matrix (8.37) act as integral operators:

$$\widehat{P}_{12}^{(1,2)}\eta(x, t) = \int_{-\infty}^{\infty} p_{12}^{(1,2)}(t, \tau)\eta(x, \tau)d\tau, \tag{8.41}$$

$$\widehat{P}_{21}^{(1,2)}\xi(x, t) = \int_{-\infty}^{\infty} p_{21}^{(1,2)}(t, \tau)\xi(x, \tau)d\tau, \tag{8.42}$$

with the kernels

$$p_{12}^{(1,2)}(t, \tau) = \mp\frac{1}{4\pi}\int_{-\infty}^{\infty} a(\omega)\exp(i\omega t - i\omega\tau)d\omega, \tag{8.43}$$

$$p_{21}^{(1,2)}(t, \tau) = \mp\frac{1}{4\pi}\int_{-\infty}^{\infty} \frac{1}{a(\omega)}\exp(i\omega t - i\omega\tau)d\omega. \tag{8.44}$$

As alternative notation, we define them via the operators $\widehat{a}$ and its inverse:

$$\begin{aligned} \widehat{a}\eta(x, t) &= \frac{1}{2\pi}\int_{-\infty}^{\infty}\left[\eta(x, \tau)\int_{-\infty}^{\infty} a(\omega)\exp(i\omega(t - \tau))d\omega\right]d\tau, \\ \widehat{a}^{-1}\xi(x, t) &= \frac{1}{2\pi}\int_{-\infty}^{\infty}\left[\xi(x, \tau)\int_{-\infty}^{\infty} \frac{1}{a(\omega)}\exp(i\omega(t - \tau))d\omega\right]d\tau, \end{aligned} \tag{8.45}$$

hence its action is equivalent to

$$\widehat{P}_{12}^{(1,2)}\eta(x,\,t) = \mp\widehat{a}\eta(x,\,t), \tag{8.46}$$

$$\widehat{P}_{21}^{(1,2)}\xi(x,\,t) = \mp\widehat{a}^{-1}\xi(x,\,t). \tag{8.47}$$

8.4 Separated equations and definition for left and right waves

Let us introduce the shorthands:

$$\partial_t \equiv \frac{\partial}{\partial t},\ \partial_x \equiv \frac{\partial}{\partial x}. \tag{8.48}$$

In t-representation, the matrix operator equation (8.35) takes the form:

$$\partial_x\Psi = \widehat{L}\Psi, \tag{8.49}$$

$$\widehat{L} = \begin{pmatrix} 0 & -\partial_t\widehat{a^2} \\ -\partial_t & 0 \end{pmatrix}, \tag{8.50}$$

where

$$\Psi = \begin{pmatrix} \Psi_1 \\ \Psi_2 \end{pmatrix}. \tag{8.51}$$

One can check that the operator $\widehat{a^2}$, defined as:

$$\widehat{a^2}\psi(x,\,t) = \frac{1}{2\pi}\int_{-\infty}^{\infty} a^2(\omega)\exp(i\omega t - i\omega\tau)\psi(x,\,\tau)d\omega d\tau, \tag{8.52}$$

acts as square of $\widehat{a}$, defined by equation (8.45).

Making the similar calculations, we can find that $\widehat{a}^2$ is expressed as a product of the operators in equation (8.22)

$$\widehat{a}^2 = \widehat{\varepsilon}\widehat{\mu},$$

that commute

$$\widehat{\varepsilon}\widehat{\mu}\psi(x,\,t) = \widehat{\mu}\widehat{\varepsilon}\psi(x,\,t). \tag{8.53}$$

We note, that this relation is true only if operators $\widehat{\varepsilon}$ and $\widehat{\mu}$ are convolution type integrals. We also prove the commutation of operators ∂_x and $\widehat{a}^2$ for the further operations.

Acting the operator $\widehat{\mathbf{P}}^{(1)}$ defined by equation (8.39) on the equation (8.49) we find:

$$\widehat{\mathbf{P}}^{(1)}\partial_x\Psi = \widehat{\mathbf{P}}^{(1)}\widehat{L}\tilde{\Psi}. \tag{8.54}$$

We can commute $\widehat{\mathbf{P}}^{(1)}$ and ∂_x, because projectors doesn't depend on x. Using also the proven relations, write

$$\partial_x \widehat{\mathbf{P}}^{(1)}(t)\Psi = \widehat{\mathbf{P}}^{(1)}(t)\widehat{L}\Psi = \widehat{L}\widehat{\mathbf{P}}^{(1)}(t)\Psi. \tag{8.55}$$

After substituting Ψ and $\widehat{L}$ (8.51, 8.50) and $\widehat{\mathbf{P}}^{(1)}$ (8.39) we find:

$$\partial_x \begin{pmatrix} \frac{1}{2}\Psi_1 + \frac{1}{2}\widehat{a}\Psi_2 \\ \frac{1}{2}\widehat{a}^{-1}\Psi_1 + \frac{1}{2}\Psi_2 \end{pmatrix} = \begin{pmatrix} -\frac{1}{2}\widehat{a}\,\partial_t\Psi_1 - \frac{1}{2}\widehat{a}^2\partial_t\Psi_2 \\ -\frac{1}{2}\partial_t\Psi_1 - \frac{1}{2}\widehat{a}^{-1}\widehat{a}^2\partial_t\Psi_2 \end{pmatrix}. \tag{8.56}$$

Conventionally projecting the vector Ψ of (8.51), we introduce new variables Π and Λ as the operator combination of fields components $\Psi_1 = B$, $\Psi_2 = E$:

$$\Lambda \equiv \frac{1}{2}(B - \widehat{a}E), \tag{8.57}$$

as well as

$$\Pi \equiv \frac{1}{2}(B + \widehat{a}E). \tag{8.58}$$

From equation (8.56) we get two equations that determine evolution with respect to x of the boundary regime (8.25):

$$\begin{aligned} \partial_x\Pi(x, t) &= -\widehat{a}\,\partial_t\Pi, \\ \partial_x\Lambda(x, t) &= \widehat{a}\,\partial_t\Lambda. \end{aligned} \tag{8.59}$$

These variables are left (Λ) and right (Π) waves variables, as follows from the equation (8.59) structure. When a dispersion is absent, a is the velocity of waves propagation.

Using relations (8.58) and (8.57) from equation (8.25) we derive boundary regime conditions for left and right waves:

$$\begin{aligned} \Lambda(0, t) &= \frac{1}{2}(B(0, t) - \widehat{a}E(0, t)) = \frac{1}{2}(k(t) - \widehat{a}j(t)), \\ \Pi(0, t) &= \frac{1}{2}(B(0, t) + \widehat{a}E(0, t)) = \frac{1}{2}(k(t) + \widehat{a}j(t)). \end{aligned} \tag{8.60}$$

As the result, we have the complete statement of problem that is the system of operator equations for time-domain dispersion of left and right waves in the linear case (8.59) with the boundary condition (8.60).

8.5 Nonlinearity account

Let us consider a nonlinear problem. We start again from Maxwell's equations (8.6) with material relations, that contain now the nonlinear terms:

$$D = \hat{\varepsilon} E + P_{NL},$$
$$B = \hat{\mu} H + M_{NL}, \tag{8.61}$$

P_{NL}—nonlinear part of polarization, M_{NL}—the part for magnetization. The linear parts of polarization and magnetization already have been taken into account by the term switch operators $\hat{\varepsilon}, \hat{\mu}$. In the time-domain, a closed nonlinear version of equation (8.24) is:

$$\partial_t(\hat{\varepsilon} E) + \partial_t P_{NL} = - \partial_x \hat{\mu}^{-1} B - \partial_x \hat{\mu}^{-1} M_{NL},$$
$$\frac{\partial B}{\partial t} = - \frac{\partial E}{\partial x}. \tag{8.62}$$

Action of operator $\hat{\mu}$ on the first equation of system (8.62) and using the same notations Ψ and $\hat{L}$ from equations (8.50) and (8.51) once more, we obtain an analogue of the matrix equation (8.49) with nonlinear terms:

$$\partial_x \Psi - \hat{L} \Psi = -\begin{pmatrix} \partial_x M_{NL} \\ 0 \end{pmatrix} - \begin{pmatrix} \hat{\mu}\, \partial_t P_{NL} \\ 0 \end{pmatrix}. \tag{8.63}$$

The vector at the rhs of the Ampère equation (8.62)

$$\mathbb{N}(E, B) = \begin{pmatrix} \partial_x M_{NL} + \partial_t \hat{\mu}\, P_{NL} \\ 0 \end{pmatrix} \tag{8.64}$$

contains all possible nonlinear terms of polarization and magnetization origin: In such notations the matrix equation (8.63) gets the following form:

$$\partial_x \Psi - \hat{L} \Psi = -\mathbb{N}(E, B). \tag{8.65}$$

Next, acting by the operators $\hat{\mathbf{P}}^{(1)}$ (8.39) and $\hat{\mathbf{P}}^{(2)}$ on equation (8.65) we derive the system projected to the left/right waves subspaces:

$$\partial_x \Pi + \hat{a}\, \partial_t \Pi = \mathbb{N}_1(\hat{a}^{-1}(\Pi - \Lambda), \Pi + \Lambda),$$
$$\partial_x \Lambda - \hat{a}\, \partial_t \Lambda = - \mathbb{N}_1(\hat{a}^{-1}(\Pi - \Lambda), \Pi + \Lambda), \tag{8.66}$$

where

$$\mathbb{N}_1(E, B) \equiv \frac{1}{2}(\partial_x M_{NL} + \partial_t \hat{\mu}\, P_{NL}). \tag{8.67}$$

We have obtained an important result, the system of equations of left and right wave interaction that arise from the account of *arbitrary nonlinearity with general temporal dispersion.*

8.6 Wave propagation in a metamaterial within the lossless Drude dispersion and Kerr nonlinearity

8.6.1 Drude model for dispersion

The Drude model represents a situation that is typical for plasma, or a good conductor, see section 6.3.2. For such a case we use relations from [31], accounting for equation (8.3):

$$\varepsilon(\omega) = \left(1 - \frac{\omega_{pe}^2}{\omega^2}\right), \tag{8.68}$$

$$\mu(\omega) = \left(1 - \frac{\omega_{pm}^2}{\omega^2}\right), \tag{8.69}$$

where ω_{pe} and ω_{pm}—parameters, that are dependent on the density, charge, and mass of the prevailing charge carrier. These parameters are named as the electric and magnetic plasma frequencies [31]. Many sources, e.g. [16, 17], address such a model to describe the material relations of a metamaterial.

The relations (8.68) and (8.69) give positive values of energy density (8.1) so, that

$$\begin{aligned} W &= \varepsilon_0 \frac{d(\omega\varepsilon(\omega))}{d\omega} E^2 + \mu_0 \frac{d(\omega\mu(\omega))}{d\omega} \\ W &= \varepsilon_0 \left(1 + \frac{\omega_{pe}^2}{\omega^2}\right) E^2 + \mu_0 \left(\frac{\omega_{pm}^2}{\omega^2} + 1\right) H^2 > 0. \end{aligned} \tag{8.70}$$

The kernel $a(\omega)$ of the integral operator $\hat{a}$ is:

$$a(\omega) = c^{-1} \sqrt{\left(1 - \frac{\omega_{pe}^2}{\omega^2}\right)\left(1 - \frac{\omega_{pm}^2}{\omega^2}\right)}. \tag{8.71}$$

After expansion $a(\omega)$ in the Taylor series in condition $\omega \ll \omega_{pe}, \omega_{pm}$ in the vicinity of $\omega = 0$, one gets:

$$\hat{a}\eta(t) \approx c^{-1}\left[\omega_{pe}\omega_{pm}\partial_t^{-2} - \frac{1}{2}\frac{\omega_{pe}^2 + \omega_{pm}^2}{\omega_{pe}\omega_{pm}} + \left(\frac{1}{2\omega_{pe}\omega_{pm}} + \frac{1}{8}\frac{(-\omega_{pe}^2 - \omega_{pm}^2)^2}{\omega_{pe}^3\omega_{pm}^3}\right)\partial_t^2\right]\eta(t). \tag{8.72}$$

The operator of integration acts as

$$\partial_\alpha^{-1} f(\alpha) = \int_0^\alpha f(\beta)d\beta. \tag{8.73}$$

From figure 8.1 it is visible, that in the range of frequencies $\omega < 0.5\omega_{pe}$, we may restrict the expansion by the first term within the 0.005% relative error. The relative error for frequencies till $0.9\omega_{pe}$ is less than 10% as is seen in figure 8.1.

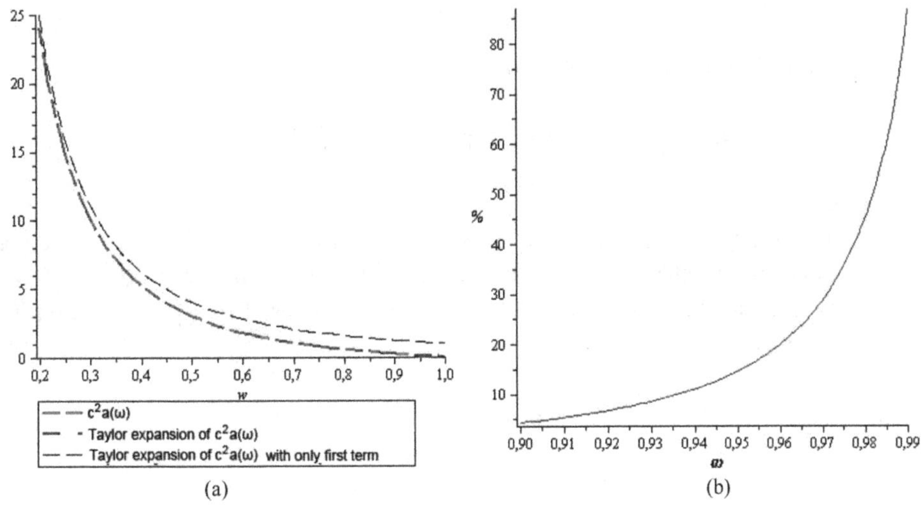

Figure 8.1. Taylor expansion (8.72) and exact form (8.71) at range $[0.2, 1.0]\omega_{pe}$ (a) and relative error (in percent) of Taylor expansion at the interval $[0.9, 0.99]\omega_{pe}$ (b), $\omega_{pm} = 1.2\omega_{pe}$.

A simplification $\omega_{pm} = \omega_{pe}$ we go to:

$$ca(\omega) = \sqrt{\left(1 - \frac{\omega_{pe}^2}{\omega^2}\right)\left(1 - \frac{\omega_{pe}^2}{\omega^2}\right)} = \sqrt{\left(1 - \frac{\omega_{pe}^2}{\omega^2}\right)^2} = \left(1 - \frac{\omega_{pe}^2}{\omega^2}\right), \tag{8.74}$$

that already has algebraic form, that allows to take approximation by differential operators.

Further, we mark ω_{pe} as p, and ω_{pm} as q for compactness. We leave the only term in the relation (8.72), in the frame of all estimations. Hence, this minimal version of equation (8.72) we plug in the system (8.59) to obtain the system:

$$\begin{aligned} \partial_x \Pi &= -\, c^{-1} pq \partial_t^{-1} \Pi, \\ \partial_x \Lambda &= c^{-1} pq \partial_t^{-1} \Lambda. \end{aligned} \tag{8.75}$$

Differentiating equations of this system by t once more, we write the resulting independent equations for the right and left wave which hybrid fields are completely separated

$$\begin{aligned} \partial_{xt} \Pi &= -\, c^{-1} pq \Pi, \\ \partial_{xt} \Lambda &= c^{-1} pq \Lambda, \end{aligned} \tag{8.76}$$

and describe the waves propagation and dispersion.

8.6.2 Interaction of left and right waves with Kerr effect

For nonlinear Kerr materials [32], the third-order nonlinear part of polarization [21]:

$$P_{NL} = \chi^{(3)} E^3$$

enter equation (8.67), then we find $\mathbb{N}_1$:

$$\mathbb{N}_1 \equiv \frac{1}{2}\hat{\mu}(\hat{\mu}^{-1}\partial_x M_{NL} + \partial_t P_{NL}) = \frac{\chi^{(3)}}{2}\hat{\mu}\,\partial_t E^3, \tag{8.77}$$

for $M_{NL} = 0$.

The effect of negative permeability was demonstrated at the THz range, see e.g. [29]. The integral operator $\hat{\mu}$ for the chosen model is approximated as $\mu_0(1 - q^2\partial_t^{-2})$, we also see, that in the THz range the term $q^2\partial_t^{-2}$ prevails. Finally, from equation (8.66) we derive:

$$\begin{aligned} c\partial_x\Pi - pq\partial_t^{-1}\Pi &= -\frac{\chi^{(3)}}{2}\mu_0 q^2\partial_t^{-1}[\hat{a}^{-1}(\Pi - \Lambda)]^3, \\ c\partial_x\Lambda + pq\partial_t^{-1}\Lambda &= \frac{\chi^{(3)}}{2}\mu_0 q^2\partial_t^{-1}[\hat{a}^{-1}(\Pi - \Lambda)]^3. \end{aligned} \tag{8.78}$$

The operator $\hat{a}^{-1}$ within same approximation is written as:

$$\hat{a}^{-1}\eta(x, t) \approx \frac{c}{pq}\partial_t^2\eta(x, t). \tag{8.79}$$

We differentiate the equations, and substitute it to the system (8.78), in unified notations it is the system:

$$\begin{aligned} c\Pi_{xt} + pq\Pi &= -\frac{\mu_0\chi^{(3)}c^3}{2p^3q}[(\Pi - \Lambda)_{tt}]^3, \\ c\Lambda_{xt} - pq\Lambda &= \frac{\mu_0\chi^{(3)}c^3}{2p^3q}[(\Pi - \Lambda)_{tt}]^3. \end{aligned} \tag{8.80}$$

This system may be marked as the *main result* of this chapter.

With some rescaling and differentiation by time, changing the fields as $\Pi_{tt} = \alpha\pi$, $\Lambda_{tt} = \alpha\lambda$, $x = \beta\zeta$ with the new parameters $\alpha = \sqrt{\frac{2p^4q^2}{\mu_0\chi^3c^3}}$, $\beta = \frac{c}{pq}$ introduction. The new form of the system

$$\begin{aligned} \pi_{\zeta t} + \pi &= -\,\partial_t^2(\pi - \lambda)^3, \\ \lambda_{\zeta t} - \lambda &= \partial_t^2(\pi - \lambda)^3, \end{aligned} \tag{8.81}$$

needs extra boundary conditions for the performed t-derivatives.

Consider the unidirectional case of equation (8.80) with $\lambda = 0$, that corresponds boundary regime specifications from equation (8.60): $(k(t) - \hat{a}j(t)) = 0$. Such neglecting the interaction in the system is valid till the effect of the left wave generation would be noticeable. So, we have

$$\pi_{\zeta t} + \pi = -\partial_t^2\pi^3, \tag{8.82}$$

It is instructive to compare the equation with the Schafer–Wayne equation [24], we find, that left and right waves for ordinary materials change their direction compared to metamaterials.

8.6.3 Stationary problem solutions

Linear case

We introduce the change of variables in the system (8.80):

$$\eta = x,\ \xi = x - vt;\ x = \eta,\ t = \frac{\eta - \xi}{v}, \tag{8.83}$$

v has dimension of velocity. Then the transformations read:

$$\begin{aligned} \partial_t &= -v\partial_\xi,\ \partial_x = \partial_\eta + \partial_\xi, \\ \partial_{xt} &= -v\left[\partial_{\eta\xi} + \partial_\xi^2\right], \\ \Pi(x,\ t) &\to \mathsf{R}(\eta,\ \xi),\ \Lambda(x,\ t) \to \mathsf{L}(\eta,\ \xi) \end{aligned} \tag{8.84}$$

The linear equations in new variables are

$$\begin{aligned} -v\left[\partial_{\eta\xi} + \partial_\xi^2\right]\mathsf{R} &= -pqc^{-1}\mathsf{R}, \\ -v\left[\partial_{\eta\xi} + \partial_\xi^2\right]\mathsf{L} &= pqc^{-1}\mathsf{L}. \end{aligned} \tag{8.85}$$

We can take the only R wave in absence of interaction (nonlinearity). For the stationary wave the independence of R on η we choose as a definition of stationary state:

$$\partial_\eta \mathsf{R} = 0,$$

having the second order ordinary differential equation:

$$c^{-1}pq\mathsf{R} = v\partial_\xi^2\mathsf{R}. \tag{8.86}$$

The dimension of the lhs is the dimension of $\mathsf{k}\omega$:

$$\mathsf{k}\omega = \frac{pq}{c}. \tag{8.87}$$

Also we found v:

$$v = \frac{\omega}{\mathsf{k}}. \tag{8.88}$$

Let's return to equation (8.86) and rewrite it with accounting equations (8.87) and (8.88):

$$\partial_\xi^2\mathsf{R} - \mathsf{k}^2\mathsf{R} = 0. \tag{8.89}$$

For a specified decaying boundary regime the solution is:

$$\mathsf{R} = A\Theta(\xi)\exp\left(\mathsf{k}(x - vt)\right). \tag{8.90}$$

For the L-wave the equation differs only by the sign from equation (8.89):

$$\partial_\xi^2 \mathsf{L} + \mathsf{k}^2 \mathsf{L} = 0, \tag{8.91}$$

that gives the oscillating solution, with the odd function choice at $x = 0$

$$\mathsf{L} = B \sin(\mathsf{k}(x - vt)). \tag{8.92}$$

As we see, the negative value for μ drastically changes the character of propagation of the waves R and L, whose definition is given by equations (8.57) and (8.58).

Nonlinear case

Equation (8.82) in variables (8.83) and in the case of stationary condition

$$c\Pi_{\xi\xi} + pq\Pi = -\alpha[\Pi_{\xi\xi}]^3, \tag{8.93}$$

is nonlinear ODE, where

$$\alpha \equiv \frac{\mu_0 \chi^{(3)} c^3 v^6}{2p^3 q}.$$

The cubic equation with respect to the second derivative $\Pi_{\xi\xi}$: of the field, that may be solved by Cardano formula

$$[\Pi_{\xi\xi}]^3 + \frac{c}{\alpha}\Pi_{\xi\xi} + \frac{pq}{\alpha}\Pi = 0. \tag{8.94}$$

The discriminant of the cubic equation:

$$Q = \left(\frac{c}{3\alpha}\right)^3 + \left(\frac{pq}{2\alpha}\Pi\right)^2 > 0, \tag{8.95}$$

defines solutions of the cubic equation (8.94) as:

$$\begin{aligned}
(\Pi_{\xi\xi})_1 &= \sqrt[3]{-\frac{pq}{2\alpha}\Pi + \sqrt{Q}} + \sqrt[3]{-\frac{pq}{2\alpha}\Pi - \sqrt{Q}},\\
(\Pi_{\xi\xi})_{2,3} &= -\frac{1}{2}\left[\left(-\frac{pq}{2\alpha}\Pi + \sqrt{Q}\right)^{1/3} + \left(-\frac{pq}{2\alpha}\Pi - \sqrt{Q}\right)^{1/3}\right]\\
&\pm \frac{i}{2}\left[\left(-\frac{pq}{2\alpha}\Pi + \sqrt{Q}\right)^{1/3} - \left(-\frac{pq}{2\alpha}\Pi - \sqrt{Q}\right)^{1/3}\right],
\end{aligned} \tag{8.96}$$

We consider only the real root $(\Pi_{\xi\xi})_1$. Its Taylor series, the first two terms look:

$$(\Pi_{\xi\xi})_1 \approx \frac{-pq}{c}\Pi + \frac{p^3 q^3 \alpha}{c^4}\Pi^3 + \cdots, \tag{8.97}$$

accounting only the first term, we return to the linear case:

$$(\Pi_{\xi\xi})_1 = \frac{-pq}{c}\Pi, \tag{8.98}$$

with respect to equation (8.87) we rewrite it as equation (8.89)

$$\Pi_{\xi\xi} = -\mathrm{k}\omega\Pi. \tag{8.99}$$

If we take account of the second term of expansion (8.97), we find

$$\Pi_{\xi\xi} = -\mathrm{k}\omega\Pi + \frac{(\mathrm{k}\omega)^3\alpha}{c}\Pi^3. \tag{8.100}$$

Multiplying all the equation on Π_ξ:

$$\Pi_\xi\Pi_{\xi\xi} = -\mathrm{k}\omega\Pi_\xi\Pi + \frac{(\mathrm{k}\omega)^3\alpha}{c}\Pi^3\Pi_\xi, \tag{8.101}$$

and rearrange as

$$\frac{1}{2}\partial_\xi(\Pi_\xi^2) = -\mathrm{k}\omega\Pi_\xi\Pi + \frac{(\mathrm{k}\omega)^3\alpha}{c}\Pi^3\Pi_\xi, \tag{8.102}$$

after integrating, yields

$$\Pi_\xi^2 + \mathrm{k}\omega\Pi^2 - \frac{(\mathrm{k}\omega)^3\alpha}{2c}\Pi^4 = \mathbb{E}, \tag{8.103}$$

where

$$\mathbb{E} = \mathrm{const.}$$

The classical form for the equation for elliptic functions

$$\Pi_\xi^2 = -\mathrm{k}\omega\Pi^2 + \frac{(\mathrm{k}\omega)^3\alpha}{2c}\Pi^4 + \mathbb{E}, \tag{8.104}$$

allows to proceed, factorizing the rhs as:

$$\begin{aligned}\frac{(\mathrm{k}\omega)^3\alpha}{2c}\Pi^4 - \mathrm{k}\omega\Pi^2 + \mathbb{E} &= \frac{(\mathrm{k}\omega)^3\alpha}{2c}\left[\Pi^4 - \frac{2c}{(\mathrm{k}\omega)^2\alpha}\Pi^2 + \frac{\mathbb{E}2c}{(\mathrm{k}\omega)^3\alpha}\right] \\ &= A(\Pi - P_1)(\Pi - P_2)(\Pi - P_3)(\Pi - P_4),\end{aligned} \tag{8.105}$$

where $A = \frac{(\mathrm{k}\omega)^3\alpha}{2c}$ P_i, $i = \overline{1, 4}$ are roots of the characteristic equation:

$$P^4 - \frac{2c}{(\mathrm{k}\omega)^2\alpha}P^2 + \frac{2c\mathbb{E}}{(\mathrm{k}\omega)^3\alpha} = 0, \tag{8.106}$$

that is biquadratic one. Following the textbook of Davis [33] we introduce new variables:

$$
\begin{aligned}
y^2 &= \frac{(P_2 - P_4)(\Pi - P_1)}{(P_1 - P_4)(\Pi - P_2)},\\
K^2 &= \frac{(P_2 - P_3)(P_1 - P_4)}{(P_1 - P_4)(P_2 - P_4)},\\
M^2 &= \frac{(P_2 - P_4)(P_1 - P_3)}{4}.
\end{aligned}
\tag{8.107}
$$

Hence:

$$
y = \mathrm{sn}(A^{1/2}M\xi,\ K). \tag{8.108}
$$

Substituting to equation (8.107) and solving it with respect to Π: we obtain finally

$$
\Pi = \left(P_2 y^2 \frac{(P_1 - P_4)}{(P_2 - P_4)} - P_1\right)\left(y^2 \frac{(P_1 - P_4)}{(P_2 - P_4)} - 1\right)^{-1}. \tag{8.109}
$$

See exercise 1. Another, more compact and direct approach is given and illustrated in [27]. See also figure 8.2.

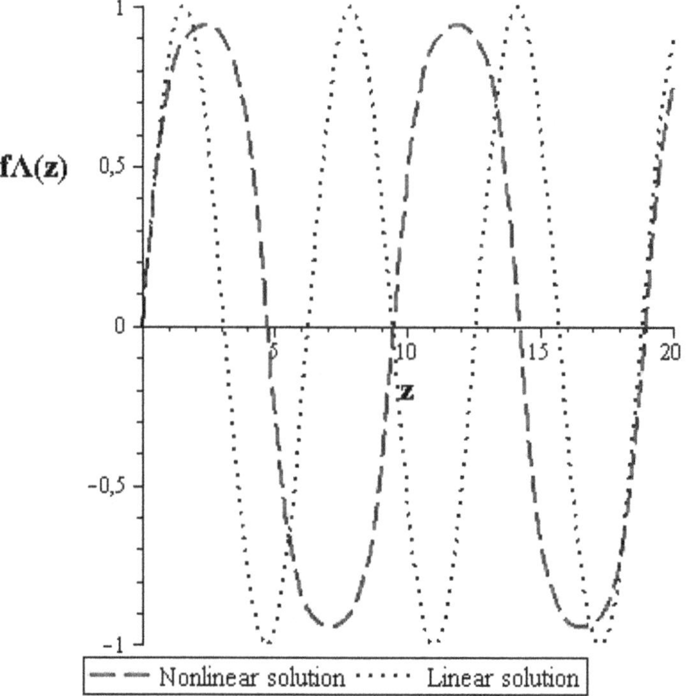

Figure 8.2. Comparison of linear and nonlinear waves, $f^2 = 1/(I\omega v)$, $\omega = 0.8p$, $I = 1$ GHz cm^{-2}. Reprinted from [27], Copyright (2016), with permission from Elsevier.

8.7 Discussion and conclusion

Using the projection operators approach we derived the general system of equations (8.66), that describes interaction between opposite directed waves propagating in 1D-metamaterial, with Kerr nonlinearity [29]. The system is specified for a lossless Drude model as (8.80). The results may be used in experiments that investigate amplitude dependence of reflected waves from a metamaterial layer [30].

Let us compare our results with Kinsler's method of derivation of a wave equation for left and right waves in time domain and scalar form [26]:

$$\partial_x \Pi(x, t) = \partial_t \alpha_r \beta_r \Pi + \frac{1}{2} \alpha_c \beta_r (\Pi - \Lambda), \tag{8.110}$$

where α, β—from expanding ε and μ:

$$\begin{aligned} \varepsilon(\omega) &= \varepsilon_r(\omega) + \varepsilon_c(\omega) = \alpha_r^2(\omega) + \alpha_r(\omega)\alpha_c(\omega), \\ \mu(\omega) &= \mu_r(\omega) + \mu_c(\omega) = \beta_r^2(\omega) + \beta_r(\omega)\beta_c(\omega). \end{aligned} \tag{8.111}$$

Index c is called the correction parameter, which 'represents the discrepancy between the true values and the reference' [26]. Correction parameters are depended only from nonlinearity. Index r indicates 'reference' values, close to the true medium properties, typically by including all the dispersive properties. As we see, the projection operators method introduces left and right waves in a more transparent manner (see also equation (8.110)).

Exercise 1. Calculate the parameters P_i and, next, the parameter M, K, constructing and plot the solution Π with different values of $\mathbb{E}$, in conditions of the paper [27]: $p = 1$ GHz, $q = 0.8p$, $I = 1$ GW cm^{-2} = 10^{13} W m^{-2}, $\chi^{(3)} = 10^{-20}$ m^2 V^{-2} for two cases: (a) $\omega = 0.8p$, (b) $\omega = 50p$.

Exercise 2. Solve and plot the linear equations (8.85) with Gaussian boundary conditions.

References

[1] Bose J C 1898 On the rotation of plane of polarisation of electric waves by a twisted structure *Proc. R. Soc.* **63** 146–52

[2] Kock W E 1948 Metallic delay lenses *Bell Syst. Tech. J.* **27** 58–82

[3] Brown J 1960 Artificial dielectrics *Progress in Dielectrics* vol 2 ed J B Birks (New York: Wiley) pp 193–225

[4] Veselago V G 1968 The electrodynamics of substances with simultaneously negative values of ε and μ *Sov. Phys. Usp.* **10** 509–14

[5] Malyuzhinets G D 1951 A note on the radiation principle *Z. Tech. Fiz.* **21** 940–2 (in Russian. English translation in Sov. Phys. Tech. Phys.)

[6] Sihvola A 2007 Metamaterials in electromagnetics *Metamaterials* **1** 2–11

[7] Smith D R, Padilla W J, Vier D C, Nemat-Nasser S C and Schultz S 2000 Composite medium with simultaneously negative permeability and permittivity *Phys. Rev. Lett.* **84** 4184–7

[8] Shelby R A, Smith D R and Schultz S 2001 Experimental verification of a negative index of refraction *Science* **292** 77–9
[9] Lipworth G, Mrozack A, Hunt J, Marks D L, Driscoll T, Brady D and Smith D R 2013 Metamaterial apertures for coherent computational imaging on the physical layer *J. Opt. Soc. Am.* A **30** 1603–12
[10] Ziolkowski R W and Kipple A 2003 Causality and double-negative metamaterials *Phys. Rev.* E **68** 026615
[11] Milonni P W 2005 *Fast Light, Slow Light and Left-Handed Light* (Philadelphia, PA: Institute of Physics Publishing) pp 184–5
[12] Wen S, Xiang Y, Dai X, Tang Z, Su W and Fan D 2007 Theoretical models for ultrashort electromagnetic pulse propagation in nonlinear metamaterials *Phys. Rev.* A **75** 033815
[13] Rotman W 1962 Plasma simulation by artificial dielectrics and parallel plate media *IRE Trans. Antennas Propag.* **10** 82–95
[14] Ziolkowski R W and Auzanneau F 1997 Passive artificial molecule realizations of dielectric materials *J. Appl. Phys.* **82** 3195–8
[15] Pendry J B, Holden A J, Stewart W J and Youngs I 1996 Extremely low frequency plasmons in metallic mesostructures *Phys. Rev. Lett.* **76** 4773–6
[16] Ziolkowski R 2003 Pulsed and CW Gaussian beam interactions with double negative metamaterial slabs *Opt. Soc. Am.* **11** 662–81
[17] Zhao Y, Argyropoulos C and Hao Y 2008 Full-wave finite-difference time-domain simulation of electromagnetic cloaking structures *Opt. Soc. Am.* **16** 6717–30
[18] Fleck J A 1970 Ultrashort-pulse generation by q-switched lasers *Phys. Rev.* B **1** 84
[19] New G H C, Kinsler P and Radnor B P 2005 Theory of direction pulse propagation *Phys. Rev.* A **72** 063807
[20] Amiranashvili S and Demircan A 2011 Ultrashort optical pulse propagation in terms of analytic signal *Adv. Opt. Technol.* **2011** 989515
[21] Kuszner M and Leble S 2014 Ultrashort opposite directed pulses dynamics with Kerr effect and polarization account *J. Phys. Soc. Jpn* **83** 034005
[22] Leble S 1990 *Nonlinear Waves in Waveguides* (Heidelberg: Springer)
[23] Leble S and Perelomova A 2018 *Dynamical Projectors Method in Hydro- and Electrodynamics* (Boca Raton, FL: CRC Press)
[24] Chung Y, Jones C K R T, Schäfer T and Wayne C E 2005 Ultra-short pulses in linear and nonlinear media *Nonlinearity* **18** 1351–74
[25] Kuszner M and Leble S 2011 Directed electromagnetic pulse dynamics: projecting operators method *J. Phys. Soc. Jpn* **80** 024002
[26] Kinsler P 2010 *Phys. Rev.* A **81** 023808
[27] Ampilogov D and Leble S 2016 Directed electromagnetic wave propagation in 1D metamaterial: projecting operators method *Phys. Lett.* A **380** 2271–8
[28] Boyd R W 1992 *Nonlinear Optics* (Boston, MA: Academic)
[29] Fedotov V 2017 Metamaterials *Springer Handbook of Electronic and Photonic Materials* ed S Kasap and P Capper (Cham: Springer)
[30] Pravdin K V and Popov I Y 2015 Layered system with metamaterials *J. Phys.: Conf. Ser.* **661** 012025
[31] Wen S, Xiang Y, Dai X, Tang Z, Su W and Fan D 2007 Theoretical models for ultrashort electromagnetic pulse propagation in nonlinear metamaterials *Phys. Rev.* A **75** 033815

[32] Argyropoulos C *et al* 2013 Negative refraction, gain and nonlinear effects in hyperbolic metamaterials *Opt. Express* **21** 15037–47
Argyropoulos C *et al* 2012 Enhanced nonlinear effects in metamaterials and plasmonics *Adv. Electromagn.* **1** 46–51
[33] Davis H T 1962 *Introduction to Nonlinear Differential and Integral Equations* (New York: Dover)

IOP Publishing

Chapter 9

Problems of electromagnetism in a piecewise continuous matter

9.1 Electro- and magneto-statics

There are known problems in physics that ignore 'dynamics', dependence of the medium and field variables on time. We do not concentrate on such problems because our main direction of investigation is electrodynamics, but mention them, in the context of class of adjacent problems, that may be a limiting case or a basis of the time-dependent fields description. The Maxwell equations for time-independent fields and free charges density ρ and free current density $\vec{j}_{sw}$ distribution may be split into two groups, the first is used for the electrostatics:

$$\nabla \cdot \vec{D} = 4\pi\rho, \tag{9.1}$$

$$\nabla \times \vec{E} = 0. \tag{9.2}$$

While for the magneto-statics we write:

$$\nabla \cdot \vec{B} = 0, \tag{9.3}$$

$$\text{rot}\vec{H} = \frac{4\pi}{c}\vec{j}_{sw}, \tag{9.4}$$

It is seen that the first group of equations contains only electric field, while the second one only magnetic field. Such problems are generally formulated for bodies, restricted in space, hence with boundaries. A transition through a boundary needs the formulation of a kind of the fields continuity conditions, based on a solid fundamental relation. It is conventionally the Maxwell's equations in integral form, i.e. such that is naturally chosen because of the absence of derivatives in its formulations, that are poorly-defined at matter discontinuities, i.e. boundaries.

There are many excellent descriptions of the static problems in textbooks [1; 2], we send a reader to these books, here we only touch some questions of magneto-statics and magneto-dynamics.

9.2 Boundary conditions

9.2.1 Absence of surface charges and currents case

If the medium is not continuous at some surfaces, we should use boundary conditions on these borders. One can derive the conditions using equation (9.1), but it is more convenient to use the integral form of Maxwell equations of section 5.1.3. Let us divide the medium to sub-volumes choosing one of them as a principal, denoting it as the V with the corresponding boundary surface. We put for simplicity zero charge and current in a vicinity of a point under consideration $\vec{r}_0$, so we take the first (Coulomb) one

$$\int_S \vec{D}\, d\vec{S} = 0, \tag{9.5}$$

and apply it to a thin (of small height h) cylindric volume with the center at $\vec{r}_0$ as it is shown in the figure 9.1. The integral naturally splits into three ones

$$\int = \int_{S_w} + \int_{S_z} + \int_{S_b} = 0,$$

over the internal, external and lateral sides of the cylinder surface. We, as before, suppose that the inhomogeneity scale of the fields $\vec{D}$, ... is much bigger than the cylinder base diameter. We also suppose that the normal unit vector $\vec{n}$ of this closed surface is directed out, as at the figure. Therefore we can write approximately

$$\int_{S_z} \vec{D}\, d\vec{S} = D_{nz} S, \quad \int_{S_w} \vec{D}\, d\vec{S} = -D_{nw} S,$$

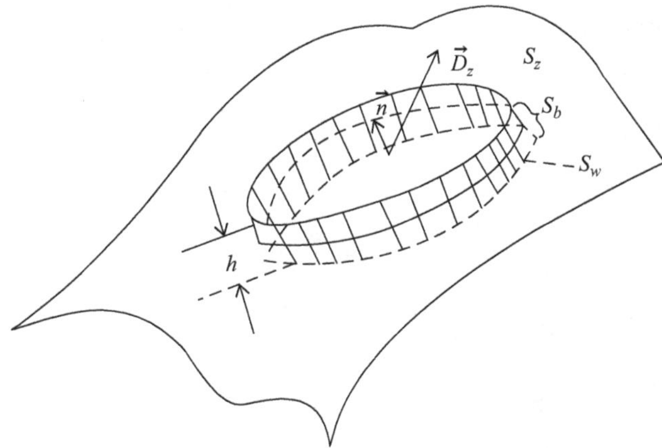

Figure 9.1. Cylinder, as a volume for equation (9.5) integration, bases of the surface S are S_z, outside the volume and S_w inside. The lateral surface marked as S_b.

where $D_{nz} = (\vec{D}_z \cdot \vec{n})$ and $D_{nw} = (\vec{D}_w \cdot \vec{n})$ are the normal projections of the vector $\vec{D}$ values in the vicinity of the central base points of the cylinders inside and outside the matter we consider. Next we tend to zero the height h of the cylinder that yields

$$0 = \int \vec{D}\, d\vec{S} \rightarrow (D_{nz} - D_{nw})S,$$

that finally gives the condition of continuity of the normal component of the vector $\vec{D}$:

$$D_{nz} - D_{nw} = 0. \tag{9.6}$$

Quite similar from the second Maxwell equation (5.12b)

$$\int \vec{B} \cdot d\vec{S} = 0,$$

the continuity of the normal component of the magnetic induction vector

$$B_{nz} - B_{nw} = 0, \tag{9.7}$$

is derived.

The next boundary condition is obtained if one applies the Faraday equation (5.12c)

$$\int_L \vec{E} \cdot d\vec{l} + \frac{1}{c}\int_S \frac{\partial \vec{B}}{\partial t} \cdot d\vec{S} = 0,$$

to the rectangular contour L of the normal dimension a, that is shared between the domains V_w and V_z, divided by the surface S as at figure 9.2. The first integral is divided as:

$$\int_L = \int_{L_w} + \int_{L_z} + \int_a, \tag{9.8}$$

the choice of the contour position allows to scratch its width $a \rightarrow 0$ so that the sum of integrals by the vertical sides $\int_a$ of the rectangle tends to zero.

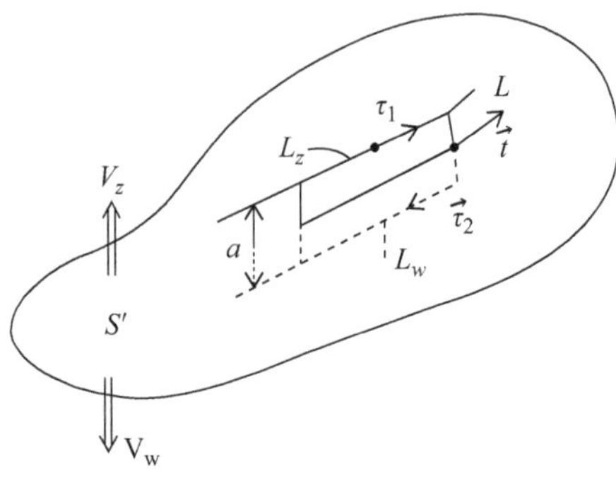

Figure 9.2. Contour L.

$$\int_S \frac{\partial \vec{B}}{\partial t} \cdot d\vec{S} \to 0$$

and, marking the $L_{z,w}$ length as L, the whole equation results in

$$\int_{L_w} \vec{E}\, d\vec{l} + \int_{L_z} \vec{E}\, d\vec{l} = E_{tw}L - E_{tz}L = 0,$$

relying on the approximation of the integrals (13.49) in the Faraday equation (5.12c). The vector identity $\vec{\tau_1} \cdot \vec{t} = -\vec{\tau_2} \cdot \vec{t}$ for the projections on the tangent directions $\vec{\tau}_{1,2}$ is used, so, that $\vec{E} \cdot \vec{t} = E_t$. Then go to the limit, so that the transverse dimension a of the circuit in figure 9.2 was much smaller than the longitudinal (tangential).

Finally

$$E_{tw} - E_{tz} = 0, \tag{9.9}$$

that means the continuity of the tangential components of the vector $\vec{E}$. It should be mentioned that the tangential component of a vector is two-dimensional. After a choice of convenient basic vectors $\vec{t_1}$, $\vec{t_2}$ it reads as

$$E_{twi} - E_{tzi} = 0, i = 1, 2. \tag{9.10}$$

Now let us add the last (Ampère–Maxwell equation)

$$\int_L \vec{H} \cdot d\vec{L} - \frac{1}{c} \int_S \frac{\partial \vec{D}}{\partial t} \cdot d\vec{S} = \frac{4\pi}{c} I_{sw}, \tag{9.11a}$$

and apply it to the same contour, neglecting surface currents, that yields

$$H_{twi} - H_{tzi} = 0, i = 1, 2, \tag{9.12}$$

in quite a similar way.

Remark. The boundary conditions are directly generalized to account for the surface charge and current density. It is enough to leave the correspondent term at the Coulomb and Ampère–Maxwell equations, see section 9.2.4.

9.2.2 Conditions for magnetic and electric moments

In problems of magnetic moment evolution, such as propagation of domain walls (DW) in wires or thin layers, a mathematical statement of problem includes boundary conditions for magnetic moment density. Let the boundary we consider, divide media with different magnetic properties, for example magnetics and dielectrics.

We introduce the magnetic moment $\vec{M}$ as

$$\vec{B} = \vec{H} + 4\pi\vec{M}. \tag{9.13}$$

The continuity of the normal component of the vector $\vec{B}$ yields

$$H_{nw} + 4\pi M_{nw} = H_{nz} + 4\pi M_{nz}. \tag{9.14}$$

Suppose the magnetic moment of the medium, that is labeled by z, is zero, then

$$H_{nw} + 4\pi M_{nw} = H_{nz}. \tag{9.15}$$

Similarly, from the continuity of the tangent component of the vector $\vec{H}$ we derive that

$$H_{tw} = B_{tw} - 4\pi M_{tw} = H_{tz} = B_{tz}. \tag{9.16}$$

For an isotropic paramagnetic $\vec{B}_w = \mu_w \vec{H}_w$, while for a dielectric $\vec{B} = \vec{H}$, hence

$$\mu_w H_{tw} - 4\pi M_{tw} = H_{tz} = H_{tw}, \tag{9.17}$$

therefore

$$\frac{4\pi}{\mu_w - 1} M_{tw} = H_{tw} = H_{tz}, \tag{9.18}$$

or

$$M_{tw} = \frac{\mu_w - 1}{4\pi} H_{tz}. \tag{9.19}$$

For the normal component, similar

$$B_{nz} = B_{nw} = H_{nw} + 4\pi M_{nw} = \mu_w H_{nw} = H_{nz}, \tag{9.20}$$

or

$$M_{nw} = (\mu_w - 1) H_{nw} = \frac{\mu_w - 1}{4\pi\mu_w} H_{nz}. \tag{9.21}$$

The relations (9.19) and (9.21) pave the way to boundary conditions for magnetic moment formulation.

For electric moment $\vec{P}$, introduced via

$$\vec{D} = \vec{E} + 4\pi\vec{P} = \varepsilon\vec{E}, \tag{9.22}$$

for isotropic dielectric, one arrives at expressions (9.13), (9.19) and (9.21) for a border between dielectric and vacuum:

$$P_{nw} = \frac{\varepsilon_w - 1}{4\pi\varepsilon_w} E_{nz}. \tag{9.23}$$

$$P_{tw} = \frac{\varepsilon_w - 1}{4\pi} E_{tz}. \tag{9.24}$$

9.2.3 Magnetic moment: spins contribution

The spin variables and related magnetic moments have relativistic quantum origin. Its account, that is necessary in a para- and ferromagnetic medium, is possible via Ampère–Maxwell and the absence-of-magnetic-charges equation. Generally, for a dynamic problem,

$$\frac{1}{c}\frac{\partial \vec{D}}{\partial t} = \operatorname{rot}\vec{H} - \frac{4\pi}{c}\vec{j} = \operatorname{rot}\left(\vec{B} - 4\pi\vec{M}\right) - \frac{4\pi}{c}\vec{j},$$
$$\operatorname{div}\vec{B} = \operatorname{div}\vec{H} + 4\pi\operatorname{div}\vec{M} = 0.$$

Let us consider a magnetostatic problem (see section 9.1). For the stationary case without currents, taking only the magnetization vector $M = M_s$ originated from spins we have

$$\begin{aligned} \operatorname{rot}\left(\vec{B} - 4\pi\vec{M}_s\right) &= 0, \\ \operatorname{div}\vec{H} + 4\pi\operatorname{div}\vec{M}_s &= 0. \end{aligned} \tag{9.25}$$

It follows that, if

$$\vec{H} = -\operatorname{grad}\psi, \tag{9.26}$$

whence the first of equation (9.25) holds identically, while the second yields

$$-\operatorname{div}\operatorname{grad}\psi + 4\pi\operatorname{div}\vec{M}_s = 0,$$

or, in the Poisson form

$$\Delta\psi = 4\pi\operatorname{div}\vec{M}_s.$$

There is a form for solution, built by means of the Green formula, applied to the domain shown at figure 9.3. In condition of continuous ψ at boundary, taking the boundary condition for the magnetic field (9.13) into account, we arrive at

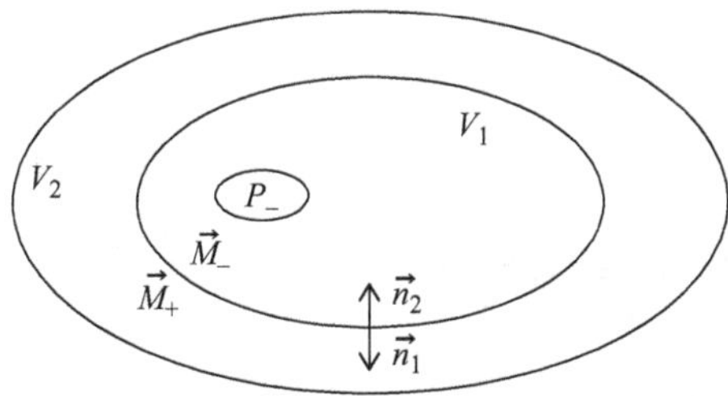

Figure 9.3. The domains for a solution by the Green formula are shown, see also [4].

$$\psi = -\int_V \frac{\operatorname{div}'\vec{M}_s}{|\vec{r}-\vec{r}'|}dV' + \frac{1}{4\pi}\int_S \frac{M_{n-} - M_{n+}}{|\vec{r}-\vec{r}'|}dS', \tag{9.27}$$

where V is any volume that contains the magnetic body and $M_{n\pm}$ are normal components of the magnetization vector outside (+) and inside (−) as limits to the surface of magnetics S. The identity

$$-\frac{\operatorname{div}'\vec{M}_s}{|\vec{r}-\vec{r}'|} = -\operatorname{div}'\frac{\vec{M}_s}{|\vec{r}-\vec{r}'|} - \left(\vec{M}_s, \operatorname{grad}'\frac{1}{|\vec{r}-\vec{r}'|}\right), \tag{9.28}$$

allows to simplify the expression for ψ if S lies outside the magnetic, by the Ostrogradskij–Gauss theorem $\int_V \operatorname{div}'\frac{\vec{M}_S}{|\vec{r}-\vec{r}'|}dV' = \int_{S'}\cdot\frac{\vec{M}_S}{|\vec{r}-\vec{r}'|}dS' = 0$. Plugging it in equation (9.27) gives

$$\psi = -\int_V \left(\vec{M}_s, \operatorname{grad}'\frac{1}{|\vec{r}-\vec{r}'|}\right)dV + \frac{1}{4\pi}\int_S \frac{M_{n-} - M_{n+}}{|\vec{r}-\vec{r}'|}dS. \tag{9.29}$$

By the ψ definition (9.26)

$$\begin{aligned}\vec{H} = -\operatorname{grad}\psi = \operatorname{grad}\int_V \left(\vec{M}_s, \operatorname{grad}'\frac{1}{|\vec{r}-\vec{r}'|}\right)dV \\ -\frac{1}{4\pi}\operatorname{grad}\int_S \frac{M_{n-} - M_{n+}}{|\vec{r}-\vec{r}'|}dS.\end{aligned} \tag{9.30}$$

The second term for a long cylindric isotropic magnetic (the ends contribution is small, otherwise see section 9.3) gives zero because the magnetization $\vec{M}$ is parallel to the cylinder axis.

The first term for unique point dipole $\vec{M}_s = \vec{m}\delta(\vec{r}-\vec{r}')$ gives

$$\psi = \left(\vec{m}, \operatorname{grad}\frac{1}{|\vec{r}|}\right) = -\frac{\left(\vec{m}, \frac{\vec{r}}{|\vec{r}|}\right)}{|\vec{r}|^2},$$

that produces the field by the definition (9.26)

$$\vec{H} = -\operatorname{grad}\psi = \operatorname{grad}\frac{\left(\vec{m}, \frac{\vec{r}}{|\vec{r}|}\right)}{|\vec{r}|^2}.$$

9.2.4 Surface charges and currents

The first Maxwell equation (9.5) in its complete version gains the charge of the internal volume inside S at rhs

$$\int_S \vec{D}\,d\vec{S} = 4\pi q. \tag{9.31}$$

Application to the volume of figure 9.1 modifies the boundary condition (9.9) as

$$E_{tw} - E_{tz} = 4\pi\sigma, \tag{9.32}$$

where σ is *surface charge density*. The next equation that needs a modification arises from the full Ampère–Maxwell's (9.11a), it is equation (9.12)

$$H_{twi} - H_{tzi} = \frac{4\pi}{c} j_{ti}, \; i = 1, 2, \tag{9.33}$$

where j_{ti} are the components of the *surface current density*. In some textbooks the second term of the field (9.30) is interpreted as generated by the surface 'effective magnetic charges' [2] with the density M_n.

9.3 Demagnetization field

9.3.1 Instructive example

Consider a finite cylinder made of a magnetic body. As it was noticed above, for a cylinder in constant field as in figure 5.5, the magnetization inside is constant, hence the div $\vec{M} = 0$ and the first term in (9.27) disappears

$$\vec{H} = -\text{grad}\,\psi = -\frac{1}{4\pi}\,\text{grad} \int_S \frac{M_{n-} - M_{n+}}{|\vec{r} - \vec{r}\,'|} dS. \tag{9.34}$$

If, following the general thesis on the symmetry, we pose the normal component at the lateral surface of the cylinder equal to zero, only the integrals over both bases S_1, S_2 of the body survive [1], then

$$\vec{H} = -\frac{1}{4\pi}\,\text{grad} \int_{S_1} \frac{M_{n-} - M_{n+}}{|\vec{r} - \vec{r}\,'|} dS - \frac{1}{4\pi}\,\text{grad} \int_{S_2} \frac{M_{n-} - M_{n+}}{|\vec{r} - \vec{r}\,'|} dS. \tag{9.35}$$

If magnetization outside the body is zero, we simplify equation (9.35) as

$$\vec{H} = -\frac{1}{4\pi}\,\text{grad} \int_{S_1} \frac{M_{n-}}{|\vec{r} - \vec{r}\,'|} dS - \frac{1}{4\pi}\,\text{grad} \int_{S_2} \frac{M_{n-}}{|\vec{r} - \vec{r}\,'|} dS. \tag{9.36}$$

Both expressions look as the magnetic poles with opposite magnetic charges.

The direction of such field is opposite with respect to the direction of the field, whose origin is defined by the first term of (9.27) i.e. by the internal magnetization, or as sometimes it is written in textbooks, the 'magnetic charges' $q_m = \text{div}\,\vec{M}$.

9.3.2 General demagnetization

Generally the surface contribution inside the magnetic body is given by equation (9.34)

$$\vec{H}_d = \frac{1}{4\pi}\int_S \frac{\sigma_m(\vec{r}-\vec{r}\,')}{|\vec{r}-\vec{r}\,'|^4}dS, \tag{9.37}$$

where $\sigma_m = M_{n-} - M_{n+}$ is named as surface magnetic charges. The value of this field depends on the form of the body, as it already shown in the case of the example of the section 9.3.1.

It is also marked by the term demagnetizing field, for its tendency to diminish a magnetization scale by a demagnetizing factor [2]. In ferromagnets it is responsible for a shape anisotropy. If the magnetization is excited by an external magnetic field H_e, inside a magnetic body we have the magnetic field stress $H_e + H_d$. If $H_d = -\nu M$, the factor ν is named the demagnetizing factor.

9.4 Stray fields

9.4.1 On definition

To illustrate applications of the derived formula for the magnetic field strength (9.30) originated from the second, the surface term of equation (9.27), we address the reader to the important problem of domain wall (DW) dynamics theory, within the example of such phenomenon in a microwire [3]. The DW in wires as a phenomena is presented in many papers, e.g. in [4], such interest is supported by an existence of isolated DW with, generally, rather simple behavior. In stationary conditions of homogeneous wire the wall moves with constant velocity, a small acceleration appears in conditions of defects presence [5]. Here in this book, the theory is given in chapter 16, particularly, the LLG equation theory in section 16.2. Note, that the surface layer of atoms have lesser number of the closest neighbors compared to the 'bulk' ones, hencefore its magnetic properties are as for paramagnetic ones even for a ferromagnetic wire as a whole. So, condition (9.21) is valid for the case. This part of the magnetic field reads

$$\vec{H} = -\frac{1}{4\pi}\int_S M_{n-}\text{grad}\,\frac{1}{|\vec{r}-\vec{r}\,'|}dS' = \frac{1}{4\pi}\int_S M_{n-}\frac{\vec{r}-\vec{r}\,'}{|\vec{r}-\vec{r}\,'|^3}dS'. \tag{9.38}$$

For $\vec{r}$ outside the body we will name it the stray field [6].

9.4.2 Landau–Lifshits–Gilbert (LLG) equation and domain wall (DW) motion model

Cylindric coordinates

Let the magnetization vector be rescaled as $\vec{M} = M_s\vec{m}$; M_s—magnetization of saturation, having the unit magnetization vector $\vec{m}$. To describe the surface phenomena at a microwire, let us go to cylindric coordinates z, ρ, ϕ, the components of the unit magnetization vector are:

$$\begin{aligned} m_z &= m_z, \\ m_\rho &= m_x \cos\phi + m_y \sin\phi, \\ m_\phi &= m_x \sin\phi + m_y \cos\phi, \end{aligned} \tag{9.39}$$

so that $\vec{M} = M_s\vec{m}$; M_s is magnetization of saturation, see figure 9.4. Having in mind the cylindrical surface of a body, similar to [7] we rewrite equation (9.38) as

$$\vec{H} = \frac{M_s}{4\pi}\int_S \frac{m_{x-}\cos\phi + m_{y-}\sin\phi}{|\vec{r} - \vec{r}\,'|^2}\vec{n}\, dS', \tag{9.40}$$

where $\vec{n} = \frac{\vec{r} - \vec{r}\,'}{|\vec{r} - \vec{r}\,'|}$.

Nonlinear transformations

Following [8], it is useful to introduce a complex variable $\Omega(\vec{r}, t)$, such that

$$\begin{aligned} \exp(\Omega) &= \frac{m^x + im^y}{1 + m^z}, \quad \text{or inversely, } m^x + im^y \\ &= \frac{2\exp(\Omega)}{1 + \exp(\Re\Omega)}, \quad m^z = \frac{1 - \exp(\Re\Omega)}{1 + \exp(\Re\Omega)}. \end{aligned} \tag{9.41}$$

In order to build a procedure of finding particular solutions of the system and obtain the general structure of magnetization field we rely upon the mentioned form of nonlinear transforms (16.22), while the space–time dependence on arguments of

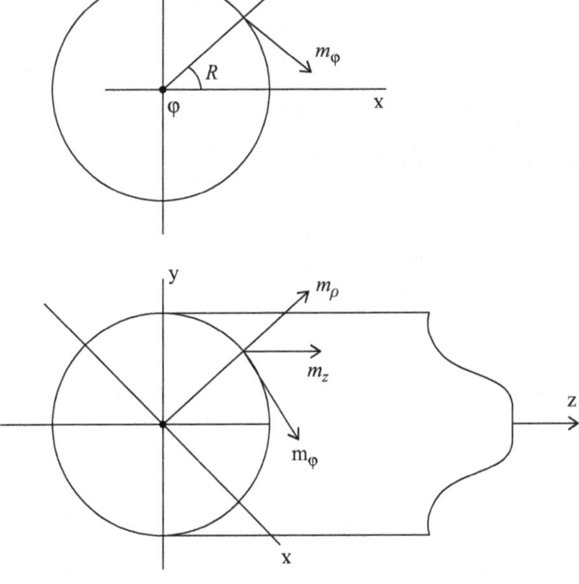

Figure 9.4. The vector $\vec{m}$ components in cylindrical coordinates at a wire surface.

equation (16.22) is calculated from the following equation after plugging equation (16.37) into the basic LLG equation [9], see chapter 16, section 16.2 for the form and the parameters definition:

$$(i + \alpha)\Omega_t - \gamma H + J\nabla^2\Omega + (K - J(\nabla\Omega)^2)\tanh\left(\frac{\Omega + \Omega^*}{2}\right) = 0. \tag{9.42}$$

Here, in equation (9.42), H is magnetic field stress z-component, J—exchange stiffness (linked to exchange integral), γ is the (positive) gyromagnetic ratio, K is an effective easy-axis anisotropy coefficient, a is the Gilbert loss parameter. The variable $\Omega(\vec{r}, t))$ is also convenient to formulate an initial-boundary problem of a DW evolution, via an inversion of the transform (16.22).

$$\begin{aligned} m^x &= \frac{\exp(\Omega) + \exp(\Omega^*).}{1 + \exp[2\Re\Omega].}, \\ m^y &= -i\frac{\exp(\Omega) - \exp(\Omega^*).}{1 + \exp[2\Re\Omega].} \end{aligned} \tag{9.43}$$

where the superscript '*' stands for complex conjugation.

9.5 Microwire: DW and observations

9.5.1 Wire DW model

The exact solutions of equation (9.42) are found in [9], we take the simplest one, corresponding to the 'planar' DW model.

$$\Omega(\vec{r}, t) = \Sigma(\vec{r}) - \frac{i\gamma}{1 - i\alpha}Ht, \tag{9.44}$$

where, the function $\Sigma(\vec{r})$ can be written in the following form:

$$\Sigma = \beta' z + \rho[A\cos(\nu\varphi) + B\sin(\nu\varphi)], \tag{9.45}$$

where β', is real, $A = A' + iA''$, $B = iB''$ and A', A'', B'' are real. In particular, as limitation of boundary condition, we assume the magnetization to be radially directed at points $\varphi = 0, \pi/2, \pi, -\pi/2$ [10]. This condition leads to the following values of parameters: $A'' = 0$, $B'' = \pi/(2R)$, where R is the radius of the inner core of the wire. Hence, the real part of Σ simplified as

$$\Sigma = \beta' z + \rho[A'\cos(\varphi) + B'\sin(\varphi)]. \tag{9.46}$$

This leads to the only relation between coefficients for the planar DW

$$(A')^2 + (\beta')^2 = \frac{K}{J} + \frac{\pi^2}{4R^2}, \tag{9.47}$$

that contains two parameters A' and β', while for the flexural planar DW we have two relations [11]. DW propagation velocity V of DW is visible from linear structure of the Ω; that reads as

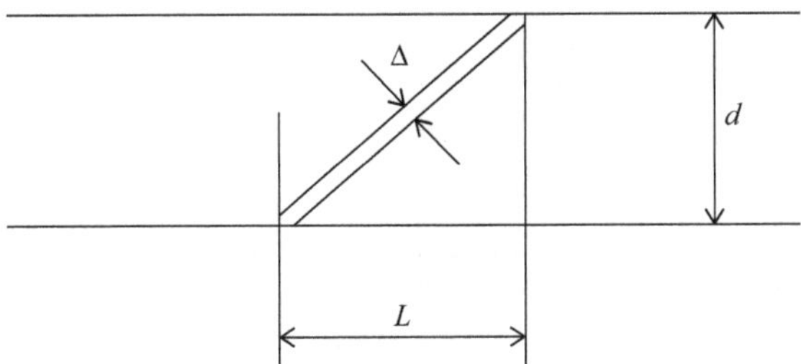

Figure 9.5. The cross-section of the wire at $f = 0$, the DW length is shown.

$$S = -\frac{\gamma\alpha}{1+\alpha^2}\frac{1}{\beta'}, \tag{9.48}$$

where $S = dV/dH$ is the mobility of DW. The explicit formula for the DW length L is extracted from equation (9.46) directly, e.g. at $\phi = 0$

$$L = 2R\frac{A'}{\beta'}. \tag{9.49}$$

see the figure 9.5. A more complicated 'flexural' solution is described in [3; 11] and in chapter 16.6. The cross-section of the wire at $t = 0$, the DW length is shown.

9.6 The stray field of the planar DW

At the wire surface, for the stationary solution, taking $t = 0$, without loss of generality, we write

$$\Sigma = \beta' z + RA'\cos(\varphi), \tag{9.50}$$

For the real Σ, equations (9.43) and (9.50) give the projections of the unit magnetization vector

$$m^x = \frac{2\exp(\beta' z + RA'\cos(\varphi))}{1+\exp[2\beta' z + RA'\cos(\varphi)].} = \frac{1}{\cosh[\beta' z + RA'\cos(\varphi)].}, \tag{9.51}$$
$$m^y = 0,$$

the form of the hyperbolic cosine in equation (9.51) shows the width of the DW:

$$\Delta = 1/\beta'. \tag{9.52}$$

that is the measure of the trace of the wall at the wire lateral surface. Finally, the stray field is evaluated by equation (9.51) with equation (9.40) account

$$\vec{H} = \frac{M_s}{4\pi}\int_S \frac{\cos\varphi'}{\cosh\left[\frac{z'}{\Delta} + RA'\cos(\varphi')\right]}\frac{\vec{n}}{|\vec{r} - \vec{r}'|^2}dS', \tag{9.53}$$

where $\vec{n} = \frac{\vec{r} - \vec{r}'}{|\vec{r} - \vec{r}'|}$, $\vec{r}' = (R\cos\varphi', R\sin\varphi', z')$. The parameters A'; β' are found from the mobility (9.48) and the link (9.47). Plugging it in equation (9.53) yields

$$\vec{H} = \frac{M_s}{4\pi} \int_S \frac{\cos\varphi'}{\cosh\left[\frac{z'}{\Delta} + RA'\cos(\varphi')\right]} \times \frac{\vec{r} - \vec{r}\,'}{[(x - R\cos\varphi')^2 + (y - R\sin\varphi')^2 + (z - z')^2]^{3/2}} d\varphi' dz'. \tag{9.54}$$

For the integration we put $x = 0$, $y = R + \delta$, $z = 0$, see figure 9.4 and choose $\frac{z'}{\Delta} + RA'\cos(\varphi') = \zeta$, $z' = \Delta(\zeta - RA'\cos(\varphi'))$, $dz' = \Delta d\zeta$. Then, for the y-component it is obtained

$$H_y = \frac{M_s\Delta}{4\pi} \int_S \frac{\cos\varphi'}{\cosh\zeta} \times \frac{R + \delta - R\sin\varphi'}{[R^2\cos^2\varphi' + (R + \delta - R\sin\varphi')^2 + (\Delta(\zeta - RA'\cos(\varphi')))^2]^{3/2}} d\varphi' d\zeta. \tag{9.55}$$

We arrive at the following formula to compute the radial component of the stray field,

$$H_\rho = \frac{M_s\Delta}{4R^2} \int_0^{2\pi} \int_{-1/2}^{1/2} d\varphi d\zeta \times \frac{\cos\varphi(1 + \delta - \sin\varphi)}{\cosh(\zeta)[\cos^2\varphi + (1 + \delta - \sin\varphi)^2 + (\Delta(\zeta - A\cos\varphi))^2]^{3/2}}, \tag{9.56}$$

where $\delta' = \delta/R$ is distance from the surface of the wire, $\Delta' = \Delta/R$—width of the DW, $L' = L/R$, measured in radius of the wire units. Primes are omitted. The resulting numerical calculations have been performed for two microwires with nucleus made of $Fe_{77;5}Si_{17;5}B_{15}$ and geometric parameters $d = 9$ μm and 12 μm, respectively, where d—diameter of metallic nucleus [11]. The results are depicted in figure 9.6. For this case, from equation (9.48), neglecting α^2 in the denominator, we have $\Delta = S/(\gamma\alpha) = 50/(0.016 \times 1.8 \times 10^{11}) = 1.7 \times 10^{-2}$ mkm, that is about half-width, see (9.51).

The DW length is evaluated as $L = 2RA'/\beta' = 2RA'\Delta$. To express A' via Δ we take equation (9.38), that gives

$$A' = \sqrt{\frac{K}{J} + \frac{\pi^2}{4R^2} - \Delta^{-2}}. \tag{9.57}$$

Finally,

$$L = 2R\Delta\sqrt{\frac{K}{J} + \frac{\pi^2}{4R^2} - \Delta^{-2}} = \sqrt{\frac{4KR^2\Delta^2}{J} - 4R^2 + \pi^2\Delta^2}. \tag{9.58}$$

Numerical values for the data shown above looks as follows.

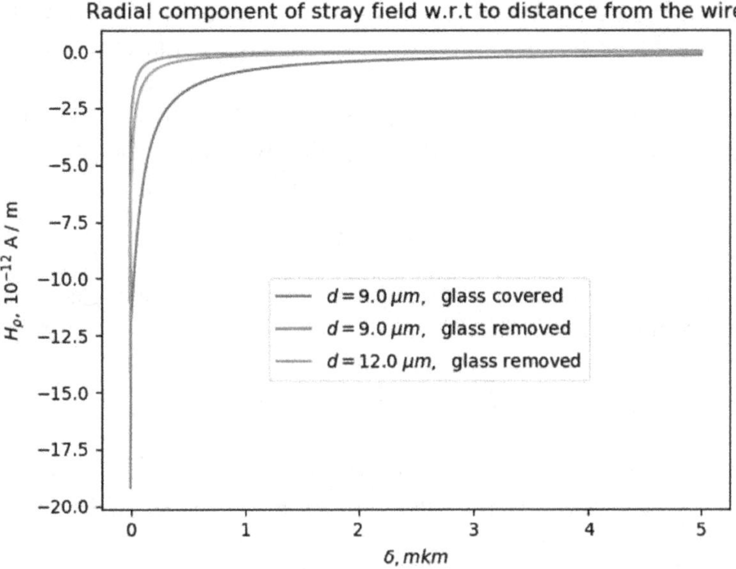

Figure 9.6. The dependence of the radial component of magnetic field H_ρ on the distance from the wire.

9.6.1 Observation of domain walls in ferromagnetic microwires

An interesting and important example of practical applications of the basic electrodynamics is observation of moving domains of magnetization. Such an object we have introduced in the preceded sections, the DW. To determine the parameters of the DW included in the formulas of the stray fields, it is necessary to carry out the following measurements. We will start with equation (9.38) for a round coil with a radius of R_c, integrating over the plane of the coil, as shown in figure 9.4.

Otherwise, considering the electromotive force as an integral of elementary work round L, surrounding the surface S, we have

$$\varepsilon = \oint_L \vec{E}\, \vec{d}\, l = -\frac{1}{c}\frac{\partial \Phi}{\partial t} = -\frac{1}{c}\frac{\partial}{\partial t}\oint_S \vec{B}\, d\varphi dz, \tag{9.59}$$

this gives the *III Maxwell's law in integral form.*

The vector $\vec{B}$ would be estimated via

$$\vec{H} = -\mathrm{grad}\,\psi = \mathrm{grad}\int_V \frac{\mathrm{div}'\vec{M}}{|\vec{r}-\vec{r}\,'|}dV' = \int_V \mathrm{div}'\vec{M}\,\mathrm{grad}\frac{1}{|\vec{r}-\vec{r}\,'|}dV'. \tag{9.60}$$

As the first application of the general theory of stray fields theory, we choose the conditions of a ferromagnetic microwire for several reasons. Firstly, there is rich experimental information about the values of the parameter H_0, called 'critical propagation' field excitation of a DW pair by several coils [4] or as a 'nucleation field' [13]; the second is cylindrical symmetry and a small cross section, which simplifies the theoretical relationships. Thirdly, there are publications that offer a good zoo of DW forms, their parameters and dynamics [10–12]. We will focus our main efforts in this section on the case of several DC excitation coils far from the ends of the wire.

DW observation is performed using measuring coils located close to a generating one [4, 12]. Recovery $\vec{M}(\vec{r})$ directly from the expression for $\vec{H}$ obtained with equation (14.9a) is difficult. However, in the dipole approximation

$$\vec{M} \sim \vec{m},$$

the relation (9.60) is written as if the wire and the position of the coils allowed an approximate condition

$$\vec{H} = \frac{3\vec{r}(\vec{r}\cdot\vec{m}) - r^2\vec{m}}{r^5}.$$

The ability to approximate the magnetization vector by the dipole moment follows from its small spatial length in comparison with the radius of the exciting coil c and the radius of the microwire a: $a \ll c$.

This leads to a working formula that expresses $\vec{m}$ as in [12]. Next we expect that the geometry of the wire and the position of the coils allows an approximate ratio

$$(\vec{r}\cdot\vec{m}) = 0,$$

because the planes of the receiving and generating coils are almost orthogonal to the axis of the wire and, therefore, to $\vec{m}$, finally, for the modules we have a connection:

$$H = \frac{m}{r^3}. \tag{9.61}$$

The signal theory is based on the Faraday effect caused by the movement of the domain system with $\vec{m}(\vec{r}, t)$, which is described in [12].

The theory of scattering field estimation is based on the general relations of electrodynamics, and the shape of the DW is selected on the basis of a concrete solution of the LLG equation. The radial field component in the vicinity of the surface of the wire is calculated by approximate integration in the resulting formula.

References

[1] Novozhilov Y V and Yappa Y A 1986 *Electrodynamics* (Mir: Moscow) [Transl. from Russian by Kisin V I]

[2] Jackson J D 1998 *Classical Electrodynamics* 3rd edn (New York: Wiley)

[3] Leble S 2019 *Waveguide Propagation of Nonlinear Waves. Impact of Inhomogeneity and Accompanying Effects* (*Springer Series on Atomic, Optical, and Plasma Physics* vol. 109) (Switzerland: Springer)

[4] Rodioniova V *et al* The defects influence on domain wall propagation in bistable glass-coated microwires *Physica* B **407** 1446–9

[5] Leble S B and Rodionova V V 2020 Dynamics of domain walls in a cylindrical amorphous ferromagnetic microwire with magnetic inhomogeneities *Theor. Math. Phys.* **202** 252–64

[6] Bregermann F-J, Normanna N and Mendea H H 1983 *J. Magn. Magn. Mater.* **38** 325–30

[7] Derby N and Olbert S 2010 Cylindrical magnets and ideal solenoids *Am. J. Phys.* **78** 229
[8] Lakshmanan M and Nakamura K 1984 Landau–Lifshitz equation of ferromagnetism: exact treatment of the Gilbert damping *Phys. Rev. Lett.* **53** 2497–9
[9] Vereshchagin M 2018 Structure of domain wall in cylindrical amorphous microwire *Physica B: Condens. Matter* **549** 91–3
[10] Janutka A and Gawroński P 2015 Structure of magnetic domain wall in cylindrical microwire *IEEE Trans. Magn.* **51** 1–6
[11] Vereshchagin M, Baraban I, Leble S and Rodionova V 2020 Structure of head-to-head domain wall in cylindrical amorphous ferromagnetic microwire and a method of anisotropy coefficient estimation *J. Magn. Magn. Mater.* **504** 166446
[12] Gudoshnikov S A, Grebenshchikov Y B, Ljubimov B Y, Palvanov P S, Usov N A, Ipatov M, Zhukov A and Gonzalez J 2009 Ground state magnetization distribution and characteristic width of head to head domain wall in fe-rich amorphous microwire *Status Solidi* A **206** 613–7
[13] Pátek K and Tomáý I 1985 Quantitative characterization of magnetic defects by domain nucleation *Phys. Stat. Sol. A* **89** 595

IOP Publishing

Practical Electrodynamics with Advanced Applications

Sergey Leble

Chapter 10

Reflection and refraction of electromagnetic waves at a boundary

10.1 Reflection and transmission of a plane wave on a border

10.1.1 General textbook relations

Let us consider a problem of wave propagation in a stratified medium. It is implied that an electromagnetic wave passes through a layered medium with different ε, μ and σ (see the definitions in chapter 5, section 5.1.7) divided by plane boundaries $z = z_i$ at Cartesian coordinates x, y, z. There is a long history of problems of electromagnetic field interaction with dielectric and conducting bodies, see e.g. [1]. There are different aspects of such phenomenon from diffraction (see e.g. [2]) to focusing in different frequency ranges up to recent x-rays [3–5] problems. Modern electrodynamics involves the whole complex plane for conductivity, permeability and permittivity coefficients, e.g. when studying metamaterials [6].

Textbooks on electrodynamics conventionally contain the results on reflection and refraction of electromagnetic waves in a form of generalized Snell and Fresnel formulas. The famous Stratton book [4], basing on [7], starts with complex Snell's law equalizing projections of the incident wave vector $\vec{k}$ and the transmitted wave vector $\vec{k}''$ on the x-axis, span along the matters' interface, $k_x = k''_x$ in adjacent media one of which is supposed to be conducting. The classic Snell's relation follows directly from the electric field boundary condition $k \sin \alpha = k'' \sin \beta$ and trigonometry in the case of zero conductivity and absorption (see figure 10.1 below), but the conductivity account automatically implies complex k'', while zero conductance yields real $\vec{k}$. It means that such generalization of Snell's law leads to complex angles as well as necessity of its interpretation. The next question relates to the direction of propagation of the wave which, in [4], is chosen as orthogonal to constant phase plane. In the textbooks [4] the relations between complex $\vec{k}''$ and $\vec{k}$ are plugged

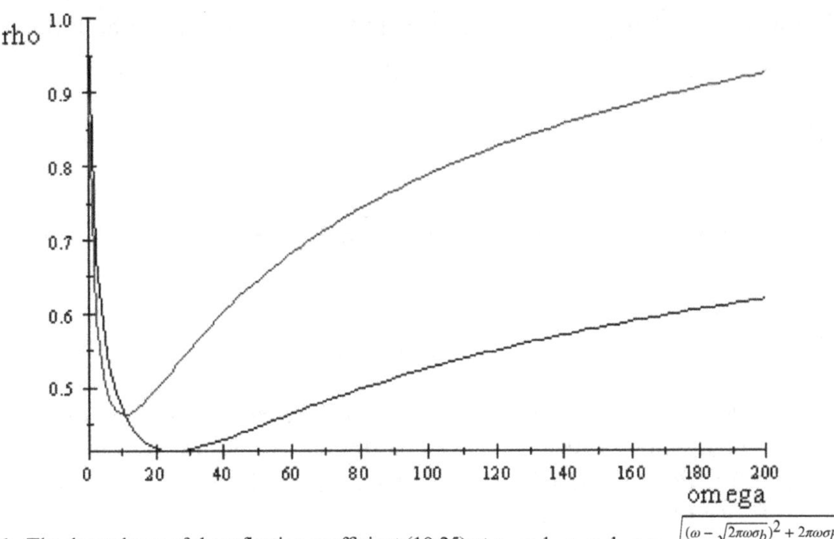

Figure 10.1. The dependence of the reflection coefficient (10.25) at $\mu_b = 1$, $\varepsilon_a = 1$, $\rho = \sqrt{\frac{(\omega - \sqrt{2\pi\omega\sigma_b})^2 + 2\pi\omega\sigma_b}{(\omega + \sqrt{2\pi\omega\sigma_b})^2 + 2\pi\omega\sigma_b}}$ on frequency in the range 0–200 rad s^{-1}, $\sigma_b = 2$—black. The case $\mu_b = 1$, $\varepsilon_a = 2$ at the same range—red.

directly into Fresnel formulas, derived in conditions of real $\vec{k}''$ that should be approved. Similar definitions use authors of papers [3, 8].

In this textbook we base on the direct solution of the complex dispersion equation (similar to textbook [9]) but do not use the real unit vector of propagation direction determined via phase front. We rely upon the definition of propagation direction by means of Pointing vector (as mentioned in the book [6]), with the time averaging afterwards and systematically solving boundary conditions for field components equalizing real and imaginary parts. We restrict ourselves to the case of incident wave propagating in a dielectric medium. Similar to the cited [11], a plane boundary is posed in the plane $z = 0$ and an isotropic conductor at $z < 0$ with a given dielectric constant is considered. The peculiarities of the case (wave damping) imply modification of the problem formulation. The conductivity is introduced via a direct account of Ohm's law in the Maxwell–Ampère equation, while the dielectric permittivity is supposed to be a complex function of frequency, the magnetic permeability is supposed to be a constant.

The main task of this chapter is to describe an eventual symbolic computing base of rather complicated reflexion and refraction theory of a plane monochrome electromagnetic wave in a general case of complex permittivity and permeability with conductivity account.

10.1.2 Ampère–Maxwell equation for a wave with given frequency

The equation of Ampère–Maxwell for matter, that accounts Ohm's law we write now as

$$\frac{1}{c}\frac{\partial \vec{D}}{\partial t} = \nabla \times \vec{H} - \frac{4\pi}{c}\sigma \vec{E}, \tag{10.1}$$

taking material relations for $\vec{D}$ and $\vec{H}$ as for isotropic media without dispersion,

$$\vec{D} = \varepsilon\vec{E}, \tag{10.2a}$$

$$\vec{H} = \frac{1}{\mu}\vec{B}. \tag{10.2b}$$

For a harmonic wave with real frequency ω we write

$$\vec{E} = \vec{E}_0(\vec{r})e^{-i\omega t} + c.c, \tag{10.3a}$$

$$\vec{B} = \vec{B}_0(\vec{r})e^{-i\omega t} + c.c. \tag{10.3b}$$

The expansion guarantees the reality of the fields, while integration over ω would mimic arbitrary time dependence at the boundary. So we will use the complex amplitude functions $\vec{E}_0$, $\vec{B}_0$. Plugging the material relations (10.2) as well as (10.3) into (10.1), we obtain

$$\nabla \times \vec{B}_0 = \frac{\mu}{c}(-i\varepsilon\omega + 4\pi\sigma)\vec{E}_0. \tag{10.4}$$

Let us evaluate the rotation part of the Faraday equation, e.g. as equation (5.11b), and plug the result in equation (10.4). We hence deduce the Helmholtz equation with a complex coefficient (compare for equation (3.62)):

$$\Delta\vec{E}_0 + \frac{i\omega\mu}{c^2}(-i\varepsilon\omega + 4\pi\sigma)\vec{E}_0 = 0. \tag{10.5}$$

Next we go to study a particular case of plane wave with a complex z-component of the wave vector $k_z = k$ and real x, y components of a wave vector $\vec{k} = (\vec{k}_\perp, k)$; $\vec{k}_\perp = (k_x, k_y)$. The Fourier method implementation is natural via supposing that the complex amplitude function $\vec{E}_0$ has the form:

$$\vec{E}_0 = \vec{A}e^{i\vec{k}\vec{r}}. \tag{10.6}$$

The substitution of equation (10.6) into (10.5) yields the dispersion equation

$$k^2 + k_\perp^2 - \frac{i\omega\mu}{c^2}(-i\varepsilon\omega + 4\pi\sigma) = 0. \tag{10.7}$$

It is seen that the z-component of the wave vector $\vec{k}$ is complex that determine a damping decrement along z. Therefore

$$k = \frac{\omega}{c}\sqrt{\varepsilon\mu + \frac{4\pi i\sigma\mu}{\omega} - k_\perp^2}. \tag{10.8}$$

A boundary conditions for the fields $\vec{E}$, $\vec{B}$ at $z = 0$ (see chapter 9, section 9.2.1) and the normal unit vector choice $\vec{n} = \{1, 0, 0\}$ leads to

$$(\varepsilon E)_{za} = (\varepsilon E)_{zb}, \tag{10.9a}$$

$$B_{za} = B_{zb}, \tag{10.9b}$$

$$\mathbf{n} \times (\vec{E}_b - \vec{E}_a) = 0, \tag{10.9c}$$

$$\mathbf{n} \times \left(\left(\frac{1}{\mu} \vec{B} \right)_b - \left(\frac{1}{\mu} \vec{B} \right)_a \right) = 0, \tag{10.9d}$$

the indices a, b denote two adjacent media. A convenient expression for the complex amplitude of magnetic field B_0 is obtained directly from the Faraday equation (5.11b):

$$\vec{B}_0 = \frac{ic}{\omega} \mathrm{rot} \vec{E}_0. \tag{10.10}$$

Plugging equation (10.3) into (10.9) gives

$$\varepsilon_a E_{0za} = \varepsilon_b E_{0zb}, \tag{10.11a}$$

$$B_{0za} = B_{0zb}, \tag{10.11b}$$

$$\mathbf{n} \times (\vec{E}_{0b} - \vec{E}_{0a}) = 0, \tag{10.11c}$$

$$\mathbf{n} \times \left(\frac{1}{\mu_b} \vec{B}_{0b} - \frac{1}{\mu_a} \vec{B}_{0a} \right) = 0. \tag{10.11d}$$

Linear independence of $e^{i\omega t}$, $e^{-i\omega t}$ is taken into account.

10.1.3 Reflection and transmission of wave propagating orthogonally to a boundary

Problem formulation, its scattering form

Let us take the simplest case of a plane wave with a wavefront parallel to the plane $z = 0$, that means $\vec{E}_a \cdot \vec{n} = 0$, $(\vec{k} \cdot \vec{n}) = k$, $\vec{k} \cdot \vec{r} = kz$, and choose the polarization as $E_{ay} = 0$, $E_{ax} \neq 0$. Hence from the boundary condition (10.11c) it follows that $E_{by} = E_{ay} = 0$. In the case of linear polarization of the electromagnetic wave along x, the only nonzero component of magnetic field (see relation (10.10)) is expressed as

$$B_{0y} = \frac{ic}{\omega} \frac{\partial E_{0x}}{\partial z}. \tag{10.12}$$

The problem, we do consider, may be formulated as a scattering one, the electromagnetic field amplitudes are expressed as

$$\begin{aligned} E_{0a} &= P e^{ik_a z} + R e^{-ik_a z}, \\ E_{0b} &= Q e^{ik_b z}. \end{aligned} \tag{10.13}$$

It means, that the amplitudes of electromagnetic field of the reflected wave R and the passed one Q at both media could be expressed in terms of the initial wave amplitude P. The boundary conditions (10.11) yield

$$P + R = Q,\ k_a'(P - R) = k_b'Q,$$

that results in

$$\begin{aligned} R &= \frac{k_a' - k_b'}{k_a' + k_b'}P, \\ Q &= \frac{2k_a'}{k_a' + k_b'}P; \end{aligned} \tag{10.14}$$

where the rescaled wave vector components $k_{a,\,b}' = k_{a,b}/\mu_{a,b}$ are introduced. The wavevectors are defined by equation (10.8)

$$k_{a,\,b}' = \frac{1}{c\sqrt{\mu_{a,b}}}\sqrt{\omega^2\varepsilon_{a,b} + 4\pi i\omega\sigma_{a,b}}\,. \tag{10.15}$$

A substitution of these expressions into relation (10.14) gives expressions for the coefficient of reflection and transmission as a function of matter parameters and frequency. Namely, let the complex coefficient $R = |R|\exp[i\phi] = ReR + \imath ImR$, $\tan\phi = \frac{ImR}{ReR}$ be inserted into the full expression of the reflected wave (10.13):

$$\begin{aligned} E_R &= \rho P\exp(i\phi)e^{-i\omega t}e^{-ik_a z} + c.c \\ &= \rho P(e^{-i(\omega t + k_a z - i\phi)} + e^{-i(\omega t + k_a z - i\phi)}) \\ &= \rho P\cos(\omega t + k_a z - i\phi). \end{aligned} \tag{10.16}$$

The real and imaginary parts of the amplitude R are defined by

$$R = R_1 + iR_2 = \frac{k_a' - k_b'}{k_a' + k_b'}P = \frac{(k_a' - k_b')(k_a' + k_b')^*}{|k_a' + k_b'|^2}P, \tag{10.17}$$

the incident wave amplitude is chosen as a real positive $P = |P|$. So the module

$$\rho = |R|/P = \sqrt{R_1^2 + R_2^2}/P, \quad \tan\phi = \frac{R_1}{R_2}$$

of the amplitude coefficient R defines the reflection coefficient while ϕ is the phase change with respect to the incident wave. Finally

$$\rho = \frac{|k_a' - k_b'|}{|k_a' + k_b'|}, \tag{10.18}$$

is evaluated via complex wavevectors (10.15).

A boundary dielectric-metal

Consider the case when the electromagnetic wave propagates from a dielectric ($\sigma_a = 0$, $\mu_a = 1$) to a metal ($\omega^2\varepsilon_b \ll 4\pi\omega\sigma_b$) falling on a boundary at $z = 0$. The wave vector for the dielectric is

$$k_a' = \frac{\omega}{c}\sqrt{\varepsilon_a}, \tag{10.19}$$

and for the conductor it is

$$k_b' = \frac{\sqrt{4\pi i\omega\sigma_b}}{c\sqrt{\mu_b}}, \tag{10.20}$$

hence the wave decrease as

$$|\exp[ik_bz]| = \exp\left[\frac{-Im(\sqrt{i})}{c}\sqrt{4\pi\omega\mu_b\sigma_b z}\right] = \exp\left[-\frac{z}{\delta}\right],$$

where

$$\delta = \frac{c}{\sqrt{2\pi\omega\mu_b\sigma_b}}, \tag{10.21}$$

is named as 'skin layer thickness'.

The reflection and transmission amplitudes are

$$\begin{aligned} R &= \frac{\omega\sqrt{\varepsilon_a\mu_b} - \sqrt{4\pi i\omega\sigma_b}}{\omega\sqrt{\varepsilon_a\mu_b} + \sqrt{4\pi i\omega\sigma_b}}P, \\ Q &= \frac{2\sqrt{\varepsilon_a\mu_b}}{\sqrt{\varepsilon_a\mu_b} + \sqrt{\frac{4\pi i\sigma_b}{\omega}}}P. \end{aligned} \tag{10.22}$$

In the case of the reflection coefficient evaluation, the relation (10.22) yields

$$\begin{aligned} \rho &= |R|/P \\ &= \frac{\left|\left(\omega\sqrt{\varepsilon_a\mu_b} - i^{1/2}\sqrt{4\pi\omega\sigma_b}\right)\left(\omega\sqrt{\varepsilon_a\mu_b} + i^{1/2}\sqrt{4\pi\omega\sigma_b}\right)^*\right|}{\left|\omega\sqrt{\varepsilon_a\mu_b} + i^{1/2}\sqrt{4\pi\omega\sigma_b}\right|^2}. \end{aligned} \tag{10.23}$$

The identity $i^{1/2} = \exp[i\pi/4] = \cos(\pi/4) + i\sin(\pi/4) = \sqrt{2}/2 + i\sqrt{2}/2$ should be taken into account. Then

$$\rho = \frac{\left|\left(\omega\sqrt{\varepsilon_a\mu_b} - \sqrt{2\pi\omega\sigma_b} - i\sqrt{2\pi\omega\sigma_b}\right)\right|}{\left|\omega\sqrt{\varepsilon_a\mu_b} + \sqrt{2\pi\omega\sigma_b} - i\sqrt{2\pi\omega\sigma_b}\right|}. \tag{10.24}$$

Using the identity for arbitrary complex numbers a, b, $|ab| = |a||b|$ yields

$$\rho = \sqrt{\frac{(\omega\sqrt{\varepsilon_a \mu_b} - \sqrt{2\pi\omega\sigma_b})^2 + 2\pi\omega\sigma_b}{(\omega\sqrt{\varepsilon_a \mu_b} + \sqrt{2\pi\omega\sigma_b})^2 + 2\pi\omega\sigma_b}}. \tag{10.25}$$

The frequency dependence is shown in figure 10.1

10.2 Problem of a plane wave with fixed frequency refraction

Now we will consider the inclined wavefront. The electric field vector now also lies within the wavefront plane and its direction defines polarization.

Changing notations for basic fields, we present here the Maxwell equations for a medium without space charge in the Lorentz–Heaviside's unit system (c—the velocity of light in vacuum)

$$\nabla \cdot \mathfrak{D} = 0, \tag{10.26a}$$

$$\nabla \times \mathfrak{E} = -\frac{1}{c}\frac{\partial \mathfrak{B}}{\partial t}, \tag{10.25b}$$

$$\nabla \cdot \mathfrak{B} = 0, \tag{10.26c}$$

$$\nabla \times \mathfrak{H} = \frac{1}{c}\frac{\partial \mathfrak{D}}{\partial t} + \frac{4\pi}{c}\mathfrak{J}, \tag{10.26d}$$

where the standard set of electric $\mathfrak{E}$, $\mathfrak{D}$ and magnetic $\mathfrak{B}$, $\mathfrak{H}$ fields is used. The material relations (of state) are assumed as for isotropic medium (see previous section 10.1.1; or a textbook, e.g. [10]). We, following the physical circumstances of, e.g. synchrotron radiation, also do restrict ourselves by fixed frequency of incident and, hence, scattered wave:

$$\mathfrak{H} = \frac{1}{\mu}\mathfrak{B}, \tag{10.27a}$$

where the dielectric permittivity ε, and magnetic permeability μ are supposed constant. The current density for this isotropic case is given by the simple version of Ohm's law:

$$\mathfrak{J} = \sigma\mathfrak{E}. \tag{10.28}$$

We, for textbook purposes, study a real σ case, implying consumption of its imaginary part in the real part of ε, if exists.

Introducing the complex fields functional amplitudes $\mathbf{E}(\vec{r})$, $\mathbf{B}(\vec{r})$, … , marked by the boldface characters:

$$\mathfrak{E} = \frac{1}{2}[\mathbf{E}(\vec{r})\exp(i\omega t) + \mathbf{E}^{*}(\vec{r})\exp(-i\omega t)], \tag{10.29a}$$

$$\mathfrak{D} = \frac{1}{2}[\mathbf{D}(\vec{r})\exp(i\omega t) + \mathbf{D}^*(\vec{r})\exp(-i\omega t)], \tag{10.29b}$$

$$\mathfrak{B} = \frac{1}{2}[\mathbf{B}(\vec{r})\exp(i\omega t) + \mathbf{B}^*(\vec{r})\exp(-i\omega t)], \tag{10.29c}$$

$$\mathfrak{H} = \frac{1}{2}[\mathbf{H}(\vec{r})\exp(i\omega t) + \mathbf{H}^*(\vec{r})\exp(-i\omega t)], \tag{10.29d}$$

$$\mathfrak{J} = \frac{1}{2}[\mathbf{J}(\vec{r})\exp(i\omega t) + \mathbf{J}^*(\vec{r})\exp(-i\omega t)], \tag{10.29e}$$

where the material relations are written in terms of these amplitudes:

$$\mathbf{D}(\vec{r}) = \varepsilon\mathbf{E}(\vec{r}), \tag{10.30}$$

$$\mathbf{H}(\vec{r}) = \frac{1}{\mu}\mathbf{B}(\vec{r}), \tag{10.31}$$

$$\mathbf{J}(\vec{r}) = \sigma\mathbf{E}(\vec{r}). \tag{10.32}$$

It reduces the problem formulation to the couple of vectors $\mathbf{E}(\vec{r})$, $\mathbf{B}(\vec{r})$, with the known links that follow from the time-independent Maxwell equations (10.26a) and (10.26c), that yields next reduction to four field components.

We would mark the field and medium parameters by primes for reflected waves (in the first medium $\varepsilon' = \varepsilon$, $\mu' = \mu$ are real),

$$\vec{k}^2 = \vec{k}'^2 = \frac{\omega^2\varepsilon\mu}{c^2}, \tag{10.33}$$

having $k_x = \frac{\omega \sin\alpha\sqrt{\varepsilon\mu}}{c}$. We mark by double primes the parameters of the second medium and refracted waves. We also suppose that for the second one $\varepsilon'' = \varepsilon_1'' + i\varepsilon_2''$ is complex, as for metals, and μ'' is real. Plugging equations (10.27)–(10.29) into the Maxwell's equations (10.26), the following equations for amplitudes inside the second medium are obtained:

$$\nabla\cdot\varepsilon''\mathbf{E}'' = 0, \tag{10.34a}$$

$$\nabla\times\mathbf{E}'' = -\frac{i\omega}{c}\mathbf{B}'', \tag{10.34b}$$

$$\nabla\cdot\mathbf{B}'' = 0, \tag{10.34c}$$

$$\nabla\times\mathbf{B}'' = \left(\frac{i\omega\mu''(\varepsilon''_1 - i\cdot\varepsilon''_2)}{c} + \frac{4\pi\mu''}{c}\sigma\right)\mathbf{E}'', \tag{10.34d}$$

as well as the corresponding conjugate ones. Excluding $\mathbf{B}''$ by plugging equation (10.34b) in equation (10.34a) it yields the Helmholtz equation

$$-\Delta\mathbf{E}'' = k''^2\mathbf{E}'', \tag{10.35}$$

where $(\frac{\omega^2\mu''\varepsilon''}{c^2} - i\frac{4\pi\omega\mu''}{c^2}\sigma)$ is denoted as k''^2. We base on solutions of equation (10.35) and link equation (10.34) within a statement of standard scattering problem.

Next, for a generalized harmonic wave, we write for the refracted one

$$\mathbf{E}'' = \mathbf{E}_0''e^{i\vec{k}''\vec{r}}, \tag{10.36}$$

$$\mathbf{B}'' = \mathbf{B}_0''e^{i\vec{k}''\vec{r}}, \tag{10.37}$$

as, similar, for both incoming and reflected ones. Substituting equations (10.36) and (10.37) into equation (10.34) yields links for complex amplitudes

$$\mathbf{E}_0'' \cdot \vec{k}'' = 0, \tag{10.38}$$

$$\vec{k}'' \times \mathbf{E}_0'' = -\frac{\omega}{c}\mathbf{B}_0'', \tag{10.39}$$

$$\mathbf{B}_0'' \cdot \vec{k}'' = 0, \tag{10.40}$$

$$\nabla \times \mathbf{B}_0''e^{i\vec{k}''\vec{r}} = \left(\frac{i\omega\mu''\varepsilon''}{c} + \frac{4\pi\mu''}{c}\sigma\right)\mathbf{E}_0''e^{i\vec{k}''\vec{r}}. \tag{10.41}$$

Let us put equation (10.39) into equation (10.41), having

$$\begin{aligned} i\vec{k}'' \times \mathbf{B}'' = i\vec{k}'' \times \mathbf{B}_0''e^{i\vec{k}''\vec{r}} &= -i\vec{k}'' \times \frac{c}{\omega}\left(\vec{k}'' \times \mathbf{E}_0''\right)e^{i\vec{k}''\vec{r}} \\ &= \left(\frac{i\omega\mu''\varepsilon''}{c} + \frac{4\pi\mu''}{c}\sigma\right)\mathbf{E}_0''e^{i\vec{k}\vec{r}}, \end{aligned} \tag{10.42}$$

or

$$-i\vec{k}'' \times \frac{c}{\omega}(\vec{k}'' \times \mathbf{E}_0'') = \left(\frac{i\omega\mu''\varepsilon''}{c} + \frac{4\pi\mu''}{c}\sigma\right)\mathbf{E}_0''. \tag{10.43}$$

Using *bac* – *cab* formula and taking equation (10.38) into account we get the dispersion equation

$$-\frac{ic}{\omega}\left(\vec{k}''(\vec{k}'', \mathbf{E}_0'') - \mathbf{E}_0''(\vec{k}'', \vec{k}'')\right) = \left(\frac{i\omega\mu''\varepsilon''}{c} + \frac{4\pi\mu''}{c}\sigma\right)\mathbf{E}_0''. \tag{10.44}$$

Due to equation (10.38) we have

$$\frac{ic}{\omega}\mathbf{E}_0''\vec{k}''^2 = \left(\frac{i\omega\mu''\varepsilon''}{c} + \frac{4\pi\mu''}{c}\sigma\right)\mathbf{E}_0''. \tag{10.45}$$

Hence

$$k''^2 = \frac{\omega^2 \mu'' \varepsilon''}{c^2} - \frac{4i\pi\omega\mu''}{c^2}\sigma, \tag{10.46}$$

that we would name a complex dispersion relation, which explicit form for a complex $\vec{k}''$ will be given in the next section after accounting for boundary conditions.

10.3 Boundary conditions impact

10.3.1 Snell law

Let us choose the boundary plane at $x = 0$ with the normal unit vector $\vec{n}$ and the tangential one $\vec{t}$ lying in the plane $x = 0$. Normal projection at the boundary points gives us

$$\varepsilon(\vec{n} \cdot \mathbf{E}_0)e^{i\vec{k}\vec{r}} + \varepsilon(\vec{n} \cdot \mathbf{E}_0{}')e^{i\vec{k}'\vec{r}} = \varepsilon''(\vec{n} \cdot \mathbf{E}_0{}'')e^{i\vec{k}''\vec{r}}, \tag{10.47}$$

and tangential one is equal to

$$(\vec{t} \cdot \mathbf{E}_0)e^{i\vec{k}\vec{r}} + (\vec{t} \cdot \mathbf{E}_0{}')e^{i\vec{k}'\vec{r}} = (\vec{t} \cdot \mathbf{E}_0{}'')e^{i\vec{k}''\vec{r}}, \tag{10.48}$$

with similar relations for the magnetic field. As the vector $\vec{r}$ lies in the plane $x = 0$, the scalar products in the exponents do not contain z-components of the wave-vectors, hence

$$k_x'' = k_x, \; k_x' = k_x. \tag{10.49}$$

In the case of a conventional dielectric $\sigma = 0$, $\varepsilon''_2 = 0$,

$$k_x'' = \frac{\omega\sqrt{\varepsilon''\mu''}\,\sin\beta}{c} = k_x = \frac{\omega\sqrt{\varepsilon\mu}\,\sin\alpha}{c}, \tag{10.50}$$

yields the Snell law. Note $n = \sqrt{\varepsilon\mu}$, $n'' = \sqrt{\varepsilon''\mu''}$, the refraction indices, see figure 10.2,

$$n''\sin\beta = n\sin\alpha. \tag{10.51}$$

Therefore the wave vector $\vec{k}''$ should have the real component k_x'' and, eventually, complex $k_z'' = k_{z1}'' + i \cdot k_{z2}''$ component. Therefore the sum of its squares in equation (10.46) yields

$$\begin{aligned} k_x''^2 + k_z''^2 &= k_x''^2 + k_{z1}''^2 + 2ik_{z1}''k_{z2}'' - k_{z2}''^2 \\ &= \frac{\omega^2\mu''(\varepsilon_1'' - i \cdot \varepsilon_2'')}{c^2} - \frac{4i\pi\omega\mu''}{c^2}\sigma. \end{aligned} \tag{10.52}$$

Equalizing the real and imaginary parts of equation (10.52) gives:

$$k_x''^2 + k_{z1}''^2 - k_{z2}''^2 = \frac{\mu''\varepsilon_1''\omega^2}{c^2}, \tag{10.53}$$

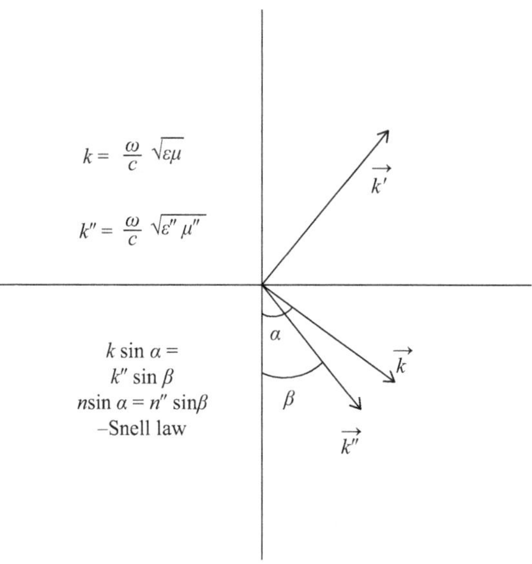

Figure 10.2. The geometry of the wavevectors and Snell's law for the case of zero conductivity $\sigma = 0$.

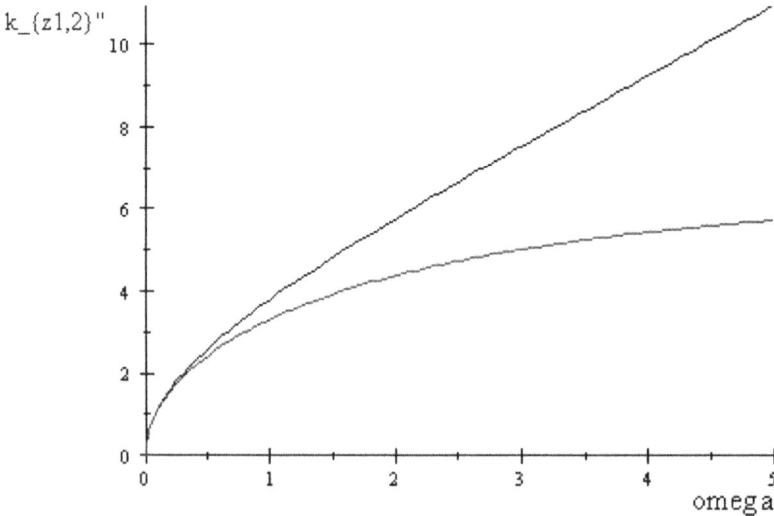

Figure 10.3. The dependence of the components k''_{z1}, k''_{z2} on frequency (10.74) in the range 100–200 rad s^{-1} at $\mu_b = 1$, $\varepsilon_a = 1$, $\sigma_b = 2$.

$$2k''_{z1}k''_{z2} = -\frac{\mu''\varepsilon''_2\omega^2}{c^2} - \frac{4\pi\mu''\omega}{c^2}\sigma = -\frac{\mu''\omega^2}{c^2}\left(\varepsilon''_2 + \frac{4\pi}{\omega}\sigma\right). \tag{10.54}$$

Next, a bit simplifying the problem, we restrict ourselves by real conductivity σ and $\varepsilon''_2 = 0$, $\varepsilon''_1 = \varepsilon_1$-omitting the index by ε. Solving this bi-quadratic system and denoting $k''_x = \frac{\omega \sin\alpha}{c}$, we get two real roots (figure 10.3)

$$k''_{z1} = \frac{\omega\sqrt{\mu\varepsilon}}{\sqrt{2}c}\sqrt{\sqrt{\frac{16\pi^2\sigma^2}{\omega^2\varepsilon^2} + \left(\frac{\varepsilon_1}{\varepsilon} - \sin^2\alpha\right)^2} + \frac{\varepsilon_1}{\varepsilon} - \sin^2\alpha}, \tag{10.55}$$

$$k''_{z2} = \frac{\omega\sqrt{\mu\varepsilon}}{\sqrt{2}c}\sqrt{\sqrt{\frac{16\pi^2\sigma^2}{\omega^2\varepsilon^2} + \left(\frac{\varepsilon_1}{\varepsilon} - \sin^2\alpha\right)^2} + \sin^2\alpha - \frac{\varepsilon_1}{\varepsilon}}. \tag{10.56}$$

The general case gives similar expressions.

10.3.2 To Fresnel formulas generalization

We write the relations (10.47) and (10.48) together with ones for magnetic field at $\vec{r}$ in the plane $x = 0$ and dropping one dimension by choosing the axes as following,

$$\varepsilon E_{0z}e^{i\vec{k}\vec{r}} + \varepsilon E'_{0z}e^{i\vec{k}'\vec{r}} = \varepsilon'' E''_{0z}e^{i\vec{k}''\vec{r}}, \tag{10.57}$$

$$E_{0x}e^{i\vec{k}\vec{r}} + E'_{0x}e^{i\vec{k}'\vec{r}} = E''_{0x}e^{i\vec{k}''\vec{r}}, \tag{10.58}$$

$$B_{0z}e^{i\vec{k}\vec{r}} + B'_{0z}e^{i\vec{k}'\vec{r}} = B''_{0z}e^{i\vec{k}''\vec{r}}, \tag{10.59}$$

$$\frac{1}{\mu}B_{0y}e^{i\vec{k}\vec{r}} + \frac{1}{\mu}B'_{0y}e^{i\vec{k}'\vec{r}} = \frac{1}{\mu''}B''_{0y}e^{i\vec{k}''\vec{r}}, \tag{10.60}$$

that fix the only polarization. It is possible only if $\vec{k}\vec{r} = \vec{k}'\vec{r} = \vec{k}''\vec{r}$, hence

$$\varepsilon E_{0z} + \varepsilon E'_{0z} = \varepsilon'' E''_{0z}, \tag{10.61}$$

$$E_{0x} + E'_{0x} = E''_{0x}, \tag{10.62}$$

$$B_{0z} + B'_{0z} = B''_{0z}, \tag{10.63}$$

$$B_{0y} + B'_{0y} = \frac{\mu}{\mu''}B''_{0y}. \tag{10.64}$$

Expressing the magnetic field components via equation (10.39) yields

$$\varepsilon E_{0z} + \varepsilon E'_{0z} = \varepsilon'' E''_{0z}, \tag{10.65}$$

$$E_{0x} + E'_{0x} = E''_{0x}, \tag{10.66}$$

$$(k_z E_{0x} - k_x E_{0z}) - (k_z E'_{0x} + k_x E'_{0z}) = \frac{\mu}{\mu''}((k''_{z1} + i\cdot k''_{z2})E''_{0x} - k''_x E''_{0z}). \tag{10.67}$$

Putting first and second equation in the third one we get

$$(k_z E_{0x} - k_x E_{0z}) - (k_z E'_{0x} + k_x E'_{0z}) = \frac{\mu}{\mu''}((k''_{z1} + i\cdot k''_{z2})(E_{0x} + E'_{0x}) - k''_x\frac{\varepsilon}{\varepsilon''}(E_{0z} + E'_{0z})). \tag{10.68}$$

Using the link between the components from (10.34a) for a reflected wave with

$$E'_{0z} = \frac{k_x}{k_z} E'_{0x}, \tag{10.69}$$

because $k'_x = k_x$, $k'_z = -k_z$.

Next, solving with respect to E'_{0x}, one has

$$E'_{0x} = E_{0x} \frac{\mu''\varepsilon'' k_x^2 + \mu''\varepsilon'' k_z^2 - k_z''\mu\varepsilon'' k_z + \mu\varepsilon k_x k_z}{\mu\varepsilon k_x^2 - \mu''\varepsilon'' k_x^2 + \mu''\varepsilon'' k_z^2 + k_z''\mu\varepsilon'' k_z}. \tag{10.70}$$

Similarly

$$E''_{0x} = E_{0x} \frac{\mu\varepsilon k_x^2 + 2\mu''\varepsilon'' k_z^2 + \mu\varepsilon k_x k_z}{\mu\varepsilon k_x^2 - \mu''\varepsilon'' k_x^2 + \mu''\varepsilon'' k_z^2 + k_z''\mu\varepsilon'' k_z}, \tag{10.71}$$

with the account for link as equation (10.69)

$$E''_{0z} = -\frac{k_x}{k''_z} E''_{0x}, \tag{10.72}$$

and

$$E''_{0z} = -E_{0x} k_x \frac{\mu\varepsilon k_x^2 + 2\mu''\varepsilon'' k_z^2 + \mu\varepsilon k_x k_z}{k_z''(\mu\varepsilon k_x^2 - \mu''\varepsilon'' k_x^2 + \mu''\varepsilon'' k_z^2 + k_z''\mu\varepsilon'' k_z)}. \tag{10.73}$$

The formulas (10.71) and (10.73) simplify for dielectric matter $\sigma = 0$, $\varepsilon_2'' = 0$, $\varepsilon_1'' = \varepsilon$. Starting from equation (10.74) we reduce it as

$$k''_{z1} = \frac{\omega\sqrt{\mu\varepsilon}}{c} \sqrt{\frac{\varepsilon_1}{\varepsilon} - \sin^2\alpha}\,, \tag{10.74}$$

$$k''_{z2} = \frac{\omega\sqrt{\mu\varepsilon}}{\sqrt{2}\,c} \sqrt{\sqrt{\left(\frac{\varepsilon_1}{\varepsilon} - \sin^2\alpha\right)^2} + \sin^2\alpha - \frac{\varepsilon_1}{\varepsilon}} = 0, \tag{10.75}$$

on a way to Fresnel formulas. To evaluate the intensities of reflected and refracted waves we will use the Pointing vectors defined in chapter 5, section 5.2.5 by equation (5.55) via vectors $\vec{k}'$, $\vec{k}''$ given by equation (10.53) and amplitudes $\vec{E}_0'$, $\vec{E}_0''$.

10.4 Energy density flux

10.4.1 Preparation: Pointing vector for a matter

To define the direction of energy transfer, let us evaluate the averaged in time Pointing vector. The instant flux density in terms of the introduced complex amplitudes (10.29) is proportional to:

$$\mathfrak{E} \times \mathfrak{B} = \frac{1}{4}\{\mathbf{E} \times \mathbf{B}\exp(2i\omega t) + \mathbf{E}^{*}\times\mathbf{B}^{*}\exp(-2i\omega t) + \mathbf{E} \times \mathbf{B}^{*} + \mathbf{E}^{*}\times\mathbf{B}\},$$

because of the proportionality of the fields $\vec{B}$ and $\vec{H}$, due to equation (10.2). The averaged flux density inside the conducting medium is expressed via Pointing vector:

$$\mathfrak{S}'' = \frac{c}{4\pi}\frac{1}{\tau}\int_0^\tau dt\mathfrak{E}'' \times \mathfrak{B}''. \tag{10.76}$$

Integrals of exponent functions over the time period $\tau = \frac{2\pi}{\omega}$ give 0. Hence:

$$\mathfrak{S} = \frac{c}{16\pi}[\mathbf{E} \times \mathbf{B}^{*}+\mathbf{E}^{*}\times\mathbf{B}]. \tag{10.77}$$

Plugging for refracted wave

$$\mathbf{E}'' = \mathbf{E}_0'' e^{i\vec{k}''\vec{r}}, \tag{10.78}$$

$$\mathbf{B}'' = \mathbf{B}_0'' e^{i\vec{k}''\vec{r}}, \tag{10.79}$$

into equation (10.77) yields

$$\mathfrak{S}'' = \frac{c}{16\pi}\left[\mathbf{E}_0'' e^{i\vec{k}\vec{r}} \times \mathbf{B}_0''^{*} e^{-i\vec{k}''^{*}\vec{r}} + \mathbf{E}_0''^{*} e^{-i\vec{k}^{*}\vec{r}} \times \mathbf{B}_0'' e^{i\vec{k}\vec{r}}\right]. \tag{10.80}$$

Extracting scalars from vector products, we have

$$\mathfrak{S}'' = \frac{c}{16\pi} e^{i[\vec{k}''-\vec{k}''^{*}]\vec{r}}[\mathbf{E}_0'' \times \mathbf{B}_0''^{*} + \mathbf{E}_0''^{*} \times \mathbf{B}_0'']. \tag{10.81}$$

Expressing B''_0 from equation (10.39)

$$\mathbf{B}_0'' = -\frac{c}{\omega}\vec{k}'' \times \mathbf{E}_0'', \tag{10.82}$$

and, plugging this equation (10.82) into equation (10.81) results in

$$\mathfrak{S}'' = -\frac{c^2}{16\pi\omega} e^{i[\vec{k}''-\vec{k}''^{*}]\vec{r}}[\mathbf{E}_0'' \times [\vec{k}''^{*}\times\mathbf{E}_0''^{*}] + \mathbf{E}_0''^{**} \times [\vec{k}'' \times \mathbf{E}_0'']]. \tag{10.83}$$

10.4.2 Polarization choice

Choosing such polarization that the field vector **E** lies inside the plane of the vector $\vec{k}$, hence in the two dimensional case via the identity bac–cab, the term of equation (10.83) is rewritten so as

$$-\mathbf{E}_0''^{*}(\mathbf{E}_0'', \vec{k}''^{*}) - \mathbf{E}_0''(\mathbf{E}''^{*}_0, \vec{k}''), \tag{10.84}$$

into more simple form, adding the opposite terms

$$-\mathbf{E}_0''^{*}(\mathbf{E}_0'', \vec{k}'' - 2i\Im\vec{k}'') - \mathbf{E}_0''(\mathbf{E}''^{*}_0, \vec{k}''^{*} + 2i\Im\vec{k}''), \tag{10.85}$$

(the symbols $\Re$ and $\Im$ denote real and imaginary parts of a complex number), expanding the expressions we arrive at

$$\begin{aligned} &-\mathbf{E}_0''^*(\mathbf{E}_0'', \vec{k}'') - \mathbf{E}_0''^*(\mathbf{E}_0'', -2i\Im\vec{k}'') \\ &-\mathbf{E}_0''(\mathbf{E}''^*_0, \vec{k}''^*) - \mathbf{E}_0''(\mathbf{E}''^*_0, 2i\Im\vec{k}''). \end{aligned} \tag{10.86}$$

Now we can apply equations (10.38), $(\mathbf{E}_0'', \vec{k}'') = (\mathbf{E}''^*_0, \vec{k}''^*) = 0$, that result in

$$\mathbf{E}_0''^*(\mathbf{E}_0'', 2i\Im\vec{k}'') - \mathbf{E}_0''(\mathbf{E}''^*_0, 2i\Im\vec{k}''). \tag{10.87}$$

Let us pick up the results in expression for $\mathfrak{S}''$, of the (10.83)

$$\mathfrak{S}'' = -\frac{c^2}{16\pi\omega} e^{i[\vec{k}''-\vec{k}''^*]\vec{r}} \times \left[2\Re\vec{k}''(\mathbf{E}_0'', \mathbf{E}_0''^*) + \mathbf{E}_0''^*(\mathbf{E}_0'', 2i\Im\vec{k}'') - \mathbf{E}_0''(\mathbf{E}_0''^*, 2i\Im\vec{k}'')\right]. \tag{10.88}$$

After some algebra of [13], reading the bac–cab identity from right to left, we finally obtain

$$4[\Im\vec{k}'' \times [\Im\mathbf{E}_0'' \times \mathbf{E}_0'']]. \tag{10.89}$$

So the Pointing vector is:

$$\mathfrak{S}'' = -\frac{c^2}{8\pi\omega} e^{i[\vec{k}''-\vec{k}''^*]\vec{r}}[\Re\vec{k}''^*(\mathbf{E}_0'', \mathbf{E}_0''^*) + 2[\Im\vec{k}'' \times [\Im\mathbf{E}_0'' \times \mathbf{E}_0'']]]= \tag{10.90}$$

$$-\frac{c^2}{8\pi\omega} e^{i[\vec{k}''-\vec{k}''^*]\vec{r}}[\Re\vec{k}''^*(\mathbf{E}_0'', \mathbf{E}_0''^*) + 2[\Im\vec{k}'' \times [\Im\mathbf{E}_0'' \times (\Re\mathbf{E}_0'' + i\Im\mathbf{E}_0'')]]]. \tag{10.91}$$

Finally

$$\mathfrak{S}'' = -\frac{c^2}{8\pi\omega} e^{i[\vec{k}''-\vec{k}''^*]\vec{r}}[\Re\vec{k}''^*(\mathbf{E}_0'', \mathbf{E}_0''^*) + 2[\Im\vec{k}'' \times [\Im\mathbf{E}_0'' \times \Re\mathbf{E}_0'']]]. \tag{10.92}$$

We arrive at a compact formula for the Pointing vector for refracted wave:

$$\mathfrak{S}'' = -\frac{c^2}{8\pi\omega} e^{-2(\Im\vec{k}'', \vec{r})}(\Re\vec{k}''(\mathbf{E}_0'', \mathbf{E}_0''^*) + 2[\Im\vec{k}'' \times [\Im\mathbf{E}_0'' \times \Re\mathbf{E}_0'']]). \tag{10.93}$$

It is important to note, that the expression (10.93) contains only the real parameters, hence it is real as a whole!

10.4.3 The generalized Snell law

To formulate the generalized Snell law, that may be composed from two relations as follows. For the incident wave we put

$$\left(\frac{\vec{S}}{S}, \vec{i}\right) = \sin\alpha, \tag{10.94}$$

and, for the refracted one

$$\left(\frac{\vec{S}''}{S''}, \vec{i}\right) = \sin\beta, \tag{10.95}$$

where $\vec{S}''$ is proportional to (10.93). So, the ratio of $\frac{\sin\beta}{\sin\alpha}$ (relative refraction index) is completely defined by the Pointing vectors of the waves. At $z = 0$ the expression (10.95) is written as

$$\sin\beta = \frac{\Re k_x''(\mathbf{E}_0'', \mathbf{E}_0''^*) - E_{0x}''^*(\mathbf{E}_0'', \vec{k}''^*) - E_{0x}''(\mathbf{E}_0''^*, \vec{k}'')}{|\Re\vec{k}''(\mathbf{E}_0'', \mathbf{E}_0''^*) - \mathbf{E}_0''^*(\mathbf{E}_0'', \vec{k}''^*) - \mathbf{E}_0''(\mathbf{E}_0''^*, \vec{k}'')|}. \tag{10.96}$$

Now we can plug the results in equation (10.96) to derive the generalized Snell law and Fresnel formulas. To evaluate the intensities of reflected and refracted waves we will use the Pointing vectors defined by equation (10.93) via vectors $\vec{k}'$, $\vec{k}''$ given by equation (10.53) and amplitudes $\vec{E}_0'$, $\vec{E}_0''$. It simplifies as

$$\mathfrak{S}'' = -\frac{c^2}{16\pi\omega}e^{-2\Im k_z''z}[\Re\vec{k}''(\mathbf{E}_0'', \mathbf{E}_0''^*) - \mathbf{E}_0''^*(\mathbf{E}_0'', \vec{k}''^*) - \mathbf{E}_0''(\mathbf{E}_0''^*, \vec{k}'')]. \tag{10.97}$$

From this it follows that the parameter $\Im k_z''$ should be positive. ·It defines the decrement of attenuation. The first term is equal to

$$\begin{aligned}\Re\vec{k}''(\mathbf{E}_0'', \mathbf{E}_0''^*) &= \Re\vec{k}''(E_{0x}''E_{0x}''^* + E_{0z}''E_{0z}''^*)\\ &= \Re\vec{k}''\left(1 + \frac{k_x^2}{|k_z''|^2}\right)E_{0x}''E_{0x}''^*,\end{aligned} \tag{10.98}$$

while the rest are expanding as

$$-\mathbf{E}_0''^*(\mathbf{E}_0'',\vec{k}''^*) - \mathbf{E}_0''(\mathbf{E}_0''^*, \vec{k}'') = 2i\mathbf{E}_0''^*(\mathbf{E}_0'',\Im\vec{k}_z'') - 2i\mathbf{E}_0''(\mathbf{E}_0''^*,\Im\vec{k}_z''). \tag{10.99}$$

For the x-component we obtain

$$\begin{aligned}2i\mathbf{E}_{0x}''^*\mathbf{E}_{0z}''\Im k_z'' - 2i\mathbf{E}_{0x}''\mathbf{E}_{0z}''^*\Im k_z'' &= -2ik_x\left(\frac{k_z''^*}{|k_z''|^2} - \frac{k_z''}{|k_z''|^2}\right)E_{0x}''^*E_{0x}''\Im k_z''\\ &= -4k_x\frac{(\Im k_z'')^2}{|k_z''|^2}E_{0x}''^*E_{0x}'',\end{aligned} \tag{10.100}$$

while for the z-component we have zero $2i\mathbf{E}_{0z}''^*\mathbf{E}_{0z}''\Im k_z'' - 2i\mathbf{E}_{0z}''\mathbf{E}_{0z}''^*\Im k_z'' = 0$. We should plug it into equation (10.96). Doing the transformations of the vector module, one has

$$\left|\left(\Re \vec{k}''(\mathbf{E}_0'', \mathbf{E}_0''^{*}) - \mathbf{E}_0''^{*}(\mathbf{E}_0'', \vec{k}''^{*}) - \mathbf{E}_0''(\mathbf{E}_0''^{*}, \vec{k}'')\right)\right| =$$

$$E_{0x}''^{*}E_{0x}''\sqrt{\left(k_x\left(1 + \frac{k_x^2}{|k_z''|^2}\right) - 4k_x\frac{(\Im k_z'')^2}{|k_z''|^2}\right)^2 + \left(\Re k_z''\left(1 + \frac{k_x^2}{|k_z''|^2}\right)\right)^2}. \tag{10.101}$$

Finally the Snell law is written as

$$\sin\beta = \frac{-4k_x(\Im k_z'')^2 + k_x(|k_z''|^2 + k_x^2)}{\sqrt{\left(k_x(|k_z''|^2 + k_x^2) - 4k_x(\Im k_z'')^2\right)^2 + \left(\Re k_z''(|k_z''|^2 + k_x^2)\right)^2}}, \tag{10.102}$$

or

$$\sin\beta = \frac{1 - \frac{4(\Im k_z'')^2}{|k_z''|^2 + k_x^2}}{\sqrt{\left(1 - \frac{4(\Im k_z'')^2}{|k_z''|^2 + k_x^2}\right)^2 + \left(\frac{\Re k_z''}{k_x}\right)^2}}, \tag{10.103}$$

From equation (10.74) we evaluate $\Re k_z''$ and the imaginary part $\Im k_z''$ as functions of the matter parameters and the incident angle α.

In the case of $\Im k_z'' = 0$, $k_x = \frac{\sqrt{\varepsilon\mu}\,\omega \sin\alpha}{c}$ we arrive at the conventional Snell's law for a boundary between dielectrics ($\mu = 1$)

$$\sin\beta = \frac{\frac{\sqrt{\varepsilon}\,\omega}{c}\sin\alpha}{\sqrt{\frac{\varepsilon''\omega^2}{c^2}}} = \frac{\sqrt{\varepsilon}\sin\alpha}{\sqrt{\varepsilon''}}, \tag{10.104}$$

after account for

$$k_x^2 + k_{z1}^{2}{}'' = \frac{\varepsilon''\omega^2}{c^2}. \tag{10.105}$$

For the Fresnel formulas, based on explicit formulas (19.8), (10.70) and (10.71), there is a program written by means of symbolic computation [13].

10.5 Discussion

By Bergmann [7] (see also [4]), the Snell law for a conductive medium is expressed in rather an ambiguous way (notations of book [4] are used). The real angle is determined by

$$\sin\theta_1 = \sin\beta = \frac{\alpha_2 \sin\theta_0}{\sqrt{q^2 + \alpha_2^2\sin^2\theta_0}}, \tag{10.106}$$

where θ_0 is the angle of incident wave, $\alpha_2 = \omega\sqrt{\mu_2\varepsilon_2} = \omega/c_2$, and

$$q = \rho(\alpha_1\cos\gamma - \beta_1\sin\gamma),$$

is defined via *complex* $\cos\theta_1 = \rho e^{i\gamma}$. The index '1' marks the conducting medium, and

$$k_1^2 = \omega^2\mu_1\varepsilon_1 + i\omega\sigma_1\mu_1 \\ = (\vec{k}'', \vec{k}''), \tag{10.107}$$

the rhs is written via the notation of this article with

$$k_1 = \alpha_1 + i\beta_1.$$

The expression via the medium parameters are given by the formula (10.60) of [4]

$$q^2 = \frac{1}{2}\left[\alpha_1^2 - \beta_1^2 - \alpha_2^2 \sin^2\theta_0 + \sqrt{4\alpha_1^2\beta_1^2 + \left(\alpha_1^2 - \beta_1^2 - \alpha_2^2 \sin^2\theta_0\right)^2}\,\right].$$

The solution with respect to α_1, β_1 gives

$$\mu_1 = \sqrt{\frac{\omega^2\mu_1\varepsilon_1}{2} + \sqrt{\left(\frac{\omega^2\mu_1\varepsilon_1}{2}\right)^2 + \left(\frac{\omega\sigma\mu_1}{2}\right)^2}},$$

$$\varepsilon_1 = \frac{\omega\sigma\mu_1}{2\sqrt{\frac{\omega^2\mu_1\varepsilon_1}{2} + \sqrt{\left(\frac{\omega^2\mu_1\varepsilon_1}{2}\right)^2 + \left(\frac{\omega\sigma\mu_1}{2}\right)^2}}}.$$

Relations between notations of equations (10.103) and (10.106) are extracted from equations (10.46) and (10.107)

$$k_1^2 = \alpha_1^2 + 2i\alpha_1\beta_1 - \beta_1^2 = k_x^2 + (k_{1z}'')^2 + 2i\Re k_z''\Im k_z'' - (\Im k_z'')^2.$$

Equalizing real and imaginary parts yields the system

$$\alpha_1^2 - \beta_1^2 = k_x^2 + (\Re k_z'')^2 - (\Im k_z'')^2,$$

$$\alpha_1\beta_1 = \Re k_z''\Im k_z''.$$

The system is a bi-quadratic equation with respect to $\Re k_z''$, $\Im k_z''$.

For a test let us take the case

$$\mu_1 = \mu_2 = \varepsilon_2 = 1,\ \varepsilon_1 = 2,\ \sigma_1 = 3.$$

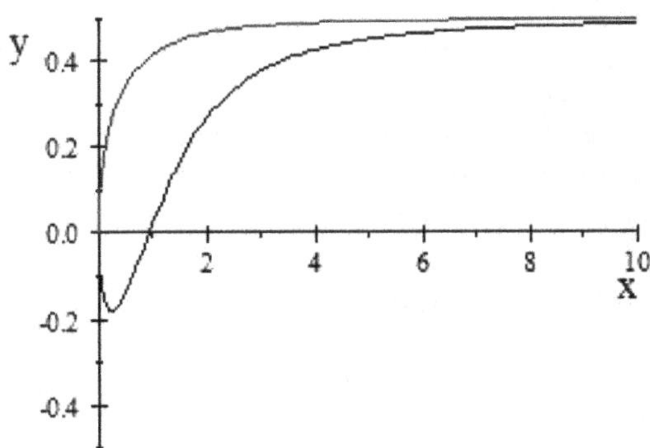

Figure 10.4. The dependence of $\sin\beta$, evaluated by the expressions (10.106) (conventional) and one by (10.103) is plotted as a function of frequency at range [0, 10].

Then $\alpha_2 = \frac{\omega}{c}$.

The evaluation of $\sin\beta$ by the expressions (10.106) (conventional) and (10.103) is plotted as a function of frequency below in figure 10.4.

The expression for the sine of refraction angle is derived via Pointing vector as a function of the attenuation parameter $\Im k_z''$ and wave vector components of the incident wave and the conducting medium parameters ε'', σ, μ''. The resulting formula is compared to the conventional one obtained by means of complex angle introduction and its interpretation in terms of phase front propagation. There is a difference between the results that are plotted as a function of frequency at some exemplary values of parameters. There are a lot of possible applications of this paper results in Rentgen optics [5] and, after a slight development, in the theory of electromagnetic waves propagation in metamaterials [6].

Exercise 1. Consider the generalization of relations (10.74) to the case of $\varepsilon''_2 \neq 0$, $\varepsilon''_1 = 0$ [12].

Exercise 2. Reproduce the algebra of [13], to obtain the Pointing vector (10.90).

Exercise 3. On the basis of the resulting formulas for field amplitudes (19.8), (10.70) and (10.71), derive the conventional Fresnel formula for non-conducting medium ($\sigma = 0$).

References

[1] Guggenheim E A 1967 *Thermodynamics An Advanced Treatment for Chemists and Physicists* (Amsterdam: Elsevier)

[2] Fock V A and Armstrong J C 1967 Electromagnetic diffraction and propagation problems. International series of monographs in electromagnetic waves, vol 1 *J. Phys.* **35** 362

[3] Fock V 1948 The Fresnel reflection law and diffraction laws *UFN* **36** 308–27

[4] Stratton J A 1941 *Electromagnetic Theory* (New York: McGraw-Hill)

[5] Aristov V V and Shabelnikov L G 2008 Recent advances in x-ray refractive optics *Physics: Uspekhi* **51** 57–77

[6] Schurig D and Smith D R 2005 Negative index lenses Negative-refraction metamaterials: Fundamental Principles and Applications ed G I Eleftheriades and K G Balmain (New York: Wiley)

[7] Bergmann J 1932 Die Erzeugung zirkular polarislerter elektrischer Wellen durcb einmalige Totalreflexion *Phys. Z.* **33** 582

[8] De Roo R and Tai C-T 2003 Plane wave reflection and refraction involving a finitely conducting medium *IEEE Antennas Propag. Mag.* **45** 54–61

[9] Suffczinski M 1965 *Elektrodynamika* (Warszawa: PWN)

[10] Jackson J D 1998 *Classical Electrodynamics* 3rd edn (New York: Wiley)

[11] Keam R B 1994 Plane wave excitation of an infinite dielectric rod *IEEE Microwave Guided Wave Lett.* **4** 326

[12] Agrawal G P 2000 *Nonlinear fiber optics Nonlinear Science at the Dawn of the 21st Century* (Berlin: Springer) pp 195–211

[13] Leble S and Vereshchagin S 2016 On reflection and refraction of plane electromagnetic waves at a conducting matter surface *Task Q.* **20** 195–205

IOP Publishing

Practical Electrodynamics with Advanced Applications

Sergey Leble

Chapter 11

New dielectric guides techniques

11.1 Planar waveguides

This chapter is devoted to waveguides, such formed by a medium inhomogeneity, that can capture electromagnetic wave in some domain. The capture is realized by means of a sharp change of matter parameters that we model by boundary condition. The guides are conventionally used as components in integrated optical circuits, as the long distances transmission medium for light wave communications, or for biomedical imaging [1]. The simplest waveguides in 2D and 3D electrodynamics are planar. In the first section of this chapter we follow the paper from [2] and others follow [3] by the textbook adaptation.

11.1.1 Novel experiments in dielectric guides

The search for a new type of dielectric waveguides for millimeter-wave range, which has a number of advantages over other previously available waveguide structures integrated circuits, is described. Dispersion characteristics and the field distributions in the waveguide are calculated using the concept of effective dielectric constant. Field distributions have been measured in the tenth GHz range in order to check the accuracy of the analytical results. This measurement has been done using a novel experimental technique, which should also be applicable to many other millimeter-wave waveguides and components [4]. More attention is given to numerical modeling of a waveguide geometry and matter, this way, doing a search of appropriate dielectric materials. The scale of transistors and capacitors in electronics is reduced to less than a few nanometers, that accompanied by a growth of leakage currents that pose a serious problem. To overcome this dilemma, high-κ (high dielectric constant, kappa) materials that exhibit a larger permittivity and band gap are introduced as gate dielectrics. Currently, HfO2 is widely used as a high-κ dielectric; however, a higher-κ material remains desired for further enhancement. To find new high-κ materials, the authors conduct a high-performance *ab initio* calculation for band gap and permittivity [8].

A new leaky nonradiative dielectric guide with a double-layer dielectric slab is presented in [7], for the millimeter-wave range. It combines a multimode theory with a mode-matching method. Numerical results show that proper selection of the permittivities can vary the leakage constant over a wide range. A leakage constant larger than 10% may be obtained easily without entering the cutoff region (in which the transistor behaves as an open switch). Leaky-wave antennas with shorter length and medium gain can thus be developed. In addition, the present leaky guide possesses the advantages of simple fabrication and physical stability.

Another important area of investigation is related to the search of compensation dispersion and nonlinearity. A phenomenon in which the effect is realized is named soliton-like behavior [3]. In the optic wave range, the paper [6] entitled *Observation of Manakov Spatial Solitons in AlGaAs Planar Waveguides* appeared in 1995. As implied by the theory, the specific conditions for a coupled nonlinear Schrödinger (CNS) equation (11.31) and (11.32), section 11.1.2, that lead to the possibility of this observation are:

- unit ratio between self-phase modulation and cross-phase modulation, that defines the relative magnitudes of the components of the tensor χ (for the definition see below, section 11.1.2, with $n_1 \approx n_2$, $n_3 \approx n_4$ in equation (11.31)), and
- the four-wave mixing terms must be negligible.

The exact implementation of the conditions is very specific, needing to search for the materials with such properties. For a strongly birefringent material, that was used in the experiment of [6]; the first condition is fulfilled approximately and the second one is not valid, but due to fast phase interchanges between polarization components is not visible. The approximate character of theoretical description in solitonics is a usual circumstance and, if a deviation of real matter from one described by integrable system stands, the result of observations depends on whether the effect is accrued with time. We would add, that an effect of spatial soliton dragging is observed by the same group [23]. Unfortunately in the paper [23], the link with the Manakov paper [24], where the case of CNLS integrability was established, was not mentioned, but a natural reference to the earlier pioneering work [25] was given. In particular, this paper initiated the investigation of self-focusing phenomena from the standpoint of the theory of integrable systems [26]. The article [23] already announced the possibility of using such spatial solitons, as the tool for focusing in the direction, orthogonal to propagation direction. It opens a way to all-optical switch construction. The dragging of bright temporal solitons should be noted [22], but its high phase sensibility, perhaps, not allow practical applications for switching devices. A vector soliton, dragging in a planar glass waveguide [27], was experimentally observed and may be an alternative approach to an optical inverter or NOT gate, a logic gate. The results of numerical simulation raise the question of integrability of the general equation applied to an isotropic matter, see also [28] on soliton-like pulses collisions in a glass medium. The next step of the idea development is directly related to polariton-plasmonics ideas [12].

A soliton vector pulses of electromagnetic field of two types has been observed recently in the fiber laser guide [30]. One was a polarization-locked vector soliton, that appears as a result of compensations produced while the soliton propagates due to randomly varying birefringence medium [29]. The collision property [31] also exhibits features that are typical of Manakov-like solitons to integrable systems.

Summarizing, the three different types of solitons for the CNS can be specified as follows

- *polarization instability of soliton* [31], closest to the conventional one (NS-nonlinear Schrödinger), with two versions:
  - periodic wave [32],
  - solitonic wave [33].

If the Kerr coefficient is positive (as in optical fibers), the fast axis becomes unstable in both cases.

- *Manakov soliton*, the classical integrable case [24]; space ones in plates, a temporal version is the pulse in fibers [14], the spatial one is observed in specially engineered planar anisotropic guides where the ratio of the non-linear constants is close to unity [6, 27].
- *Polarization-locked vector soliton*. This has elliptic polarization, stabilized by cross-phase modulation and self-phase modulation to compensate for the differing phase velocities' of the orthogonal components [29].

The coupled nonlinear Schrödinger (CNS) equation has the general form, but in a frame with the averaged linear group velocities c_i have opposite signs. We should mention the recent communication about hybrid vector spatial plasmon solitons in a Kerr slab embedded between metal plates, analyzed with a modified nonlinear Schrödinger (MNS) equation [34].

11.1.2 Dielectric slab as a waveguide

11.1.2.1 Main equations

Suppose a slab of a dielectric fills the interval $y \in [-h, h]$, see figure 11.1, and its dimension along x, z is much bigger than $2h$. We shall call such plate a planar waveguide, in accordance with most publications exposition (e.g. [1, 6]). The general theory may be applied to such geometry, following details of modifications to the results in the book [38], section 3.4. In the first section we follow [2] again, that is a development of the consideration on [38]. The matter model includes the linear isotropic dielectric with constant permeability ε inside the interval $y \in [-h, h]$ and unity outside.

Paper [6] describes how the Manakov soliton was observed and the output laser beam transversal to propagation along z profiles were directly demonstrated. The conditions of the experiments involve nonlinear diffraction in a direction (say, x) that mimics the 1D soliton evolution, orthogonal to the transverse slab coordinate y. Hence, developing the results of [38] in the weak nonlinearity range account, we first summarize the generalities in order to specify the choice of notation within the reference frame marked at figure 11.1.

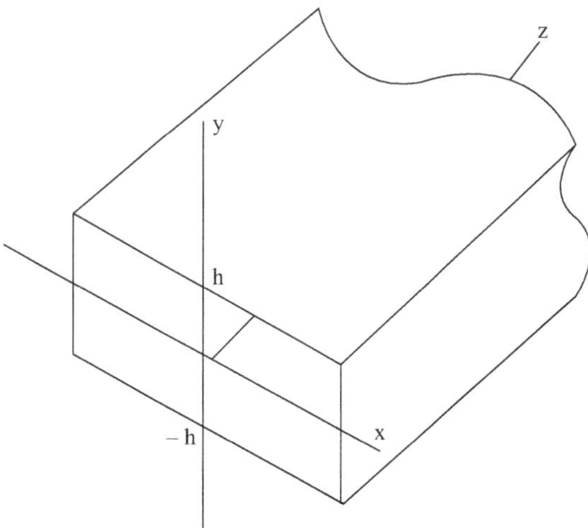

Figure 11.1. This geometry and coordinates of a planar waveguide position.

Maxwell's equations in Gauss units are most convenient for theoretical study, cf chapter 10, section 10.1.1, they form the system:

$$\frac{1}{c}\frac{\partial \mathbf{D}}{\partial t} = \text{rot}\mathbf{H}, \quad \frac{1}{c}\frac{\partial \mathbf{B}}{\partial t} = -\text{rot}\mathbf{E}, \quad \text{div}\,\mathbf{D} = 0, \quad \text{div}\,\mathbf{B} = 0. \tag{11.1}$$

The equations are chosen to account for the averaged constants of the dielectric layer ε and μ within a narrow frequency range, separating the nonlinearity and anisotropy in the definition of the polarization vector **P**.

As the dielectric matter is conventionally used, we arrange experiments so that no currents or charges $\rho = 0, \vec{j} = 0$, are taken into account. We also restrict ourselves to the following form of the permittivities :

$$\mathbf{D} = \varepsilon\mathbf{E} + 4\pi\mathbf{P}, \qquad \mathbf{B} = \mu\mathbf{H}, \tag{11.2}$$

where, the following relations introduce permittivity tensors χ^i_{ik}

$$P_i = \chi^0_{ik}E_k + \chi^1_{ik}|\mathbf{E}|^2 E_k + \cdots, \tag{11.3}$$

that separate the linear anisotropic and nonlinear parts. Repetition of indices implies summation and only Kerr nonlinearity is taken into account.

Combining the Maxwell equations (11.1) and using the definition of the D'Alembert operator $\Box = \frac{\mu\varepsilon}{c^2}\frac{\partial^2}{\partial t^2} - \Delta$, we obtain the basic closed set of wave equations for the electric field in the vector form

$$\Box\mathbf{E} = -4\pi[\frac{\mu}{c^2}\mathbf{P}_{tt} + \nabla(\nabla\mathbf{P})]. \tag{11.4}$$

The function $Y(y)$ stands for a transverse waveguide mode.

11.1.2.2 Waveguide modes

Let the linearized equation (11.4) for the TE mode have the solution

$$E_x = E_e(x,\, t)Y(y)\exp i(kz - \omega t) + c.c. \tag{11.5}$$

Inside the interval $y \in [-h,\, h]$ the functions

$$Y(y) = Y_0 \sin(\alpha y), \tag{11.6}$$

satisfy the equation

$$Y_{yy} = -\alpha^2 Y, \tag{11.7}$$

and describe the (transverse) orthonormal waveguide basis, with $\alpha^2 = \omega^2/c_0^2 - k^2$, $c_0^2 = c^2/\varepsilon$, for a dielectric we take $\mu = 1$. This is nothing other than the dispersion relation arising from the linear version of the basic system (11.4) after the space and time variables separation.

The spectrum of α is determined from boundary conditions that link Fourier components inside and outside the dielectric guide interval $[-h,\, h]$. We shall comment on this below. The continuity conditions follow from the integral form of Maxwell equations; those are the tangential components of electric field $\vec{E}$ and the normal component of the vector $\vec{D}$, see chapter 9, section 9.2.1. For the upper boundary the continuity condition for the tangential components reads

$$E_t|_{h-0} = E_t|_{h+0}, \tag{11.8}$$

the index marks the limit $E_t|_{h-0} = \lim_{y\to h} E_t$ from inside of the slab, similar for the rhs, and for the lower limit, from outside it is

$$E_t|_{-h-0} = E_t|_{-h+0}, \tag{11.9}$$

then, outside the slab, the solution exponentially decays

$$Y(y) = \begin{cases} Y_0 \exp[-py],\ y \in [h,\, \infty) \\ -\ Y_0 \exp[py],\ y \in (-\infty,\, -h]. \end{cases} \tag{11.10}$$

The single-mode linear matching of Y and $\exp[-py]$, $y > 0$ at both boundaries $y = \pm h$ implies

$$\sin(\alpha h) = \exp[-ph]. \tag{11.11}$$

The relations (11.10) account these conditions (11.8) and (11.9), while the linearized wave equation (11.4) gives

$$p^2 = k^2 - \omega^2/c^2,\ \alpha^2 = \omega^2/c_0^2 - k^2. \tag{11.12}$$

Next, excluding k, write:

$$p^2 = -\alpha^2 + (\varepsilon - 1)\omega^2/c^2. \tag{11.13}$$

For the normal component the upper boundary condition reads

$$D_n|_{h-0} = \varepsilon E_n|_{h-0} = E_n|_{h+0}, \tag{11.14}$$

and, for the lower one,

$$D_n|_{-h+0} = \varepsilon E_n|_{-h+0} = E_n|_{-h-0}. \tag{11.15}$$

We would extract the normal component, that is $E_n = E_y$, in the context of this section, from the basic

$$\operatorname{div} \vec{E} = 0, \tag{11.16}$$

that, for constant ε, is valid inside the slab and in outer space points.

We restrict ourselves by the only TE component presence, putting $E_z = 0$, so equation (11.16) reads

$$\frac{\partial E_y}{\partial y} = -\frac{\partial E_x}{\partial x}, \tag{11.17}$$

the solution (11.5) for $y \in [-h, h]$, is expressed by

$$E_y = \frac{\partial E_e}{\partial x} \frac{\cos[\alpha y]}{\alpha} \exp i(kz - \omega t) + c.c. \tag{11.18}$$

And, for $y \in [-\infty, -h]$,

$$E_y = -\frac{\partial E_e}{\partial x} \frac{\exp(py)}{p} \exp i(kz - \omega t) + c.c., \tag{11.19}$$

while, for upper half plane $y \in [h, \infty]$,

$$E_y = \frac{\partial E_e}{\partial x} \frac{\exp(-py)}{p} \exp i(kz - \omega t) + c.c. \tag{11.20}$$

Plugging equations (11.18) and (11.19) into equation (11.14)

$$\varepsilon E_y|_{h-0} = E_y|_{h+0}, \tag{11.21}$$

yields, at the upper boundary

$$\varepsilon Y_0 \left[\frac{\cos(\alpha h)}{\alpha} \right] = Y_0 \frac{\exp[-ph]}{p}, \tag{11.22}$$

or

$$\varepsilon \frac{\cos(\alpha h)}{\alpha} = \frac{1}{p} \exp(-ph). \tag{11.23}$$

Dividing equation (11.11) by equation (11.22) relates the wave number α with k and ω. Boundary conditions yield

$$\alpha \tan(\alpha h) = p\varepsilon, \qquad \alpha^2 + p^2 = \omega^2(\varepsilon - 1)/c^2. \tag{11.24}$$

Multiplying the equalities (11.24) by h, h^2 introduces dimensionless variables ph and αh, so that

$$hp = \frac{1}{\varepsilon}\alpha h \tan(\alpha h), \qquad (h\alpha)^2 + (hp)^2 = h^2\omega^2(\varepsilon - 1)/c^2. \tag{11.25}$$

Equations (11.25) in the domain $ph > 0$ have compatible solutions at the points of intersection of tangent curves in the intervals $n\pi < \alpha h < n\pi + \pi/2$, $n = 0, 1, \ldots$ with the circumferences, defined by the second equality of equation (11.25). For a choice of the parameters, see caption, the solution of the dimensionless system (11.25), $ph \to p$, $h\alpha \to \alpha$, is shown in the figure 11.2, in this case the only positive eigenvalue is observed. It means that only the guide mode exists with such parameters h, ω, ε values.

11.1.2.3 Coupled NS system

Let us illustrate the algorithm for deriving the weak nonlinear system of the modes interaction, that turns out to be the coupled nonlinear Schrödinger equation (CNS). Let us start from linear terms, the action of the operator □ (named often as 'quabla') on the TE wave (11.5) gives

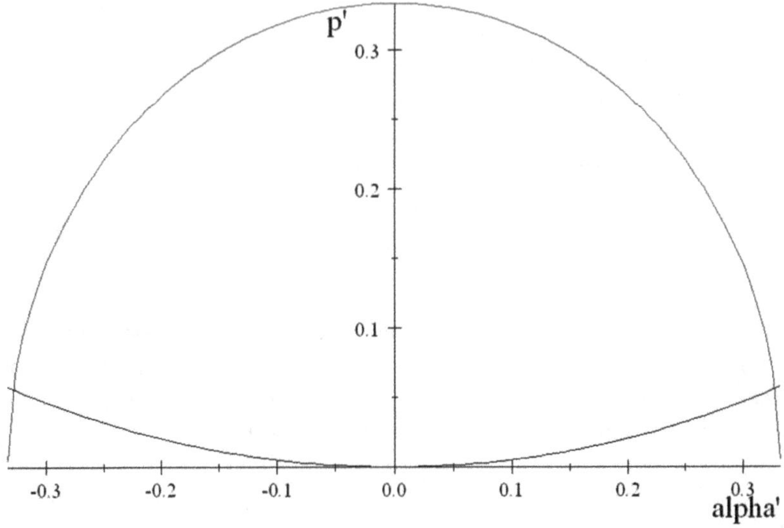

Figure 11.2. The plots of the $ph = p'$ as a function of $\alpha h = \alpha'$ are given for $\epsilon = 2$, $h = 10^{-2}$ m, $\omega = 10^{10}$ Hz, $c = 3 \times 10^8$ m s^{-1}. The first of equation (11.25) is drawn in black, the second—in red. The intersection points projection to the axis αh give the eigenvalues.

$$\exp[-i(kz - \omega t)]\Box E_x = \exp[-i(kz - \omega t)]\Box$$
$$[E_e(z, x, t)Y(y)\exp i(kz - \omega t) + c.c.]$$
$$= (-\omega^2/c_0^2 + k^2 + \alpha^2)YE_e + \left[\frac{-2\iota\omega}{c_0^2}\frac{\partial}{\partial t} - 2ik\frac{\partial}{\partial z} \right. \tag{11.26}$$
$$\left. - \Delta_\perp + \frac{1}{c_0^2}\frac{\partial^2}{\partial t^2}\right]YE_e,$$

with $c_0 = c/\sqrt{\mu\varepsilon}$, where

$$\Delta_\perp = \frac{\partial^2}{\partial x^2} + \frac{\partial^2}{\partial y^2}.$$

The relation between the scales in the x and z directions depends on the pulse duration and the laser beam cross-section. It defines the terms that should be taken into account in the linear part of the final model equation. We easily recognize propagation and dispersion operators in equation (11.26).

We describe the geometry and physics of the experiment in [6]. The two parallel beams of the TE mode and TM mode inputs are chosen orthogonal, and given by the expressions

$$E_{x,y} = E_{e,m}(z, x, t)Y(y)\exp i(kz - \omega t) + c.c. \tag{11.27}$$

These are the simplest one-mode wave trains which account for the 2+1 nature of evolution along both the axis of propagation and the orthogonal axis. We now plug the relation $\nabla P = -\varepsilon\nabla\mathbf{E}$, obtained from equation (11.2), substituted into the third equation of (11.1).

If the amplitude functions depend weakly on the variables, the substitution of P_i with account for equations (11.27) and (11.26) (and similar for E_y) into equation (11.4), after accounting for the dispersion relation $\omega^2 = c_0^2(k^2 + \alpha^2)$, and multiplying by $\exp[-i(kz - \omega t)]$, gives for the first (x) component

$$\left[\frac{-2\iota\omega}{c_0^2}\frac{\partial}{\partial t} - 2ik\frac{\partial}{\partial z} - \Delta_\perp + \frac{1}{c_0^2}\frac{\partial^2}{\partial t^2}\right]YE_e = -4\pi\left[\frac{\mu}{c^2}\mathbf{P}_{1tt} - \varepsilon\frac{\partial}{\partial x}(\nabla\mathbf{E})\right]. \tag{11.28}$$

This equation is the 2+1 evolution of a TE pulse, launched at the boundary, due to propagation, nonlinearity, diffraction along x and dispersion along z action. After transformation to the moving frame, for a relatively long pulse, i.e. if $\lambda_x \ll \lambda_z$, and if $\lambda_{x,z}$ are the corresponding scales along x and z, we arrive at

$$Y\left\{\frac{\partial E_e}{\partial z} + \frac{1}{2\iota k}\frac{\partial^2 E_e}{\partial x^2}\right\} = -Y\frac{4\iota\pi\omega^2\mu}{c^2}\left[\overset{0}{\chi}_{2k}E_k + \overset{1}{\chi}_{2k}|\mathbf{E}|^2E_k + \cdots\right] + \varepsilon\frac{\partial}{\partial y}\frac{\partial E_k}{\partial x_k}, \tag{11.29}$$

and similarly,

$$Y\left\{\frac{\partial E_m}{\partial z} + \frac{1}{2\imath k}\frac{\partial^2 E_m}{\partial x^2}\right\} = -Y\frac{4\pi\imath\omega^2\mu}{c^2}\left[\chi^0_{3k}E_k + \chi^1_{3k}|\mathbf{E}|^2 E_k + \cdots\right] + \varepsilon\frac{\partial}{\partial z}\frac{\partial E_k}{\partial x_k}. \quad (11.30)$$

The equation for the third (z) component does not in fact contribute, due to the zero input value of the TE mode electric field.

Multiplying by the function Y and integrating equations (11.29) and (11.30) across the slab, we obtain the integrals defining the nonlinear and dispersion constants as in [38]. See chapter 20 for orthogonality and normalization conditions formulation. Finally, for one mode, we arrive at the nonlinear Schrödinger equation. We write the system in dimensionless form by rescaling the electric field components ($E_e \mapsto e_+$, $E_m \mapsto e_-$), coordinates, and time ($x \mapsto \tau$, $z \mapsto \zeta$), as in the case of the Maxwell–Bloch equations (see equation (4.2) in [2], section 4.1 for the fields).

The resulting evolution equation is the coupled nonlinear Schrödinger (CNS) system

$$\imath e_{-,\zeta} = \frac{1}{2}e_{-,\tau\tau} + (n_1|e_-|^2 + n_2|e_+|^2)e_-, \quad (11.31)$$

$$\imath e_{+,\zeta} = \frac{1}{2}e_{-,\tau\tau} + (n_3|e_-|^2 + n_4|e_+|^2)e_+. \quad (11.32)$$

There are more terms (linear and nonlinear) that may be neglected under certain special conditions (see, e.g. [6]). The Manakov integrable case occurs when the nonlinear terms in both equations (11.31) and (11.32) enter with approximately equal weights ($n_1 \approx n_2$, $n_3 \approx n_4$). More general systems are studied in [49].

11.1.2.4 Nonlocal dispersion term

A more exact matching condition is more complicated: the relation (11.25) that connects α with wave parameters ω, k is valid only for the divided variables, otherwise some interval of the spectrum contributes (see [38], Sect. 3.4). Supposing that the 'linear' excitation conditions of a single mode contribution is expressed by a deviation from the single-mode state Π by weak nonlinearity and interaction with the environment of the dielectric layer. We choose an initial state as the only TE component presents and such that the dispersion parameter is small, introducing a correction Π.

$$E_x = E_e(z, x, t)Y(y)\exp i(kz - \omega t) + c.c. + \Pi. \quad (11.33)$$

Then inside the dielectric layer the Helmholtz equation

$$\Delta\Pi = \varepsilon\frac{\omega^2}{c^2}\Pi, \quad (11.34)$$

is valid and outside, one has the analogous equation with $\varepsilon = 1$.

To account for this more general matching, we introduce the Green function for a half plane that describes this correction Π. Finally, it should be decomposed into the components for different polarizations to fit the boundary condition (in y). The evolution equation gains an additional term which, after integration across the slab, yields the extra dispersion term in integral form. The resulting equation may be written along these lines as in [38].

This corresponds to the likely overlapping excited inputs, bringing us back again to the experiment in [6]. Note also that internal waves in inhomogeneous water layers (thermoclyne or pycnoclyne) were successfully described with nonlocal analogues of the KdV equations: most close to the Joseph one [39], and their two-dimensional generalizations [40]. Some mathematical aspects of propagating modes in planar waveguides were presented in the recent publication [1].

11.2 Cylindrical dielectric waveguides

11.2.1 On the problem

The form of a waveguide choice naturally implies its study in cylindrical coordinates and the transverse coordinates basic functions preference. It is also supposed that the waveguide material has a dispersion and nonlinearity such as self-action and interaction.

A novel (for 1980) dielectric waveguide is a thin dielectric tube containing an internal high-dielectric-constant gas. Authors account the attenuation constant and radiation loss of this waveguide are calculated and compared with those of a conventional O-guide (a dielectric tube waveguide). The advantage of the gas-confined guide is that the attenuation constant is decreased. Experimental investigations are carried out to verify the low-loss property of the gas-confined guide. The case of Lorentz–Drude dispersion and Kerr nonlinearity is specified in the study and resulting equation. The guide modes dispersion is naturally combined with the material one [5].

The problem of an electromagnetic wave propagation in dielectric cylinder started from the celebrated Hondroz–Debye paper [41]. Important applications of the linear version of the theory and its nonlinear generalizations were intensely studied in connection with the optical fiber technique [33]. Its solitonic stage [42] introduced new ideas as related to conventional nonlinear Schrödinger solitons [43] as to dispersion managed (DM) ones [10] with hopes and realizations in communication lines [11] and optical devices control [44]. The search of possibilities for increasing transmission rate density challenged the theory development related to more short pulses description by means of either generalizations of NS equation [34] or derivation of a new type (e.g. Shafer–Wayne) equation [35]. Their integrability motivates investigation of correspondent pulses stability [18] together with the multimode case and the nonlinearity parameters calculation [21].

In [37], the theoretical aspects of the problem of developing the multimode fiber systems description and their numerical integration [36] were studied, that support the idea of real pulse excitation which stability arises from quasisolitonic behavior in the coupled NS equations dynamics range [37]. Its adjustment to the photonic crystals [19] was delivered by means of modeling by the fiber parameters choice [20].

In [21, 37] the resulting multimode system was obtained by direct use of the expansion of electromagnetic field components in the Hondroz–Debye basis and combining of the equations 'by hands', forming the nonlinear terms.

In this section we develop and used a more advanced approach of dynamic projecting operators technique [45] that allow to do the mentioned combinations as well as equations combination in a systematic way. It, after the waveguide cylindrical functions basis expansion, leads to (hybrid) fields with polarization account. An important feature of this study is accounting for both waveguide and matter dispersion. The limit of a zero-range potential model applied to photonic crystal theory can be found in [46].

11.2.2 Linear problem

We start directly with the Maxwell system, chapter 10, section 10.1.1, in cylindrical coordinates ρ, φ, z, see figure 11.3. Bessel functions satisfy orthogonality conditions hence form convenient basis. To complete the statement of the problem, we define the boundary problem with conditions at the point $z = 0$ that fix the transversal components (polarizations) of the electric field as a function of time.

The procedure of the projecting is properly described in [45], firstly we expand basic fields in series with respect to the orthogonal waveguide Hondros–Debye basis

$$N_{nl}^{-1}J_l(\alpha_{nl}\rho)e^{il\varphi}, \tag{11.35}$$

N_{nl} are the normalization constants for the transverse modes nl,

$$E_z(\rho, \varphi, z, t) = \sum\nolimits_{l,n} A_{ln}(z, t)J_l(\alpha_{nl}\rho)e^{il\varphi} + c.c., \tag{11.36a}$$

$$\begin{aligned} E_r(\rho, \varphi, z, t) = \sum\nolimits_{l,n} \mathcal{B}_{ln}(z, t)\frac{il}{\rho}J_l(\alpha_{ln}\rho)e^{il\varphi} \\ + \mathcal{C}_{ln}(z, t)\frac{\partial}{\partial\rho}J_l(\alpha_{ln}\rho)e^{il\varphi} + c.c., \end{aligned} \tag{11.36b}$$

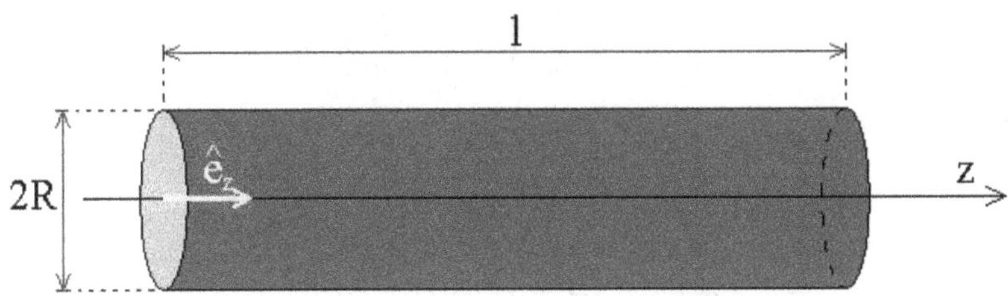

Figure 11.3. Geometry and coordinates of a dielectric waveguide line segment.

$$E_\varphi(\rho,\,\varphi,\,z,\,t) = \sum\nolimits_{l,n} \mathcal{D}_{ln}(z,\,t)\frac{il}{\rho}J_l(\alpha_{ln}\rho)e^{il\varphi} + \sum\nolimits_{l,n} \mathcal{F}_{ln}(z,\,t)\frac{\partial}{\partial\rho}J_l(\alpha_{ln}\rho)e^{il\varphi} + c.c.\,, \tag{11.36c}$$

$$B_z(\rho,\,\varphi,\,z,\,t) = \sum\nolimits_{l,n} \mathcal{K}_{ln}(z,\,t)J_l(\alpha_{ln}\rho)e^{il\varphi} + c.c.\,, \tag{11.36d}$$

$$B_r(\rho,\,\varphi,\,z,\,t) = \sum\nolimits_{l,n} \mathcal{L}_{ln}(z,\,t)\frac{il}{\rho}J_l(\alpha_{ln}\rho)e^{il\varphi} + \sum\nolimits_{l,n} \mathcal{M}_{ln}(z,\,t)\frac{\partial}{\partial\rho}J_l(\alpha_{ln}\rho)e^{il\varphi} + c.c.\,, \tag{11.36e}$$

$$B_\varphi(\rho,\,\varphi,\,z,\,t) = \sum\nolimits_{l,n} \mathcal{R}_{ln}(z,\,t)\frac{il}{\rho}J_l(\alpha_{ln}\rho)e^{il\varphi} + \sum\nolimits_{l,n} \mathcal{S}_{ln}(z,\,t)\frac{\partial}{\partial\rho}J_l(\alpha_{ln}\rho)e^{il\varphi} + c.c. \tag{11.36f}$$

We substitute these expansions to the Maxwell equations for matter (11.1), that in cylindrical coordinates have the form

$$\frac{\partial}{\rho\partial\rho}(\rho D_\rho) + \frac{\partial D_\varphi}{\rho\partial\varphi} + \frac{\partial D_z}{\partial z} = 0, \tag{11.37a}$$

$$\frac{\partial}{\rho\partial\rho}(\rho B_\rho) + \frac{\partial B_\varphi}{\rho\partial\varphi} + \frac{\partial B_z}{\partial z} = 0, \tag{11.37b}$$

$$\left(\frac{\partial E_z}{\rho\partial\varphi} - \frac{\partial E_\varphi}{\partial z}\right) = -\frac{1}{c}\frac{\partial B_\rho}{\partial t}, \tag{11.37c}$$

$$\left(\frac{\partial E_\rho}{\partial z} - \frac{\partial E_z}{\partial\rho}\right) = -\frac{1}{c}\frac{\partial B_\varphi}{\partial t}, \tag{11.37d}$$

$$\frac{1}{\rho}\left(\frac{\partial(\rho E_\varphi)}{\partial\rho} - \frac{\partial E_\rho}{\partial\varphi}\right) = -\frac{1}{c}\frac{\partial B_z}{\partial t}, \tag{11.37e}$$

$$\left(\frac{\partial H_z}{\rho\partial\varphi} - \frac{\partial H_\varphi}{\partial z}\right) = \frac{1}{c}\frac{\partial D_\rho}{\partial t}, \tag{11.37f}$$

$$\left(\frac{\partial H_\rho}{\partial z} - \frac{\partial H_z}{\partial\rho}\right) = \frac{1}{c}\frac{\partial D_\varphi}{\partial t}, \tag{11.37g}$$

$$\frac{1}{\rho}\left(\frac{\partial(\rho H_{\varphi})}{\partial\rho} - \frac{\partial H_{\rho}}{\partial\varphi}\right) = \frac{1}{c}\frac{\partial D_z}{\partial t}. \tag{11.37h}$$

Plugging the electromagnetic field components (11.36) into this system of equation (11.37), yields the following system of differential equations, that, after calculating the scalar product of the basic equation (11.35) with each equation from (11.37) gives the system of equations for the coefficients of expansion (11.36):

$$\mathcal{B}_{ln}(z, t) + \mathcal{F}_{ln}(z, t) = 0, \quad \mathcal{L}_{ln}(z, t) + \mathcal{S}_{ln}(z, t) = 0, \tag{11.38a}$$

$$\mathcal{C}_{ln}(z, t) - \mathcal{D}_{ln}(z, t) = 0, \quad \mathcal{M}_{ln}(z, t) - \mathcal{R}_{ln}(z, t) = 0, \tag{11.38b}$$

$$\frac{\partial}{\partial z}\mathcal{A}_{ln}(z, t) + \mathcal{C}_{ln}(z, t)\alpha_{ln}^2 = 0, \quad \frac{\partial}{\partial z}\mathcal{K}_{ln}(z, t) + \mathcal{M}_{ln}(z, t)\alpha_{ln}^2 = 0, \tag{11.38c}$$

$$\begin{aligned} \mathcal{A}_{ln}(z, t) - \frac{\partial}{\partial z}\mathcal{C}_{ln}(z, t) &= -\frac{1}{c}\frac{\partial}{\partial t}\mathcal{L}_{ln}(z, t), \\ \mathcal{K}_{ln}(z, t) - \frac{\partial}{\partial z}\mathcal{M}_{ln}(z, t) &= \frac{\varepsilon\mu}{c}\frac{\partial}{\partial t}\mathcal{B}_{ln}(z, t), \end{aligned} \tag{11.38d}$$

$$\frac{\partial}{\partial z}\mathcal{B}_{ln}(z, t) = -\frac{1}{c}\frac{\partial}{\partial t}\mathcal{M}_{ln}(z, t), \quad \frac{\partial}{\partial z}\mathcal{L}_{ln}(z, t) = \frac{\varepsilon\mu}{c}\frac{\partial}{\partial t}\mathcal{C}_{ln}(z, t), \tag{11.38e}$$

$$\alpha_{ln}^2\mathcal{B}_{ln}(z, t) = \frac{1}{c}\frac{\partial}{\partial t}\mathcal{K}_{ln}(z, t), \quad \alpha_{ln}^2\mathcal{L}_{ln}(z, t) = -\frac{\varepsilon\mu}{c}\frac{\partial}{\partial t}\mathcal{A}_{ln}(z, t). \tag{11.38f}$$

Some of the equations are equivalent to each other. If exclude ones, the system of six equations is obtained

$$\frac{\varepsilon\mu}{c}\frac{\partial}{\partial t}\mathcal{A}_{ln}(z, t) = -\alpha_{ln}^2\mathcal{L}_{ln}(z, t), \tag{11.39a}$$

$$\frac{\varepsilon\mu}{c}\frac{\partial}{\partial t}\mathcal{B}_{ln}(z, t) = \mathcal{K}_{ln}(z, t) - \frac{\partial}{\partial z}\mathcal{M}_{ln}(z, t), \tag{11.39b}$$

$$\frac{\varepsilon\mu}{c}\frac{\partial}{\partial t}\mathcal{C}_{ln}(z, t) = \frac{\partial}{\partial z}\mathcal{L}_{ln}(z, t), \tag{11.39c}$$

$$\frac{1}{c}\frac{\partial}{\partial t}\mathcal{K}_{ln}(z, t) = \alpha_{ln}^2\mathcal{B}_{ln}(z, t) \tag{11.39d}$$

$$\frac{1}{c}\frac{\partial}{\partial t}\mathcal{L}_{ln}(z, t) = -\mathcal{A}_{ln}(z, t) + \frac{\partial}{\partial z}\mathcal{C}_{ln}(z, t), \tag{11.39e}$$

$$\frac{1}{c}\frac{\partial}{\partial t}\mathcal{M}_{ln}(z, t) = -\frac{\partial}{\partial z}\mathcal{B}_{ln}(z, t). \tag{11.39f}$$

Considering a process of construction of projection operator method, by its direct view, there will be six projection operators for such a system. However, by the physical sense of the Maxwell system, it would be reasonable to reduce the system to four equations. Then it will be possible to associate two new variables to electric component with left and right direction of wave propagation, that differ by polarization, and two variables with magnetic component with left and right direction of wave propagation as well. Hence, from the system

$$\frac{\partial}{\partial t}\mathcal{A}_{ln}(z,\, t) = -\frac{\alpha_{ln}^2 c}{\varepsilon\mu}\mathcal{L}_{ln}(z,\, t), \tag{11.40a}$$

$$\frac{\partial}{\partial t}\mathcal{K}_{ln}(z,\, t) = c\alpha_{ln}^2\mathcal{B}_{ln}(z,\, t), \tag{11.40b}$$

$$\frac{\partial}{\partial z}\mathcal{L}_{ln}(z,\, t) = \frac{\varepsilon\mu}{c}\frac{\partial}{\partial t}\mathcal{C}_{ln}(z,\, t), \tag{11.40c}$$

$$\frac{\partial}{\partial z}\mathcal{M}_{ln}(z,\, t) = -\frac{\varepsilon\mu}{c}\frac{\partial}{\partial t}\mathcal{B}_{ln}(z,\, t) + \mathcal{K}_{ln}(z,\, t), \tag{11.40d}$$

$$\frac{\partial}{\partial z}\mathcal{C}_{ln}(z,\, t) = \frac{1}{c}\frac{\partial}{\partial t}\mathcal{L}_{ln}(z,\, t) + \mathcal{A}_{ln}(z,\, t), \tag{11.40e}$$

$$\frac{\partial}{\partial z}\mathcal{B}_{ln}(z,\, t) = -\frac{1}{c}\frac{\partial}{\partial t}\mathcal{M}_{ln}(z,\, t), \tag{11.40f}$$

the first two equations

$$\frac{\partial}{\partial t}\frac{\varepsilon\mu}{\alpha_{ln}^2 c}\mathcal{A}_{ln}(z,\, t) = -\,\mathcal{L}_{ln}(z,\, t), \tag{11.41a}$$

$$\frac{\partial}{\partial t}\frac{1}{c\alpha_{ln}^2}\mathcal{K}_{ln}(z,\, t) = \mathcal{B}_{ln}(z,\, t), \tag{11.41b}$$

will be used to reduce the system of six equations to a system of four first order in derivatives of z ones.

$$\frac{\partial}{\partial z}\mathcal{A}_{ln}(z,\, t) = -\,\alpha_{ln}^2\mathcal{C}_{ln}(z,\, t), \tag{11.42a}$$

$$\frac{\partial}{\partial z}\mathcal{C}_{ln}(z,\, t) = \left(-\frac{\partial^2}{\partial t^2}\frac{\varepsilon\mu}{\alpha_{ln}^2 c^2} + 1\right)\mathcal{A}_{ln}(z,\, t), \tag{11.42b}$$

$$\frac{\partial}{\partial z}\mathcal{K}_{ln}(z,\, t) = -\,\alpha_{ln}^2\mathcal{M}_{ln}(z,\, t), \tag{11.42c}$$

$$\frac{\partial}{\partial z}\mathcal{M}_{ln}(z, t) = \left(-\frac{\varepsilon\mu}{\alpha_{ln}^2 c^2}\frac{\partial^2}{\partial t^2} + 1\right)\mathcal{K}_{ln}(z, t). \tag{11.42d}$$

With this complete set of four equations it is possible to present the system in a form of matrix evolution equation by recipe of [4]

$$\partial_z \Psi(z, t) - \hat{L}\Psi(z, t) = 0. \tag{11.43}$$

Following the procedure presented in [45] and, shortly, in chapter 20, section 20.3, a form of evolution equation defines the projection operators explicit form. The evolution operator $\hat{L}$ has been constructed and applied to the evolution equation

$$\partial_z \begin{pmatrix} \mathcal{A}_{ln}(z, t) \\ \mathcal{C}_{ln}(z, t) \\ \mathcal{K}_{ln}(z, t) \\ \mathcal{M}_{ln}(z, t) \end{pmatrix} = \begin{pmatrix} 0 & -\alpha_{ln}^2 & 0 & 0 \\ -\frac{\partial^2}{\partial t^2}\frac{\varepsilon\mu}{\alpha_{ln}^2 c^2} + 1 & 0 & 0 & 0 \\ 0 & 0 & 0 & -\alpha_{ln}^2 \\ 0 & 0 & -\frac{\partial^2}{\partial t^2}\frac{\varepsilon\mu}{\alpha_{ln}^2 c^2} + 1 & 0 \end{pmatrix} \times \begin{pmatrix} \mathcal{A}_{ln}(z, t) \\ \mathcal{C}_{ln}(z, t) \\ \mathcal{K}_{ln}(z, t) \\ \mathcal{M}_{ln}(z, t) \end{pmatrix}. \tag{11.44}$$

11.2.3 Transformation to frequency domain

Fourier transformation from time to frequency domain provides the solution for the system and leads to the current form

$$\frac{\partial}{\partial z}\tilde{\mathcal{A}}_{ln}(z, \omega) = -\tilde{\alpha}_{ln}^2\tilde{\mathcal{C}}_{ln}(z, \omega), \tag{11.45a}$$

$$\frac{\partial}{\partial z}\tilde{\mathcal{C}}_{ln}(z, \omega) = \left(\omega^2\frac{\varepsilon\mu}{\tilde{\alpha}_{ln}^2 c^2} + 1\right)\tilde{\mathcal{A}}_{ln}(z, \omega), \tag{11.45b}$$

$$\frac{\partial}{\partial z}\tilde{\mathcal{K}}_{ln}(z, \omega) = -\tilde{\alpha}_{ln}^2\tilde{\mathcal{M}}_{ln}(z, \omega), \tag{11.45c}$$

$$\frac{\partial}{\partial z}\tilde{\mathcal{M}}_{ln}(z, \omega) = \left(\frac{\varepsilon\mu}{\tilde{\alpha}_{ln}^2 c^2}\omega^2 + 1\right)\tilde{\mathcal{K}}_{ln}(z, \omega). \tag{11.45d}$$

This form of the ordinary differential equations with constant coefficients allows us to propose a solution $\tilde{\mathcal{A}}_{ln}(z, \omega) = \breve{\mathcal{A}}_{ln}(k, \omega)e^{ikz}$, which will be adopted to other variables. Applying this solution to our system, the result

$$ik\breve{\mathcal{A}}_{ln}(k, \omega) = -\tilde{\alpha}_{ln}^2\breve{\mathcal{C}}_{ln}(k, \omega), \tag{11.46a}$$

$$ik\breve{\mathcal{C}}_{ln}(k, \omega) = \left(\omega^2\frac{\varepsilon\mu}{\tilde{\alpha}_{ln}^2c^2} + 1\right)\breve{\mathcal{A}}_{ln}(k, \omega), \tag{11.46b}$$

$$ik\breve{\mathcal{K}}_{ln}(k, \omega) = -\tilde{\alpha}_{ln}^2\breve{\mathcal{M}}_{ln}(k, \omega), \tag{11.46c}$$

$$ik\breve{\mathcal{M}}_{ln}(k, \omega) = \left(\frac{\varepsilon\mu}{\tilde{\alpha}_{ln}^2c^2}\omega^2 + 1\right)\breve{\mathcal{K}}_{ln}(k, \omega) \tag{11.46d}$$

is obtained in the form of an eigenvalue problem, note that the spectral values $\tilde{\alpha}_{ln} = \alpha_{ln}$ in both domains. The condition of nonzero solution existence reads as

$$\det\begin{pmatrix} ik & -\alpha_{ln}^2 & 0 & 0 \\ \left(\omega^2\frac{\varepsilon\mu}{\alpha_{ln}^2c^2} + 1\right) & ik & 0 & 0 \\ 0 & 0 & ik & -\alpha_{ln}^2 \\ 0 & 0 & \left(\frac{\varepsilon\mu}{\alpha_{ln}^2c^2}\omega^2 + 1\right) & ik \end{pmatrix} = 0, \tag{11.47}$$

$$\begin{aligned} &ik\det\begin{pmatrix} ik & 0 & 0 \\ 0 & ik & -\alpha_{ln}^2 \\ 0 & \left(\frac{\varepsilon\mu}{\alpha_{ln}^2c^2}\omega^2 + 1\right) & ik \end{pmatrix} \\ &-\alpha_{ln}^2\det\begin{pmatrix} \left(\omega^2\frac{\varepsilon\mu}{\alpha_{ln}^2c^2} + 1\right) & 0 & 0 \\ 0 & ik & -\alpha_{ln}^2 \\ 0 & \left(\frac{\varepsilon\mu}{\alpha_{ln}^2c^2}\omega^2 + 1\right) & ik \end{pmatrix} = 0. \end{aligned} \tag{11.48}$$

Expanding the determinants in (11.48) gives the biquadratic equation

$$-k^2\left(-k^2 + \left(\frac{\varepsilon\mu}{c^2}\omega^2 + \alpha_{ln}^2\right)\right) - \left(\omega^2\frac{\varepsilon\mu}{c^2} + \alpha_{ln}^2\right)\left(-k^2 + \left(\frac{\varepsilon\mu}{c^2}\omega^2 + \alpha_{ln}^2\right)\right) = 0, \tag{11.49}$$

its factoring yields

$$\left(-k^2 - \left(\omega^2\frac{\varepsilon\mu}{c^2} + \alpha_{ln}^2\right)\right)\left(-k^2 + \left(\frac{\varepsilon\mu}{c^2}\omega^2 + \alpha_{ln}^2\right)\right) = 0. \tag{11.50}$$

This equation naturally splits:

$$-k^2 - \left(\omega^2\frac{\varepsilon\mu}{c^2} + \alpha_{ln}^2\right) = 0, \; - k^2 + \left(\frac{\varepsilon\mu}{c^2}\omega^2 + \alpha_{ln}^2\right) = 0. \tag{11.51}$$

Above calculations, returning to notations with indexes, lead to the dispersion relation

$$k_{nl} = \pm\sqrt{\frac{\varepsilon\mu}{c^2}\omega^2 + \alpha_{ln}^2}. \tag{11.52}$$

In equation (11.52) we have a positive and negative value of the coefficient. We, by physical reason, choose the positive value of k_{nl}^2 as the frequency should have the real value. The positive value of k_{nl} is connected to the wave propagating to the right direction and negative is connected to the wave propagating to the left. With use of the dispersion relation the variables of state vector can be presented as

$$\breve{\mathcal{A}}_{ln}(k, \omega) = i\alpha_{ln}^2\sqrt{\frac{\varepsilon\mu}{c^2}\omega^2 + \alpha_{ln}^2}^{(-1)}\breve{\mathcal{C}}_{ln}(k, \omega), \tag{11.53a}$$

$$\breve{\mathcal{C}}_{ln}(k, \omega) = \frac{-i}{\alpha_{ln}^2}\sqrt{\frac{\varepsilon\mu}{c^2}\omega^2 + \alpha_{ln}^2}\,\breve{\mathcal{A}}_{ln}(k, \omega), \tag{11.53b}$$

$$\breve{\mathcal{K}}_{ln}(k, \omega) = i\sqrt{\frac{\varepsilon\mu}{c^2}\omega^2 + \alpha_{ln}^2}^{(-1)}\alpha_{ln}^2\breve{\mathcal{M}}_{ln}(k, \omega), \tag{11.53c}$$

$$\breve{\mathcal{M}}_{ln}(k, \omega) = \frac{-i}{\alpha_{ln}^2}\sqrt{\frac{\varepsilon\mu}{c^2}\omega^2 + \alpha_{ln}^2}\,\breve{\mathcal{K}}_{ln}(k, \omega). \tag{11.53d}$$

Taking into consideration that ε is dependent on ω, from that moment it will be treated as an operator $\hat{\varepsilon}$. Hence the wave vector $\hat{k}_{nl}$, as the function of $\hat{\varepsilon}$ will also an operator. To simplify calculations a variable $\hat{k}_{nl} = \sqrt{\omega^2\frac{\hat{\varepsilon}\mu}{c^2} + \alpha_{ln}^2}$ will be used in the projection operators explicit expressions, having for the left TM wave

$$P_{11} = \frac{1}{2}\begin{pmatrix} 1 & i\alpha_{ln}^2\left(\sqrt{\omega^2\frac{\varepsilon\mu}{c^2} + \alpha_{ln}^2}\right)^{-1} & 0 & 0 \\ \frac{-i}{\alpha_{ln}^2}\left(\sqrt{\omega^2\frac{\varepsilon\mu}{c^2} + \alpha_{ln}^2}\right) & 1 & 0 & 0 \\ 0 & 0 & 0 & 0 \\ 0 & 0 & 0 & 0 \end{pmatrix}, \tag{11.54a}$$

for the TE one

$$P_{21} = \frac{1}{2}\begin{pmatrix} 1 & -i\alpha_{ln}^2\left(\sqrt{\omega^2\frac{\varepsilon\mu}{c^2} + \alpha_{ln}^2}\right)^{-1} & 0 & 0 \\ \frac{i}{\alpha_{ln}^2}\left(\sqrt{\omega^2\frac{\varepsilon\mu}{c^2} + \alpha_{ln}^2}\right) & 1 & 0 & 0 \\ 0 & 0 & 0 & 0 \\ 0 & 0 & 0 & 0 \end{pmatrix}, \tag{11.54b}$$

for the right TE,

$$P_{12} = \frac{1}{2}\begin{pmatrix} 0 & 0 & 0 & 0 \\ 0 & 0 & 0 & 0 \\ 0 & 0 & 1 & i\alpha_{ln}^2\left(\sqrt{\omega^2\frac{\varepsilon\mu}{c^2} + \alpha_{ln}^2}\right)^{-1} \\ 0 & 0 & \frac{-i}{\alpha_{ln}^2}\left(\sqrt{\omega^2\frac{\varepsilon\mu}{c^2} + \alpha_{ln}^2}\right) & 1 \end{pmatrix}, \tag{11.54c}$$

and, finally for the right TM

$$P_{22} = \frac{1}{2}\begin{pmatrix} 0 & 0 & 0 & 0 \\ 0 & 0 & 0 & 0 \\ 0 & 0 & 1 & -i\alpha_{ln}^2\left(\sqrt{\omega^2\frac{\varepsilon\mu}{c^2} + \alpha_{ln}^2}\right)^{-1} \\ 0 & 0 & \frac{i}{\alpha_{ln}^2}\left(\sqrt{\omega^2\frac{\varepsilon\mu}{c^2} + \alpha_{ln}^2}\right) & 1 \end{pmatrix}. \tag{11.54d}$$

The corresponding eigenvectors are

$$\Psi_{11} = \frac{1}{2}\begin{pmatrix} -i\alpha_{ln}^2\left(\sqrt{\omega^2\frac{\varepsilon\mu}{c^2} + \alpha_{ln}^2}\right) \\ 1 \\ 0 \\ 0 \end{pmatrix}\check{\mathcal{C}}_{ln}(k, \omega), \tag{11.55}$$

$$\Psi_{12} = \frac{1}{2}\begin{pmatrix} 0 \\ 0 \\ -i\alpha_{ln}^2\left(\sqrt{\omega^2\frac{\varepsilon\mu}{c^2} + \alpha_{ln}^2}\right) \\ 1 \end{pmatrix}\check{\mathcal{M}}_{ln}(k, \omega), \tag{11.56}$$

$$\Psi_{21} = \frac{1}{2}\begin{pmatrix} i\alpha_{ln}^2\left(\sqrt{\omega^2\frac{\varepsilon\mu}{c^2} + \alpha_{ln}^2}\right) \\ 1 \\ 0 \\ 0 \end{pmatrix}\breve{C}_{ln}(k, \omega), \tag{11.57}$$

$$\Psi_{22} = \frac{1}{2}\begin{pmatrix} 0 \\ 0 \\ i\alpha_{ln}^2\left(\sqrt{\omega^2\frac{\varepsilon\mu}{c^2} + \alpha_{ln}^2}\right) \\ 1 \end{pmatrix}\breve{\mathcal{M}}_{ln}(k, \omega). \tag{11.58}$$

The projecting actions on equation (11.45) result as

$$\partial_z P_{11}\Psi(z, \omega) - \hat{L}P_{11}\Psi(z, \omega) = 0. \tag{11.59}$$

Recall the we still use the notation $\hat{k}_{nl} = \sqrt{\omega^2\frac{\hat{\varepsilon}\mu}{c^2} + \alpha_{ln}^2}$ the mode vectors already in t-domain are as follows

$$\hat{P}_{11}\Psi(z, \omega) = \begin{pmatrix} \mathcal{A}_{ln}(z, t) - i\alpha_{ln}^2\hat{k}_{nl}^{-1}\mathcal{C}_{ln}(z, t) \\ \frac{i}{\alpha_{ln}^2}\hat{k}_{nl}\mathcal{A}_{ln}(z, t) + \mathcal{C}_{ln}(z, t) \\ 0 \\ 0 \end{pmatrix} = \begin{pmatrix} -i\alpha_{ln}^2\hat{k}_{nl}^{-1}\Lambda_1(z, t) \\ \Lambda_1(z, t) \\ 0 \\ 0 \end{pmatrix}, \tag{11.60}$$

$$\hat{P}_{12}\Psi(z, \omega) = \begin{pmatrix} 0 \\ 0 \\ \mathcal{K}_{ln}(z, t) - i\alpha_{ln}^2\hat{k}_{nl}^{-1}\mathcal{M}_{ln}(z, t) \\ \frac{i}{\alpha_{ln}^2}\hat{k}_{nl}\mathcal{K}_{ln}(z, t) + \mathcal{M}_{ln}(z, t) \end{pmatrix} = \begin{pmatrix} 0 \\ 0 \\ -i\alpha_{ln}^2\hat{k}_{nl}^{-1}\Lambda_2(z, t) \\ \Lambda_2(z, t) \end{pmatrix}, \tag{11.61}$$

$$\hat{P}_{21}\Psi(z, \omega) = \begin{pmatrix} \mathcal{A}_{ln}(z, t) + i\alpha_{ln}^2\hat{k}_{nl}^{-1}\mathcal{C}_{ln}(z, t) \\ -\frac{i}{\alpha_{ln}^2}\hat{k}_{nl}\mathcal{A}_{ln}(z, t) + \mathcal{C}_{ln}(z, t) \\ 0 \\ 0 \end{pmatrix} = \begin{pmatrix} -i\alpha_{ln}^2\hat{k}_{nl}^{-1}\Pi_1(z, t) \\ \Pi_1(z, t) \\ 0 \\ 0 \end{pmatrix}, \tag{11.62}$$

$$\hat{P}_{22}\Psi(z, \omega) = \begin{pmatrix} 0 \\ 0 \\ \mathcal{K}_{ln}(z, t)i\alpha_{ln}^2\hat{k}_{nl}^{-1}\mathcal{M}_{ln}(z, t) \\ \frac{-i}{\alpha_{ln}^2}\hat{k}_{nl}\mathcal{K}_{ln}(z, t) + \mathcal{M}_{ln}(z, t) \end{pmatrix} = \begin{pmatrix} 0 \\ 0 \\ -i\alpha_{ln}^2\hat{k}_{nl}^{-1}\Pi_2(z, t) \\ \Pi_2(z, t) \end{pmatrix}. \tag{11.63}$$

It means, that Λ_1 can be associated to polarized electric wave propagating to the left. Λ_2 is associated to the polarized magnetic wave propagating in the same direction. Waves propagating in the opposite direction are described by Π_1 associated to the electric wave and Π_2 associated to the magnetic one.

11.2.4 Projection operators in time domain

In the further analysis only the first projection operator P_{11} will be used. Acting on the state vector $\Psi(\mathcal{A}, \mathcal{C}, \mathcal{K}, \mathcal{M})$ with P_{11},

$$\partial_z P_{11}\Psi(\mathcal{A}, \mathcal{C}, \mathcal{K}, \mathcal{M}) - \hat{L}P_{11}\Psi(\mathcal{A}, \mathcal{C}, \mathcal{K}, \mathcal{M}) = 0, \tag{11.64}$$

the t-domain evolution equation describing electric wave (TE mode) propagating in left direction is derived

$$\partial_z\Psi(\Lambda_1(z, t)) - \hat{L}\Psi(\Lambda_1(z, t)) = 0. \tag{11.65}$$

The dielectric susceptibility coefficient $\varepsilon(\omega)$ either originated from the quantum version of the Lorentz formula (see e.g. [47, 48]) or directly from phenomenology [35]. The approximation may be obtained dependent on the frequency range; for example at high frequency, as the Taylor expansion at small ω^{-2}:

$$\varepsilon(\omega) \approx 1 + 4\pi\chi(\omega) \approx 1 - 4\pi\chi_0\omega^{-2}. \tag{11.66}$$

Including the ε dependence over ω (11.66) it is natural to rewrite the dispersion relation (11.52)

$$k_{nl} = \omega\sqrt{\frac{(1 - 4\pi\chi_0\omega^{-2})\mu}{c^2} + \frac{\alpha_{nl}^2}{\omega^2}}. \tag{11.67}$$

The right-hand side of the expression is expanded in Taylor series over $\omega^{-2} \to 0$

$$k_{nl} = \omega\sqrt{\frac{\mu}{c^2} + \frac{\alpha_{nl}^2 - \mu c^{-2}4\pi\chi_0}{\omega^2}} = \omega\left(\sqrt{\frac{\mu}{c^2}} + \frac{\alpha_{ln}^2 - \mu c^{-2}4\pi\chi_0}{2\sqrt{\frac{\mu}{c^2}}}\frac{1}{\omega^2}\right). \tag{11.68}$$

The equation (11.59) takes the form

$$\partial_z\Lambda(z, \omega) - i\omega\left(\sqrt{\frac{\mu}{c^2}} + \frac{\alpha_{nl}^2 - \mu c^{-2}4\pi\chi_0}{2\sqrt{\frac{\mu}{c^2}}}\frac{1}{\omega^2}\right)\Lambda(z, \omega) = 0. \tag{11.69}$$

After the inverse Fourier transformation the dispersion operator (11.52) in z, t representation with dielectric susceptibility coefficient (11.66) included, takes the form

$$\hat{k}_{nl} = \sqrt{\frac{\mu}{c^2}}\partial_t + \frac{\alpha_{nl}^2 - \mu c^{-2}4\pi\chi_0}{2\sqrt{\frac{\mu}{c^2}}}\partial_t^{-1}. \tag{11.70}$$

With the approximate dispersion relation in time-representation it is possible to present the projection operators (11.54) as below

$$\hat{P}_{11} = \frac{1}{2}\begin{pmatrix} 1 & -i\alpha_{ln}^{2}\hat{k}_{nl}^{-1} & 0 & 0 \\ i\alpha_{ln}^{-2}\hat{k}_{nl} & 1 & 0 & 0 \\ 0 & 0 & 0 & 0 \\ 0 & 0 & 0 & 0 \end{pmatrix} \quad \hat{P}_{21} = \frac{1}{2}\begin{pmatrix} 1 & i\alpha_{ln}^{2}\hat{k}_{nl}^{-1} & 0 & 0 \\ -i\alpha_{ln}^{-2}\hat{k}_{nl} & 1 & 0 & 0 \\ 0 & 0 & 0 & 0 \\ 0 & 0 & 0 & 0 \end{pmatrix}$$
$$\hat{P}_{12} = \frac{1}{2}\begin{pmatrix} 0 & 0 & 0 & 0 \\ 0 & 0 & 0 & 0 \\ 0 & 0 & 1 & i\alpha_{ln}^{2}\hat{k}_{nl}^{-1} \\ 0 & 0 & -i\alpha_{ln}^{-2}\hat{k}_{nl}^{-1} & 1 \end{pmatrix} \quad \hat{P}_{22} = \frac{1}{2}\begin{pmatrix} 0 & 0 & 0 & 0 \\ 0 & 0 & 0 & 0 \\ 0 & 0 & 1 & -i\alpha_{ln}^{2}\hat{k}_{nl}^{-1} \\ 0 & 0 & i\alpha_{ln}^{-2}\hat{k}_{nl}^{-1} & 1 \end{pmatrix}. \tag{11.71}$$

11.3 Including nonlinearity

In this section we would include the nonlinear part of the vector $\vec{D}$, equation (11.2), and explicit form of the polarization vector (11.3) to the Maxwell system, that will have the form

$$\frac{\partial}{r\partial r}(r(D_L)_r) + \frac{\partial(D_L)_\varphi}{r\partial\varphi} + \frac{\partial(D_L)_z}{\partial z} + \frac{\partial}{r\partial r}(r(P_{NL})_r) + \frac{\partial(P_{NL})_\varphi}{r\partial\varphi} + \frac{\partial(P_{NL})_z}{\partial z} = 0, \tag{11.72}$$

$$\frac{\partial}{r\partial r}(rB_r) + \frac{\partial B_\varphi}{r\partial\varphi} + \frac{\partial B_z}{\partial z} = 0, \tag{11.73}$$

$$\left(\frac{\partial E_z}{r\partial\varphi} - \frac{\partial E_\varphi}{\partial z}\right) = -\frac{1}{c}\frac{\partial B_r}{\partial t}, \tag{11.74}$$

$$\left(\frac{\partial E_r}{\partial z} - \frac{\partial E_z}{\partial r}\right) = -\frac{1}{c}\frac{\partial B_\varphi}{\partial t}, \tag{11.75}$$

$$\frac{1}{r}\left(\frac{\partial(rE_\varphi)}{\partial r} - \frac{\partial E_r}{\partial\varphi}\right) = -\frac{1}{c}\frac{\partial B_z}{\partial t}, \tag{11.76}$$

$$\left(\frac{\partial H_z}{r\partial\varphi} - \frac{\partial H_\varphi}{\partial z}\right) = \frac{1}{c}\frac{\partial(D_L)_r}{\partial t} + \frac{1}{c}\frac{\partial(P_{NL})_r}{\partial t}, \tag{11.77}$$

$$\left(\frac{\partial H_r}{\partial z} - \frac{\partial H_z}{\partial r}\right) = \frac{1}{c}\frac{\partial(D_L)_\varphi}{\partial t} + \frac{1}{c}\frac{\partial(P_{NL})_\varphi}{\partial t}, \tag{11.78}$$

$$\frac{1}{r}\left(\frac{\partial(rH_\varphi)}{\partial r} - \frac{\partial H_r}{\partial\varphi}\right) = \frac{1}{c}\frac{\partial(D_L)_z}{\partial t} + \frac{1}{c}\frac{\partial(P_{NL})_z}{\partial t}, \tag{11.79}$$

where the linear part of the vector $\vec{D}$ accounts the linear dispersion

$$\mathbf{D}_L = \int_{-\infty}^{\infty} \int_{-\infty}^{\infty} \varepsilon(\omega) e^{-i\omega(t-t')} d\omega \mathbf{E}(t') dt' = \hat{\varepsilon}\mathbf{E}. \tag{11.80}$$

The further manipulations with the system are divided into two stages of projecting procedure: the first—to waveguide modes, the second—projecting to subspace with definite polarization and propagation direction. As the first step the electromagnetic field coefficients (11.36) will be presented with the use of derived relation (11.42),

$$E_z(r, \varphi, z, t) = \sum\nolimits_{l,n} \mathcal{A}_{ln}(z, t) J_l(\alpha_{nl} r) e^{il\varphi} + c.c., \tag{11.81a}$$

$$\begin{aligned} E_r(r, \varphi, z, t) = &\sum\nolimits_{l,n} \frac{\partial}{\partial t} \frac{1}{c\alpha_{ln}^2} \mathcal{K}_{ln}(z, t) \frac{il}{r} J_l(\alpha_{ln} r) e^{il\varphi} \\ &+ \sum\nolimits_{l,n} \mathcal{C}_{ln}(z, t) \frac{\partial}{\partial r} J_l(\alpha_{ln} r) e^{il\varphi} + c.c., \end{aligned} \tag{11.81b}$$

$$\begin{aligned} E_\varphi(r, \varphi, z, t) = &\sum\nolimits_{l,n} \mathcal{C}_{ln}(z, t) \frac{il}{r} J_l(\alpha_{ln} r) e^{il\varphi} \\ &- \sum\nolimits_{l,n} \frac{\partial}{\partial t} \frac{1}{c\alpha_{ln}^2} \mathcal{K}_{ln}(z, t) \frac{\partial}{\partial r} J_l(\alpha_{ln} r) e^{il\varphi} + c.c., \end{aligned} \tag{11.81c}$$

$$B_z(r, \varphi, z, t) = \sum\nolimits_{l,n} \mathcal{K}_{ln}(z, t) J_l(\alpha_{nl} r) e^{il\varphi} + c.c., \tag{11.81d}$$

$$\begin{aligned} B_r(r, \varphi, z, t) = &\sum\nolimits_{l,n} \mathcal{M}_{ln}(z, t) \frac{\partial}{\partial r} J_l(\alpha_{ln} r) e^{il\varphi} \\ &- \sum\nolimits_{l,n} \frac{\partial}{\partial t} \frac{\varepsilon\mu}{\alpha_{ln}^2 c} \mathcal{A}_{ln}(z, t) \frac{il}{r} J_l(\alpha_{ln} r) e^{il\varphi} + c.c., \end{aligned} \tag{11.81e}$$

$$\begin{aligned} B_\varphi(r, \varphi, z, t) = &\sum\nolimits_{l,n} \mathcal{M}_{ln}(z, t) \frac{il}{r} J_l(\alpha_{ln} r) e^{il\varphi} \\ &+ \sum\nolimits_{l,n} \frac{\partial}{\partial t} \frac{\varepsilon\mu}{\alpha_{ln}^2 c} \mathcal{A}_{ln}(z, t) \frac{\partial}{\partial r} J_l(\alpha_{ln} r) e^{il\varphi} + c.c. \end{aligned} \tag{11.81f}$$

The nonlinear part $(\mathbf{P}_{NL})_i$ is given by formula (11.3) and expressed in Cartesian coordinates. Each coordinate $(\mathbf{P}_{NL})_i$ should be expressed in cylindrical coordinates. To simplify the calculations, one can use the third order susceptibility tensor relation, similar to [47]. The cylindrical coordinates of $(\mathbf{P}_{NL})_i$ have the form

$$\begin{aligned} (\mathbf{P}_{NL})_r = \chi_{zzzz}(&E_r^3(3\cos^3\varphi + \sin^2\varphi\cos\varphi) - E_r^2 E_\varphi(7\sin\varphi\cos^2\varphi + \sin^3\varphi) \\ &+ E_r E_\varphi^2(7\sin^2\varphi\cos\varphi + \cos^3\varphi) \\ &- E_\varphi^3(3\sin^3\varphi + \sin\varphi\cos^2\varphi) + E_z^2(E_r\cos\varphi - E_\varphi\sin\varphi)), \end{aligned} \tag{11.82}$$

$$(\mathbf{P}_{NL})_\varphi = \chi_{zzzz}\Big(E_r^3(3\sin^3\varphi + \sin\varphi\cos^2\varphi) + E_r^2E_\varphi(7\sin^2\varphi\cos\varphi + \cos^3\varphi) \\ +E_\varphi^2E_r(7\cos^2\varphi\sin\varphi + \sin^3\varphi) + E_\varphi^3(3\cos^3\varphi + \sin^2\varphi\cos\varphi) \\ +E_z^2(E_r\sin\varphi + E_\varphi\cos\varphi)\Big), \tag{11.83}$$

$$(\mathbf{P}_{NL})_z = \chi_{zzzz}\Big(3E_z^3 + (E_r^2 + E_\varphi^2)E_z\Big). \tag{11.84}$$

At this point the nonlinear part can be included to the Maxwell system (11.72). Presented in our basis (11.81) the Maxwell system will be reduced to a form which after simplification can be described as the system of equations

$$\sum\nolimits_{l,n}\hat{\varepsilon}(-\mathcal{C}_{ln}(z,t)\alpha_{ln}^2 + \frac{\partial}{\partial z}\mathcal{A}_{ln}(z,t))J_l(\alpha_{ln}r)e^{il\varphi} + \frac{\partial}{r\partial r}(r(P_{NL})_r) + \frac{\partial(P_{NL})_\varphi}{r\partial\varphi} \\ + \frac{\partial(P_{NL})_z}{\partial z} + c.c. = 0, \tag{11.85a}$$

$$\sum\nolimits_{l,n}(\mathcal{M}_{ln}(z,t)\alpha_{ln}^2 + \frac{\partial}{\partial z}\mathcal{K}_{ln}(z,t))J_l(\alpha_{nl}r)e^{il\varphi} + c.c. = 0, \tag{11.85b}$$

$$\sum\nolimits_{l,n}(\mathcal{A}_{ln}(z,t) - \frac{\partial}{\partial z}\mathcal{C}_{ln}(z,t) - \frac{\partial^2}{\partial t^2}\frac{\varepsilon\mu}{\alpha_{ln}^2c^2}\mathcal{A}_{ln}(z,t))\frac{il}{r}J_l(\alpha_{ln}r)e^{il\varphi} + c.c. \\ = \sum\nolimits_{l,n}\frac{1}{c}\frac{\partial}{\partial t}(\frac{\partial}{\partial z}\frac{1}{\alpha_{ln}^2}\mathcal{K}_{ln}(z,t) + \mathcal{M}_{ln}(z,t))\frac{\partial}{\partial r}J_l(\alpha_{ln}r)e^{il\varphi} + c.c., \tag{11.85c}$$

$$\sum\nolimits_{l,n}\left(\frac{1}{c\alpha_{ln}^2}\frac{\partial}{\partial z}\frac{\partial}{\partial t}\mathcal{K}_{ln}(z,t) + \frac{1}{c}\frac{\partial}{\partial t}\mathcal{M}_{ln}(z,t)\right)\frac{il}{r}J_l(\alpha_{ln}r)e^{il\varphi} + c.c. \\ = \sum\nolimits_{l,n}\left(\mathcal{A}_{ln}(z,t) + \frac{\partial^2}{\partial t^2}\frac{\hat{\varepsilon}\mu}{\alpha_{ln}^2c}\mathcal{A}_{ln}(z,t) - \frac{\partial}{\partial z}\mathcal{C}_{ln}(z,t)\right)\frac{\partial}{\partial r}J_l(\alpha_{ln}r)e^{il\varphi} + c.c., \tag{11.85d}$$

$$\sum\nolimits_{l,n}\left(\mathcal{K}_{ln}(z,t) - \frac{\partial}{\partial z}\mathcal{M}_{ln}(z,t) - \frac{\hat{\varepsilon}\mu}{c}\frac{\partial^2}{\partial t^2}\frac{1}{c\alpha_{ln}^2}\mathcal{K}_{ln}(z,t)\right)\frac{il}{r}J_l(\alpha_{ln}r)e^{il\varphi} + c.c. \\ = \sum\nolimits_{l,n}\frac{\partial}{\partial t}\left(\frac{\partial}{\partial z}\frac{\varepsilon\mu}{\alpha_{ln}^2c}\mathcal{A}_{ln}(z,t) + \frac{\hat{\varepsilon}\mu}{c}\mathcal{C}_{ln}(z,t)\right)\frac{\partial}{\partial r}J_l(\alpha_{ln}r)e^{il\varphi} + \frac{\mu}{c}\frac{\partial(P_{NL})_r}{\partial t} + c.c. \tag{11.85e}$$

$$\sum\nolimits_{l,n}\left(\hat{\varepsilon}\mu\frac{\partial^2}{\partial t^2}\frac{1}{c^2\alpha_{ln}^2}\mathcal{K}_{ln}(z,t) + \frac{\partial}{\partial z}\mathcal{M}_{ln}(z,t) - \mathcal{K}_{ln}(z,t)\right)\frac{\partial}{\partial r}J_l(\alpha_{nl}r)e^{il\varphi} + c.c. \\ = \sum\nolimits_{l,n}\left(\frac{\partial}{\partial z}\frac{\partial}{\partial t}\frac{\hat{\varepsilon}\mu}{\alpha_{ln}^2c}\mathcal{A}_{ln}(z,t) + \frac{1}{c}\frac{\partial}{\partial t}\hat{\varepsilon}\mu\mathcal{C}_{ln}(z,t)\right)\frac{il}{r}J_l(\alpha_{ln}r)e^{il\varphi} + c.c. + \frac{\mu}{c}\frac{\partial}{\partial t}(P_{NL})_\varphi \tag{11.85f}$$

$$\sum\nolimits_{l,n}\frac{\partial}{\partial t}\frac{\hat{\varepsilon}}{\alpha_{ln}^2 c}(\mathcal{A}_{ln}(z,\,t)\alpha_{ln}^2 J_l(\alpha_{ln}r))e^{il\varphi} - \frac{1}{c}\frac{\partial}{\partial t}\hat{\varepsilon}\sum\nolimits_{l,n}\mathcal{A}_{ln}(z,\,t)J_l(\alpha_{nl}r)e^{il\varphi} + c.c. = \frac{1}{c}\frac{\partial (P_{NL})_z}{\partial t}. \tag{11.85g}$$

From that system, four equations will emerge

$$\sum\nolimits_{l,n}(\mathcal{M}_{ln}(z,\,t)\alpha_{ln}^2 + \frac{\partial}{\partial z}\mathcal{K}_{ln}(z,\,t))J_l(\alpha_{nl}r)e^{il\varphi} + c.c. = 0, \tag{11.86a}$$

$$\sum\nolimits_{l,n}(\mathcal{A}_{ln}(z,\,t) - \frac{\partial}{\partial z}\mathcal{C}_{ln}(z,\,t) - \frac{\partial^2}{\partial t^2}\frac{\varepsilon\mu}{\alpha_{ln}^2 c^2}\mathcal{A}_{ln}(z,\,t))\frac{il}{r}J_l(\alpha_{ln}r)e^{il\varphi} + c.c. = 0 \tag{11.86b}$$

$$\sum\nolimits_{l,n}\left(\mathcal{K}_{ln}(z,\,t) - \frac{\partial}{\partial z}\mathcal{M}_{ln}(z,\,t) - \frac{\hat{\varepsilon}\mu}{c}\frac{\partial^2}{\partial t^2}\frac{1}{c\alpha_{ln}^2}\mathcal{K}_{ln}(z,\,t)\right)\frac{il}{r}J_l(\alpha_{ln}r)e^{il\varphi} + c.c. \tag{11.86c}$$

$$=\sum\nolimits_{l,n}\frac{\partial}{\partial t}\left(\frac{\partial}{\partial z}\frac{\varepsilon\mu}{\alpha_{ln}^2 c}\mathcal{A}_{ln}(z,\,t) + \frac{\hat{\varepsilon}\mu}{c}\mathcal{C}_{ln}(z,\,t)\right)\frac{\partial}{\partial r}J_l(\alpha_{ln}r)e^{il\varphi} + \frac{\mu}{c}\frac{\partial}{\partial t}(P_{NL})_r + c.c. \tag{11.86d}$$

$$\sum\nolimits_{l,n}\left(\hat{\varepsilon}\mu\frac{\partial^2}{\partial t^2}\frac{1}{c^2\alpha_{ln}^2}\mathcal{K}_{ln}(z,\,t) + \frac{\partial}{\partial z}\mathcal{M}_{ln}(z,\,t) - \mathcal{K}_{ln}(z,\,t)\right)\frac{\partial}{\partial r}J_l(\alpha_{nl}r)e^{il\varphi} + c.c. \tag{11.86e}$$

$$= \sum\nolimits_{l,n}\left(\frac{\partial}{\partial z}\frac{\partial}{\partial t}\frac{\hat{\varepsilon}\mu}{\alpha_{ln}^2 c}\mathcal{A}_{ln}(z,\,t) + \frac{1}{c}\frac{\partial}{\partial t}\hat{\varepsilon}\mu\mathcal{C}_{ln}(z,\,t)\right)\frac{il}{r}J_l(\alpha_{ln}r)e^{il\varphi} + \frac{\mu}{c}\frac{\partial}{\partial t}(P_{NL})_\varphi + c.c. \tag{11.86f}$$

which will lead to the evolution equation $\partial_z\Psi(z,\,t) - \hat{L}\Psi(z,\,t) = \mathcal{N}(\Psi)$. However to present that equation the third and fourth equation have to be rebuilt with the use of standard Bessel functions properties. To form a nonlinear part $\mathcal{N}(\Psi)$ it is crucial to find the proper relation between variables $\mathcal{A}_{ln}$, $\mathcal{C}_{ln}$ and variables $\mathcal{K}_{ln}$, $\mathcal{M}_{ln}$, so the nonlinear part is well defined.

$$\sum\nolimits_{l,n}\left(\mathcal{M}_{ln}(z,\,t)\alpha_{ln}^2 + \frac{\partial}{\partial z}\mathcal{K}_{ln}(z,\,t)\right)J_l(\alpha_{nl}r)e^{il\varphi} + c.c. = 0, \tag{11.87a}$$

$$\sum\nolimits_{l,n}\left(\frac{\varepsilon\mu}{\alpha_{ln}^2 c^2}\frac{\partial^2}{\partial t^2}\mathcal{A}_{ln}(z,\,t) - \mathcal{A}_{ln}(z,\,t) + \frac{\partial}{\partial z}\mathcal{C}_{ln}(z,\,t)\right)\frac{il}{r}J_l(\alpha_{ln}r)e^{il\varphi} + c.c. = 0, \tag{11.87b}$$

$$\sum_{l,n} -i\left(\frac{\hat{\varepsilon}\mu}{c^2\alpha_{ln}^2}\frac{\partial^2}{\partial t^2}\mathcal{K}_{ln}(z,t) - \mathcal{K}_{ln}(z,t) + \frac{\partial}{\partial z}\mathcal{M}_{ln}(z,t)\right)$$
$$\frac{1}{2}(J_{l-1}(\alpha_{nl}r) + J_{l+1}(\alpha_{nl}r))e^{il\varphi} + c.c.$$
$$= \sum_{l,n} \frac{\hat{\varepsilon}\mu}{c}\frac{\partial}{\partial t}\left(\frac{\partial}{\partial z}\frac{1}{\alpha_{ln}^2}\mathcal{A}_{ln}(z,t) + \mathcal{C}_{ln}(z,t)\right)\frac{1}{2}(J_{l-1}(\alpha_{nl}r) - J_{l+1}(\alpha_{nl}r))e^{il\varphi} \tag{11.87c}$$
$$+ \frac{\mu}{c}\frac{\partial}{\partial t}(P_{NL})_r + c.c.\,,$$

$$\sum_{l,n}\left(\hat{\varepsilon}\mu\frac{\partial^2}{\partial t^2}\frac{1}{c^2\alpha_{ln}^2}\mathcal{K}_{ln}(z,t) - \mathcal{K}_{ln}(z,t) + \frac{\partial}{\partial z}\mathcal{M}_{ln}(z,t)\right)$$
$$\frac{1}{2}(J_{l-1}(\alpha_{nl}r) - J_{l+1}(\alpha_{nl}r))e^{il\varphi} + c.c.$$
$$= \sum_{l,n} \frac{i\hat{\varepsilon}\mu}{c}\frac{\partial}{\partial t}\left(\frac{1}{\alpha_{ln}^2}\frac{\partial}{\partial z}\mathcal{A}_{ln}(z,t) + \mathcal{C}_{ln}(z,t)\right)\frac{1}{2}(J_{l-1}(\alpha_{nl}r) + J_{l+1}(\alpha_{nl}r))e^{il\varphi} \tag{11.87d}$$
$$+ \frac{\mu}{c}\frac{\partial}{\partial t}(P_{NL})_\varphi + c.c.\,.$$

With simple algebraic operations Bessel functions $J_{l-1}(\alpha_{nl}r)$ and $J_{l+1}(\alpha_{nl}r)$ will be separated,

$$\sum_{l,n} -i\left(\frac{\hat{\varepsilon}\mu}{c^2\alpha_{ln}^2}\frac{\partial^2}{\partial t^2}\mathcal{K}_{ln}(z,t) - \mathcal{K}_{ln}(z,t) + \frac{\partial}{\partial z}\mathcal{M}_{ln}(z,t)\right)$$
$$J_{l-1}(\alpha_{ln}r)e^{il\varphi} + c.c.$$
$$= \sum_{l,n} \frac{\hat{\varepsilon}\mu}{c}\frac{\partial}{\partial t}\left(\frac{\partial}{\partial z}\frac{1}{\alpha_{ln}^2}\mathcal{A}_{ln}(z,t) + \mathcal{C}_{ln}(z,t)\right)J_{l-1}(\alpha_{nl}r)e^{il\varphi} \tag{11.88}$$
$$+ \frac{\mu}{c}\frac{\partial(P_{NL})_r}{\partial t} - \frac{i\mu}{c}\frac{\partial}{\partial t}(P_{NL})_\varphi,$$

$$\sum_{l,n} -i\left(\frac{\hat{\varepsilon}\mu}{c^2\alpha_{ln}^2}\frac{\partial^2}{\partial t^2}\mathcal{K}_{ln}(z,t) - \mathcal{K}_{ln}(z,t) + \frac{\partial}{\partial z}\mathcal{M}_{ln}(z,t)\right)$$
$$J_{l+1}(\alpha_{ln}r)e^{il\varphi} + c.c.$$
$$= -\sum_{l,n} \frac{\hat{\varepsilon}\mu}{c}\frac{\partial}{\partial t}\left(\frac{\partial}{\partial z}\frac{1}{\alpha_{ln}^2}\mathcal{A}_{ln}(z,t) + \mathcal{C}_{ln}(z,t)\right)J_{l+1}(\alpha_{ln}r)e^{il\varphi} \tag{11.89}$$
$$+ \frac{\mu}{c}\frac{\partial(P_{NL})_r}{\partial t} + \frac{i\mu}{c}\frac{\partial}{\partial t}(P_{NL})_\varphi.$$

Now, doing the first stage projecting, the result can be integrated over φ

$$
\begin{aligned}
&\int_0^{2\pi} \sum\nolimits_{l,n} - i\left(\frac{\hat{\varepsilon}\mu}{c^2\alpha_{ln}^2}\frac{\partial^2}{\partial t^2}\mathcal{K}_{ln}(z, t) - \mathcal{K}_{ln}(z, t) + \frac{\partial}{\partial z}\mathcal{M}_{ln}(z, t)\right) \\
&J_{l-1}(\alpha_{nl}r)e^{il\varphi}e^{-il'\varphi}d\varphi + c.c. \\
&= \int_0^{2\pi} \sum\nolimits_{l,n} \frac{\hat{\varepsilon}\mu}{c}\frac{\partial}{\partial t}\left(\frac{\partial}{\partial z}\frac{1}{\alpha_{ln}^2}\mathcal{A}_{ln}(z, t) + \mathcal{C}_{ln}(z, t)\right)J_{l-1}(\alpha_{ln}r)e^{il\varphi}e^{-il'\varphi}d\varphi \\
&+ \int_0^{2\pi} e^{-il'\varphi}\left(\frac{\mu}{c}\frac{\partial(P_{NL})_r}{\partial t} - \frac{i\mu}{c}\frac{\partial}{\partial t}(P_{NL})_\varphi\right)d\varphi,
\end{aligned}
\tag{11.90}
$$

$$
\begin{aligned}
&-\int_0^{2\pi} \sum\nolimits_{l,n} i\left(\frac{\hat{\varepsilon}\mu}{c^2\alpha_{ln}^2}\frac{\partial^2}{\partial t^2}\mathcal{K}_{ln}(z, t) - \mathcal{K}_{ln}(z, t) + \frac{\partial}{\partial z}\mathcal{M}_{ln}(z, t)\right) \\
&J_{l+1}(\alpha_{ln}r)e^{il\varphi}e^{-il'\varphi}d\varphi + c.c. \\
&+ \int_0^{2\pi} \sum\nolimits_{l,n} \frac{\hat{\varepsilon}\mu}{c}\frac{\partial}{\partial t}\left(\frac{\partial}{\partial z}\frac{1}{\alpha_{ln}^2}\mathcal{A}_{ln}(z, t) + \mathcal{C}_{ln}(z, t)\right)J_{l+1}(\alpha_{ln}r)e^{il\varphi}e^{-il'\varphi}d\varphi \\
&= \int_0^{2\pi} e^{-il\varphi}\left(\frac{\mu}{c}\frac{\partial(P_{NL})_r}{\partial t} + \frac{i\mu}{c}\frac{\partial}{\partial t}(P_{NL})_\varphi\right)d\varphi.
\end{aligned}
\tag{11.91}
$$

Within the same stage of projecting, both equations will be multiplied by the Bessel function $rJ_{l\pm1}(\alpha_{ln'}r)$ and integrate over $r \in [0, r_0]$

$$
\begin{aligned}
&\int_0^{r_0} \sum\nolimits_{n} - i\left(\frac{\hat{\varepsilon}\mu}{c^2\alpha_{ln}^2}\frac{\partial^2}{\partial t^2}\mathcal{K}_{ln}(z, t) - \mathcal{K}_{ln}(z, t) + \frac{\partial}{\partial z}\mathcal{M}_{ln}(z, t)\right) \\
&rJ_{l-1}(\alpha_{ln}r)J_{l-1}(\alpha_{ln'}r)dr + c.c. \\
&- \int_0^{r_0} \sum\nolimits_{n} \frac{\hat{\varepsilon}\mu}{c}\frac{\partial}{\partial t}\left(\frac{\partial}{\partial z}\frac{1}{\alpha_{ln}^2}\mathcal{A}_{ln}(z, t) + \mathcal{C}_{ln}(z, t)\right)J_{l-1}(\alpha_{ln}r)rJ_{l-1}(\alpha_{ln'}r)dr \\
&= \int_0^{r_0}\int_0^{2\pi} e^{-il\varphi}\left(\frac{\mu}{c}\frac{\partial(P_{NL})_r}{\partial t} - \frac{i\mu}{c}\frac{\partial}{\partial t}(P_{NL})_\varphi\right)rJ_{l-1}(\alpha_{ln'}r)drd\varphi,
\end{aligned}
\tag{11.92}
$$

$$
\begin{aligned}
&\int_0^{r_0} \sum\nolimits_{n} - i\left(\frac{\hat{\varepsilon}\mu}{c^2\alpha_{ln}^2}\frac{\partial^2}{\partial t^2}\mathcal{K}_{ln}(z, t) - \mathcal{K}_{ln}(z, t) + \frac{\partial}{\partial z}\mathcal{M}_{ln}(z, t)\right) \\
&rJ_{l+1}(\alpha_{ln}r)J_{l+1}(\alpha_{ln'}r)dr + c.c. \\
&+ \int_0^{r_0} \sum\nolimits_{n} \frac{\hat{\varepsilon}\mu}{c}\frac{\partial}{\partial t}\left(\frac{\partial}{\partial z}\frac{1}{\alpha_{ln}^2}\mathcal{A}_{ln}(z, t) + \mathcal{C}_{ln}(z, t)\right)rJ_{l+1}(\alpha_{ln}r)J_{l+1}(\alpha_{ln'}r)dr \\
&= \int_0^{r_0}\int_0^{2\pi} e^{-il\varphi}\left(\frac{\mu}{c}\frac{\partial(P_{NL})_r}{\partial t} + \frac{i\mu}{c}\frac{\partial}{\partial t}(P_{NL})_\varphi\right)rJ_{l+1}(\alpha_{ln'}r)drd\varphi.
\end{aligned}
\tag{11.93}
$$

Based on the relation of Bessel functions orthogonality

$$\int_0^{r_0} J_{l\pm 1}(\alpha_{ln}r)rJ_{l\pm 1}(\alpha_{ln'}r)dr = N_{l\pm 1}\delta_{n,n'} \tag{11.94}$$

the obtained equations can be simplified to this form

$$\begin{aligned}&-i\left(\frac{\hat{\varepsilon}\mu}{c^2\alpha_{ln}^2}\frac{\partial^2}{\partial t^2}\mathcal{K}_{ln}(z,t) - \mathcal{K}_{ln}(z,t) + \frac{\partial}{\partial z}\mathcal{M}_{ln}(z,t)\right)N_{l-1} + c.c.\\&-\frac{\hat{\varepsilon}\mu}{c}\frac{\partial}{\partial t}\left(\frac{\partial}{\partial z}\frac{1}{\alpha_{ln}^2}\mathcal{A}_{ln}(z,t) + \mathcal{C}_{ln}(z,t)\right)N_{l-1}\\&= \int_0^{r_0}\int_0^{2\pi} e^{-il\varphi}\left(\frac{\mu}{c}\frac{\partial (P_{NL})_r}{\partial t} - \frac{i\mu}{c}\frac{\partial}{\partial t}(P_{NL})_\varphi\right)rJ_{l-1}(\alpha_{ln'}r)drd\varphi,\end{aligned} \tag{11.95}$$

$$\begin{aligned}&-i\left(\frac{\hat{\varepsilon}\mu}{c^2\alpha_{ln}^2}\frac{\partial^2}{\partial t^2}\mathcal{K}_{ln}(z,t) - \mathcal{K}_{ln}(z,t) + \frac{\partial}{\partial z}\mathcal{M}_{ln}(z,t)\right)N_{l+1} + c.c.\\&+\frac{\hat{\varepsilon}\mu}{c}\frac{\partial}{\partial t}\left(\frac{\partial}{\partial z}\frac{1}{\alpha_{ln}^2}\mathcal{A}_{ln}(z,t) + \mathcal{C}_{ln}(z,t)\right)N_{l+1}\\&= \int_0^{r_0}\int_0^{2\pi} e^{-il\varphi}\left(\frac{\mu}{c}\frac{\partial (P_{NL})_r}{\partial t} + \frac{i\mu}{c}\frac{\partial}{\partial t}(P_{NL})_\varphi\right)rJ_{l+1}(\alpha_{ln'}r)drd\varphi.\end{aligned} \tag{11.96}$$

To separate $\frac{\hat{\varepsilon}\mu}{c^2\alpha_{ln}^2}\frac{\partial^2}{\partial t^2}\mathcal{K}_{ln}(z,t) - \mathcal{K}_{ln}(z,t) + \frac{\partial}{\partial z}\mathcal{M}_{ln}(z,t)$ and $\frac{\partial}{\partial z}\frac{1}{\alpha_{ln}^2}\mathcal{A}_{ln}(z,t) + \mathcal{C}_{ln}(z,t)$ from the above system it is necessary to divide each equation by normalization parameter $N_{l\pm 1}$ and sum by side both equations. The result is

$$\begin{aligned}&-i\left(\frac{\hat{\varepsilon}\mu}{c^2\alpha_{ln}^2}\frac{\partial^2}{\partial t^2}\mathcal{K}_{ln}(z,t) - \mathcal{K}_{ln}(z,t) + \frac{\partial}{\partial z}\mathcal{M}_{ln}(z,t)\right) + c.c.\\&= \frac{1}{2N_{l-1}}\int_0^{r_0}\int_0^{2\pi} e^{-il\varphi}\left(\frac{\mu}{c}\frac{\partial (P_{NL})_r}{\partial t} - \frac{i\mu}{c}\frac{\partial}{\partial t}(P_{NL})_\varphi\right)rJ_{l-1}(\alpha_{ln'}r)drd\varphi\\&+\frac{1}{2N_{l+1}}\int_0^{r_0}\int_0^{2\pi} e^{-il\varphi}\left(\frac{\mu}{c}\frac{\partial (P_{NL})_r}{\partial t} + \frac{i\mu}{c}\frac{\partial}{\partial t}(P_{NL})_\varphi\right)rJ_{l+1}(\alpha_{ln'}r)drd\varphi,\end{aligned} \tag{11.97}$$

$$\begin{aligned}&\frac{\hat{\varepsilon}\mu}{c\alpha_{ln}^2}\frac{\partial}{\partial t}\left(\frac{\partial}{\partial z}\mathcal{A}_{ln}(z,t) + \alpha_{ln}^2\mathcal{C}_{ln}(z,t)\right) + c.c.\\&= -\frac{1}{2N_{l-1}}\int_0^{r_0}\int_0^{2\pi} e^{-il\varphi}\left(\frac{\mu}{c}\frac{\partial (P_{NL})_r}{\partial t} - \frac{i\mu}{c}\frac{\partial}{\partial t}(P_{NL})_\varphi\right)rJ_{l-1}(\alpha_{ln'}r)drd\varphi\\&+\frac{1}{2N_{l+1}}\int_0^{r_0}\int_0^{2\pi} e^{-il\varphi}\left(\frac{\mu}{c}\frac{\partial (P_{NL})_r}{\partial t} + \frac{i\mu}{c}\frac{\partial}{\partial t}(P_{NL})_\varphi\right)rJ_{l+1}(\alpha_{ln'}r)drd\varphi.\end{aligned} \tag{11.98}$$

At this moment we arrive at the system of four equations

$$\frac{\partial}{\partial z}\mathcal{A}_{ln}(z, t) + \alpha_{ln}^2\mathcal{C}_{ln}(z, t) = \mathcal{N}_1 \tag{11.99}$$

$$\frac{\partial}{\partial z}\mathcal{C}_{ln}(z, t) - \left(-\frac{\partial^2}{\partial t^2}\frac{\hat{\varepsilon}\mu}{\alpha_{ln}^2c^2}\mathcal{A}_{ln}(z, t) + \mathcal{A}_{ln}(z, t)\right) = 0 \tag{11.100}$$

$$\frac{\partial}{\partial z}\mathcal{K}_{ln}(z, t) - (-\alpha_{ln}^2\mathcal{M}_{ln}(z, t)) = 0, \tag{11.101}$$

$$\frac{\partial}{\partial z}\mathcal{M}_{ln}(z, t) - \left(-\frac{\hat{\varepsilon}\mu}{c^2\alpha_{ln}^2}\frac{\partial^2}{\partial t^2}\mathcal{K}_{ln}(z, t) + \mathcal{K}_{ln}(z, t)\right) = \mathcal{N}_2 \tag{11.102}$$

where

$$\begin{aligned}\mathcal{N}_1 &= -\frac{1}{N_{l-1}}\int_0^{r_0}\int_0^{2\pi} e^{-il\varphi}\left(\frac{\mu}{c}\frac{\partial(P_{NL})_r}{\partial t} - \frac{i\mu}{c}\frac{\partial}{\partial t}(P_{NL})_\varphi\right)rJ_{l-1}(\alpha_{ln'}r)drd\varphi\\ &+\frac{1}{N_{l+1}}\int_0^{r_0}\int_0^{2\pi} e^{-il\varphi}\left(\frac{\mu}{c}\frac{\partial(P_{NL})_r}{\partial t} + \frac{i\mu}{c}\frac{\partial}{\partial t}(P_{NL})_\varphi\right)rJ_{l+1}(\alpha_{ln'}r)drd\varphi,\end{aligned} \tag{11.103}$$

$$\begin{aligned}\mathcal{N}_2 &= \frac{1}{N_{l-1}}\int_0^{r_0}\int_0^{2\pi} e^{-il\varphi}\left(\frac{\mu}{c}\frac{\partial(P_{NL})_r}{\partial t} - \frac{i\mu}{c}\frac{\partial}{\partial t}(P_{NL})_\varphi\right)rJ_{l-1}(\alpha_{ln'}r)drd\varphi,\\ &+\frac{1}{N_{l+1}}\int_0^{r_0}\int_0^{2\pi} e^{-il\varphi}\left(\frac{\mu}{c}\frac{\partial(P_{NL})_r}{\partial t} + \frac{i\mu}{c}\frac{\partial}{\partial t}(P_{NL})_\varphi\right)rJ_{l+1}(\alpha_{ln'}r)drd\varphi,\end{aligned} \tag{11.104}$$

can be presented in a form $\partial_z\Psi(z, t) - \hat{L}\Psi(z, t) = \mathcal{N}(\Psi)$, where

$$\Psi(z, t) = \begin{pmatrix}\mathcal{A}_{ln}(z, t)\\ \mathcal{C}_{ln}(z, t)\\ \mathcal{K}_{ln}(z, t)\\ \mathcal{M}_{ln}(z, t)\end{pmatrix},$$

$$\hat{L} = \begin{pmatrix} 0 & -\alpha_{ln}^2 & 0 & 0\\ -\frac{\partial^2}{\partial t^2}\frac{\hat{\varepsilon}\mu}{\alpha_{ln}^2c^2} + 1 & 0 & 0 & 0\\ 0 & 0 & 0 & -\alpha_{ln}^2\\ 0 & 0 & -\frac{\partial^2}{\partial t^2}\frac{\hat{\varepsilon}\mu}{\alpha_{ln}^2c^2} + 1 & 0\end{pmatrix}, \tag{11.105}$$

$$\mathcal{N}(\Psi) = \begin{pmatrix}\mathcal{N}_1\\ 0\\ 0\\ \mathcal{N}_2\end{pmatrix}.$$

With the zero nonlinear part the system reduces to equation (11.43).

11.3.1 Application of projection operators

As all components to the equation are given, it is possible to proceed with the procedure of projection operators application. The P_{11} operator is chosen as the exemplary one to the presented equation

$$\frac{\partial}{\partial z}P_{11}\begin{pmatrix} \mathcal{A}_{ln}(z,t) \\ \mathcal{C}_{ln}(z,t) \\ \mathcal{K}_{ln}(z,t) \\ \mathcal{M}_{ln}(z,t) \end{pmatrix} - P_{11}\begin{pmatrix} 0 & -\alpha_{ln}^2 & 0 & 0 \\ -\frac{\partial^2}{\partial t^2}\frac{\hat{\varepsilon}\mu}{\alpha_{ln}^2 c^2} + 1 & 0 & 0 & 0 \\ 0 & 0 & 0 & -\alpha_{ln}^2 \\ 0 & 0 & -\frac{\partial^2}{\partial t^2}\frac{\hat{\varepsilon}\mu}{\alpha_{ln}^2 c^2} + 1 & 0 \end{pmatrix} \tag{11.106}$$

$$\begin{pmatrix} \mathcal{A}_{ln}(z,t) \\ \mathcal{C}_{ln}(z,t) \\ \mathcal{K}_{ln}(z,t) \\ \mathcal{M}_{ln}(z,t) \end{pmatrix} = P_{11}\begin{pmatrix} \mathcal{N}_1 \\ 0 \\ 0 \\ \mathcal{N}_2 \end{pmatrix}.$$

As the result of projection the equation is acquired

$$\frac{\partial}{\partial z}\begin{pmatrix} -i\alpha_{ln}^2\hat{k}_{nl}^{-1}\Lambda_1(z,t) \\ \Lambda_1(z,t) \\ 0 \\ 0 \end{pmatrix} - \begin{pmatrix} 0 & -\alpha_{ln}^2 & 0 & 0 \\ -\frac{\partial^2}{\partial t^2}\frac{\varepsilon\mu}{\alpha_{ln}^2 c^2} + 1 & 0 & 0 & 0 \\ 0 & 0 & 0 & -\alpha_{ln}^2 \\ 0 & 0 & -\frac{\partial^2}{\partial t^2}\frac{\varepsilon\mu}{\alpha_{ln}^2 c^2} + 1 & 0 \end{pmatrix} \tag{11.107}$$

$$\begin{pmatrix} -i\alpha_{ln}^2\hat{k}_{nl}^{-1}\Lambda_1(z,t) \\ \Lambda_1(z,t) \\ 0 \\ 0 \end{pmatrix} = \frac{1}{2}\begin{pmatrix} \mathcal{N}_1 \\ \frac{-i}{\alpha_{ln}^2}\hat{k}_{nl}\mathcal{N}_1 \\ 0 \\ 0 \end{pmatrix}.$$

Comparing the dispersion relation (11.52) with $\hat{L}$ operator it is possible to replace $-\frac{\partial^2}{\partial t^2}\frac{\hat{\varepsilon}\mu}{\alpha_{ln}^2 c^2} + 1$ with $\alpha_{ln}^{-2}\hat{k}_{nl}^2$. As a result the system of two equations

$$\begin{aligned} &\frac{\partial}{\partial z}(-i\alpha_{ln}^2\hat{k}_{nl}^{-1}\Lambda_1(z,t)) - (-\alpha_{ln}^2)\Lambda_1(z,t) = \frac{1}{2}\mathcal{N}_1 \\ &\frac{\partial}{\partial z}\Lambda_1(z,t) + i\hat{k}_{nl}\Lambda_1(z,t) = \frac{1}{2}(-i\alpha_{ln}^{-2}\hat{k}_{nl})\mathcal{N}_1, \end{aligned} \tag{11.108}$$

is given. Equations are linearly depended. Hence the final general equation describing the propagation of the polarized, directed electromagnetic wave to the left side has the form

$$\frac{\partial}{\partial z}\Lambda_1(z, t) + i\hat{k}_{nl}\Lambda_1(z, t) = \frac{1}{2}(-i\alpha_{ln}^{-2}\hat{k}_{nl})\mathcal{N}_1, \tag{11.109}$$

which can be rewritten with the explicit form of operators

$$\begin{aligned} &\frac{\partial}{\partial z}\Lambda_1(z, t) + i\left(-\frac{\partial^2}{\partial t^2}\frac{\hat{\varepsilon}\mu}{\alpha_{ln}^2 c^2} + 1\right)^{1/2}\Lambda_1(z, t) \\ &= -i\frac{1}{2}\alpha_{ln}^{-2}\left(-\frac{\partial^2}{\partial t^2}\frac{\hat{\varepsilon}\mu}{\alpha_{ln}^2 c^2} + 1\right)^{1/2} \\ &\left(\frac{1}{N_{l-1}}\int_0^{r_0}\int_0^{2\pi} e^{-il\varphi}\left(\frac{\mu}{c}\frac{\partial (P_{NL})_r}{\partial t} - \frac{i\mu}{c}\frac{\partial}{\partial t}(P_{NL})_\varphi\right) rJ_{l-1}(\alpha_{ln'}r)\right. \\ &\left. dr d\varphi + \frac{1}{N_{l+1}}\int_0^{r_0}\int_0^{2\pi} e^{-il\varphi}\left(\frac{\mu}{c}\frac{\partial (P_{NL})_r}{\partial t} + \frac{i\mu}{c}\frac{\partial}{\partial t}(P_{NL})_\varphi\right) rJ_{l+1}(\alpha_{ln'}r) dr d\varphi\right). \end{aligned} \tag{11.110}$$

The way to solutions and its numerical realization is presented in [3, 36], see also the general exposition in [50]

References

[1] Gómez-Correa J E, Balderas-Mata S E, Garza-Rivera A, Jaimes-Nájera A, Trevino J P, Coello V, Rogel-Salazar J and Chávez-Cerda S 2019 *Mathematical-Physics of Propagating Modes in Planar Waveguides* (arXiv: 1905.09650v1)

[2] Leble S B 2003 Nonlinear waves in optical waveguides and soliton theory applications *Optical Solitons* (Berlin: Springer)

[3] Leble S 2019 *Waveguide Propagation of Nonlinear Waves Impact of Inhomogeneity and Accompanying Effects* (Berlin: Springer)

[4] Itoh T 1976 Inverted strip dielectric wave guide for millimeter wave integrated circuits December *IEEE Trans. Microwave Theory Tech.* **24** 821–7

[5] Yamamoto A novel low-loss dielectric waveguide for millimeter and submillimeter wavelengths *IEEE Trans. Microw. Theory Tech.* **28** 580–5

[6] Kang J U, Stegeman G I, Atchison J S and Akhmediev N 1996 *Phys. Rev. Lett.* **76** 3699–702

[7] Zeng X-Y, Luk K-M and Xu S-J 2001 A novel leaky NRD guide with a double-layer dielectric slab *IEEE Trans. Microw. Theory Tech.* **49** 585–8

[8] Yim K, Yong Y, Lee J, Lee K, Nahm H-H, Yoo J, Lee C, Hwang C S and Han S 2015 Novel high-κ dielectrics for next-generation electronic devices screened by automated *ab initio* calculations *NPG Asia Mater.* **7** 190

[9] Haus H A and Wong W S 1996 *Rev. Mod. Phys.* **68** 423

[10] Turitsyn S K, Balea B G and Fedoruk M P 2012 Dispersion-managed solitons in fibre systems and lasers *Phys. Rep.* **521** 135–203

[11] Lee S C J, Breyer F, Randel S, van den Boom H P A and Koonen A M J 2008 High-speed transmission over multimode fiber using discrete multitone modulation *J. Opt. Netw.* **7** 183

[12] Harder M and Hu C-M 2018 Cavity spintronics: an early review of recent progress in the study of magnon–photon level repulsion *Solid State Phys.* **69** 47–121

[13] Turitsyn S, Doran N, Nijhof J, Mezentsev V, Schäfer T and Forysiak W 1999 Dispersion managed solitons *Les Houches Lectures* Lecture 7 p 93

[14] Cautaerts V, Kodama Y, Maruta A and Sugavara H 1999 Nonlinear pulses in ultra-fast communications *Les Houches Lectures* Lecture 9 p 147

[15] Seve E, Millot G and Trillo S 2000 *Phys. Rev.* E **61** 3139–50

[16] Leble S 1998 *Comput. Math. Appl.* **35** 73–81
Leble S 2000 *Theor. Math. Phys.* **122** 239–50

[17] Wabnitz S, Pitois S and Millot G 2001 Nonlinear polarization dynamics of counter-propagating waves in an isotropic optical fiber: theory and experiments *J. Opt. Soc. Am.* B **18** 432–43

[18] Blow K J, Doran N J and Wood D 1988 Generation and stabilization of short soliton pulses in the amplified nonlinear Schrödinger equation *J. Opt. Soc. Am.* B **5** 381

[19] Russell P St J 2003 Photonic crystal fibers *Science* **299** 358

[20] Reichel B and Blow K J 2008 Approximation of photonic crystal fibres with large air holes by the step index fibre model *J. Mod. Opt.* **55** 1479

[21] Leble S B and Reichel B 2008 Mode interaction in few-mode optical fibers with Kerr effect *J. Mod. Opt.* **55** 3653–66

[22] Islam M N 1990 *Opt. Lett.* **15** 417

[23] Kang J U, Stegeman G I and Atchison J S 1996 *Opt. Lett.* **21** 189–91

[24] Manakov S V 1973 On the theory of self-focusing stationary self-focusing of electromagnetic waves *J. Exp. Theor. Phys.* **65** 505–16

[25] Zakharov V E and Shabat A B 1972 Exact theory of two-dimensional self-focusing and one-dimensional self-modulation of waves in nonlinear media *Sov. Phys. JETP* **34** 62

[26] Mollenauer L F, Stolen R H and Gordon J P 1980 Experimental observation of picosecond pulse narrowing and solitons in optical fibers *Phys. Rev. Lett.* **45** 1095

[27] Bertolotti M *et al* 1999 *Opt. Commun.* **168** 399–403

[28] Delqué M, Fanjoux G and Sylvestre T 2008 Collision between scalar and vector spatial solitons in Kerr media *Opt. Quantum Electron.* **40** 281–91

[29] Cundiff S T, Collings B C, Achmediev N N, Soto-Crespo J M, Bergman K and Knox W H 1999 *Phys. Rev. Lett.* **2** 3988–91

[30] Yan M, Hao Q, Shen X and Zeng H 2018 Experimental study on polarization evolution locking in a stretched-pulse fiber laser *Opt. Express* **26** 16086

[31] Anastassiu C *et al* 1999 *Phys. Rev. Lett.* **83** 2332–5

[32] Winful H G 1986 Polarization instabilities in birefringent nonlinear media: application to fiber-optic devices *Opt. Lett.* **11** 33

[33] Blow K J, Doran N J and Wood D 1987 Polarization instabilities for solitons in birefringent fibers *Opt. Lett.* **12** 202

[34] Doktorov E V and Shchesnovich V S 1995 Modified nonlinear Schrödinger equation: Spectral transform and N-soliton solution *J. Math. Phys.* **36** 7009

[35] Schäfer T and Wayne C E 2004 Propagation of ultra-short optical pulses in cubic nonlinear media *Physica* D **196** 90–105

[36] Reichel B and Leble S 2008 On convergence and stability of a numerical scheme of coupled nonlinear schrödinger equations *Comput. Math. Appl.* **55** 745

[37] Leble S and Reichel B 2009 Coupled nonlinear Schrodinger equations in optical fibers theory: from general aspects to solitonic ones *Eur. Phys. J. Spec. Top.* **173** 5–55

[38] Leble S 1991 *Nonlinear Waves in Waveguides* (Berlin: Springer) p 164

[39] Joseph R J 1977 Solitary waves in a finite depth fluid *J. Phys. A: Math. Gen.* **10** 1225–7

[40] Leble S 1982 On the Kadomtsev–Petviashvili equation analogue in the internal waves theory *Izv. Akad. Nauk SSSR, Fiz. Atm.Okean* **20** 1199–1205

[41] Hondros D and Debye P 1910 Elektromagnetische wellen an dielektrischen drahten *Ann. Phys.* **32** 465

[42] Hasegawa A and Matsumoto M 2003 *Optical Solitons in Fibers, Springer Series in Photonics* (Berlin: Springer)

[43] Zakharov V E 1968 Stability of periodic waves of finite amplitude on the surface of a deep fluid *J. Appl. Mech. Tech. Phys.* **9** 190–4

[44] Chow K K and Lin C 2008 Photonic crystal fibers for nonlinear signal processing *Opti- cal Fiber Communication Conference and Exposition and The National Fiber Optic Engineers Conference, OSA Technical Digest (CD)* (Washington, DC: Optical Society of America) paper OMP6

[45] Leble S and Perelomova A 2018 *Dynamical Projectors Method in Hydro- and Electrodynamics* (Boca Raton, FL: CRC Press)

[46] Popov I Yu 1999 Zero-range potentials model for planar waveguide in photonic crystal *Tech. Phys. Lett.* **25** 45–9

[47] Boyd R W 1992 *Nonlinear Optics* (Boston, MA: Academic)

[48] Fock V 1976 *Nachala Kvantovoj mechaniki (Foundations of Quantum Mechanics)* 2nd rev edn (Moscow: Nauka) 374

[49] Ma W X 2020 Long-time asymptotics of a three-component coupled nonlinear Schrödinger system *J. Geom. Phys.* **153** 103669

[50] Ma W X, Huang T W and Zhang Y 2010 A multiple exp-function method for nonlinear differential equations and its application *Phys. Scr.* **82** 065003

Chapter 12

Propagation of electromagnetic waves in exclusive dispersive media such as metamaterials

This chapter is a direct continuation and advanced development of chapter 8, this textbook.

12.1 Electromagetic waves in metamaterial

12.1.1 On dispersion in 1D metamaterial

A concept of a metamaterial from a formal point of view is directly based on dispersion relation [1], that develops on the Drude model, see chapter 6, subsection 6.3.2 as a particular case of Lorentz model. The term '*meta*material' itself has an interesting history [1]. One of the important applications of such a model relates to metamaterials that are characterized by negative values of the parameters ε and μ, that must be anomalous dispersive, i.e. their permittivity and permeability must be frequency dependent, otherwise they would not be causal [2]. The two-time derivative Lorentz material (2TDLM) model encompasses the metamaterial models most commonly discussed; it has the frequency domain susceptibility [3]:

$$\chi_e = \frac{\omega_p^2\chi_\alpha + i\omega_p\chi_\beta\omega - \chi_\gamma\omega^2}{\omega_0^2 + i\omega\Gamma - \omega^2}, \tag{12.1}$$

its particular case is the 2TDLM model, which produces a resonant response at $\omega = \omega_0$ when $\Gamma = 0$. The natural 1D metamaterials construction originated from layered medium [4].

The nonlinear behavior of electromagnetic (EM) wave propagation depends on relations between the field and induced polarization. It is obvious that it is necessary to use either a numerical scheme or approximations to obtain an analytical solution of a nonlinear problem. The first successful approach of such a reduction was the use of a set of slowly varying envelopes. The simplest model scalar equation for a

doi:10.1088/978-0-7503-2576-9ch12

directed wave propagation, based on this approach, have a form of the nonlinear Schrödinger equation, derived by Zakharov in 1968 [5, 6]. Its integrability [7] made the model very attractive because of the rich 'zoo' of the equation explicit solutions [8].

A natural step of integrable generalization of such a model lies in a plane of better approximations of dispersion, dissipation [9] and nonlinearity (modified nonlinear Schrödinger (MNS) equation, see e.g. [8]), that allows to extend pulse durations down to picoseconds.

There are plenty of alternative ideas on the few-cycle pulse soliton-type description in different media [10]

The next step of this movement to the ultrashort pulses description that maintains integrability is made in the works of Shäfer–Wayne [11, 12]. The short pulse equation (SPE) again relates to unidirectional propagation for which a special kind of dispersion law and nonlinearity action was accounted for in a rescaled evolution. In 2017 Z Zhaqilao *et al* derived an N-fold Darboux transformation from the Lax pair of the two-component short pulse system that give an interesting class of solutions with loops [13].

A generalization that allows to include the description and interaction of opposite directed waves is connected with the idea of joint account of the correspondent spaces of 'hybrid' electric–magnetic amplitudes [14, 15]. The projecting operators (PO) method [14] works directly at arbitrary dispersion and weak nonlinearity. Strong nonlinearity needs the method modification [16]. Similar universality demonstrates the method of [17]. The PO technique gives systematic transition to hybrid fields with simultaneous superposition of nonlinear terms, that effectively approximate weak nonlinearity, arriving at the mentioned celebrated model equations at the subspaces of directed waves [18, 19]. The field hybridization may account for *ab initio* dispersion and dissipation [9] and nonlinearity by iteration procedure [16].

Next, natural for electromagnetic field step accounts polarization and leads to double component vector equations [19]. Both direction and polarization are studied theoretically and, of importance, experimentally in [20].

Its continued systematic application of projecting approach, originated from [14] for 1D metamaterial with account of both polarizations of the EM wave. The technique and results of the work [21, 22] on nonlinear evolution equations of opposite directed waves with one polarization in Drude 1D-metamaterial have been developed. The application of projecting operators method for the metamaterials is widen to include electromagnetic wave polarization, similar to [19]. The nonlinear equation for Drude metamaterial is derived and the result is compared with our previous results and vector SPE. On the basis of resulting equations, the wave packets for linear and nonlinear cases are studied in [23], where for the case with equal amplitudes, the approximate solutions are derived.

For metamaterial the situation is different, as shown in [22] for unique polarization, the change of arrangements of wave modes is recovered. Now, the questions to be answered: what happens with dispersion and polarization in a waveguide? And how do interactions of all four modes in a waveguide filled with a metamaterial look?

12.2 Directed modes in rectangular waveguides: polarization, dispersion, nonlinearity

We would start from the problem of an electromagnetic wave in rectangular waveguides formulation, based on the author's book [24], doing simplifications and comments for didactic purposes. A wave motion of electromagnetic field in a metal rectangular waveguide filled with matter with given dispersion and non-linearity is described on the basis of Maxwell's equations with appropriate boundary conditions. For metamaterials applications we choose the Drude models for both permittivities as dispersion relations and Kerr nonlinearity.

12.2.1 Maxwell's equations for matter inside a waveguide

For such a problem, for points inside a waveguide, more close to engineering, we would present the Maxwell system in SI units as:

$$\operatorname{div} \vec{D} = 0, \tag{12.2}$$

$$\operatorname{div} \vec{B} = 0, \tag{12.3}$$

$$\operatorname{rot}\vec{E} = -\frac{\partial \vec{B}}{\partial t}, \tag{12.4}$$

$$\operatorname{rot}\vec{H} = \frac{\partial \vec{D}}{\partial t}, \tag{12.5}$$

that is formulated for the case of absence of charges and currents. The material operator relations for an isotropic dispersive medium in a waveguide we write as in section 13.2 from [25] in SI units:

$$\vec{D} = \varepsilon_0 \hat{\varepsilon} \vec{E},\ \vec{H} = \frac{1}{\mu_0} \hat{\mu}^{-1} \vec{B}.$$

Here the operators $\hat{\varepsilon}$ and $\hat{\mu}$ act as integral ones

$$\hat{\varepsilon}\vec{E} = \int_{-\infty}^{\infty} \tilde{\varepsilon}(t-s)\vec{E}(\tau)d\tau,\ \hat{\mu}^{-1}\vec{B} = \frac{1}{\sqrt{2\pi}} \int_{-\infty}^{\infty} \tilde{\mu}^{-1}(t-\tau)\vec{B}(\tau)d\tau, \tag{12.6}$$

where

$$\tilde{\varepsilon}(t-s) = \frac{1}{2\pi} \int_{-\infty}^{\infty} \varepsilon(\omega)\exp(i\omega(t-s))d\omega, \tag{12.7}$$

$$\tilde{\mu}(t-\tau) = \frac{1}{\sqrt{2\pi}} \int_{-\infty}^{\infty} \mu(\omega)\exp(i\omega(t-s))d\omega, \tag{12.8}$$

Here: ε_0, μ_0 are vacuum SI constants and $\varepsilon(\omega)$, $\mu(\omega)$ are functional parameters that define wave dispersion of the medium of propagation, see e.g. [26]. The Colombian equation (12.2) goes to

$$\operatorname{div} \hat{\varepsilon}\vec{E} = 0. \tag{12.9}$$

Note, that the operator div and $\hat{\varepsilon}$ commute. The Ampère equation (12.5) of the Maxwell system is transformed as

$$\operatorname{rot}\vec{B} = \frac{\hat{\varepsilon}\hat{\mu}}{c^2}\frac{\partial \vec{E}}{\partial t}. \tag{12.10}$$

The equations (12.3), (12.4), (12.9) and (12.10) form the closed system for two vectors, the problem $\vec{E}$, $\vec{B}$. The equations of the system are however not independent, as was shown in chapter 2, section 2.3. The complete problem formulation for a wave in a waveguide needs boundary conditions at waveguide walls and a boundary regime at an end. These are formulated in the next section.

12.3 Boundary conditions: the transversal waveguide modes evolution

Suppose the geometry of a metal rectangular tube is taken in such a form, that its axis lies along the x, that marks the direction of the waves propagation. In this chapter, in didactic purposes we, compared with [24] simplify the waveguide geometry, take the square cross-section with dimension a. It, in final relations simplifies the coefficients form. We suppose that on the metal walls of a rectangular waveguide the following boundary conditions hold [24].

$$E_{x,y}\big|_{z=0,a} = 0, \quad E_{x,z}\big|_{y=0,a} = 0, \tag{12.11}$$

$$B_z\big|_{z=0,a} = 0, \quad B_y\big|_{y=0,a} = 0. \tag{12.12}$$

The transversal waveguide modes we use have the standard form [27].

Starting from the x-component of the electric field vector, we write the expansion

$$E_x = \sum_{n,m} \varphi_{nm}(x,\, t)\sin\frac{\pi m}{a}y \,\sin\frac{\pi n}{a}z. \tag{12.13}$$

It by construction satisfies the boundary conditions (12.11) at the waveguide walls. The rest components of the electric field are chosen as

$$E_y = \sum_{n,m} \chi_{nm}(x,\, t)\cos\frac{\pi m}{a}y \,\sin\frac{\pi n}{a}z, \tag{12.14}$$

$$E_z = \sum_{n,m} \psi_{nm}(x,\, t)\sin\frac{\pi m}{a}y \,\cos\frac{\pi n}{a}z, \tag{12.15}$$

to adjust them to the boundary conditions (12.11). The magnetic field expansions should account for the Faraday equation (12.4) and, naturally fit the boundaries (12.12).

$$B_x = \sum_{n,m} \alpha_{nm}(x,\, t)\cos\frac{\pi m}{a}y\,\cos\frac{\pi n}{a}z, \tag{12.16}$$

$$B_y = \sum_{n,m} \beta_{nm}(x,\, t)\sin\frac{\pi m}{a}y\,\cos\frac{\pi n}{a}z, \tag{12.17}$$

$$B_z = \sum_{n,m} \gamma_{nm}(x,\, t)\cos\frac{\pi m}{a}y\,\sin\frac{\pi n}{a}z. \tag{12.18}$$

Plugging the expansions (12.13), (12.14) and (12.16) into the Maxwell system (12.2)–(12.5) and omitting indices yields

$$\frac{\partial\varphi}{\partial x} - \frac{\pi m}{a}\chi - \frac{\pi n}{a}\psi = 0, \tag{12.19}$$

$$\frac{\partial\alpha}{\partial x} + \frac{\pi m}{a}\beta + \frac{\pi n}{a}\gamma = 0, \tag{12.20}$$

and, for dynamic equations (12.4) for magnetic field it is:

$$\begin{aligned} &\frac{\pi m}{a}\psi - \frac{\pi n}{a}\chi + \frac{\partial\alpha}{\partial t} = 0,\\ &\frac{\pi n}{a}\varphi - \frac{\partial\psi}{\partial x} + \frac{\partial\beta}{\partial t} = 0,\\ &\frac{\partial\chi}{\partial x} - \frac{\pi m}{a}\varphi + \frac{\partial\gamma}{\partial t} = 0. \end{aligned} \tag{12.21}$$

The electric field evolution is described by equation (12.10) that gives

$$\begin{aligned} &\frac{\pi m}{a}\gamma - \frac{\pi n}{a}\beta - \frac{\hat{\varepsilon}\hat{\mu}}{c^2}\frac{\partial\varphi}{\partial t} = 0,\\ &\frac{\pi n}{a}\alpha - \frac{\partial\gamma}{\partial x} - \frac{\hat{\varepsilon}\hat{\mu}}{c^2}\frac{\partial\chi}{\partial t} = 0,\\ &\frac{\partial\beta}{\partial x} + \frac{\pi m}{a}\alpha - \frac{\hat{\varepsilon}\hat{\mu}}{c^2}\frac{\partial\psi}{\partial t} = 0. \end{aligned} \tag{12.22}$$

The equations (12.21) and (12.22) differ from the ones from [24] only by hats over ε and μ because we account a material dispersion now.

Excluding ψ, γ by the relations (12.22) and (12.21) as

$$\frac{a}{\pi n}\frac{\partial\varphi}{\partial x} - \frac{m}{n}\chi = \psi, \tag{12.23}$$

$$-\frac{a}{\pi n}\frac{\partial \alpha}{\partial x} - \frac{m}{n}\beta = \gamma, \tag{12.24}$$

Let us introduce the constants that are expressed in terms of the waveguide dimension parameters and the transverse mode number m, n:

$$p \equiv \frac{m}{n},\ q \equiv \frac{\pi}{a}\left(\frac{m^2}{n} + n\right),\ r \equiv \frac{\pi n}{a}, \tag{12.25}$$

we arrive at:

$$\frac{\partial \alpha}{\partial t} = q\chi - p\frac{\partial \varphi}{\partial x}, \tag{12.26}$$

$$\frac{\partial \beta}{\partial t} = \frac{1}{r}\frac{\partial^2 \varphi}{\partial x^2} - r\varphi - p\frac{\partial \chi}{\partial x}, \tag{12.27}$$

$$\frac{\hat{\varepsilon}\hat{\mu}}{c^2}\frac{\partial \varphi}{\partial t} = -p\frac{\partial \alpha}{\partial x} - q\beta, \tag{12.28}$$

$$\frac{\hat{\varepsilon}\hat{\mu}}{c^2}\frac{\partial \chi}{\partial t} = r\alpha + r^{-1}\frac{\partial^2 \alpha}{\partial x^2} + p\frac{\partial \beta}{\partial x}, \tag{12.29}$$

compare with [25]. The last equations from the (12.21) and (12.22) go to identities. Developing the content of the corresponding study in [25], let us take now the case with matter dispersion account and concentrate our efforts on peculiarities of the waveguide excitation in the next section. The linear Cauchy problem may be reformulated as the system of partial differential equations of the first order time derivatives. Introducing

$$\frac{\hat{\varepsilon}\hat{\mu}}{c^2} = \hat{a}^2, \tag{12.30}$$

we rewrite (12.26) in standard matrix operator form (see section 20.3)

$$\frac{\partial}{\partial t}\begin{pmatrix} \alpha \\ \beta \\ \varphi \\ \chi \end{pmatrix} = \begin{pmatrix} 0 & 0 & -p\partial_x & q \\ 0 & 0 & \frac{1}{r}\partial_{xx} & -p\partial_x - r \\ -p\hat{a}^{-2}\partial_x & -q\hat{a}^{-2} & 0 & 0 \\ \hat{a}^{-2}\left(r + \frac{1}{r}\partial_{xx}\right) & p\hat{a}^{-2} & 0 & 0 \end{pmatrix}\begin{pmatrix} \alpha \\ \beta \\ \varphi \\ \chi \end{pmatrix}, \tag{12.31}$$

with the notation

$$\partial_x = \frac{\partial}{\partial x}.$$

The system determines a Cauchy problem with the appropriate initial conditions for $\alpha(x, 0)$, $\beta(x, 0)$, $\varphi(x, 0)$, $\chi(x, 0)$ that may be extracted from the waveguide mode expansions as equation (13.71). Each such equation determines evolution of the specific mode with fixed direction of propagation and polarization. That is obtained by projecting operators as in section 20.3. The presence of waveguide dispersion, however, enlarges the order of derivatives with respect to x. The presence of matter dispersion is determined by integral operators $\hat{\varepsilon}$, $\hat{\mu}$.

12.3.1 The boundary regime problem formulation for the transversal modes

The geometry of a waveguide and physics of its excitation imply rather a boundary regime formulation than a Cauchy problem: one supposes the excitation is realized in the vicinity of one or both ends of the waveguide. Therefore we need a reformulation of the problem as in section 8.3.2. The algorithm of transformation of equations (12.31) is described in detail in [24], it is also presented in chapter 20 as an exercise.

The results of transformations is the x-evolution system, denote, $\nu = 1/(p(rp^2 - q))$

$$\alpha_x = -\frac{1}{p}\hat{a}^2\varphi_t - \frac{q}{p}\beta, \tag{12.32}$$

$$\chi_x = \nu\hat{a}^2\varphi_{tt} - \nu q r\varphi - \frac{r}{p}\varphi - \frac{1}{p}\beta_t, \tag{12.33}$$

$$\beta_x = -\nu r^2 p^2 \alpha - \nu\hat{a}^2\alpha_{tt} + \nu(q + rp^2)\hat{a}^2\chi_t, \tag{12.34}$$

$$\varphi_x = \frac{q}{p}\chi - \frac{1}{p}\alpha_t. \tag{12.35}$$

Going to matrix operators notations, as is used through this book, we write

$$\frac{\partial\Psi}{\partial x} = L\Psi, \tag{12.36}$$

with the standard Ψ as the column

$$\Psi = \begin{pmatrix} \alpha \\ \chi \\ \beta \\ \varphi \end{pmatrix}. \tag{12.37}$$

The matrix operator L reads:

$$L = \begin{pmatrix} 0 & 0 & -\frac{q}{p} & -\frac{1}{p}\hat{a}^2\partial_t \\ 0 & 0 & -\frac{1}{p}\partial_t & \nu\hat{a}^2\partial_{tt} - \nu qr - \frac{r}{p} \\ -\nu r^2p^2 - \nu\hat{a}^2\partial_{tt} & \nu(q + rp^2)\hat{a}^2\partial_t & 0 & 0 \\ -\frac{1}{p}\partial_t & \frac{q}{p} & 0 & 0 \end{pmatrix}. \tag{12.38}$$

Delivering the Fourier transformation for the four basic variables, $\varphi(x, t)$, $\psi(x, t)$, $\alpha(x, t)$ and $\beta(x, t)$, it presents as:

$$\varphi(x, t) = \frac{1}{\sqrt{2\pi}} \int \widehat{\varphi}(\omega, x)\exp(i\omega t)d\omega, \tag{12.39}$$

$$\chi(x, t) = \frac{1}{\sqrt{2\pi}} \int \widehat{\chi}(\omega, x)\exp(i\omega t)d\omega, \tag{12.40}$$

$$\alpha(x, t) = \frac{1}{\sqrt{2\pi}} \int \widehat{\alpha}(\omega, x)\exp(i\omega t)d\omega, \tag{12.41}$$

$$\beta(x, t) = \frac{1}{\sqrt{2\pi}} \int \widehat{\beta}(\omega, x)\exp(i\omega t)d\omega. \tag{12.42}$$

The Fourier transform of the vector-column (12.37), looks now as $\tilde{\Psi}$

$$\tilde{\Psi} = \begin{pmatrix} \widehat{\alpha} \\ \widehat{\chi} \\ \widehat{\beta} \\ \widehat{\varphi} \end{pmatrix}. \tag{12.43}$$

Recall the definition of the operator $\hat{a}^2$, its ω-representation and denote it $a(\omega)^2$, hence the matrix $\widehat{L}$ represents the operator (12.38):

$$\widehat{L} = \begin{pmatrix} 0 & 0 & -\frac{q}{p} & -i\omega\frac{a(\omega)^2}{p} \\ 0 & 0 & -\frac{i\omega}{p} & -\nu\omega^2a^2 - \nu qr - \frac{r}{p} \\ -\nu r^2p^2 + \nu\omega^2a^2 & i\nu(q + rp^2)\omega a^2 & 0 & 0 \\ -i\frac{\omega}{p} & \frac{q}{p} & 0 & 0 \end{pmatrix}. \tag{12.44}$$

The condition of a nonzero eigenvectors of the matrix (12.44)

$$\tilde{\Psi}_i = \begin{pmatrix} \widehat{\alpha_i} \\ \widehat{\chi_i} \\ \widehat{\beta_i} \\ \widehat{\varphi} \end{pmatrix}. \tag{12.45}$$

Existence reads:

$$\det(\widehat{L} - \lambda\widehat{I}) = 0. \tag{12.46}$$

That is the algebraic equation with respect to λ with the roots λ_i.

The evolution operator (12.44) in ω-representation is the block-matrix:

$$\widehat{L} = \begin{pmatrix} 0 & L \\ M & 0 \end{pmatrix}, \tag{12.47}$$

with 2×2 matrices L, M. The vectors $\tilde{\Psi}_i$ with four components $\tilde{\Psi}_{ij}$, $j = 1, 2, 3, 4$ we correspondingly divide in two-component vectors,

$$\tilde{\Psi} = \begin{pmatrix} \widehat{A} \\ \widehat{B} \end{pmatrix}. \tag{12.48}$$

Then it is convenient to introduce two-component columns. For $\widehat{A}$ it is

$$\widehat{A} = \begin{pmatrix} 1 \\ \gamma \end{pmatrix}, \tag{12.49}$$

that guarantees the standard normalization as in chapter 20, see also [25], while the column $\widehat{B}$ has components to be expressed in terms of $\widehat{A}$. Hence the eigen problem

$$\begin{pmatrix} 0 & L \\ M & 0 \end{pmatrix}\begin{pmatrix} \widehat{A} \\ \widehat{B} \end{pmatrix} = \lambda\begin{pmatrix} \widehat{A} \\ \widehat{B} \end{pmatrix}, \tag{12.50}$$

may be reformulated as the system

$$\begin{aligned} L\widehat{B} &= \lambda\widehat{A} \\ M\widehat{A} &= \lambda\widehat{B}, \end{aligned} \tag{12.51}$$

or, having the direct link

$$\widehat{B} = \lambda^{-1}M\widehat{A}, \tag{12.52}$$

we write

$$LM\widehat{A} = \lambda^2\widehat{A}. \tag{12.53}$$

Denoting the 2×2 matrix as $C = LM$ and its eigenvalue as $m = \lambda^2$, we write the eigen problem as a conventional (temporal use of a, b, c, d)

$$\begin{pmatrix} a & b \\ c & d \end{pmatrix}\begin{pmatrix} 1 \\ \gamma \end{pmatrix} = m\begin{pmatrix} 1 \\ \gamma \end{pmatrix}, \tag{12.54}$$

or

$$\begin{aligned} a + b\gamma &= m, \\ c + d\gamma &= m\gamma. \end{aligned} \tag{12.55}$$

So, we arrive at expression for the component γ and the spectral condition.

$$\begin{aligned} \gamma &= b^{-1}(m - a), \\ bc + d(m - a) &= m(m - a). \end{aligned} \tag{12.56}$$

Finally, the eigenvalues of the whole problem are evaluated as

$$\begin{aligned} \lambda_i^{\pm} &= \pm\sqrt{m_i}, \\ m_i &= \frac{a + d}{2} \pm \sqrt{\frac{(a + d)^2}{4} - \det C}. \end{aligned} \tag{12.57}$$

The matrix C (12.54) in explicit form defined by equation (12.44) is calculated as the product

$$LM = \begin{pmatrix} -\frac{1}{p^2}(-q\nu p^3 r^2 + q\nu p a^2\omega^2 + a^2\omega^2) & -\frac{i}{p^2}qa^2\omega(r\nu p^3 + q\nu p + 1) \\ \frac{i}{p^2}r\omega(r\nu p^3 + q\nu p + 1) & -\frac{1}{p^2}r(-\nu p^3 a^2\omega^2 + \nu pq^2 + q) \end{pmatrix}. \tag{12.58}$$

Its eigenvalues are

$$\begin{aligned} m_1 &= pr\nu(a^2\omega^2 + qr) \\ m_2 &= -(a^2\omega^2 + qr)\frac{pq\nu + 1}{p^2}. \end{aligned} \tag{12.59}$$

It yields eigenvectors

$$\beta_{i,\pm} = \lambda_{i,\pm}^{-1}M\alpha_i = \lambda_{i,\pm}^{-1}\begin{pmatrix} a^2\nu\omega^2 - p^2r^2\nu + ia^2\gamma_i\nu\omega(rp^2 + q) \\ \frac{1}{p}q\gamma_i - \frac{i}{p}\omega \end{pmatrix}, \tag{12.60}$$

where γ is found from (12.56)

$$\gamma_i = i\frac{p^2}{qa^2\omega}\frac{m_i + \frac{1}{p^2}(a^2\omega^2 - p^3qr^2\nu + pqa^2\nu\omega^2)}{pq\nu + p^3r\nu + 1}. \tag{12.61}$$

Plugging the eigenvalues (12.59) to (12.61) wonderfully simplifies

$$\begin{aligned}\gamma_1 &= \frac{i}{q}\omega,\\ \gamma_2 &= -i\frac{r}{a^2\omega}.\end{aligned} \tag{12.62}$$

The eigenvectors are constructed by two-component columns. For the four-component eigenvectors $\psi_i^{\pm}$ it is

$$\psi_i^{\pm} = \begin{pmatrix} \widehat{A}_i \\ \frac{1}{\lambda_i^{\pm}}M\widehat{A}_i \end{pmatrix}, \tag{12.63}$$

where $\lambda_{i,\pm} = \pm\sqrt{m_i}$ and

$$\widehat{A}_i = \begin{pmatrix} 1 \\ \gamma_i \end{pmatrix}, \tag{12.64}$$

that are equal for the cases $\pm$.

So, to determine all the necessary ingredients of the algorithm for the projecting operators calculation, we start from the 2×2 submatrices L, M and the matrix LM, arriving at the eigenvalues formula (12.90) constructing eigenvectors by the formulas (12.63) and (12.64).

12.3.2 Projecting operators

Here we conventionally build the projecting matrix with operator matrix elements for the evolution operator of the form, typical for electrodynamics, see also applications for cylindric fibers in section 11.2.4. During calculations we will use temporary notations for shorthand, rewriting the result in terms of the matrices of the section 12.3.1. Along the general scheme consider the spectral problem for the evolution operator $\widehat{L}$ from equation (12.44),

$$\widehat{L}\Omega = \Omega\Lambda, \tag{12.65}$$

for the matrix Ω, built from the eigen columns Ψ_i, $i = 1, 2, 3, 4$, and the diagonal matrix of eigenvalues Λ.

Let us write the eigenvector matrix in explicit form, starting from notations of equations (12.63) and (12.64):

$$\Omega = \begin{pmatrix} 1 & 1 & 1 & 1 \\ \gamma_1 & \gamma_1 & \gamma_2 & \gamma_2 \\ \lambda_{1,+}^{-1}(M\widehat{A_1})_1 & \lambda_{1,-}^{-1}(M\widehat{A_1})_1 & \lambda_{2,+}^{-1}(M\widehat{A_2})_1 & \lambda_{2,-}^{-1}(M\widehat{A_2})_1 \\ \lambda_{1,+}^{-1}(M\widehat{A_1})_2 & \lambda_{1,-}^{-1}(M\widehat{A_1})_2 & \lambda_{2,+}^{-1}(M\widehat{A_2})_2 & \lambda_{2,-}^{-1}(M\widehat{A_2})_2 \end{pmatrix}, \tag{12.66}$$

it gives matrix elements of projecting operators, by the conventional formula [25]

$$P_{ik}^{(j)} = \Omega_{ij}\Omega_{jk}^{-1},\ i, j.\ k = 1, 2, 3, 4. \tag{12.67}$$

The explicit expression is very complicated, its application may be a bit simplified if applying the following interpretation. The projector number j is the direct product of the j column of Ω and the j row of the matrix Ω^{-1}. Its action may be made directly in such form, as shown below.

Let us introduce the notations:

$$\Omega = \begin{pmatrix} 1 & 1 & 1 & 1 \\ \gamma_1 & \gamma_1 & \gamma_2 & \gamma_2 \\ c & -c & e & -e \\ c_1 & -c_1 & e_1 & -e_1 \end{pmatrix}, \tag{12.68}$$

with obvious relations

$$\begin{aligned} c &= \lambda_{1,+}^{-1}(M\widehat{A_1})_1, \\ c_1 &= \lambda_{1,+}^{-1}(M\widehat{A_1})_2, \\ e &= \lambda_{2,+}^{-1}(M\widehat{A_2})_1, \\ e_1 &= \lambda_{2,+}^{-1}(M\widehat{A_2})_2. \end{aligned} \tag{12.69}$$

The inverse to Ω matrix is

$$\Omega^{-1} = \frac{1}{2}\begin{pmatrix} -\dfrac{\gamma_2}{\gamma_1 - \gamma_2} & \dfrac{1}{\gamma_1 - \gamma_2} & -\dfrac{e_1}{c_1 e - c e_1} & \dfrac{e}{c_1 e - c e_1} \\ -\dfrac{\gamma_2}{\gamma_1 - \gamma_2} & \dfrac{1}{\gamma_1 - \gamma_2} & \dfrac{e_1}{c_1 e - c e_1} & -\dfrac{e}{c_1 e - c e_1} \\ \dfrac{\gamma_1}{\gamma_1 - \gamma_2} & -\dfrac{1}{\gamma_1 - \gamma_2} & \dfrac{c_1}{c_1 e - c e_1} & -\dfrac{c}{c_1 e - c e_1} \\ \dfrac{\gamma_1}{\gamma_1 - \gamma_2} & -\dfrac{1}{\gamma_1 - \gamma_2} & -\dfrac{c_1}{c_1 e - c e_1} & \dfrac{c}{c_1 e - c e_1} \end{pmatrix}. \tag{12.70}$$

Hence the direct product of the first column of Ω and the 1st row of the matrix Ω^{-1} yields the first projecting operator

$$\tilde{P}^{(1)} = \frac{1}{2}\begin{pmatrix}1\\ \gamma_1\\ c\\ c_1\end{pmatrix}\begin{pmatrix}-\frac{\gamma_2}{\gamma_1-\gamma_2} & \frac{1}{\gamma_1-\gamma_2} & -\frac{e_1}{c_1e-ce_1} & \frac{e}{c_1e-ce_1}\end{pmatrix}$$
$$= \frac{1}{2}\begin{pmatrix}-\frac{\gamma_2}{\gamma_1-\gamma_2} & \frac{1}{\gamma_1-\gamma_2} & -\frac{e_1}{c_1e-ce_1} & \frac{e}{c_1e-ce_1}\\ -\frac{\gamma_1\gamma_2}{\gamma_1-\gamma_2} & \frac{\gamma_1}{\gamma_1-\gamma_2} & -\frac{\gamma_1e_1}{c_1e-ce_1} & \frac{\gamma_1e}{c_1e-ce_1}\\ -\frac{c\gamma_2}{\gamma_1-\gamma_2} & \frac{c}{\gamma_1-\gamma_2} & -\frac{ce_1}{c_1e-ce_1} & \frac{ce}{c_1e-ce_1}\\ -\frac{c_1\gamma_2}{\gamma_1-\gamma_2} & \frac{c_1}{\gamma_1-\gamma_2} & -\frac{c_1e_1}{c_1e-ce_1} & \frac{ec_1}{c_1e-ce_1}\end{pmatrix}, \tag{12.71}$$

similar for the third column/row we built the third projecting operator:

$$\tilde{P}^{(3)} = \frac{1}{2}\begin{pmatrix}1\\ \gamma_2\\ e\\ e_1\end{pmatrix}\begin{pmatrix}\frac{\gamma_1}{\gamma_1-\gamma_2} & -\frac{1}{\gamma_1-\gamma_2} & -\frac{c_1}{-c_1e+ce_1} & \frac{c}{-c_1e+ce_1}\end{pmatrix}$$
$$= \frac{1}{2}\begin{pmatrix}\frac{\gamma_1}{\gamma_1-\gamma_2} & -\frac{1}{\gamma_1-\gamma_2} & \frac{c_1}{ec_1-ce_1} & -\frac{c}{ec_1-ce_1}\\ \gamma_1\frac{\gamma_2}{\gamma_1-\gamma_2} & -\frac{\gamma_2}{\gamma_1-\gamma_2} & \gamma_2\frac{c_1}{ec_1-ce_1} & -c\frac{\gamma_2}{ec_1-ce_1}\\ \gamma_1\frac{e}{\gamma_1-\gamma_2} & -\frac{e}{\gamma_1-\gamma_2} & c_1\frac{e}{ec_1-ce_1} & -c\frac{e}{ec_1-ce_1}\\ \gamma_1\frac{e_1}{\gamma_1-\gamma_2} & -\frac{e_1}{\gamma_1-\gamma_2} & e_1\frac{c_1}{ec_1-ce_1} & -c\frac{e_1}{ec_1-ce_1}\end{pmatrix}. \tag{12.72}$$

The identity

$$\sum_{i=1}^{4}\tilde{P}^{(i)} = I,$$

states the completeness of the projecting operators set. It realizes transition to hybrid variables, that is convenient to determine ones by the first component

$$(\tilde{P}^{(i)}\tilde{\Psi})_1 = \Pi^i, \tag{12.73}$$

in the frequency domain. For example,

$$(\tilde{P}^{(i)}\tilde{\Psi})_1 = \left(-\frac{\gamma_2}{\gamma_1-\gamma_2} \quad \frac{1}{(\gamma_1-\gamma_2)} \quad \frac{e_1}{-c_1e+ce_1} \quad -\frac{e}{-c_1e+ce_1}\right)\begin{pmatrix}\widehat{\alpha}\\ \widehat{\chi}\\ \widehat{\beta}\\ \widehat{\varphi}\end{pmatrix}$$
$$= \frac{1}{\gamma_1-\gamma_2}\widehat{\chi} - \frac{\gamma_2}{\gamma_1-\gamma_2}\widehat{\alpha} + \frac{e}{ec_1-ce_1}\widehat{\varphi} - \frac{e_1}{ec_1-ce_1}\widehat{\beta} = \Pi^1. \tag{12.74}$$

From equation (12.73), that is linear system with respect to $\widehat{\alpha}, \widehat{\chi}, \widehat{\beta}, \widehat{\varphi}$ the original variables may be expressed in terms of Π^i.

Such formalism may be applied in practice by direct application of a symbolic computations program. In the example calculations we used the SWP program. The dynamic projection operators for original equations are obtained by transition to t-domain by Fourier transformation, prolongating the functions to the whole t- and ω-axis similar to previous sections.

Considering waveguides, it is natural in linear theory and first approximation of a nonlinear to restrict the description by unidirectional waves. It may be realized by simple conditions:

$$(\tilde{P}^{(2)}\tilde{\Psi})_1 = 0, \; (\tilde{P}^{(4)}\tilde{\Psi})_1 = 0. \tag{12.75}$$

It gives two algebraic constraints, that allow to express all four variables as combinations of two. Namely, taking the form as for $P^{(1,3)}$, we evaluate

$$P^{(2)} = \frac{1}{2}\begin{pmatrix} -\frac{\gamma_2}{\gamma_1-\gamma_2} & \frac{1}{\gamma_1-\gamma_2} & \frac{e_1}{c_1e-ce_1} & -\frac{e}{c_1e-ce_1} \\ -\gamma_1\frac{\gamma_2}{\gamma_1-\gamma_2} & \frac{\gamma_1}{\gamma_1-\gamma_2} & \gamma_1\frac{e_1}{c_1e-ce_1} & -\gamma_1\frac{e}{c_1e-ce_1} \\ c\frac{\gamma_2}{\gamma_1-\gamma_2} & -\frac{c}{\gamma_1-\gamma_2} & -c\frac{e_1}{c_1e-ce_1} & c\frac{e}{c_1e-ce_1} \\ \gamma_2\frac{c_1}{\gamma_1-\gamma_2} & -\frac{c_1}{\gamma_1-\gamma_2} & -e_1\frac{c_1}{c_1e-ce_1} & c_1\frac{e}{c_1e-ce_1} \end{pmatrix},$$

and

$$P^{(4)} = \frac{1}{2}\begin{pmatrix} \frac{\gamma_1}{\gamma_1-\gamma_2} & -\frac{1}{\gamma_1-\gamma_2} & -\frac{c_1}{ec_1-ce_1} & \frac{c}{ec_1-ce_1} \\ \gamma_1\frac{\gamma_2}{\gamma_1-\gamma_2} & -\frac{\gamma_2}{\gamma_1-\gamma_2} & -\gamma_2\frac{c_1}{ec_1-ce_1} & c\frac{\gamma_2}{ec_1-ce_1} \\ -\gamma_1\frac{e}{\gamma_1-\gamma_2} & \frac{e}{\gamma_1-\gamma_2} & c_1\frac{e}{ec_1-ce_1} & -c\frac{e}{ec_1-ce_1} \\ -\gamma_1\frac{e_1}{\gamma_1-\gamma_2} & \frac{e_1}{\gamma_1-\gamma_2} & e_1\frac{c_1}{ec_1-ce_1} & -c\frac{e_1}{ec_1-ce_1} \end{pmatrix}.$$

Applying the operators to the field vector (12.48) and plugging the results in equation (12.75) we arrive at the equations

$$\frac{\gamma_1}{\gamma_1 - \gamma_2}\widehat{\alpha} - \frac{1}{\gamma_1 - \gamma_2}\widehat{\chi} + \frac{c}{c_1 e - c e_1}\widehat{\varphi} - \frac{c_1}{c_1 e - c e_1}\widehat{\beta} = 0, \tag{12.76}$$

$$\frac{1}{\gamma_1 - \gamma_2}\widehat{\chi} - \frac{\gamma_2}{\gamma_1 - \gamma_2}\widehat{\alpha} + \frac{e}{e c_1 - c e_1}\widehat{\varphi} - \frac{e_1}{e c_1 - c e_1}\widehat{\beta} = 0. \tag{12.77}$$

As it is seen from the basic electric field components expansion (12.13) and (12.14), the coefficients by the transversal modes are marked as φ, χ, ψ. The last variable is expressed via equation (12.23) as

$$\psi = \frac{a}{\pi n}\frac{\partial \varphi}{\partial x} - \frac{m}{n}\chi, \tag{12.78}$$

that simplifies for the 'TM' mode ($\varphi = 0$, see e.g. [38]):

$$\psi = -\frac{m}{n}\chi. \tag{12.79}$$

12.4 Rectangular waveguide filled with metamaterial: nonlinearity account

The basic x-evolution equation (12.35) from this chapter 12, section 12.3.1 for a nm transversal mode vector (indices omitted), that accounts for the material dispersion (via the operator $\hat{a}$) and waveguide dispersion via geometry parameters p, q, r, ν, see equation (12.25). The projecting onto the directed TM wave subspace with equations (12.69) and (12.61) and the nonlinear term account is based on the Maxwell system (12.5), but material relation (12.6) should be modified to include the nonlinear part of the polarization vector as in equation (11.3) and, next account it as in equation (11.2). In a linear problem the system of equations separated by the linear independence the transverse modes basis. In the nonlinear case, we should form the basis and realize the projecting procedure by means of scalar product application, that is defined via integration across the waveguide, for arbitrary functions:

$$(A, B) = \int_0^a \int_0^a A(y, z)B(y, z)dydz. \tag{12.80}$$

For example for E_y component (12.14), the basic projection is proportional to

$$\cos\frac{\pi m}{a}y \sin\frac{\pi n}{a}z, \tag{12.81}$$

such functions with different m, n are orthogonal. With $m = n$ one has

$$\begin{aligned}\left(\cos\frac{\pi n}{a}y\sin\frac{\pi n}{a}z,\ \cos\frac{\pi n}{a}y\sin\frac{\pi n}{a}z\right) &= \int_0^a \cos^2\frac{\pi n}{a}ydy\int_0^a \sin^2\frac{\pi n}{a}zdz \\ = \frac{1}{4}\int_0^a\left(\cos\frac{2\pi n}{a}y + 1\right)dy\int_0^a\left(1 - \cos\frac{2\pi n}{a}z\right)dz &= \frac{a^2}{4}.\end{aligned} \tag{12.82}$$

It is convenient to introduce the normalized basic functions, proportional to equation (12.81):

$$S^{nm} = \frac{2}{a}\cos\frac{\pi m}{a}y\sin\frac{\pi n}{a}z. \tag{12.83}$$

Projecting by scalar product separates the linear part of the multimode system, but the nonlinear one describes the modes interaction, having the system of nonlinear equations

$$\frac{\pi n}{a}\frac{e_1 + c_1}{\gamma_1 e_1 + \gamma_2 c_1}\chi^{nm} - \frac{n}{m}\frac{ec_1 - ce_1}{\gamma_1 e_1 + \gamma_2 c_1}\frac{\partial\chi^{nm}}{\partial x} - \frac{\hat{\varepsilon}\hat{\mu}}{c^2}\frac{\partial\chi^{nm}}{\partial t} = \left(S^{nm}, \mathbf{P}_y^{NL}\right). \tag{12.84}$$

The system is generally infinite, but the dimension of the waveguide may restrict the number of modes to be excited. It is rewritten as

$$\begin{aligned}\frac{\partial\chi^{nm}}{\partial x} + \frac{m}{a}\frac{\gamma_1 e_1 + \gamma_2 c_1}{ec_1 - ce_1}\frac{\hat{\varepsilon}\hat{\mu}}{c^2}\frac{\partial\chi^{nm}}{\partial t} - \frac{\pi m}{a}\frac{e_1 + c_1}{\gamma_1 e_1 + \gamma_2 c_1}\chi^{nm} \\ = -\frac{m}{n}\frac{\gamma_1 e_1 + \gamma_2 c_1}{ec_1 - ce_1}\left(S^{nm}, \mathbf{P}_y^{NL}\right).\end{aligned} \tag{12.85}$$

That gives the group velocity of propagation

$$\frac{1}{c_g} = \frac{m}{a}\frac{\gamma_1 e_1 + \gamma_2 c_1}{ec_1 - ce_1}\frac{\tilde{\varepsilon}(\omega)\tilde{\mu}(\omega)}{c^2},$$

as a function of a carrying frequency ω and waveguide dispersion coefficient

$$d = \frac{\pi m}{a}\frac{e_1 + c_1}{\gamma_1 e_1 + \gamma_2 c_1}.$$

Finally, equation (12.85) reads

$$\frac{\partial\chi^{nm}}{\partial x} + \frac{1}{c_g}\frac{\partial\chi^{nm}}{\partial t} - d\chi^{nm} = -\frac{m}{n}\frac{\gamma_1 e_1 + \gamma_2 c_1}{ec_1 - ce_1}\left(S^{nm}, \mathbf{P}_y^{NL}\right). \tag{12.86}$$

The constants are sequentially defined as

$$\begin{aligned} c &= \lambda_{1,+}^{-1}(M\widehat{A_1})_1, \\ c_1 &= \lambda_{1,+}^{-1}(M\widehat{A_1})_2, \\ e &= \lambda_{2,+}^{-1}(M\widehat{A_2})_1, \\ e_1 &= \lambda_{2,+}^{-1}(M\widehat{A_2})_2. \end{aligned} \tag{12.87}$$

The eigenvalues $\lambda_i^\pm$ are expressed in terms of m_i as

$$\begin{aligned} \lambda_i^\pm &= \pm\sqrt{m_i}, \\ m_1 &= pr\nu(\tilde{a}^2\omega^2 + qr), \\ m_2 &= -(\tilde{a}^2\omega^2 + qr)\frac{pq\nu + 1}{p^2}. \end{aligned} \tag{12.88}$$

Reproducing equation (12.62)

$$\begin{aligned} \gamma_1 &= \frac{i}{q}\omega, \\ \gamma_2 &= -i\frac{r}{\tilde{a}^2\omega}, \end{aligned} \tag{12.89}$$

gives the group velocity

$$c_g = \frac{n}{m}\frac{ec_1 - ce_1}{\gamma_1 e_1 + \gamma_2 c_1}\frac{c^2}{\tilde{\varepsilon}\tilde{\mu}} \tag{12.90}$$

that define material dispersion via equation (12.1) and similar formula for χ_m

$$\tilde{a}^2 = \tilde{\varepsilon}\tilde{\mu} = c^2(1 + \chi_e)(1 + \chi_m). \tag{12.91}$$

For 2TLDM metamaterial, that generalize the Drude model used in section 8.6.1, acting in ω-representation, in an SI unit system $\tilde{\mu}(\omega)$ and $\tilde{\varepsilon}(\omega)$ defined as [2]:

$$\tilde{\mu}(\omega) = \mu_0\left(1 - \frac{\omega_m^2}{\omega(\omega + i\Gamma_m)}\right), \tag{12.92}$$

$$\tilde{\varepsilon}(\omega) = \varepsilon_0\left(1 - \frac{\omega_e^2}{\omega(\omega + i\Gamma_e)}\right). \tag{12.93}$$

That's a particular case out of resonance. For $\left|\frac{\omega_m^2}{\omega(\omega + i\Gamma_m)}\right| \gg 1$:

$$\tilde{\mu}(\omega) \approx -\mu_0\frac{\omega_m^2}{\omega(\omega + i\Gamma_m)}, \tag{12.94}$$

Similarly,

$$\tilde{\varepsilon}(\omega) \approx -\varepsilon_0\frac{\omega_e^2}{\omega(\omega + i\Gamma_e)}. \tag{12.95}$$

The result for the $\tilde{a}$ and $\tilde{a}^2$ is shown in figure 12.1 for the case $\Gamma_e = \Gamma_m$ and in figure 12.2 for the different Γ. Let us note, that the imaginary part of $\tilde{a}$ changes the sign inside the range we consider. There are cases when the model parameters for the electric and magnetic Drude models may be chosen as identical [3]:

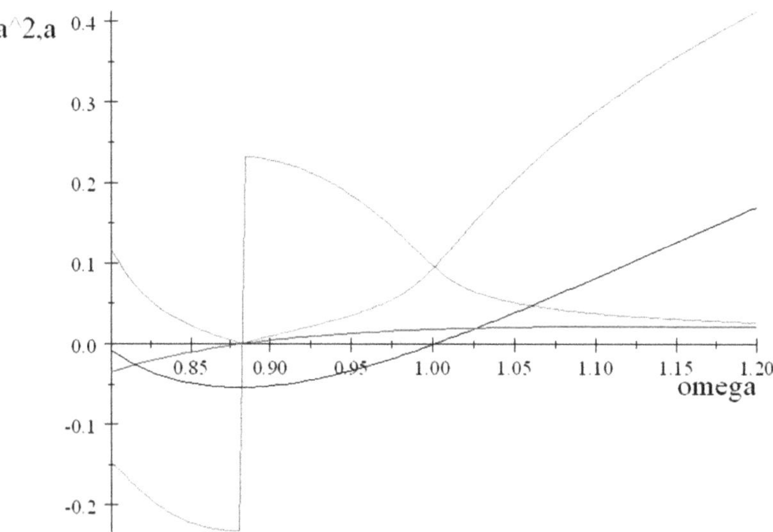

Figure 12.1. The functions $\tilde{a}(\omega)$ (Re—green, Im—brown) and $\tilde{a}^2(\omega)$ (Re—black, Im—red) are plotted in the vicinity of $\omega = \omega_e = 1$ GHz with the parameters $\omega_m = 0.8\omega_e$ and $\Gamma_e = \Gamma_m = 0.05\omega_e$.

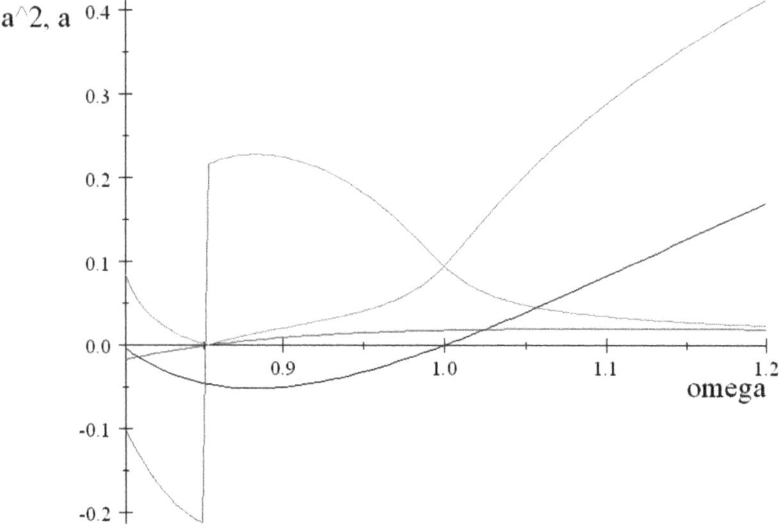

Figure 12.2. The functions $\tilde{a}(\omega)$ (Re—green, Im—brown) and $\tilde{a}^2(\omega)$ (Re—black, Im—red) are plotted in the vicinity of $\omega = \omega_e = 1$ GHz with the parameters $\omega_m = 0.8\omega_e$ and $\Gamma_e = 0.05\omega_e$, $\Gamma_m = 0.025\omega_e$.

$$\tilde{a}(\omega) = c^{-1}\frac{\omega_m\omega_e}{\omega}\sqrt{\frac{1}{(\omega + i\Gamma)^2}} = c^{-1}\frac{\omega_m\omega_e}{\omega(\omega + i\Gamma)}, \tag{12.96}$$

see figure 12.3. For this case in t-domain the operator $\widehat{a}$ may be expressed in terms of integration operator:

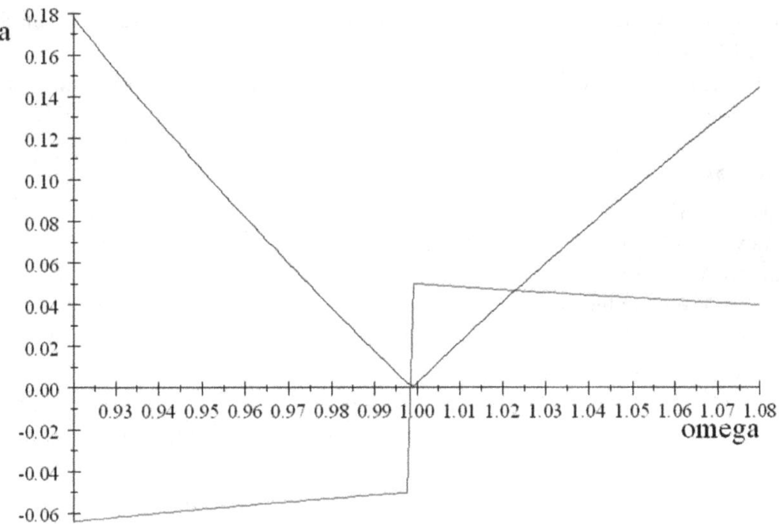

Figure 12.3. The functions $\tilde{a}(\omega)$ (Re—black, Im—red) are plotted in the vicinity of $\omega = \omega_e = 1$ GHz with the parameters $\omega_m = \omega_e$ and $\Gamma_e = \Gamma_m = 0.05\omega_e$.

$$\hat{a} = c^{-1}\omega_m\omega_{pe}(\partial_t(\partial_t + i\Gamma))^{-1} = c^{-1}\omega_m\omega_{pe}e^{-i\Gamma t}\partial_t^{-1}e^{i\Gamma t}\partial_t^{-1}. \tag{12.97}$$

Recall, that the conventional cubic nonlinearity y-component of nonlinear polarization vector with the parameter $\chi_{zzzz} = N$ is evaluated as in equation (12.98), after use of equation (12.4) and advanced notations:

$$\begin{aligned} \mathbf{P}_y^{NL} &= NE_y^3, \\ &\left(S^{nm}, \mathbf{P}_y^{NL}\right) \\ = N' &\sum_{n'''m''',n'm',n''m''} \chi^{nm}\chi^{n'm'}\chi^{n''m''}(S^{mn}, S^{n'm'}S^{n''m''}S^{m'''n'''}), \end{aligned} \tag{12.98}$$

with N' containing normalization constants. The case of one-mode waveguide with $n = m = 1$, $\chi^{11} = \chi$, $S^{11} = S$ simplifies as

$$\frac{\partial\chi}{\partial x} + \frac{1}{c_g}\frac{\partial\chi}{\partial t} - d\chi = N''\chi^3, \tag{12.99}$$

where $N'' = N'(S, S^3)$, that, after a weak dispersion account, leads to the nonlinear Schrödinger equations for a wave packet as in chapter 11, section 11.1.2. Account of few modes leads to coupled nonlinear Schrödinger equations with more rich mathematics [28].

References

[1] Fedotov V 2017 Metamaterials *Springer Handbook of Electronic and Photonic Materials* ed S Kasap and P Capper (Cham: Springer)

[2] Ziolkowski R W and Kipple A 2003 Causality and double-negative metamaterials *Phys. Rev.* E **68** 026615
[3] Ziolkowski R W and Auzanneau F 1997 Passive artificial molecule realizations of dielectric materials *J. Appl. Phys.* **82** 3195–8
[4] Pravdin K V and Popov I Y 2015 Layered system with metamaterials *J. Phys.: Conf. Ser.* **661** 012025
[5] Zakharov V E 1972 Collapse of Langmuir waves *J. Exp. Theor. Phys.* **62** 1745–55
[6] Talanov V I 1965 *ZhETF Pis. Red.* **2** 223 [1965 *JETP Lett.* **2** 141]
Zakharov V E 1968 *J. Appl. Mech. Tech. Phys.* **9** 190
[7] Zakharov V E and Shabat A B 1971 Exact theory of two-dimensional self- focusing and one-dimensional modulation of waves in nonlinear media *Zhurn. Eksp. Teor. Fiz.* **61** 118–34 [1972 *Sov. Phys. JETP* 34 62–69]
[8] Doktorov E and Leble S B 2007 *Dressing Method in Mathematical Physics* (Berlin: Springer)
[9] Perelomova A A 2000 Projectors in nonlinear evolution problem: acoustic solitons of bubbly liquid *Appl. Math. Lett.* **13** 93–8
Perelomova A A 1998 Nonlinear dynamics of vertically propagating acoustic waves in a stratified atmosphere *Acta Acust.* **84** 1002–6
[10] Sazonov S V and Ustinov N V 2006 New class of extremely short electromagnetic solitons *Pis'ma v Zh. Eksper. Teoret. Fiz.* **83** 573–8 General class of the traveling waves propagating in a nonlinear oppositely-directional coupler
[11] Schäfer T and Wayne C E 2004 Propagation of ultra-short optical pulses in cubic nonlinear media *Physica* D **196** 90–105
[12] Chung Y, Jones C K R T, Schäfer T and Wayne C E 2005 Ultra-short pulses in linear and nonlinear media *Nonlinearity* **18** 1351–74
[13] Zhaqilao Z, Hu Q and Qiao Z 2017 Multi-soliton solutions and the Cauchy problem for a two-component short pulse system *Nonlinearity* **30** 3773
[14] Leble S B 1988 *Waveguide Propagation of Nonlinear Waves in Stratified Media* (in Russian) (Leningrad: Leningrad University Press) (Extended Ed. in Springer, Berlin 1990)
[15] Kinsler P 2010 *Phys. Rev.* A **81** 023808
[16] Perelomova A 2006 Development of linear projecting in studies of non-linear flow. Acoustic heating induced by non-periodic sound *Phys. Lett.* A **357** 42–7
[17] Belov V V, Dobrokhotov S Y and Tudorovskiy T Y 2006 Operator separation of variables for adiabatic problem in quantum and wave mechanic *J. Eng. Math.* **55** 183–237
[18] Kuszner M and Leble S 2011 Directed electromagnetic pulse dynamics: projecting operators method *J. Phys. Soc. Jpn.* **80** 024002
[19] Kuszner M and Leble S 2014 Ultrashort opposite directed pulses dynamics with Kerr effect and polarization account *J. Phys. Soc. Jpn.* **83** 034005
[20] Pitois S, Millot G and Wabnitz S 2001 Nonlinear polarization dynamics of counterpropagating waves in an isotropic optical fiber: theory and experiments *J. Opt. Soc. Am.* B **18**
[21] Leble S and Ampilogov D 2016 Directed electromagnetic wave propagation in 1D metamaterial: projecting operators method *Phys. Lett.* A **380** 2271–8
[22] Ampilogov D and Leble S 2016 General equation for directed electromagnetic wave propagation in 1D metamaterial: projecting operator method *TASK* Q **20**
[23] Ampilogov D and Leble S 2018 *Interaction of orthogonal-polarized waves in 1D metamaterial with Kerr nonlinearity* (arXiv: 1802.09523)

Ampilogov D 2017 Interaction of orthogonal-polarized waves in 1D-metamaterial *TASK* Q **21** 605–19

[24] Leble S 2019 *Waveguide Propagation of Nonlinear Waves Impact of Inhomogeneity and Accompanying Effects* (Berlin: Springer)

[25] Leble S and Perelomova A 2018 *Dynamical Projectors Method in Hydro- and Electrodynamics* (Boca Raton, FL: CRC Press)

[26] Agrawal G P 2000 *Nonlinear fiber optics Nonlinear Science at the Dawn of the 21st Century* (Berlin: Springer) pp 195–211

[27] Jackson J D 1998 *Classical Electrodynamics* 3rd edn (New York: Wiley)

[28] Ma W X 2020 Long-time asymptotics of a three-component coupled nonlinear Schrodinger system *J. Geom. Phys.* **153** 103669

Chapter 13

Plasma basic equations, waveguide formation

Such medium as plasma is, by definition, the multicomponent one, containing charged and neutral particles. So, the mathematical apparatus should admit this property account. From the other side, the classical (non-quantum) electromagnetic field is described as a continuous medium. So a joint description should include an averaging procedure, that we used in definition of the matter characteristics (chapter 5, section 5.2). It, compared to mentioned chapters, should naturally have an advanced form, to account for the particles distribution in space. Such a tool suggests the application of statistical physics, where the distribution function and averaging procedure is the important part of the formalism.

In this chapter (see also chapter 7) we sketch the basic mathematical equations, adding general relations and background plasma inhomogeneity or boundary conditions as a reason for a guide formation for plasma, illustrating them in simple examples.

13.1 Maxwell-kinetic system

13.1.1 Joint electrodynamics—particles kinetics description

We restrict ourselves, to a classical description (neglecting spin variables) so that we may use the ion distribution functions and calculate densities of charge and current by integrating these functions over momenta and summing over the charge types. The macroscopic state of the plasma is described by a distribution function $f_a(\vec{r}_a, \vec{p}_a, t)$ that is a function of phase space $\vec{r}_a, \vec{p}_a$ for each type of ion a. The functions are normalized by conditions

$$\int f_a d\vec{p} = n_a, \tag{13.1}$$

doi:10.1088/978-0-7503-2576-9ch13

It determines directly by averaging the charge density

$$\rho = \sum_a e_a \int f_a(\vec{r}_a, \vec{p}_a) d\vec{p}_a, \tag{13.2}$$

and current density

$$\vec{j} = \sum_a e_a \int d\vec{p}_a f_a \vec{v}_a, \tag{13.3}$$

the complete description of the plasma is visibly closed, if one adds an equation for the distribution functions. The material equation is generally nonlocal due to spatial dispersion. The closed system of equations for the electric field components that are linked with the equation for distribution functions $f_a(\vec{r}_a, \vec{p}_a)$. The description, however is two-level. First, for fields $\vec{E}$, $\vec{B}$ it may be named as hydrodynamic. The second for f_a it conventionally relates to kinetic theory.

So, the Maxwell system is written as follows, for the Coulomb equation:

$$\operatorname{div} \vec{E} = 4\pi \sum_a e_a \int f_a(\vec{r}_a, \vec{p}_a) d\vec{p}_a, \tag{13.4}$$

the magnetic charges absence:

$$\operatorname{div} \vec{B} = 0, \tag{13.5}$$

Faraday equation:

$$\frac{1}{c}\frac{\partial \vec{B}}{\partial t} = -\text{rot}\vec{E}, \tag{13.6}$$

and the Ampère–Maxwell one:

$$\frac{1}{c}\frac{\partial \vec{E}}{\partial t} = \text{rot}\vec{B} - \frac{4\pi}{c}\sum_a e_a \int d\vec{p}_a f_a \vec{v}_a, \tag{13.7}$$

that are completed by material equations in the form

$$\vec{D} = \vec{E} + 4\pi \int \vec{j}(\vec{r}, \tau) d\tau, \tag{13.8}$$

where

$$\vec{j} = \sum_a e_a \int d\vec{p}_a f_a \vec{v}_a. \tag{13.9}$$

Note, that absence of magnetic charges is the direct reason to leave equations (13.5) and (13.6) unchanged.

13.1.2 Kinetic equation: Vlasov plasma

The explicit form of equation for distribution function—kinetic equation contains terms of transport in phase space, usually posed at the lhs, and a collision term. The sense of transition is explained in hydrodynamics:

$$\frac{df_a}{dt} = \frac{\partial f_a}{\partial t} + \vec{v}_a \cdot \nabla_a f_a + \vec{F}_a \cdot \nabla_{\vec{p}_a} f_a = St, \tag{13.10}$$

where classic $\vec{v}_a = \vec{p}_a/m$ is the velocity, e_a is the charge of the plasma particle of type a, while the symbol St denotes the collision term.

When the ratio of the mean free path of the particles to a typical space scale of a perturbation, e.g. wavelength is large, the collision term is estimated as small compared with $\vec{F}_a \cdot \nabla_{\vec{p}_a} f_a$, if $\vec{F}_a$ is the self-correlated Lorentz force field $\vec{F}_a = e_a(\vec{E} + [\vec{v}_a \times \vec{B}]/c)$, so we have

$$\frac{\partial f_a}{\partial t} + \vec{v}_a \cdot \nabla_a f_a + \vec{F}_a \cdot \nabla_{\vec{p}_a} f_a = 0, \tag{13.11}$$

called the Vlasov equation [1]. In such an approach, the interaction of particles is accounted for only by the action of the Lorentz force that depends on the electromagnetic field by definition, so accounts for the self-correlation.

On the basis of self-correlated field theory it is possible to build an elementary theory of cold plasma equilibrium and oscillations. It can be demonstrated that equation (13.11) is compatible with equations (13.4)–(13.7).

A cold plasma state of equilibrium and oscillations is naturally built by a perturbation theory for the distribution function that is conventionally built as the expansion

$$f = f_0 + \chi + \cdots \tag{13.12}$$

with a small parameter hidden in χ, proportional to amplitude of electromagnetic perturbation (e.g. a wave) and normalization

$$\int \chi d\vec{p} \approx 0. \tag{13.13}$$

In homogeneous collisionless plasma, in the absence of an external field the distribution function of nonperturbed state f_0 can be chosen as independent on the position coordinates.

Disturbances of plasma are oscillations that are related to the motion of electrons and ions. Indeed, a displacement of an electron from the equilibrium state leads to the appearance of a 'hole'—a positively charged domain. Between charges there appears an electric field that generates an attractive restoring force, which influences both positive and negative charges, that leads to plasma oscillations. Let us write the kinetic equation (13.11) for electrons with a linearized Vlasov term in $\vec{E}$ and the perturbation of the equilibrium distribution function χ:

$$\chi_t + (\vec{v}, \nabla)\chi = -e(\vec{E}, \nabla_{\vec{p}} f_0) = F(\vec{r}, t). \tag{13.14}$$

The electric charge density perturbation in such terms is equal to

$$e \int \chi d\vec{p}\,.$$

The Coulomb–Maxwell equation for the electric field coined to the distribution function perturbation is written as

$$\operatorname{div} \vec{E} = 4\pi e \int \chi d\vec{p}\,. \tag{13.15}$$

The Ampère–Maxwell one:

$$\frac{1}{c}\frac{\partial \vec{E}}{\partial t} = \operatorname{rot}\vec{B} - \frac{4\pi}{c}\sum_a e_a \int d\vec{p}_a \chi_a \vec{v}_a, \tag{13.16}$$

for symmetric in $\vec{p}$ equilibrium functions.

13.2 Waves in homogeneous plasma

The content of this section continues section 7.2, where the theory of plasma waves essentially rested on the electromagnetic field equations. Here we consider the propagation of plasma waves without treating all the variety of plasma oscillations [2, 3].

The main aim of this section study is the derivation and further analysis of the evolution equation describing the Langmuir, ion-acoustic and helicoidal waves in a homogeneous plasma using combined kinetic/electrodynamic theory. Its deterministic part includes electromagnetic field while the charge and current densities are included as integrals by momentum of charged particles distribution functions (section 13.1.1). It is assumed that explicit dependence on the magnetic field is excluded by a suitable equation for the field $\vec{B}$ of the Maxwell system. In similar conditions we consider the Langmuir and ion-acoustic waves interaction (section 14.3). Section 14.1 includes results on helicoidal waves and its three-wave interaction that exhibit solitonic behavior.

13.2.1 Cold plasma: general dispersion equation

We first consider the classical example of a wave in a cold plasma neglecting the ion motion. The dispersion relations conventionally used for metamaterials have common features with ones for plasma, hence we include the wave theory for such materials in this chapter, as well in section 18.3.3.

To explain the idea of the method consider the 1D perturbation, when the electromagnetic field propagates along the x-axis. Let the longitudinal component of electric field be formed as a wave-train, localized in a range $t > 0$, the dimensionless parameter $\beta \ll 1$ is introduced to characterize the scale of slow-varied amplitude A to be convenient in a perturbation theory development

$$E_x = A(\beta x, \beta t)\exp i(kx - \omega t) + c.c. \tag{13.17}$$

Equation (13.14) simplifies as

$$\chi_t + \frac{p}{m}\chi_x = F(x, t), \tag{13.18}$$

it is directly integrated by the method of characteristics

$$\chi = \int_0^t F\left(x - \frac{p}{m}(t - \tau), \tau\right)d\tau. \tag{13.19}$$

where m is the electron mass and p is the x-component of its momentum. Plugging equation (13.17) into expression for F from equation (13.14) and, next, to equation (13.19), we get in the zeroth order (in the small parameter β, tearing the slow varying amplitude A out from the integrand) approximation

$$\begin{aligned}\chi &= e\frac{\partial f_0}{\partial p}A(\beta x, \beta t)\int_0^t \exp i\left[k\left(x - \frac{p}{m}t\right) + \left(\frac{kp}{m} - \omega\right)\tau\right]d\tau + c.c.\\ &= e\frac{\partial f_0}{\partial p}A(\beta x, \beta t)\frac{\exp i[kx - \omega t]}{i\left(\omega - k\frac{p}{m}\right)} + c.c.\end{aligned} \tag{13.20}$$

Now both the current density $\vec{j}$ and the charge density can be found in the same approximation. Let us put equation (13.20) into equation (13.15)

$$\operatorname{div}\vec{E} = 4\pi e^2 A(\beta x, \beta t)\exp i[kx - \omega t]\int\left(\frac{\partial f_0}{\partial p}\frac{1}{i\left(\omega - k\frac{p}{m}\right)} + c.c.\right)d\vec{p}, \tag{13.21}$$

and equate coefficients by linearly independent exponents, having the dispersion equation:

$$k = -4\pi\frac{e^2}{\omega}\int\frac{\frac{\partial f_0}{\partial p}}{1 - \frac{pk}{m\omega}}dp dp_y dp_z. \tag{13.22}$$

Under the conditions of cold plasma we expand the integrand in a power series up to the third order $(kp/m\omega)^3$.

$$\frac{1}{1 - \frac{pk}{m\omega}} = 1 + \frac{kp}{m\omega} + \left(\frac{kp}{m\omega}\right)^2 + \left(\frac{kp}{m\omega}\right)^3 + \cdots. \tag{13.23}$$

If the distribution function f_0 is even in p and decreases at infinity quickly enough then only odd powers from equation (13.23) contribute to the integral (13.22). Due to the normalization condition of the unperturbed distribution function (see equation (13.1))

$$\int f_0 d\vec{p} = n, \tag{13.24}$$

the integrals of powers of the momentum component p are

$$\int f_0 p d\vec{p} = -n, \int p^3 f_0 d\vec{p} = -3n < p^2 >, \tag{13.25}$$

where the bracket (<...>) denotes averaging over the momentum. Therefore

$$m\omega^2 = -4\pi e^2 n\left[1 + \frac{k^2 < p^2 >}{m^2\omega^2}\right], \tag{13.26}$$

and the dispersion relation follows.

13.2.2 Maxwell distribution background: Langmuir waves

For the Maxwell distribution f_0 it has the form

$$\omega^2 = \omega_{Le}^2 + 3\kappa_B T_e \frac{k^2}{m}, \tag{13.27}$$

where

$$\omega_{Le} = 2e\sqrt{\frac{\pi n}{m}}, \tag{13.28}$$

is the Langmuir frequency of the electrons, κ_B—Boltzmann constant, and T_e is the electron temperature.

If the electron density

$$\rho = \rho_0 \exp(-i\omega t) + c.c. \tag{13.29}$$

and $\frac{< p^2 >^{1/2} k}{m\omega} \ll 1$ and the same condition is valid for ions, then equation (13.20) with ion terms account gives

$$\chi = \sum_a e_a \frac{\partial f_{0a}}{\partial p_a} A \frac{\exp i[kx - \omega t]}{i\left(\omega - k\frac{p_a}{m_a}\right)} + c.c., \tag{13.30}$$

where the x-component of ion 'a' momentum is denoted as p_a so as the ion mass is m_a.

Approximate equality occurs if there is no great difference between the concentrations n_a. For example, this is true for a single-ion quasineutral plasma. Obviously the Langmuir frequency is basic for longitudinal plasma oscillations in the long wavelength range.

For real ω the integral in equation (13.22) is singular. Therefore it is natural to assume that the frequency in equation (13.71) is a complex number and the wave amplitude decreases in time. A small imaginary part of the frequency may be found by the Sokhotsky formula

$$\int \frac{\phi dp}{\frac{m\omega}{k} - p} = v.\, p.\, \int \frac{\phi dp}{\frac{m\omega}{k} - p} - i\pi\phi\left(\frac{m\omega}{k}\right), \tag{13.31}$$

where $v.p. \int = \mathcal{P} \int$ the integral principle value notation. In the framework of the Vlasov theory the Debye radius, an important concept in plasma physics, is introduced [3]. For the Maxwell distribution, at $\omega = \omega' + i\omega''$ one gets for imaginary part:

$$\omega'' = \sqrt{\frac{\pi}{8}} \frac{\omega_{Le}}{(kr_{De})^2} \exp\left[-\frac{(\omega')^2}{2k^2 v_{Te}^2}\right], \tag{13.32}$$

where

$$r_{De}^2 = \frac{\kappa_B T_e}{4\pi e^2 n_e}, \tag{13.33}$$

is the Debye radius square for electrons. By this notation, introducing also for plotting convenience the dimensionless variable, $kr_{De} = s$ we transform equations (13.27) and (13.32) as

$$\frac{\omega^2}{\omega_{Le}^2} = 1 + \frac{3\kappa_B T_e}{4e^2 \pi n} k^2 = 1 + 3\frac{\kappa_B T_e}{4\pi e^2 n_e} k^2 = 1 + 3s^2.$$

Plugging it into the exponent of equation (13.32) as $\omega' = \omega$, yields

$$\frac{\omega''}{\omega_{Le}} = \sqrt{\frac{\pi}{8}} \frac{1}{(kr_{De})^2} \exp\left[-\left(\frac{1}{6}\frac{1}{(kr_{De})^2} + \frac{1}{2}\right)\right]$$

$$= \sqrt{\frac{\pi}{8}} \frac{1}{s^2} \exp\left[-\left(\frac{1}{6}\frac{1}{s^2} + \frac{1}{2}\right)\right].$$

The functions are compared in figure 13.1.

The wave exists in the domain $\omega'' \ll \omega'$. We note, finally, that if the wavelength of the oscillation is of the order of the values comparable with the Debye radius of electrons ($kr_{De} \approx 1$), the damping factor $\frac{\omega''}{\omega_{Le}}$ is also of the order 1. Therefore, in the electron plasma longitudinal waves can spread only with a much longer wavelength than Debye radius. The wave energy is transformed to plasma heating, that is the typical **accompanying effect** of the wave propagation. As a result of equation (13.19) the distribution function is obviously changed and the mean quadratic velocity of the electrons (and temperature) increases. Because the main contribution to the integral is for the momentum range in which $p/m = \omega/k$ the energy transport occurs via the Cherenkov effect. Collisionless relaxation, which is a result of the self-adjusting field, is known as Landau damping [4].

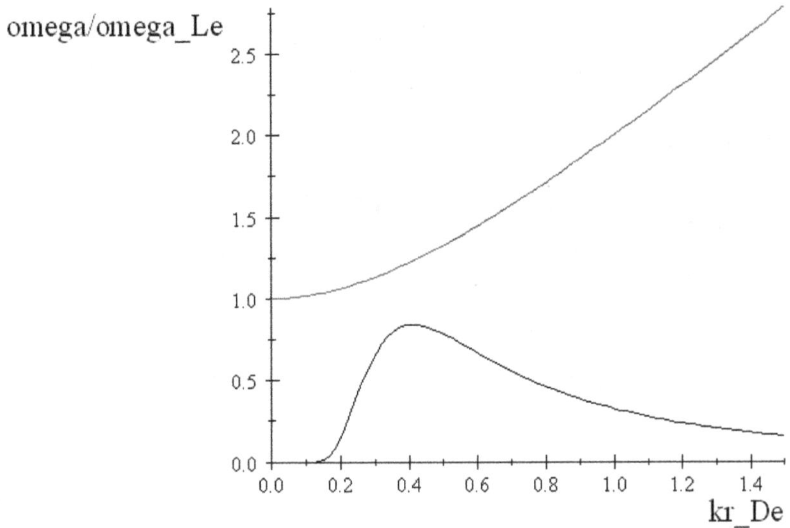

Figure 13.1. The dimensionless frequency of the Langmuir wave $\frac{\omega'}{\omega_{Le}}$ (red line) and dimensionless attenuation factor $\frac{\omega''}{\omega_{Le}}$ (black line) as functions of dimensionless wave vector kr_{De}. Reproduced from [7] (2019), with permission of Springer.

13.2.3 More roots of the dispersion equation

Accounting for the ions movement means that in addition to equation (13.14) it is necessary to introduce an analogous equation for ions (see equation (13.11)) and keep the sums in equations (13.4) and (13.7). The increase of the number of equations with time derivatives generally leads to new roots in the dispersion relation. The analysis of the dependence $\omega(k)$ and of the corresponding dispersion relation is complicated by the different functional spaces used for the kinetic and electrodynamics levels of theory. It is necessary to generalize the mathematics of splitting of waves and introduce projection operators onto subspaces with given $\omega_i(k)$. Here we use the traditional approach: we solve the kinetic equations by perturbation theory and enter the results into the electrodynamic equations.

The existence of two dispersion branches for the single-ion plasma may be demonstrated by analyzing the expression for the perturbed distribution function (13.30) generated by the wave motion. The dispersion equation can be derived from equation (13.4) or (13.15) after substitution of equation (13.30) for χ. For longitudinal waves we have

$$k = 4\pi \sum_a e_a^2 m_e \int \frac{\frac{\partial f_{0a}}{\partial p_a}}{\omega - k\frac{p_a}{m_a}} d\vec{p}_a, \tag{13.34}$$

where p_a is again the x-projection of the momentum of ion a.

By means of expansion (13.23) in equation (13.30) for both electrons and ions, we derive the expressions for real part of frequency as

$$\omega'^2 = \omega_{Li}^2 \frac{1 + 3k^2 r_{Di}^2 + 3\frac{r_{Di}^2}{r_{De}^2}}{1 + k^{-2} r_{De}^{-2}}, \tag{13.35}$$

and for the imaginary part:

$$\omega'' = -\sqrt{\frac{\pi}{8}} \frac{m_i}{m_e} \frac{\omega'^4}{k^3 v_{Te}^3} \left[1 + \sqrt{\frac{m_i}{m_e}} \left(\frac{T_e}{T_i}\right)^{3/2} \exp\left[\frac{-\omega'^2}{2k^2 v_{Ti}^2}\right]\right], \tag{13.36}$$

in the limit $\frac{\langle p^2 \rangle^{1/2} k}{m\omega} \ll 1$. The parameters r_{De}, r_{Di} are Debye radius for electrons and ions correspondingly, $v_{Te} = \sqrt{\frac{3\kappa_B T_e}{m_e}}$, $v_{Ti} = \sqrt{\frac{3\kappa_B T_i}{m_i}}$ are mean heat velocities. There is a link of the heat velocities with the Debye radii. For electrons it is

$$r_{De}^2 = \frac{3\kappa_B T_e}{m_e} \frac{m_e}{12\pi e^2 n_e} = v_{Te}^2 \frac{m_e}{12\pi e^2 n_e}, \tag{13.37}$$

quite similar for ions. Similarly to the Langmuir waves the expression (13.35) is rewritten as

$$\frac{\omega'^2}{\omega_{Li}^2} = \frac{1 + 3r_{De}^2 k^2 \frac{r_{Di}^2}{r_{De}^2} + 3\frac{r_{Di}^2}{r_{De}^2}}{1 + k^{-2} r_{De}^{-2}} = \frac{1 + 3\frac{r_{Di}^2}{r_{De}^2}(1 + s^2)}{1 + s^{-2}} = \frac{s^2(1 + 0.03(1 + s^2))}{1 + s^2}, \tag{13.38}$$

in neutral ($n_e \approx n_i$) plasma we approximate

$$\frac{r_{Di}^2}{r_{De}^2} = \frac{T_i}{T_e} = 0.01.$$

$$\omega_{Le} = 2e\sqrt{\frac{\pi n_e}{m_e}}, \quad \omega_{Li} = 2e\sqrt{\frac{\pi n_i}{m_i}}, \tag{13.39}$$

$$k^2 v_{Te}^2 = k^2 r_{De}^2 \frac{12\pi e^2 n_e}{m_e} = s^2 \frac{12\pi e^2 n_e}{m_e} = 3s^2 \omega_{Le}^2 = 3s^2 \frac{\omega_{Le}^2}{\omega_{Li}^2} \omega_{Li}^2 = 3s^2 \frac{n_e m_i}{n_i m_e} \omega_{Li}^2. \tag{13.40}$$

Next, we rewrite equation (13.36) in the form

$$\frac{\omega''}{\omega'} = -\sqrt{\frac{\pi}{24}} \left(\frac{n_i}{n_e}\right)^{3/2} \frac{\omega'^3}{3s\omega_{Li}^3} \left[\sqrt{\frac{m_e}{m_i}} + \left(\frac{T_e}{T_i}\right)^{3/2} \exp\left[\frac{-\omega'^2}{6s^2 \omega_{Ti}^2}\right]\right], \tag{13.41}$$

to determine the attenuation decrement. For the neutral ($n_e \approx n_i$) plasma we approximate $(\frac{T_e}{T_i})^{3/2} = 10^3$, $\frac{m_e}{m_i} = 10^{-3} 0.5$, having the relation to be plotted:

$$\frac{\omega''}{\omega'} = -\sqrt{\frac{\pi}{24}} \frac{\omega'^3}{3s\omega_{Li}^3} \left[\sqrt{10^{-3} 0.5} + 10^3 \exp\left[\frac{-\omega'^2}{6s^2 \omega_{Ti}^2}\right]\right]. \tag{13.42}$$

It is plotted on figure 13.3. The dependencies $\omega'(k)$ of the Langmuir and ion waves are shown in figures 13.1 and 13.2. Ion waves are possible only in non-isothermal plasma in which $T_e \gg T_i$ in the long wavelength range $kr_{De} \ll 1$ and the ion wave spectrum ([5]) is a linear function

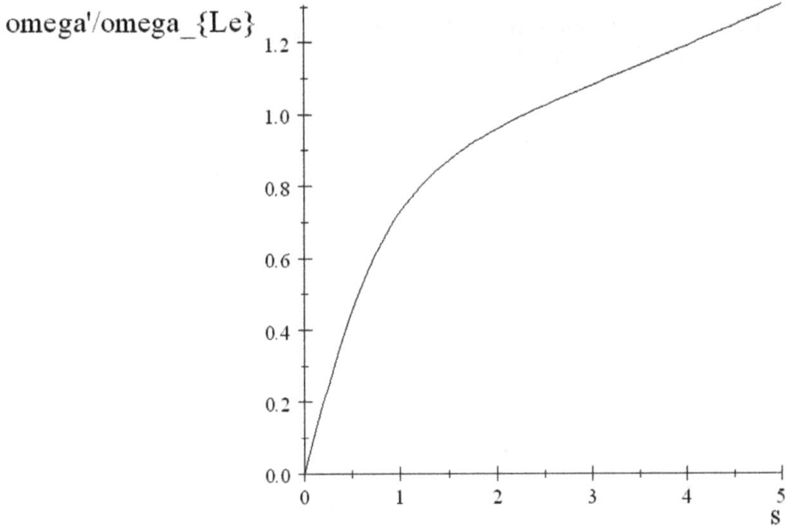

Figure 13.2. The dependence of ion-acoustic wave frequency in non-isothermal plasma on dimensionless $s = kr_{De}$.

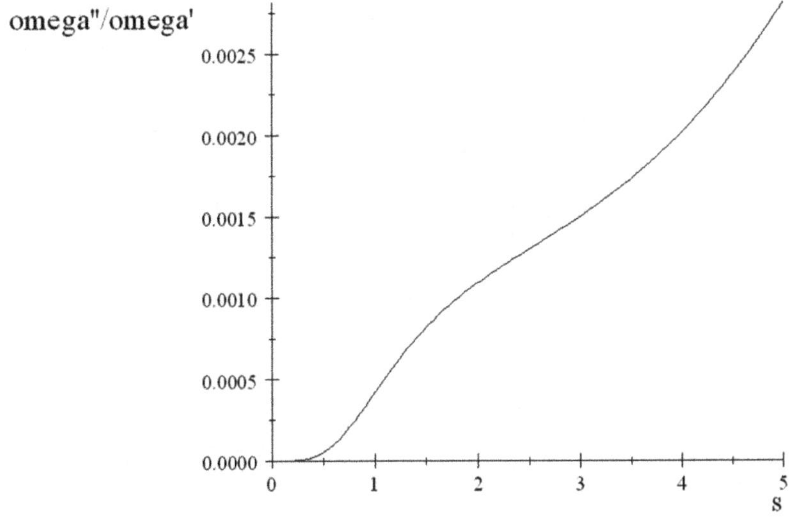

Figure 13.3. The attenuation factor for ion-acoustic wave as function of s.

$$\omega' = \pm k\sqrt{\frac{\kappa_B T_e\left(1 + 3\frac{T_i}{T_e}\right)}{m_i}} \approx s\frac{\omega_{Le}}{1 + s^2}. \tag{13.43}$$

These oscillations are known as the ion-acoustic waves and are found in collisionless magnetic hydrodynamics like the usual acoustic waves in a neutral gas [5]. It follows from equations (13.27) and (13.35) that for both types of waves there are waves propagating in opposite directions. The imaginary part of the frequency $\omega' + i\omega''$, i.e. the attenuation factor is [3]

$$\frac{\omega''}{\omega'} = -\sqrt{\frac{\pi}{8}}\frac{\omega'}{kv_{Te}}\left[\frac{1}{s^2 + 1} + \frac{v_{Te}}{v_{Ti}}\frac{\omega'^2}{k^2 v_{Ti}^2}\exp\left[\frac{-\omega'^2}{2k^2 v_{Ti}^2}\right]\right].$$

It is plotted on figure 13.4, by means of (13.40), so that

$$k^2 v_{Te}^2 = k^2 r_{De}^2\frac{12\pi e^2 n_e}{m_e} = s^2\frac{12\pi e^2 n_e}{m_e} = 3s^2\omega_{Le}^2, \tag{13.44}$$

because of equation (13.39). Plugging equation (13.38) gives

$$\frac{\omega''}{\omega'} = -\sqrt{\frac{\pi}{24}}\frac{\omega'}{s\omega_{Le}}\left[\frac{1}{s^2 + 1} + \frac{\omega'^2}{3s^2\omega_{Li}^2}\exp\left[\frac{-\omega'^2}{6s^2\omega_{Li}^2}\right]\right].$$

It is plotted on figure 13.4

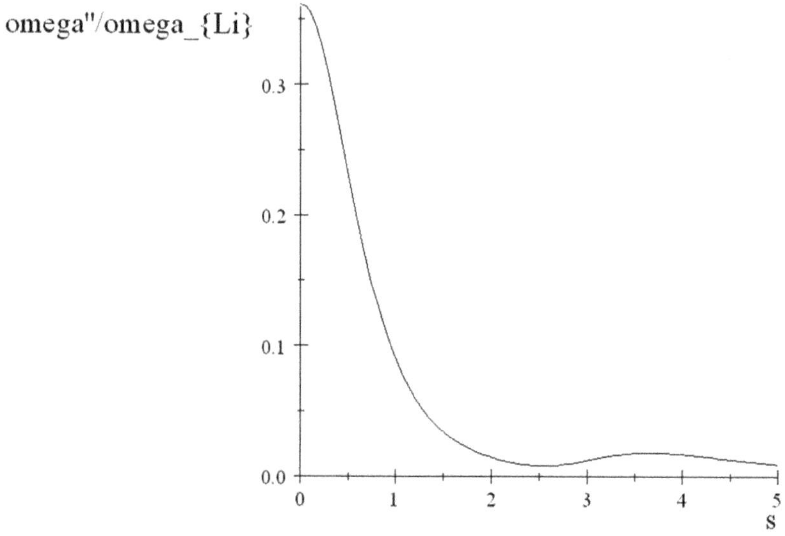

Figure 13.4. The attenuation factor for ion-acoustic wave in non-isothermal plasma as function of s.

13.3 Weakly inhomogeneous plasma

Inhomogeneity of a medium may radically change the character of wave propagation. Its impact generally does not allow to introduce the typical wave parameters such as wavelength. There is an example of so-called stratified medium, in which the basic parameters, for example, density, change along a unique direction, say along x. The case of exponential dependence of the matter parameters, however allows to transform the basic system of equations to one with constant coefficients, that allow to return to conventional wave parameters description, but with dispersion, that may be strong. The important example of such model relates to planetary or stars atmospheres [6], see also [7], that we follow here, and where the exponential atmosphere theory is reviewed.

Generally, if a scale of inhomogeneity is large compared with the wavelength, a wave description may refer to the conventional, but dependent on coordinate, parameters. It allows to introduce a small parameter as a ratio of typical scales of the plasma inhomogeneity and of a wave-like perturbation. The presence of the small parameter in the theory admits expansion with respect to it.

Here we study stratified plasma and waves whose propagation direction coincides with the x-axis. Hence, we should introduce basic plasma parameters that are functions of x: $\omega_{La}(x)$, $r_{Da}(x)$ being local. Such weak inhomogeneity of the plasma ground state described by some distribution function, now also depends on x: $f_0(\beta x)$, $\beta \ll 1$, so that the dispersion relation would be also local as, for example, in [8]. On this basis one can introduce a wave packet with the local parameters and derive the equations for amplitude function evolution in approximation with respect to the mentioned small inhomogeneity parameter. To include effects of nonlinearity we should include the amplitude parameter and waveguide transversal mode projections, restricting ourselves here by the single-mode approximation.

The approximate integration up to first order in the inhomogeneity parameter β, for its compactness we denote $p_x = p$ and set $\beta = 1$, after derivation, that gives

$$\int_0^t B\left(x - \frac{p}{m}(t - \tau)\right)\exp[in\tau]d\tau = \left\{\frac{iB}{n} + \frac{1}{n^2}\left[B_t + \frac{p}{m}B_x\right]\right\}, \tag{13.45}$$

$B_t = \frac{\partial B}{\partial t}$, $B_x = \frac{\partial B}{\partial x}$. By differentiation, the formula for an arbitrary slowly varying function $B(\beta x, \beta t)$ can be checked. We apply it as the main tool that allow to account the weak inhomogeneity of the propagation medium.

Now let us consider the wave-train ansatz as in equation (13.17). It is supposed that the amplitude function $A(\beta x, \beta t)$ in this definition as well as its derivatives are equal to zero at $t = 0$. The dispersion equation (13.22) in such approximations contains the first derivatives with respect to x, along the propagation direction and same to t:

$$ikA + A_x = 4\pi\frac{e^2}{i\omega}\int\left[\frac{A\frac{\partial f_0}{\partial p}}{\left(1-\frac{pk}{m\omega}\right)} - \frac{Ap\frac{\partial^2 f_0}{\partial p\partial x}}{m\omega\left(1-\frac{pk}{m\omega}\right)^2} - \frac{A_t\frac{\partial f_0}{\partial p}}{\omega\left(1-\frac{pk}{m\omega}\right)^2}\right.$$
$$\left. - \frac{A_x p\frac{\partial f_0}{\partial p}}{m\omega\left(1-\frac{pk}{m\omega}\right)^2}\right]dpdp_ydp_z. \tag{13.46}$$

The equation (13.46) describes the approximate evolution for the envelope amplitude $A(x,t)$. The term that contains f_{0x} indicates the locality of the distribution. It determines the relaxation process which may be associated with convective transport. If to choose for f_0 the Boltzmann distribution function and the approximation of phase velocity close to the heat one, $\omega/k \approx V_T$ yields

$$\begin{aligned}
\frac{1}{m\omega^2}\int\frac{\frac{\partial f_0}{\partial p}}{i\left(1-\frac{pk}{m\omega}\right)^2}dpdp_ydp_z &= -\frac{2kn}{m\omega^3},\\
\frac{1}{m\omega^2}\int\frac{p\frac{\partial f_0}{\partial p}}{i\left(1-\frac{pk}{m\omega}\right)^2}dpdp_ydp_z &= -\frac{n}{m\omega^2},\\
\frac{1}{m\omega^2}\int\frac{p\frac{\partial^2 f_0}{\partial p\partial x}}{i\left(1-\frac{pk}{m\omega}\right)^2}dpdp_ydp_z &= -\frac{N_x}{N}\frac{n}{m\omega^3},
\end{aligned} \tag{13.47}$$

where $N(x) = \frac{n(x)}{2\pi m\kappa_B T(x)}$. The relation (13.24) is taken into account. Plugging the formulas for momentum integrals (13.47), in the mentioned approximations, the amplitude evolution equation (13.46) takes the form

$$A_t + c_g A_x - \gamma A = 0, \tag{13.48}$$

where

$$\begin{aligned}
c_g &= \frac{\omega^3}{2\omega_L^2 k}\left[\frac{\omega^2}{\omega_L^2} - 1\right],\\
\gamma &= -\frac{\omega}{2k}\frac{N_x}{N},
\end{aligned} \tag{13.49}$$

that naturally contains the parameter c_g, that is the group velocity of the wave. Such a description appears within the collisionless Cherenkov–Landau relaxation theory of plasma, that gives the dispersion relation for complex ω (13.32) and (13.43) as well as the influence of the inhomogeneity appearance in the last term of equation (13.48). If, for example the concentration $n(x)$ and temperature depend on x

exponentially, the ratio N_x/N, does not depend on x and ω, ω_1, then equation (13.48) for the envelope $A(x, t)$ with an initial condition $A(x, 0) = \phi(x)$ is solved as

$$A(x, t) = \exp[\gamma t]\phi(x - c_g t). \tag{13.50}$$

When, $\gamma < 0$, the density of the plasma $n(x)$ decreases or the temperature increases along the wave propagation direction, and the wave amplitude attenuation at the solution to the Cauchy problem (13.50) is observed. Amplitude grows along the characteristic, i.e. the opposite effect is observed. Another problem, with a boundary condition for a wave disturbance of a plasma state of charged particles can be used [9]. Next, the flux of particles or the degree of recombination can be assigned at the boundaries [9, 10]. The presence of boundaries, as in a waveguide, determines the discrete spectrum of k and the dispersion branches $k_n(\omega)$ appear. If the wavelength is much smaller than the interval between the boundaries the discrete evolution equations have a natural transition to approximate continuous analog [11]. Let us note in addition that if the plasma density decreases in some direction upon leaving the hydrodynamical region or, on the other hand, if the contribution of the collision integral increases upon transition to the collisionless regime, then a transitional Knudsen regime interval appears. This requires special technique of a hydrodynamic and kinetic ranges matching [12]. As follows from [8] the theory may be generalized to the case of propagation directly at an angle to the density gradient if a ray description is used. Note, that the spatial nonuniformity, may cause convective energy transfer between layers of different density [13].

13.4 Plasma waveguides

13.4.1 On plasma confinement

One of the important phenomenon when plasma is captured in a finite volume, forming an eventual waveguide, is so-called *plasma confinement*. As an example, consider inhomogeneous plasma in a magnetic field oriented parallel to the z-axis and independent on time. The x coordinate we choose along the plasma density gradient. The equilibrium in this case as determined by the kinetic equation (13.11) [3] is:

$$v_x \frac{\partial f_a}{\partial x} + \frac{e_a B}{m_a c}\left(v_y \frac{\partial f_a}{\partial v_x} - v_x \frac{\partial f_a}{\partial v_y}\right) = 0. \tag{13.51}$$

The symmetry of this equation S is operator, determined by condition of zero commutator $[A, B] = AB - BA$ with the equation operator

$$\left[S,\, v_x \frac{\partial}{\partial x} + \frac{e_a B}{m_a c}\left(v_y \frac{\partial}{\partial v_x} - v_x \frac{\partial}{\partial v_y}\right)\right] = 0. \tag{13.52}$$

The natural example is given by the generator of rotation transformation in velocity space

$$R = v_y \frac{\partial}{\partial v_x} - v_x \frac{\partial}{\partial v_y},$$

with obvious invariant, defined by the condition

$$R\phi = 0,$$

which solution is

$$\phi = v_x^2 + v_y^2. \tag{13.53}$$

The next symmetry operator is

$$L = v_x \frac{\partial}{\partial x} - \Omega_a v_x \frac{\partial}{\partial v_y}, \tag{13.54}$$

where the parameter

$$\Omega_a = \frac{e_a B}{m_a c}, \tag{13.55}$$

is named a frequency of rotation (cyclotron frequency or gyrofrequency) of the particle a in the magnetic field. The equation

$$L\psi = v_x \frac{\partial \psi}{\partial x} - \Omega_a v_x \frac{\partial \psi}{\partial v_y} = 0,$$

defines the invariant

$$\psi = x + \frac{v_y}{\Omega_a},$$

as a solution.

Hence it is convenient to go from to the variables ϕ, ψ, x and search a solution of the equation (13.51) as

$$f_a = f(x, \phi, \psi). \tag{13.56}$$

Introducing a small parameter as ratio of a ion rotation radius $\rho_a = \frac{m_a \bar{v}_a}{e_a B}$, $\bar{v}_a$—mean velocity, to the mean inhomogeneity scale of plasma, we can expand the solution with respect to this parameter, arriving at

$$f_a(\vec{v}, x) \approx \left(1 + \frac{v_y}{\Omega_a} \frac{\partial}{\partial x}\right) F_a(v_z, v_x^2 + v_y^2, x). \tag{13.57}$$

In absence of electric field, the electric neutrality should take place. Having in mind equation (13.57) we write the condition

$$\sum_a e_a n_a = \sum_a e_a \int F_a(v_z, v_x^2 + v_y^2, x) dp dp_y dp_z = 0. \tag{13.58}$$

The equations of electromagnetic field give

$$\sum_a e_a \int F_a(v_z, v_x^2 + v_y^2, x) v_z dp dp_y dp_z = 0, \tag{13.59}$$

that holds, e.g. for Maxwellian distribution, while the Ampère equation after plugging (13.57) in it reads

$$\frac{dB}{dx} = -\frac{4\pi}{c}j_y = -\frac{4\pi}{c}\sum_a \frac{e_a}{\Omega_a}\frac{d}{dx}\int v_y^2 F_a(v_z, v_x^2 + v_y^2, x)dpdp_ydp_z. \tag{13.60}$$

The equality may be rewritten as, (13.55) is used,

$$\frac{d}{dx}\left(\frac{B^2}{8\pi} + P_\perp\right) = 0, \tag{13.61}$$

where

$$P_\perp = \frac{1}{2}\sum_a \int m_a(v_x^2 + v_y^2)F_a(v_z, v_x^2 + v_y^2, x)dpdp_ydp_z. \tag{13.62}$$

It follows that the sum of pressures is constant. In the practical case of low pressures

$$\beta = \frac{P_\perp}{B^2/8\pi} \ll 1. \tag{13.63}$$

The distribution function F_a may be Maxwellian,

$$F_a(v_z, v_x^2 + v_y^2, x) = \frac{n_a}{(2\pi m_a T_a(x))^{3/2}}\exp\left[-\frac{p^2}{2m_a\kappa_B T_a(x)}\right]. \tag{13.64}$$

In this case

$$\beta = \frac{\kappa_B\sum n_a T_a}{B^2/8\pi} \ll 1, \tag{13.65}$$

hence due to plasma equilibrium condition (13.51) in the direction of the magnetic field growth, the plasma pressure decreases, while in the case of homogeneous temperature the plasma density decreases. Therefore there is a possibility of the plasma confinement [14].

13.4.2 The confined plasma perturbations

Vlasov equation approach

Let now suppose that the stationary distribution function (note it here as F_a) in the confined x-layer plasma has a dynamic perturbation:

$$f_a = F_a + \chi_a, \tag{13.66}$$

that satisfy the kinetic equation, taking the dependence on invariant (13.53) into account.

$$\frac{\partial f_a}{\partial t} + v_x\frac{\partial f_a}{\partial x} + \frac{e_a E_x'}{m_a}\frac{\partial f_a}{\partial v_x} = 0. \tag{13.67}$$

The electromagnetic field E'_x, appears as a perturbation of the electric field component, that satisfy the wave equation as a direct corollary of Maxwell system (see e.g. Equation (3.3))

$$\nabla^2 E'_x = \frac{1}{c^2}\frac{\partial^2 E'_x}{\partial t^2} + \frac{4\pi}{c}\sum_a e_a \frac{\partial n_a}{\partial t} V_{ax}. \tag{13.68}$$

The hydrodynamic velocity is determined as

$$V_{ax} = \int v_{ax}\chi d\vec{v}. \tag{13.69}$$

The function F_a do not depend on time t and satisfies equation (13.51) from the previous section, hence the perturbation χ_a, similar (13.14) satisfies

$$\frac{\partial \chi_a}{\partial t} + v_x \frac{\partial \chi_a}{\partial x} = -\frac{e_a}{m_a} E'_x \frac{\partial F_a}{\partial v_x} = G_a. \tag{13.70}$$

To apply the method of section 13.3 consider the 1D perturbation, taking the electromagnetic field, that represents a wave packet propagating along the x-axis. Introduce the longitudinal component of electric field, localized in a range $t > 0$, marking $p_x = p$, $\beta \ll 1$ in the form

$$E'_x = A(\beta x, \beta t)\exp i(kx - \omega t) + c.c. \tag{13.71}$$

Without loss of generality the x-axis is chosen such that its direction of the plasma density gradient coincides with the direction of wave propagation. Equation (13.14) is directly integrated by the method of characteristics

$$\chi_a = \int_0^t G_a(x - \frac{p}{m}(t - \tau), \tau)d\tau. \tag{13.72}$$

The confinement phenomenon implies the inhomogeneity, therefore we should apply the method of section 13.3, obtaining similar results for Maxwellian (13.56) $F_a = f_{a0}$.

$$\chi_a = -\frac{e_a}{m_a}\frac{\partial F_a}{\partial v_x} A(\beta x, \beta t)\int_0^t (\exp i(k(x - \frac{p}{m}(t - \tau) - \omega\tau))d\tau + c.c. \tag{13.73}$$

The evaluation of equation (13.73) and derivation of dispersion equation is suggested to make exercise 5 in chapter 20, section 20.8.

13.4.3 Hydrodynamic equations approach: flute instability

Let us return to the hydrodynamic system, the continuity and momentum ones

$$\frac{\partial n_a}{\partial t} + \nabla \cdot (n_a \vec{u}_a) = 0, \tag{13.74a}$$

$$\frac{\partial \vec{u}_a}{\partial t} + (\vec{u}_a \cdot \nabla)\vec{u}_a = \frac{e_a}{m_a}\left\{\vec{E} + \frac{1}{c}[\vec{u}_a \times \vec{B}]\right\} + \vec{g}. \qquad (13.74b)$$

Here $\vec{u}_a$ is hydrodynamic velocity and n_a the density of plasma component 'a', $\vec{g}$ is the gravity field. The constants e_a, m_a are corresponding charge and mass. It is used in the next chapter together with the complete electrodynamic system (14.1). The equilibrium state we will mark by the index '0'. For density it will be

$$n_a^0(x),$$

with neutrality condition

$$en_e^0 + e_i n_i^0 = 0. \qquad (13.75)$$

The velocity at equilibrium is

$$u_{ay}^0 = -g/\Omega_a(x),$$

where

$$\Omega_a(x) = \frac{e_a B}{m_a c},$$

is gyrofrequency of a ion rotation in the magnetic field. The field equations give

$$\frac{d}{dx}\frac{B^2}{8\pi} = g\sum m_a n_a = \rho_m g \approx g\sum m_i n_i. \qquad (13.76)$$

The derived system defines the equilibrium state of plasma.

To study stability, the perturbations are introduced by

$$\begin{aligned} \vec{u}_a &= \vec{v}_a \exp[i\chi] + c.c., \\ n_a &= n_a^0 + \nu_a \exp[i\chi] + c.c., \\ \vec{E} &= -\nabla\varphi \exp[i\chi] + c.c., \end{aligned} \qquad (13.77)$$

for a potential (φ) electric field, $\chi = ky - \omega t$.

Having in mind a weak dependence of amplitudes v_a, ν_a, φ on x, or short waves approximation, in the first order of the inhomogeneity parameter, the equations (13.74) and Poisson equation give

$$\begin{aligned} -i(\omega - ku_{ay}^0)\nu_a + \frac{dn_a^0}{dx}v_{ax} + ikv_{ay}n_a^0 &= 0, \\ -i(\omega - ku_{ay}^0)v_{ax} - \Omega_a v_{ay} &= 0, \\ -i(\omega - ku_{ay}^0)v_{ay} + \Omega_a v_{ax} &= -ik\frac{e_a}{m_a}\varphi, \\ k^2\varphi &= 4\pi\sum_a e_a\nu_a. \end{aligned} \qquad (13.78)$$

It is the system of linear algebraic equations. Expressing all variables via the variable φ we obtain

$$\begin{aligned} v_{ax} &= -i\frac{kc}{B}\varphi, \\ v_{ay} &= -i\frac{\omega - ku_{ay}^0}{\Omega_a}v_{ax} = -\frac{\omega - ku_{ay}^0}{\Omega_a}\frac{kc}{B}\varphi, \\ \nu_a &= -\frac{kc}{B}\frac{dn_a^0}{dx}\frac{\varphi}{\omega - ku_{ay}^0} - \frac{k^2c}{B\Omega_a}n_a^0\varphi. \end{aligned} \tag{13.79}$$

Plugging v_{ax}, v_{ay}, ν_a into the last equation of the system (13.78) yields

$$\left[1 + \frac{4\pi\rho c^2}{B^2}(m_e n_e^0 + m_i n_i^0) + \frac{4\pi c^2}{kB^2}\left(\frac{\frac{edn_e^0}{dx}}{\omega - kg/\Omega_e} + \frac{\frac{e_i dn_i^0}{dx}}{\omega - kg/\Omega_i}\right)\right]\varphi = 0, \tag{13.80}$$

that gives the dispersion relation. The x-derivative of the neutrality condition (13.75)

$$e\frac{dn_e^0}{dx} = -e_i\frac{dn_i^0}{dx}. \tag{13.81}$$

simplifies equation (13.80) as

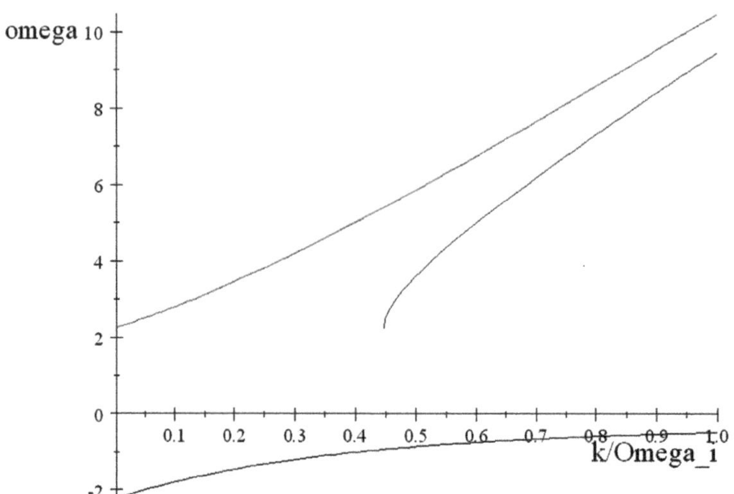

Figure 13.5. The roots of the dispersion equation $\omega(\frac{k}{\Omega_i})$ (red and black lines) for density gradient $\frac{d\ln\rho}{dx} = \frac{1}{l} = 1$ and for opposite gradient $l = -1$ (green line) as functions of $\frac{k}{\Omega_i}$. Reproduced from [7]. With permission of Springer.

$$1 + \frac{4\pi\rho c^2}{B^2} = \frac{4\pi g c^2}{B^2(\omega - kg/\Omega_e)(\omega - kg/\Omega_i)} \frac{d\rho}{dx}, \tag{13.82}$$

where $m_e n_e^0 + m_i n_i^0 \approx m_i n_i^0 \approx \rho$—mass density of the plasma.

The roots of the dispersion equation (13.82) are (figure 13.5)

$$\omega_{\pm} = \frac{1}{2}\frac{kg}{\Omega_i}\left[1 \pm \sqrt{1 + \frac{1}{lk^2 g}\frac{4\Omega_i^2}{1 + \frac{B^2}{4\pi\rho c^2}}}\right]. \tag{13.83}$$

In SI units, the elementary charge e has the value 1.6×10^{-19} C, the mass of the ion m is often given in unified atomic mass unit or Dalton: 1 u = 1 Da $\approx 1.66 \times 10^{-27}$ kg, the magnetic field B is measured in teslas, and the angular frequency ω is measured in radians per second. Taking $B = 1.4$ T gives $\Omega_i = 2.1 \times 10^7$.

Account of gravity field leads to the flute instability if the density gradient and gravity field are opposite [3] as is seen in the plot: the green curve abrupt at some value, the frequency begins to be complex.

References

[1] Vlasov A A 1938 *J. Exp. Theor. Phys.* **8** 291, reproduced in *Sov. Phys. Uspekhi* **10** 721–33

[2] Petviahsvili V I 1975 Three-dimensional solitons of extraordinary and plasma waves *Fiz. Plasmy SSSR* **I** 28, *Sov. J. Plasma Phys.* **1** 15

[3] Silin V P 1971 *Introduction to the Kinetic Gas Theory* (Moscow: Nauka)

[4] Landau L D 1946 *Zh. Eksp. Teor. Fiz.* **16** 574
Landau L D 2016 *J. Exp. Theor. Phys.* **123** 677–86

[5] Bohm D and Gross E P 1864 Theory of plasma oscillations. Excitation and damping of oscillations *Phys. Rev.* A **75** 949

[6] Erdélyi R and Mendoza-Briceno C A 2008 Waves and Oscillations in the Solar Atmosphere (IAU S247) *Heating and Magneto-Seismology. Int. Astronomical Union Symp.* (Cambridge: Cambridge University Press)

[7] Leble S 2019 *Waveguide Propagation of Nonlinear Waves Impact of Inhomogeneity and Accompanying Effects* (Berlin: Springer)

[8] Babich V M, Buldyrev V S and Molotkov I A 1985 *Space–Time Ray Method: Linear and Nonlinear Waves* (Leningrad: Leningrad University Press)

[9] Morozov A I 2012 *Introduction to Plasma Dynamics* (Boca Raton, FL: CRC Press) p 17

[10] Karpov I and Leble S 1986 The analytical theory of ionospheric effect of internal-gravity waves in the ionosphere region *Geomagn. Aeron.* **26** 234–7

[11] Piel A 2010 Plasma Physics: An Introduction to Laboratory *Space, and Fusion Plasmas* (Berlin: Springer) pp 4–5 (Archived from the original on 5 January 2016)

[12] Leble S 1990 *Nonlinear Waves in Waveguides* (Heidelberg: Springer)

[13] Silin V P and Tikhonchuk V T 1981 Parametric turbulence and Cherenkov heating of electrons in a spatially inhomogeneous plasma *J. Exp. Theor. Phys.* **81** 2039–51

[14] Kadomtsev B B 1976 (1988) *Collective Phenomena in Plasma* (Moscow: Nauka)

Chapter 14

Helicoidal and other plasma wave phenomena

In this chapter we sketch the basic mathematical equations, adding general relations and inhomogeneity or boundary conditions as a reason for a guide formation for plasma [1], illustrating them in simple examples, including the Langmuir, ion-acoustic and helicoidal waves in inhomogeneous plasma using combined kinetic/ electrodynamic theory. We also use the limiting description of an electromagnetic field interacting with plasma that is given by the hydrodynamic description. Generally, its deterministic part includes electromagnetic field while the charge and current densities are included as integrals in momentum space of charged particles distribution functions (section 13.1.1). A weak inhomogeneity is taken into account by expansion with respect to a small parameter determined as a ratio of typical scales of the plasma inhomogeneity and one for a wavelength of a wave-like perturbation, see section 13.3. In similar conditions we consider the Langmuir and ion-acoustic waves interaction as in section 14.3. Section 14.1 includes results on helicoidal waves and its three-wave interaction leading to solitons [2].

14.1 Helicoidal waves interactions

In this section we use the hydrodynamical approximation as in section 13.4.3. In such case the free path length is small compared with wavelength (the medium parameter, Knudsen number is small $Kn \ll 1$), hence the collisions should be taken into account as the principal contribution and the plasma is considered as a conductive fluid [3]. For such approach the distribution function is approximated by a local equilibrium function [7]. The method of the nonlinear wave theory hence is the same as described in sections 13.3 and 14.3 but it is more complex because the number of dynamical variables and number of basic equations increases with the number of medium components.

In the book [4] and development in [5] we have considered an interaction of Helicon and ion-acoustic waves. It is originated from the nonlinear theory of helicoidal waves interaction, published in [8]. The wave propagation is one of the

doi:10.1088/978-0-7503-2576-9ch14

most common physical phenomena, its nonlinear character manifests itself at large enough amplitude range. In matter it is ubiquitously exists as waves themselves.

The latest achievements of high technology are under growing attention to nonlinear effects in physics. A lot of research is focused on plasma phenomena, including oscillations and, naturally-waves. In the next section we consider the hydrodynamical approximation (HD) of plasma as a medium of wave propagation. In the Knudsen regime plasma may be treated then as magnetic fluid, so the Maxwell's equations may be used, including density of charge and current within the hydrodynamical approximation. In such a description, plasma is quasineutral ($n_i \approx n_e$) and energy losses are neglected.

14.1.1 Basic equations

We write the basic plasma equations that join electrodynamics and hydrodynamics, we repeat partially the introduction of equations of section 7.2 for reader convenience. The first group of equation is the modified Maxwell system (compare with those of section 13.1.1):

$$\nabla \cdot \vec{E} = 4\pi \sum_a e_a n_a, \tag{14.1a}$$

$$\nabla \cdot \vec{B} = 0, \tag{14.1b}$$

$$\frac{1}{c}\frac{\partial \vec{B}}{\partial t} = -\nabla \times \vec{E}, \tag{14.1c}$$

$$\frac{1}{c}\frac{\partial \vec{E}}{\partial t} = \nabla \times \vec{B} - \frac{4\pi}{c}\sum_a e_a n_a \vec{v}_a, \tag{14.1d}$$

while the second, hydrodynamical group consists of the continuity and the momentum balance equations, as in section 13.4.3

$$\frac{\partial n_a}{\partial t} + \nabla \cdot (n_a \vec{v}_a) = 0, \tag{14.1e}$$

$$\frac{\partial \vec{v}_a}{\partial t} + (\vec{v}_a \cdot \nabla)\vec{v}_a = \frac{e_a}{m_a}\left\{\vec{E} + \frac{1}{c}[\vec{v}_a \times \vec{B}]\right\}, \tag{14.1f}$$

where n_a stands for particles 'a' concentration, $\vec{v}_a$ is velocity of such particles, m_a and e_a are mass and charge of the particle a. Index a describes the type of ion, for unique ion type, $a = i, e$, mark ions and electrons.

Multiplying (14.1e) by the charge, yields the charge conservation law

$$\frac{\partial \rho_a}{\partial t} + \rho_a(\nabla \cdot \vec{v}_a) = 0, \tag{14.2}$$

with ρ_a is the charge density.

We will consider now the single-ion plasma, named electron one. It means, that we can omit indexes in the equations (14.1). We also suppose that the plasma movement has small acceleration $\frac{\partial v_a}{\partial t} \approx 0$, interacting with slow varying electromagnetic field. Next, let the electrons do not suffer from the energy dissipation, so we neglect the plasma resistance. In consequence we get rid of displacement term. In addition, in the first (linear) approximation, we omit the convective term $(\vec{v}, \nabla)\vec{v}$, which is visibly nonlinear.

The divergence of equation (14.1b) plugged in equation (14.2) reduces the equations number to six. Let us make next the assumption that disturbances of magnetic field $\vec{B}'$ are very small compared with homogeneous external magnetic field $\vec{B_0}$. After equation (14.1f) linearization and multiplying it vectorially by $\vec{B_0}$, we express the part of velocity:

$$\vec{v} = -\frac{c}{B_0^2}\left[\vec{B_0} \times \vec{E}\right], \tag{14.3}$$

which is named as the drift velocity of electrons across magnetic field. We can choose the direction of the vector $\vec{B_0}$ along the z-axis without loss of generality. For example,

$$v_x = \frac{c}{B_0}E_y. \tag{14.4}$$

If we substitute equation (14.3) to the Ampère equation (14.1d), the $\nabla \times \vec{B}$ will be linear with respect to $\vec{E}$, $\vec{B}$, with the velocity of the first order:

$$\nabla \times \vec{B} = -\frac{4\pi en}{B_0^2}\left[\vec{B_0} \times \vec{E}\right]. \tag{14.5}$$

Let us use the Faraday equation (14.1c) to express the derivative $\vec{B_t}$ and calculate time derivative from equation (14.5). For the components of electric field we write:

$$\Delta E_x = \frac{4\pi en}{cB_0}\frac{\partial E_y}{\partial t}, \tag{14.6a}$$

$$\Delta E_y = -\frac{4\pi en}{cB_0}\frac{\partial E_x}{\partial t}. \tag{14.6b}$$

Recall, that the magnetic field $\vec{B_0} = B_0\vec{k}$.

To derive the dispersion relations for plain waves, one use to perform the Fourier transformation of equation (14.6). We get the transition, it is determined by following: $E_i(\vec{r}, t) \to \tilde{E}_i(\vec{k}, t)$, $\Delta \to -k^2$. So equation (14.6), conventionally goes to the system of ordinary differential equations:

$$\frac{d\tilde{E}_x}{dt} - \frac{cB_0}{4\pi en}k^2\tilde{E}_y = 0 \tag{14.7a}$$

$$\frac{d\tilde{E}_y}{dt} + \frac{cB_0}{4\pi en}k^2\tilde{E}_x = 0. \tag{14.7b}$$

The equations above may be represented using schematic notation as:

$$\frac{d\psi}{dt} + L\psi = 0, \tag{14.8}$$

where ψ is a column with two components $\tilde{E}_x$, $\tilde{E}_y$ of Fourier transforms of the corresponding electromagnetic field components and the matrix L with coefficients, that are originated from the system (14.6):

$$\psi = \begin{pmatrix} \tilde{E}_x \\ \tilde{E}_y \end{pmatrix}, \tag{14.9a}$$

$$L = \begin{pmatrix} 0 & -\frac{cB_0k^2}{4\pi en} \\ \frac{cB_0k^2}{4\pi en} & 0 \end{pmatrix}. \tag{14.9b}$$

Now, using notations that figure in equations (14.9a) and (14.9b) we can find eigenvalues:

$$\lambda_\pm = \pm i\frac{cB_0k^2}{4\pi en}, \tag{14.10a}$$

and the corresponding eigenvectors

$$\varphi_\pm = \begin{pmatrix} E_1 \\ \mp iE_1 \end{pmatrix}, \tag{14.10b}$$

that open a way to the projecting procedure, see again chapter 20, section 20.3.

14.1.2 Introducing the projectors P_+ and P_-

To separate the two modes that naturally appear as eigen subspaces of the evolution matrix (14.9b) and obtaining evolutionary equations for separated modes, we use projectors, which satisfy the conventional conditions (see chapter 20):

$$(1)\ P_+ \cdot P_- = 0, \tag{14.11a}$$

$$(2)\ P_\pm^2 = P_\pm, \tag{14.11b}$$

$$(3)\ P_+ + P_- = I, \tag{14.11c}$$

where 0 and I are 'zero-matrix' and 'unit-matrix'. To derive the explicit form of projectors, let us follow the recipe of chapter 20, section 20.3, that explain how to evaluate matrix elements of $P_{\pm}$. After the prescribed calculations we arrive at:

$$P_{+} = \frac{1}{2}\begin{pmatrix} 1 & i \\ -i & 1 \end{pmatrix} \tag{14.12a}$$

$$P_{-} = \frac{1}{2}\begin{pmatrix} 1 & -i \\ i & 1 \end{pmatrix}. \tag{14.12b}$$

The projectors (14.12) are orthogonal to each other in the sense of equation (14.11a) and form the complete set as prescribed by equation (14.11c).

Any perturbation in the considered plasma-electromagnetic field system model can be presented by using the obtained projectors (14.12). The next step is to get back to 'x'-representation, let the column vector be written as

$$\Psi = \begin{pmatrix} E_1 \\ E_2 \end{pmatrix}. \tag{14.13}$$

The projectors transformed to x-domain coincide with equation (14.12) because its matrix elements do not depend on k. The completeness (14.11c) reads formally as $(P_{+} + P_{-})\Psi = \Psi$, that admits to write the mode expansion as identity. We hence have at the space of solutions two evolutionary equations. Applying the projectors to equation (14.14), we get

$$P_{+}\Psi = \frac{1}{2}\begin{pmatrix} 1 & i \\ -i & 1 \end{pmatrix}\begin{pmatrix} E_1 \\ E_2 \end{pmatrix} = \frac{1}{2}\begin{pmatrix} E_1 + iE_2 \\ -iE_1 + E_2 \end{pmatrix} = \begin{pmatrix} \Pi \\ -i\Pi \end{pmatrix}. \tag{14.14}$$

Similar we define Λ by means of the projector P_{-}. and use equation (14.6), we arrive at:

$$i\Pi_t + \frac{cB_0}{4\pi en}\Delta\Pi = 0, \tag{14.15a}$$

$$i\Lambda_t - \frac{cB_0}{4\pi en}\Delta\Lambda = 0. \tag{14.15b}$$

Both equations coincide by the form with the Schrödinger equation with zero potential and opposite dispersion (kinetic energy) terms [9, 10]. The amplitudes Π and Λ represent two waves propagating into opposite directions.

14.1.3 Model with nonlinear term: the three-waves equation

Equation (14.8) defines two kinds of modes interacting with one another via nonlinearity of general equations (14.1). We will now take the nonlinearities into account taken to the second order. For example, such a term depending on velocity of electrons $\vec{v}$ is written below. To do this, let us put equation (14.3) into the convective term of equation (14.1f):

$$(\vec{v}\cdot\nabla)\vec{v} = \frac{c^2}{B_0^2}\left(\left(E_y\frac{\partial E_y}{\partial x} - E_x\frac{\partial E_y}{\partial y}\right)\mathbf{i} - \left(E_y\frac{\partial E_x}{\partial x} - E_x\frac{\partial E_x}{\partial y}\right)\mathbf{j}\right). \qquad (14.16)$$

If we solve the momentum balance equation (14.1f) within the mentioned approximation

$$(\vec{v}\cdot\nabla)\vec{v} = \frac{e}{m}\left\{\vec{E} + \frac{1}{c}[\vec{v}\times\vec{B_0}]\right\}, \qquad (14.17)$$

with respect to $\vec{v}$, by iterations, we obtain the second order expression to the hydrodynamic velocity in terms of electric field components:

$$\vec{v}^{(2)} = \frac{c}{B_0}(E_y\mathbf{i} - E_x\mathbf{j}) + \frac{mc^3}{eB_0^3}\left(\left(E_y\frac{\partial E_y}{\partial x} - E_x\frac{\partial E_y}{\partial y}\right)\mathbf{i} + \left(E_y\frac{\partial E_y}{\partial x} - E_x\frac{\partial E_y}{\partial y}\right)\mathbf{j}\right). \qquad (14.18)$$

Let us substitute equations (14.3) and (14.18) into equation (14.1b), then together with equation (14.1c), we get for components of the electric field vector $\vec{E}$:

$$\begin{aligned}
&-\frac{\partial E_y}{\partial t} + \frac{cB_0}{8\pi en}\Delta E_x \\
&= \frac{mc^2}{2eB_0^2}\left(\frac{\partial E_y}{\partial t}\frac{\partial E_x}{\partial x} + E_y\frac{\partial^2 E_x}{\partial x\partial t} - \frac{\partial E_x}{\partial t}\frac{\partial E_x}{\partial y} - E_x\frac{\partial^2 E_x}{\partial y\partial t}\right), \\
&\frac{\partial E_x}{\partial t} + \frac{cB_0}{8\pi en}\Delta E_y \\
&= \frac{mc^2}{2eB_0^2}\left(\frac{\partial E_y}{\partial t}\frac{\partial E_y}{\partial x} + E_y\frac{\partial^2 E_y}{\partial x\partial t} - \frac{\partial E_x}{\partial t}\frac{\partial E_y}{\partial y} - E_x\frac{\partial^2 E_y}{\partial y\partial t}\right).
\end{aligned} \qquad (14.19)$$

We picked up the linear terms at lhs, it coincides with ones of the equations of the system (14.6). The nonlinear terms of the rhs account for the fields interactions.

At this moment, we can apply projectors techniques built on the operators P_+ and P_-, which were obtained in linear approximation, see equation (14.12). Simultaneously we read again the identity (14.11c) as mentioned $(P_+ + P_-)\Psi = \Psi$. It points out the transition to new variables (separate modes): $\Pi = E_x + iE_y$ and $\Lambda = E_x - iE_y$. We arrive at hybrid modes

$$\Pi = E_x + iE_y \qquad (14.20a)$$

$$\Lambda = E_x - iE_y, \qquad (14.20b)$$

and, inversely,

$$E_x = \frac{1}{2}(\Pi + \Lambda), \qquad (14.21a)$$

$$E_y = \frac{1}{2i}(\Pi - \Lambda), \qquad (14.21b)$$

Simply plugging equation (14.21) into equation (14.19), we obtain:

$$\frac{\partial}{\partial t}\Lambda + i\frac{cB_0}{8\pi en}\Delta\Lambda + i\frac{mc^2}{2eB_0^2}\left(i\frac{\partial}{\partial t}(\Pi - \Lambda)\frac{\partial}{\partial x}\Lambda + \frac{\partial}{\partial t}(\Pi + \Lambda)\frac{\partial}{\partial y}\Lambda + i(\Pi - \Lambda)\frac{\partial^2}{\partial t \partial x}\Lambda + (\Pi + \Lambda)\frac{\partial^2}{\partial t \partial y}\Lambda\right) = 0, \tag{14.22a}$$

$$\frac{\partial}{\partial t}\Pi - i\frac{cB_0}{8\pi en}\Delta\Pi - i\frac{mc^2}{2eB_0^2}\left(i\frac{\partial}{\partial t}(\Pi - \Lambda)\frac{\partial}{\partial x}\Pi + \frac{\partial}{\partial t}(\Pi + \Lambda)\frac{\partial}{\partial y}\Pi + i(\Pi - \Lambda)\frac{\partial^2}{\partial t \partial x}\Pi + (\Pi + \Lambda)\frac{\partial^2}{\partial t \partial y}\Pi\right) = 0, \tag{14.22b}$$

with visible separation of hybrid amplitudes at linear parts as in equation (14.15).

Next we represent the hybrid modes of electric field as a superposition of three wave packets with amplitudes E_i, group velocities v_i and correspondingly with frequencies ω_i.

In this moment, we should analyze the equations above, relying on a notion of nonlinear resonance. Its meaning one could understand, considering exponential factors of wave trains [11]. For the left waves we write

$$\Lambda = E_1 e^{i(k_1 x + k_2 y + \omega_1 t)} + E_2 e^{i(k_1' x + k_2' y + \omega_2 t)} + E_3 e^{i(k_1'' x + k_2'' y + \omega_3 t)} + c.c., \tag{14.23}$$

where $E_i(x, y, t)$ are slow varying amplitudes, *c.c.* represents complex conjugate parts of the expressions. For nonlinearity of the second order the typical products of exponents

$$E_1 E_2 e^{i(k_1 x + k_2 y + \omega_1 t)} e^{i(k_1' x + k_2' y + \omega_2 t)} e^{-i(k_1'' x + k_2'' y + \omega_3 t)} + c.c., \tag{14.24}$$

where the last factor came from the linear part of the equation. In this product we see the exponential term

$$e^{i(\omega_1 + \omega_2 - \omega_3)t}, \tag{14.25}$$

and similar with k_i. If $\omega_1 + \omega_2 - \omega_3 = 0$, the exponential factor does not oscillate, that may lead to the amplitude growth. Such a condition is named the *nonlinear resonance*.

Because of conjugate terms, the nonlinearity produces the products of different signs. Note, that interaction of two waves running from opposite directions is the weak effect and there is no nonlinear resonance between those waves with frequencies ω_2 and ω_3 (exercise 6 in chapter 20, section 20.8). The equation (14.22) analysis shows that the nonlinear resonance condition $\omega_1 = \omega_2 + \omega_3$ is fulfilled only in case, when waves are propagating in the same direction.

Therefore, restricting by the only left waves contribution, we plug equation (14.23) to equation (14.22) with zero Π. To continue, we note, that the second order derivatives of slow varied amplitudes E_i conventionally are supposed to be small. Analyzing and comparing terms with the same exponents and derivatives, we get the dispersion conditions.

$$\omega_1 = \pm\frac{cB_0}{8\pi en}k^2, \quad \omega_2 = \pm\frac{cB_0}{8\pi en}k'^2 \quad \text{and} \quad \omega_3 = \pm\frac{cB_0}{8\pi en}k''^2. \tag{14.26}$$

Next we apply equation (14.23) to equation (14.22), and use dispersion conditions (14.26), getting the amplitudes E_i. The three-wave equations in 2 + 1 [12, 13]

$$\begin{aligned}
&\frac{\partial E_1}{\partial t} - \frac{cB_0}{4\pi en}\left(\frac{\partial E_1}{\partial x}k_1 + \frac{\partial E_1}{\partial y}k_2\right) \\
&= \frac{mc^2}{2eB_0^2}((k_1' + k_2')\omega_3 + (k_1'' + k_2'')\omega_2)E_3E_2e^{-i(\Delta k_1 x + \Delta k_2 y)}, \\
&\frac{\partial E_2}{\partial t} - \frac{cB_0}{4\pi en}\left(\frac{\partial E_2}{\partial x}k_1' + \frac{\partial E_2}{\partial y}k_2'\right) \\
&= \frac{mc^2}{2eB_0^2}((k_1 + k_2)\omega_3 + (k_1'' + k_2'')\omega_1)E_3^*E_1e^{i(\Delta k_1 x + \Delta k_2 y)}, \\
&\frac{\partial E_3}{\partial t} - \frac{cB_0}{4\pi en}\left(\frac{\partial E_3}{\partial x}k_1'' + \frac{\partial E_3}{\partial y}k_2''\right) \\
&= \frac{mc^2}{2eB_0^2}((k_1 + k_2)\omega_2 + (k_1' + k_2')\omega_1)E_2^*E_1e^{i(\Delta k_1 x + \Delta k_2 y)}.
\end{aligned} \tag{14.27}$$

where $\Delta k_1 = k_1 + k_1' - k_1''$, $\Delta k_2 = k_2 + k_2' - k_2''$. To write the result in more compact form, we introduce the following notations: $\vec{v} = \frac{cB_0}{4\pi en}\vec{k}$, $\vec{v}\,' = \frac{cB_0}{4\pi en}\vec{k}\,'$ and $\vec{v}\,'' = \frac{cB_0}{4\pi en}\vec{k}$, $\nabla = \vec{i}\frac{\partial}{\partial x} + \vec{j}\frac{\partial}{\partial y}$. Equations (14.29) are written as:

$$\begin{aligned}
&\frac{\partial E_1}{\partial t} - (\vec{v}, \nabla)E_1 = aE_3E_2e^{-i(\vec{\Delta k}\vec{r})}, \\
&\frac{\partial \tilde{E}_2}{\partial t} - (\vec{v}\,', \nabla)E_2 = -\,bE_3^*E_1e^{i(\vec{\Delta k}\vec{r})}, \\
&\frac{\partial \tilde{E}_3}{\partial t} - (\vec{v}\,'', \nabla)E_3 = -\,cE_2^*E_1e^{i(\vec{\Delta k}\vec{r})},
\end{aligned} \tag{14.28}$$

where

$$a = \frac{mc^2}{2B_0^3 e}((k_1' + k_2')\omega_3 + (k_1'' + k_2'')\omega_2),$$
$$b = \frac{mc^2}{2eB_0^2}((k_1 + k_2)\omega_3 + (k_1'' + k_2'')\omega_1),$$
$$c = \frac{mc^2}{2eB_0^2}((k_1 + k_2)\omega_2 + (k_1' + k_2')\omega_1).$$

So, we have ordered the equations and inserted symbols for velocities $\vec{v}, \vec{v}', \vec{v}''$. The equations (14.28) and above expressions simplify in the case if *space sinchronism condition* $\overrightarrow{\Delta k} = 0$, that should be however compatible with the dispersion relations (14.26) [13]. Note, that more general N-wave systems may be derived in a similar manner, its reductions and solutions are intensely studied [15].

14.1.4 The three-wave system in 1 + 1 case

Using the dispersion conditions (14.26) and the resonance condition $\omega_1 = \omega_2 + \omega_3$, we obtain the system of three equations for $\hat{E}_i = E_i e^{\Delta kx}$, considering the amplitudes as y-independent with the only component of the wave vector k. Applying the dispersion condition (14.26), gives the following equations for the three-wave system

$$\begin{aligned} \frac{\partial E_1}{\partial t} - v\frac{\partial E_1}{\partial x} &= aE_3E_2e^{-i(\Delta kx)} \\ \frac{\partial E_2}{\partial t} - v'\frac{\partial E_2}{\partial x} &= bE_3^*E_1e^{i(\Delta kx)}, \\ \frac{\partial E_3}{\partial t} - v''\frac{\partial E_3}{\partial x} &= cE_2^*E_1e^{i(\Delta kx)}, \end{aligned} \tag{14.29}$$

with $v = \frac{cB_0}{2\pi en}k$, $v' = \frac{cB_0}{2\pi en}k'$, $v'' = \frac{cB_0}{2\pi en}k''$ standing for wave group velocities (see [8]).

14.2 Algebraic method of three-wave systems solution: solitons

In this section we touch on a class of methods, that allow to construct solutions on nonlinear integrable equations explicitly for [13]. The idea of the method is based on a possibility to represent a given nonlinear system as a compatibility condition of two linear equations, that is named Lax pair. Its algebraic realization admits to produce solutions by special procedure ('dressing') on the basis of simple joint solution of both Lax equations. Some examples are presented as exercises in chapter 20, section 20.4.

As a particular case of the dressing method we apply two-fold Darboux transformation (TfDT). The general idea of TfDT (former binary, see the discussion in [13]), using in the three-wave interactions (for N waves see the [12]), is based on two linear evolutionary equations called the Lax pair

$$D_t\Psi = -\lambda M\Psi + [H, M]\Psi, \tag{14.30a}$$

$$D_x\Psi = -\lambda N\Psi + [H, N]\Psi. \tag{14.30b}$$

In the equations above $H = H(x, t)$, the terms $[H, M]$ and $[H, N]$ have a meaning of potential. Operators D_t, D_x applying to Ψ are defined as:

$$D_t\Psi = \Psi_t + R[Y, \Psi] \tag{14.31a}$$

$$D_x\Psi = \Psi_x + S[Z, \Psi]. \tag{14.31b}$$

Existence of its joint solution is determined by the compatibility condition

$$D_tD_x\Psi = D_xD_t\Psi.$$

The compatibility condition for the Lax pair (14.30) is the nonlinear equation

$$[H_t, N] - [H_x, M] = [[H, M], [H, N]] - [Y, [RH, N]], +[Z, [SH, M]], \tag{14.32}$$

with following diagonal matrices defined as:

$$M = \mathrm{diag}\{a_1, \dots, a_3\} = \begin{pmatrix} a_1 & 0 & 0 \\ 0 & a_2 & 0 \\ 0 & 0 & a_3 \end{pmatrix},$$

$$N = \mathrm{diag}\{b_1, b_2, b_3\},$$

$R = \mathrm{diag}\{r_1, r_2, r_3\}$, $S = \mathrm{diag}\{s_1, s_2, s_3\}$, $Y = \mathrm{diag}\{y_1, y_2, y_3\}$, $Z = \mathrm{diag}\{z_1, z_2, z_3\}$. Now if we use the system of equations of (14.32), we get six equations. On the assumption that matrix H is Hermitian, reduction to three equations is obtained. This reduction gives conditions for R and S with the complex coefficients.

$$\begin{aligned} H_{ij,t} - \frac{(a_j - a_i)}{(b_j - b_i)}H_{ij,x} &= \frac{(a_k(b_j - b_i) + a_j(b_i - b_k) + a_i(b_k - b_j))}{(b_j - b_i)}H_{ik}H_{kj} \\ &+ \frac{(r_i(b_j - b_i)(y_j - y_i) - s_i(a_j - a_i)(z_j - z_i))}{(b_j - b_i)}H_{ij} \end{aligned} \tag{14.33}$$

Equations (14.33) have form of (14.29).

14.2.1 Solutions derived using dressing by two-fold Darboux transformation (TfDT)

Rewriting the equations above and taking into consideration form of matrices R, S, Y, Z, we have the system of equations for the matrix elements φ_{jk} of the simultaneous matrix solution Φ of equations (14.30):

$$\varphi_{ij,t} + r_i(y_i - y_j)\varphi_{ij} = -\lambda a_i\varphi_{ij} + (a_j - a_i)H_{ij}\varphi_{jk} \tag{14.34a}$$

$$\varphi_{ij,x} + s_i(z_i - z_j)\varphi_{ij} = -\lambda b_i\varphi_{ij} + (b_j - b_i)H_{ij}\varphi_{jk}. \tag{14.34b}$$

Let consider the system of linear equations (14.34) assuming, that $H = 0$, to get *'zero base solution'*. It is some simplification, but we are able to generate new solution by the following dressing formula

$$H[1] = H + (\lambda + \lambda^*)\chi, \tag{14.35}$$

where $\chi_{ij} = \frac{\varphi_j^* \varphi_i}{(\varphi^*, \varphi)}$. With $H = 0$ as a condition, solution of equation (14.34) is written as

$$\varphi_{ij} = D_{ij} e^{-\lambda(a_i t + b_i x) - r_i(y_i - y_j)t - s_i(z_i - z_j)x}, \tag{14.36}$$

where D_{ij} is a set of constants and φ_{ij} is a matrix element of Φ. It means, that we can take its column as

$$\varphi_i = \varphi_{i1} = D_{i1} e^{-\lambda(a_i t + b_i x) - r_i(y_i - y_1)t - s_i(z_i - z_1)x}, \tag{14.37}$$

that gives, for real a_i, b_i, r_i, s_i, y_i, z_i

$$\begin{aligned} \chi_{ik} &= \frac{\varphi_k^* \varphi_i}{(\varphi^*, \varphi)} \\ &= \frac{D_{i1} e^{-\lambda(a_i t + b_i x) - r_i(y_i - y_1)t - s_i(z_i - z_1)x} D_{k1}^* e^{-\lambda^*(a_k t + b_k x) - r_k(y_k - y_1)t - s_k(z_i - z_1)x}}{\sum_i |D_{i1}|^2 e^{-(\lambda + \lambda^*)(a_i t + b_i x) - r_i(y_i - y_1)t - s_i(z_i - z_1)x}}. \end{aligned} \tag{14.38}$$

14.2.2 Solutions plots and discussion

Thus, this method of the theoretical description of weakly nonlinear plasma disturbances leads to the canonical equations (14.28) and its mathematical notation (14.33), which are widely discussed in the literature.

In this section we have derived a system of equations, that describes interactions of unidirectional helicoidal waves with different carrying frequencies. For wave packets belonging to two types of polarizations, solutions of interactions of three waves have been obtained. We get Lax pair for 3-wave system with asynchronism ($\Delta k \neq 0$). We used the dressing method [13], more precisely-two-fold Darboux transformation to obtain particular solutions [14]. We have generated solution in the form of impulses for every of the three wave. The impulses present a good example of *solitons*, solitary waves with specific particle-like behavior [2], named for the problem under consideration as 'helicon'. The simplest soliton version is obtained by the formula (14.38) for the parameters values given at 1D figure 14.1 and 3D plot for the $u = H_{12}(x, t)$, shown at figure 14.2. The plugging of the parameters values into equation (14.38) simplifies as $\frac{e^{t+x}}{e^{-2t-2x} + e^{3t+3x} + e^{t+x}}$, that is plotted in the figure 14.1 for $x = 0$ and in the figure 14.2 as 3D plot. More complex examples of *inclined* solitons are presented in figure 14.3.

In numerical realizations, a wide range of parameters of equations were analyzed and some exemplary results are shown in figures, see figure 14.2. Considered models were not directly consulted with an experiments, but there are many possibilities for empirical research. Hypothetical medium for interactions of helicon waves may be gaseous plasma (Ar) with basics parameters: $n_e \sim 10^{10}$–10^{14} cm^{-3}, $B = 50$–1000 G [17] or in semiconductors like Bi–Sn, In and InSb. The proposed parameters for semiconductors are (at 77 K): $n_e \sim 10^{14}$ cm^{-3}, $B = 700$ G [16].

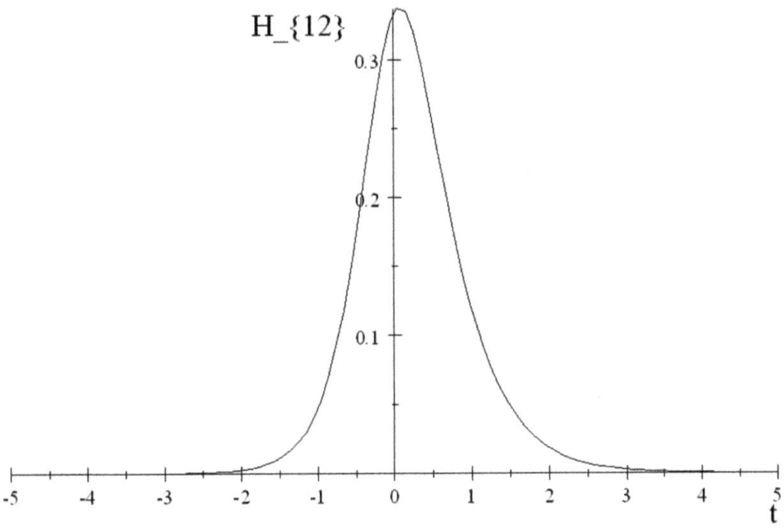

Figure 14.1. The dependence of the component H_{12} on time at $x = 0$, by equation (14.38) in the range $[-5, 5]$ s for the parameters $a_1 = 1$, $a_2 = -1$, $b_1 = 1$, $b_2 = -1$, $b_3 = 0$, $a_3 = 0$, $\lambda = 1$, $y_1 = 1$, $y_2 = 0$, $z_2 = 0$, $z_1 = 1$, $z_3 = 0$, $y_3 = 0$.

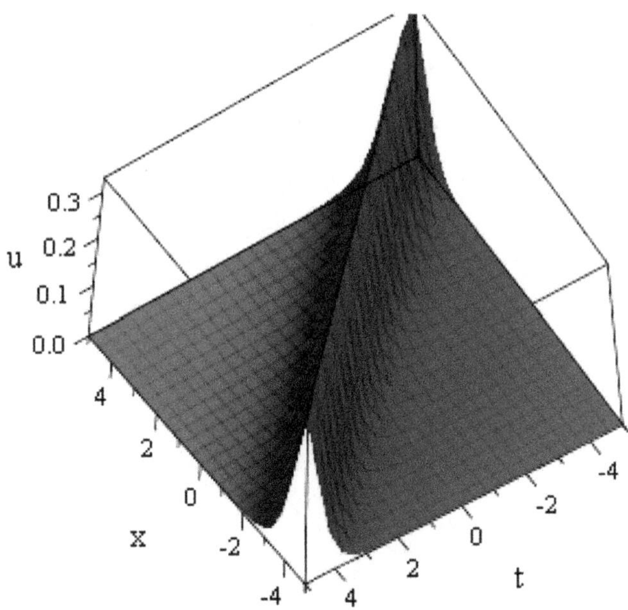

Figure 14.2. The 3D plot of the component H_{12} on x, t by equation (14.38) in the range $[-5, 5]$ s, $x \in [-5, 5]$ for the parameters $a_1 = 1$, $a_2 = -1$, $b_1 = 1$, $b_2 = -1$, $b_3 = 0$, $a_3 = 0$, $\lambda = 1$, $y_1 = 1$, $y_2 = 0$, $z_2 = 0$, $z_1 = 1$, $z_3 = 0$, $y_3 = 0$.

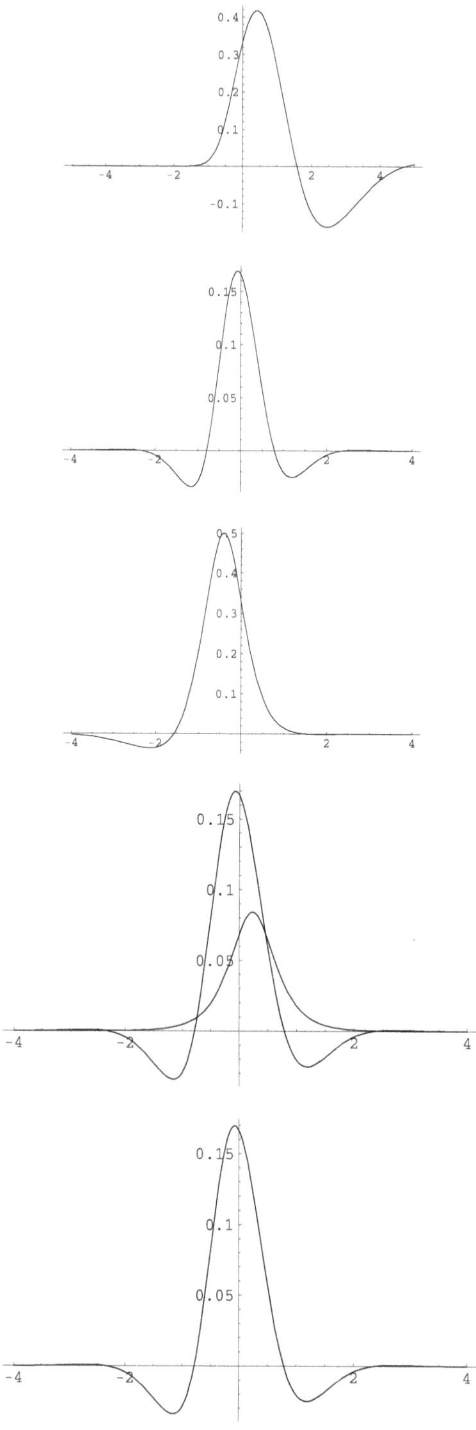

Figure 14.3. The 'inclined' solitons of a 3-wave system. Plots are given for the components H_{12}, H_{23}, H_{31} as a function of t at $y = 0$. The parameters of the equations are $a_1 = 1$, $a_2 = -1$, $b_1 = 1$, $b_2 = -1$, $b_3 = 0$, $a_3 = 0$. The last picture shows the component H_{12} values with different (real) values of the parameter k_{12}. The symmetric one has $k_{12} = 0$. The asymmetric one corresponds to $x_2 = 0.5$, $x_3 = 0.6$ [5].

14.3 Interaction of plasma waves

14.3.1 Interaction of Langmuir and ion waves

In correspondence with the announced in the title aim we should include ions into consideration, see the basic system (14.1). Under the action of coherent sources of a prescribed frequency and polarization in a plasma the wave trains of different types can be excited. These are, for example, the ionosound and Langmuir longitudinal waves discussed in section 13.2.2 and waves with the electric field $\vec{E}$ orthogonal to $\vec{k}$. Wave splitting in a simple way is impossible if the plasma is in an external magnetic field, that is, magnetized plasma [6]. In such magnetized plasma new important dispersion relation branches appear [4, 6, 13]. All possible plasma waves of sufficient amplitude interact [18, 19]. Let us consider the occurrence of an interaction in the case of longitudinal waves in an inhomogeneous plasma, for which, including an acoustic mode we derive a nonlinear system of equations. The weak nonlinearity dependence of the susceptibility on the electromagnetic field can be found by iterating equation (13.11) over the next order of amplitude parameter. To do this, one must modify the Vlasov term $f \to f_0 + \chi$. As χ is a linear function of the amplitude the expression for the distribution function will be quadratic in the field $E' = \sigma E$, σ being the amplitude parameter. So if

$$f = f^0 + \sigma\chi + \sigma^2\Phi + \cdots, \tag{14.39}$$

then

$$\frac{d\Phi}{dt} = -eE_x \int_0^t F_p\left(x - \frac{p}{m}(t - \tau),\, \tau\right) d\tau, \tag{14.40}$$

where F is determined via the rhs of equation (13.14). Since effects of the dispersion and inhomogeneity of the field are assumed to be small ($\beta \ll 1$) within the calculation of the nonlinear terms hidden in Φ we retain only contributions of the β^0 order. Therefore

$$\Phi = ie^2 \int_0^t \frac{Ae^{i\phi} + c.c}{\omega - pk/m}\left(f^0_{pp} - \frac{kf^0_p}{m\omega - pk}\right)(Ae^{i\phi} + c.c)d\tau. \tag{14.41}$$

Here ω is the carrying frequency of the wavepacket and k is the wave vector, while in correspondence with the main target we should include ions into the $\phi = kx - pkt/m + (kp/m - \omega)\tau$. Integration with respect to τ for a uniform plasma gives

$$\Phi = -\frac{e^2}{m}\frac{(A^2e^{2i\phi} + c.c)[f^0_{pp}(m\omega - pk) - kf^0_p]}{3(\omega - pk/m)^3}. \tag{14.42}$$

This contribution to the nonlinearity describes the generation of waves at multiple frequencies and may appear in the propagation of ion-acoustic waves in the range of linearity of the dispersion law (figure 13.2), see also [20, 21]. The equation for the unidirectional wave may be obtained if we substitute equation (14.42) into the Coulomb–Maxwell equation

$$\operatorname{div}\vec{E} = 4\pi\sum_a e_a\int(\chi^a + \Phi^a)d\vec{p}\,, \tag{14.43}$$

and take into account only the resonant terms adding to relation (13.48) the nonlinear terms originated from Φ^a. We get

$$A_t + c_{gi}A_x + \frac{1}{2}\frac{d^2\omega}{dk^2}A_{xx} + N_{2i}(B)^2 = 0, \tag{14.44}$$

where A, B are the amplitudes of the first and second harmonics, i marks ion and the nonlinear constant is expressed as

$$N_{2i} = \frac{2\pi e_i^3}{\omega^2}\int\frac{f_{pp}^{0i} - kf_p^{0i}/(m_i\omega - pk)}{(1 - pk/\omega m_i)}d\vec{p}\,. \tag{14.45}$$

A rather simple expression for the nonlinear constant can be written if the expansion (13.23) in the integrand is used. The latter is possible if $\omega/k \ll V_{T_i}$, V_{T_i} is the mean ion velocity, that occurs in an ion-acoustic waverange which is strongly nonisothermal $T_e \ll T_i$.

In the next approximation, interaction of Langmuir and ion-acoustic waves can appear. For the derivation it is sufficient to take into account the next order in β when χ is calculated. The contribution gives

$$e^2E_x\left[\frac{f_p^0A_t + p(AF_p^0)_x}{(kp/m - \omega)^2} + c.c.\right]_p. \tag{14.46}$$

Keeping the terms that are in resonance with long ion waves, let us write the equation for the right-hand ion-acoustic wave with slowly varying amplitude Π, having

$$\Pi_t + c_g\Pi_x = -N_{ie}|A|^2. \tag{14.47}$$

This equation can be obtained if for the electric field components, the wave equation

$$\mathrm{rotrot}\vec{E} + \vec{E}_{tt}/c^2 = -4\pi\vec{j_t}/c^2, \tag{14.48}$$

includes the current density as the functional of the distribution function

$$\vec{j} = \sum_a\frac{e_a}{m_a}\int\vec{p_a}(\sigma\chi^a + \sigma^2\Phi^a)d\vec{p}\,, \tag{14.49}$$

and projects the result on the longitudinal direction (x-axis). The second perturbation Φ can be calculated to the next order in σ if only the mean field with frequency

$\omega \sim 0$ is retained in equation (14.46). We write the corresponding contribution in the electron distribution function (14.40) denoting it as

$$\Phi_{int}^{a} = -\frac{e^2}{\omega^2}\int\left\{\left|A\left(x-\frac{p}{m}\right)\right|_{\tau_1}^{2}\left[\frac{f_p^0}{(1-pk/m\omega)^2}\right]_p + |A|_x^2\left[\frac{f_p^0}{(1-pk/m\omega)^2}\right]_p + \left[\frac{f_p^0}{1-pk/m\omega}\right]_p |A|^2\right\}d\tau. \tag{14.50}$$

The equations of the first approximation (14.45) can be read as $A_t \approx c_g A_x$ and the dependence of $A(\beta(x - c_g t), \beta^2 t)$ on the arguments is weak. To the first order in β the derivatives of A with respect to τ and τ_1 differ only by factor $A_\tau = (p/m - c_g)A_x$, $A_{\tau_1} = A_\tau(1 - p/mc_g)$. Integrating over τ and the momenta we get the nonlinear term in equation (14.43) and therefore also in equation (14.47):

$$\int \Phi_{int}^{a} d\vec{p} = -\frac{e^2}{\omega^2}\int\left\{\left|A\left(x-\frac{p}{m}\right)\right|_{\tau_1}^{2}\left[\frac{f_p^0}{(1-pk/m\omega)^2}\right]_p\right\}d\vec{p}\,. \tag{14.51}$$

In the frequency range such that $\omega/k \sim V_{Te}$, $c_g \sim V_{Te}$ the expression (14.51) simplifies. The calculation is formulated as the problem of exercise 7 in chapter 20, section 20.8.

One should remember that here ω is the carrier frequency of the Langmuir wave train $\omega \approx \omega_{Le}$.

In the same manner the nonlinear term in the equation for the amplitude function of the Langmuir wave can be derived. The term is obviously proportional to A multiplied by Π_x. The equation for the ion-acoustic wave is often written in the notations of the plasma density n [5]. The density of plasma is proportional to the derivative of Π with respect to x according to equation (13.4). Therefore in the equation for the density perturbation the term $|A|_x^2$ will stand at the right-hand side. The system of equations for interacting plasma waves can be transformed to the equation containing both left and right ion waves

$$n_{tt} - n_{xx} = (|A|^2)_{xx}, \tag{14.52}$$

and to the equation for the Langmuir wave amplitude

$$2iA_t - 3A_{xx} = nA. \tag{14.53}$$

The rescaling was made to simplify the coefficients. The system (14.52) and (14.53) has been called the Zakharov system [22]. The investigation of equations (14.52) and (14.53) shows the Lyapunov stability of the stationary solutions and the existence of the mass, energy and plasmon number conservation laws [23]. The two-dimensional generalization of equations (14.52) and (14.53) is unstable by Lyapunov stationary solutions [23]. The interaction of the Langmuir stationary waves with plasma particles and a relevant system integrability is considered in [24]. Blowing up solutions are studied in [25].

References

[1] Morozov A I 2012 *Introduction to Plasma Dynamics* (Boca Raton, FL: CRC Press) p 17

[2] Leble S B 2003 *Nonlinear Waves in Optical Waveguides and Soliton Theory Applications, Optical Solitons* (Berlin: Springer)

[3] Piel A 2010 *Plasma Physics: An Introduction to Laboratory, Space, and Fusion Plasmas* (Berlin: Springer) pp 4–5 (Archived from the original on 5 January 2016)

[4] Leble S 1990 *Nonlinear Waves in Waveguides* (Heidelberg: Springer)

[5] Leble S 2019 *Waveguide Propagation of Nonlinear Waves Impact of Inhomogeneity and Accompanying Effects* (Berlin: Springer)

[6] Bohm D and Gross E P 1864 Theory of plasma oscillations. Excitation and damping of oscillations *Phys. Rev.* A **75** 949

[7] Resibois R and de Leener M 1977 *Classical Kinetic Theory of Fluids* (New York: Wiley) pp 400

[8] Leble S B and Rohraff D W 2006 Nonlinear evolution of components of an electromagnetic field of helicoidal waves in plasma *Phys. Scr.* **123** 140–4

[9] Sato G, Oohara W and Hatakeyama R 2004 Plasma Production by Helicon Waves with Single Mode Number in Low Magnetic Fields HAL Id: hal-00001995

[10] Dmitriyev V P 2005 Helical waves on a vortex filament *Am. J. Phys.* **73** 563

[11] Whitham G B 1999 *Linear and Nonlinear Waves* (New York: Wiley)

[12] Konopelchenko B G 1992 *Introduction to Multidimensional Integrable equations: The Inverse Spectral Inverse Spectral Transform in 2+1 Dimensions* (New York: Plenum)

[13] Doktorov E and Leble S B 2007 *Dressing Method in Mathematical Physics* (Berlin: Springer)

[14] Matveev V B and Salle M A 1991 *Darboux Transformations and Solitons* (Berlin: Springer)

[15] Ma W X and Zhou Z X 2001 Binary symmetry constraints of N-wave interaction equations in 1+1 and 2+1 dimensions *J. Math. Phys.* **42** 4345–82

[16] Maxfield B W 1969 Helicon waves in solids *Am. J. Phys.* **37** 241–69

[17] Bychenkov V P, Silin V P and Uryupin S A 1988 *Phys. Rep.* **164** 119–215

[18] Gorbunov L M and Silin V P 1964 Nonlinear interaction between plasma waves *J. Exp. Theor. Phys.* **47** 203–10 1965 *Soviet Physics-JETP* **20** 135

[19] Gorbunov L M and Tumerbulatov A M 1967 Dispersion Law and Nonlinear Interaction of Langmuir Waves in a Weakly Turbulent Plasma *J. Exp. Theor. Phys.* **53** 1494–7
Gorbunov L M and Tumerbulatov A M 1968 *JETP* **26** 861

[20] Maslov V P 1983 *Mathematical Aspects of Integral Optics* (Moscow: Moscow Institute of Electronic Engineering)

[21] Dobrokhotov S Y and Maslov V P 1985 *J. Math. Sci.* **28** 91

[22] Zakharov V E 1972 SOVIET PHYSICS JETP **35** 908

[23] Talanov V I 1965 *ZhETF Pis. Red.* **2** 223
Talanov V I 1965 *JETP Lett.* **2** 141
Zakharov V E 1968 *J. Appl. Mech. Tech. Phys.* **9** 190

[24] Zakharov V E and Shabat A B 1979 *Funk. Anal. Prilozh.* **13** 13–22

[25] Cher Y, Czubak M and Sulem C 2015 Blowing up solutions to the Zakharov system for Langmuir waves *Laser Filamentation, CRM Series in Mathematical Physics* ed A D Bandrauk *et al* (Berlin: Springer)

IOP Publishing

Practical Electrodynamics with Advanced Applications

Sergey Leble

Chapter 15

Diffraction in the presence of conductivity, x-rays manipulation and focusing

15.1 General remarks

A very interesting example of electrodynamic applications relates to the very high-frequency range (see [2], named x-rays [1]). Its generation and propagation peculiarities are very important, as directly in medicine as in synchrotron radiation facilities [3]. We would study the second option, important in investigations of tiny objects in many aspects [1]. In this chapter we continue and develop the content of chapter 10, section 10.2, based mainly on two very recent publications [4] and [5].

At present, x-ray optics are developing rapidly due to the necessity of visualization of very small objects. To visualize the structure of such object, it is necessary to focus the x-rays. Currently, commonly used focusing elements are bent crystals, multilayer or mirrors, capillaries, waveguides, refractive lenses and diffraction elements, such as Fresnel's zone plates or Laue's multilayer lenses [6]. For the announced purposes, composite refractive lens are used. Since then, many experiments have been performed in which x-rays were focused. The use of lenses with a spherical or parabolic *concave* profile is one of the most popular methods of x-ray focusing now. Recently, researchers have tried using diamond lenses for x-ray focusing [6]. The materials used for x-ray focusing have some common features. First of all, they are characterized by a complex refractive index $n = 1 - \delta + i\beta$, where δ is responsible for refraction of x-rays in lenses, and β for attenuation of x-ray intensity. The ideal material for making x-ray lenses should have as large as possible value of δ and as small as possible value of the absorption coefficient β. In addition, the material should allow the technological possibility of manufacturing lenses with a small curvature radius.

In this chapter, focusing in x-ray optics using x-ray refraction mechanisms will be considered. A high-accurate method for calculating x-ray propagation is described [8]. The main idea of the approach [4], is to use a superposition of oriented Gaussian beams for the propagating wave. The direction of wave propagation in each Gaussian beam agrees with the local direction of propagation of the x-ray wave-front. Calculation of the propagation of x-ray waves through lenses, the thin lens approximation is applied. In this approximation, the wave parameters change discontinuously when the wave passed through a lens; the corresponding explicit formulas are derived. The theory is applied to high-accurate calculation focusing of x-rays by a system of many beryllium lenses [7]. A fine structure of the wave electric field on the focal plane is revealed and studied. The fine structure is formed due to diffraction of waves at the edges of the lens apertures. Tools for controlling the calculation accuracy are traced. It is shown that the amplitude of the electric field on the focal plane and the focal spot width are very sensitive to the calculation quality, while the position of the best focus can be obtained even from simple calculations.

In addition to the listed properties, the materials must be strong and inexpensive. Unfortunately, it is not easy to satisfy the latter two conditions. Beryllium is one of the most popular materials currently used for focusing x-rays; the cost of manufacturing a beryllium lens is about 5000 euros. In order to effectively focus x-rays at relatively small distances, use of complex optical systems consisting of several dozen lenses is necessary. Performing more extensive theoretical studies based on advanced mathematical modeling of interesting experiments is hence desirable, especially for a system of many lenses.

In theoretical studies in x-ray optics, use of the fast Fourier transform (FFT) to solve the paraxial wave equation [9] is popular. However, the exact solution of the paraxial equation has a form of an infinite Fourier series that has to be truncated when we move to the FFT. This leads to errors which magnitude obviously depends on circumstances, but the issue of accuracy and reliability of calculations is poorly investigated because this is a difficult theoretical question. The methods for estimating the Fourier series truncating error are well developed for functions specified analytically, and the truncation error is expressed in terms of derivatives of the decomposed function. In calculations in x-ray optics, we deal with functions digitized on a mesh, and classical methods of estimating the Fourier series truncation error are difficult to apply to functions specified in such a form. In addition, studies show that various focal spot characteristics have significantly different sensitivity to the calculation quality [7]. The function describing the electric field of the x-ray wave passed through a system of several tens of lenses becomes rapidly oscillating and requires using very dense numerical mesh for its digitization and many Fourier series terms in calculations. These circumstances show the limited possibilities of Fourier methods and stimulate the development of new methods for solving the problems of x-ray optics.

New methods should provide high accuracy of calculations in complex cases and should be equipped with tools for controlling the accuracy and reliability of calculations [7, 8] It is the mentioned method, based on the use of oriented Gaussian beams. The idea of using Gaussian beams in optics is not new. The technique of using beams in optics can be found in [10, 11]. In later works, [12, 13], Gaussian beams (a Gaussian shooting algorithm) were used in radio- location.

For the x-ray waves with exceptionally high wave frequency, the method of Gaussian beams needs to be adapted to solve the problems of x-ray optics.

In this chapter, we set the incident wave in the form of a simple coherent wavefront, and we do not touch on the problem of real sources of x-ray waves, partially coherent and non-strictly monochromatic. However, Gaussian beams can be also used to study partially coherent radiation beams, as exemplified by the Gaussian Shell Model [11].

15.2 Basic equations

In this section we study x-rays, as a very important example of electromagnetic waves in matter, that is described as wave equation (6.109) in isotropic matter from chapter 6, section 6.5. It arises as direct corollary of the Maxwell system for electromagnetic field in matter (6.1)–(6.4). Such phenomenon is characterized by the parameters ε and μ, that are either taken from experiments (phenomenology) or from a theory originated from the Drude–Lorentz model, see chapter 6. The propagation of a monochromatic electromagnetic wave with frequency ω_0 in a medium with complex refractive index $n = 1 - \delta + i\beta$ is described by the Helmholtz equation [14, 15]

$$\Delta E + k_0^2 n^2 E = 0, \tag{15.1}$$

where E is an electric field component; $\vec{r} = (x;\, y;\, z)$, $k_0 = \omega_0/c$ is wavenumber, c is the speed of light, β is an absorption coefficient and δ is a refractive decrement (δ; β are non-negative). More convenient for practical applications, a simplified wave equation can be derived from the Helmholtz equation (15.1). Let us consider the case when the wave propagates along the x-axis, and the characteristic scales l_y, l_z of the wave along the axes y and z are much larger than the characteristic scale l_x of the wave along the x-axis: $l_x \ll l_y$, $l_x \ll l_z$. In this case, equation (15.1) can be significantly simplified. If we substitute

$$E(\vec{r}) = A(\vec{r})\exp[ik_0 x], \tag{15.2}$$

into equation (15.1) we get the equation for function $A(\vec{r})$. We suppose that $A(\vec{r})$ varies slowly with variables $\vec{r}$. Neglecting the small term of a second derivative of the function $A(\vec{r})$ with respect to the x-variable, we arrive at the paraxial equation:

$$\frac{\partial A}{\partial x} + \frac{1}{2ik_0}\Delta_{\perp} A + \frac{k_0(n^2 - 1)}{2i} A = 0, \tag{15.3}$$

where

$$\Delta_{\perp} = \frac{\partial^2}{\partial x^2} + \frac{\partial^2}{\partial y^2}. \tag{15.4}$$

Equation (15.3) was first proposed in [16] for describing the propagation of a monochromatic electromagnetic wave within the paraxial approach. The Leontovich equation (15.3) is often called a paraxial equation as well.

Equations (15.1) and (15.3) describe the propagation of monochromatic waves with frequency ω_0. Considering quasimonochromatic waves characterized by a set of frequencies ω close to ω_0, we can construct the solution of such a problem in the form of a superposition of waves with different ω.

15.3 Propagation of x-rays in vacuum

Let us consider equation (15.3) for the case of vacuum (see also chapter 3, section 3.8), that is, for the refractive index $n = 1$, then equation (15.3) simplifies as

$$\frac{\partial A}{\partial x} + \frac{1}{2ik_0}\Delta_{\perp}A = 0. \tag{15.5}$$

This equation has important applications in optics in general [9, 15], where it describes the propagation of electromagnetic waves, and in acoustics [14]. Equation (15.5) has an exact particular solution in the form of Gaussian beam by Kogelnik [10]:

$$A(\vec{r}, \vec{r}_0, \sigma) = \frac{k_0\sigma^2}{k_0\sigma^2 + i(x - x_0)}\exp[-B(x)]. \tag{15.6}$$

$$B(x) = k_0\frac{\vec{\rho}^{\,2}}{k_0\sigma^2 + i(x - x_0)}. \tag{15.7}$$

Here, the two-dimensional radius-vector $\vec{\rho} = (y - y_0)\vec{j} + (z - z_0)\vec{k}$, σ is a real parameter, the Gaussian beam width. At $x = x_0$, the solution (15.7) takes the form of usual Gauss function. We can consider (15.6) with $x = 0$ as a boundary condition given at the plane $x = 0$ to equation (15.5).

Applying the formula (15.6) to formula (15.2), we obtain an approximate solution

$$E(\vec{r}, \vec{r}_0, \sigma) = \frac{k_0\sigma^2}{k_0\sigma^2 + i(x - x_0)}\exp[-B(x) + ik_0(x - x_0)], \tag{15.8}$$

of the Helmholtz equation (15.1). In our investigation, the condition $(\sigma k_0)^2 \gg 1$ is satisfied with very high accuracy ($(\sigma k_0)^2 \approx 10^9$), and this means that the function (15.8) satisfies the Helmholtz equation (15.1) with high accuracy, this is an 'almost exact' solution of the Leontovich equation (15.3). Therefore, we shall take (15.8) as a solution of the Helmholtz equation. We also shall call the solution (15.8) of the Helmholtz equation a Gaussian beam. The solution (15.8) of the Helmholtz equation (15.1) describes the wave propagation along the OX-axis; we assume that this axis coincides with the optical axis of the optical system.

Waves in x-ray optics can propagate at small angles to the OX-axis. Therefore, it is desirable to generalize the particular solution (15.8) of the Helmholtz equation (15.1) to the case when the wave propagates at an arbitrary angle to the OX-axis. Let the vector $\vec{e}_1$ indicates the direction of wave propagation and let the vector $\vec{j}$ be directed along the OZ-axis. We denote

$$\vec{e_2} = \vec{j} \times \vec{e_1}, \; \vec{e_3} = \vec{e_1} \times \vec{e_2}. \tag{15.9}$$

Then the generalization of the formula (15.8) to the case when the wave propagates along the direction of the vector $\vec{e_1}$ is written as follows

$$\begin{aligned} G(\vec{r}, \vec{r_0}, \vec{e_1}) &= \exp\left(i\vec{k}(\vec{r}-\vec{r_0})\right)\frac{k_0\,\sigma^2}{k_0\,\sigma^2 + i((\vec{r}-\vec{r_0})\,\vec{e_1})} \\ &\times \exp\left(-k_0\frac{((\vec{r}-\vec{r_0})\,\vec{e_2})^2 + ((\vec{r}-\vec{r_0})\,\vec{e_3})^2}{2(k_0\sigma^2 + i((\vec{r}-\vec{r_0})\,\vec{e_1}))}\right). \end{aligned} \tag{15.10}$$

The notations are as follows: $\vec{r_0} = (x_0, y_0, z_0)$, $\vec{k} = k_0\,\vec{e_1}$. The formula (15.10) is the principle result of the authors, Kshevetskii–Wojda [4]. In applications to synchrotron radiation only small angles to axis OX are of interest, then with high accuracy $k_x \approx k_0$, within the case.

15.4 Approximation of electromagnetic field as a superposition of Gaussian beams

15.4.1 Paraxial equation for Kshevetskii–Wojda beam

The goal of this section is to present the method of electric field $E(\vec{r})$ calculation using superposition of Gaussian beams, which will be efficient even if boundary condition $A(x_0, y, z)$ is fast oscillating function of distance from the main optical axis. In this method, the solution will be constructed with Gaussian beams propagating in direction, which coincides locally with the wavefront propagation direction.

Let's consider Helmholtz equation in vacuum. Assuming that electric field can be represented by boundary condition $E(x_0, y, z) = A(x_0, y, z)$ and using substitution

$$E(\vec{r}) = A(\vec{r})\exp(i\vec{k_0}\cdot(\vec{r} - \vec{r_0})), \tag{15.11}$$

where $\vec{r} = (x, y, z)$, $\vec{r_0} = (x_0, y, z)$, $\vec{k_0} = (k_{0,x}, k_{0,y}, k_{0,z})$, $k_0 = |\vec{k_0}|$ we get equation for function $A(\vec{r})$:

$$2\,i\vec{k_0}\cdot\vec{\nabla}A(\vec{r}) + \Delta A(\vec{r}) = 0, \tag{15.12}$$

where $\vec{\nabla} = (\partial/\partial x, \partial/\partial y, \partial/\partial z)$, Δ-Laplacian. If the direction of $\vec{k_0}$ locally coincides with the direction of wave propagation the equation (15.12) will take the following form:

$$2\,ik_0\frac{\partial A(\vec{r})}{\partial(\vec{r}\cdot\vec{e_1})} + \frac{\partial^2 A(\vec{r})}{\partial(\vec{r}\cdot\vec{e_1})^2} + \frac{\partial^2 A(\vec{r})}{\partial(\vec{r}\cdot\vec{e_2})^2} + \frac{\partial^2 A(\vec{r})}{\partial(\vec{r}\cdot\vec{e_3})^2} = 0, \tag{15.13}$$

where $\vec{e_1} = \vec{k_0}/k_0$ and vectors $\vec{e_2}$, $\vec{e_3}$ are defined by equation (15.9). Neglecting the small term of the second derivative of the function $A(\vec{r})$ with respect to the

$(\vec{r} \cdot \vec{e_1})$-variable, we arrive at the paraxial equation in the new system of coordinates:

$$2ik_0 \frac{\partial A(\vec{r})}{\partial(\vec{r} \cdot \vec{e_1})} + \frac{\partial^2 A(\vec{r})}{\partial(\vec{r} \cdot \vec{e_2})^2} + \frac{\partial^2 A(\vec{r})}{\partial(\vec{r} \cdot \vec{e_3})^2} = 0. \tag{15.14}$$

equation (15.14) has exact solution in form of the Gaussian beam (15.10), which propagates along the vector $\vec{e_1}$.

15.4.2 Superposition of Gaussian beams

We will introduce the method using superposition of Gaussian beams, which will give us approximate solution of electric field $\vec{E}(\vec{r}) = E(\vec{r})\vec{e_1}$. The function A might be a function that rapidly oscillates in the OYZ plane. On the surface **S**, which is perpendicular to the direction of propagation of electromagnetic waves, the phase of the electric field $\vec{E}$ does not change. Each surface can be approximated locally with a plane. Therefore, for a small domain of surface **S** we can select such a plane P_0, which approximates the surface **S** in the vicinity of some point $(x_0, y_0, z_0) = \vec{r_0}$ of the surface **S**. Let $\vec{e_1}^0$ be a vector, which is perpendicular to the plane P_0, then its direction locally coincides with the wave propagation direction in the vicinity of the point $\vec{r_0}$. For a sufficiently small part of surfaces **S** and P_0 set in the vicinity of point $\vec{r_0}$, accordingly: S^{part}, P_0^{part}, the following approximation can be used:

$$\iint_{S^{\text{part}}} \vec{E}(\vec{r}) \cdot d\mathbf{S} \approx \iint_{P_0^{\text{part}}} \vec{E}(\vec{r}) \cdot \vec{e_1}^0 \, d(\vec{e_2}^0 \cdot \vec{r}) d(\vec{e_3}^0 \cdot \vec{r}), \tag{15.15}$$

where vectors $\vec{e_2}^0$, $\vec{e_3}^0$ create orthonormal basis with the vector $\vec{e_1}^0$, similar to equation (15.9) and $d\mathbf{S}$ is the differential vector area of the surface **S**. The projection of the vector field $\vec{E}(\vec{r}), \vec{E}(\vec{r}) \cdot \vec{e_1}^0$ on the plane P_0 in the vicinity of the point $\vec{r_0}$ is a function, with small change of phase. For the square domain S_0 with side length h, which center is in the point $\vec{r_0}$ and that belongs to the plane P_0, following approximation can be applied:

$$\iint_{S_0} \vec{E}(\vec{r}) \cdot \vec{e_1}^0 \, d(\vec{e_2}^0 \cdot \vec{r}) d(\vec{e_3}^0 \cdot \vec{r}) \approx \vec{A_0} \exp\left(ik_0 \, \vec{e_1}^0 \cdot (\vec{r} - \vec{r_0})\right) \cdot \vec{e_1}^0 \, h^2, \tag{15.16}$$

where $\vec{A_0} = A(x_0, y_0, z_0)\vec{e_x}$ and $\vec{r}$ lies on the plane P_0, so the term $\exp(i\, k_0 \, \vec{e_1}^0 \cdot (\vec{r} - \vec{r_0})) = 1$.

15.4.3 Quasi-exact solution of the Helmholtz equation

Let us construct approximate solution of the Helmholtz equation using Gaussian beams, that we would name as 'quasi-exact':

$$C \exp\left(-a\left(\left(\vec{e_2}^0 \cdot (\vec{r} - \vec{r_0})\right)^2 + \left(\vec{e_3}^0 \cdot (\vec{r} - \vec{r_0})\right)^2\right)\right), \tag{15.17}$$

which propagates in the direction coinciding with the direction of wave propagation $\vec{e}_1^{\,0}$ and where C and a are parameters that we must determine. We determine C assuming that the electric field flux passing through the area S^{part} is approximately equal to

$$\iint_{P_0} C \exp\left(-a\left(\left(\vec{e}_2^{\,0}\cdot(\vec{r}-\vec{r}_0)\right)^2+\left(\vec{e}_3^{\,0}\cdot(\vec{r}-\vec{r}_0)\right)^2\right)\right) d\left(\vec{e}_2^{\,0}\cdot\vec{r}\right) d\left(\vec{e}_3^{\,0}\cdot\vec{r}\right). \tag{15.18}$$

This means that with high accuracy the following equality is satisfied:

$$\begin{aligned}&\vec{A}_0 \exp\left(i\,k_0\,\vec{e}_1^{\,0}\cdot(\vec{r}-\vec{r}_0)\right)\cdot\vec{e}_1^{\,0}\,h^2\\ &=\iint_{P_0} C \exp\left(-a\left(\left(\vec{e}_2^{\,0}\cdot(\vec{r}-\vec{r}_0)\right)^2+\left(\vec{e}_3^{\,0}\cdot(\vec{r}-\vec{r}_0)\right)^2\right)\right)\\ &\quad d(\vec{e}_2^{\,0}\cdot\vec{r})d(\vec{e}_3^{\,0}\cdot\vec{r}).\end{aligned} \tag{15.19}$$

It follows that

$$C=\frac{a}{\pi}\vec{A}_0\exp\left(i\,k_0\,\vec{e}_1^{\,0}\cdot(\vec{r}-\vec{r}_0)\right)\cdot\vec{e}_1^{\,0}\,h^2. \tag{15.20}$$

Both C and a are independent on variables $(\vec{e}_2^{\,0}\cdot\vec{r})$, $(\vec{e}_3^{\,0}\cdot\vec{r})$, but they can be functions of $\vec{e}_1^{\,0}\cdot(\vec{r}-\vec{r}_0)$. Additionally C depends on $\vec{A}_0$. Gaussian beam (15.17) lies on the plane P_0 perpendicular to vector $\vec{e}_1^{\,0}$ and which contains the point $\vec{r}_0$ and not necessarily on the plane $x = x_0$. If on the plane P_0 for Gaussian beam (15.17) $a(\vec{e}_1^{\,0}\cdot(\vec{r}-\vec{r}_0)) = a$ then Gaussian beams after projection on the plane $x = x_0$ have the following form:

$$\begin{aligned}&G(A_0,\vec{e}_1^{\,0},\vec{e}_2^{\,0},\vec{e}_3^{\,0},\vec{r},\vec{r}_0)=C(A_0,\vec{e}_1^{\,0},\vec{r},\vec{r}_0)\\ &\exp\left(-a(\vec{e}_1^{\,0}\cdot(\vec{r}-\vec{r}_0))\left(\left(\vec{e}_2^{\,0}\cdot(\vec{r}-\vec{r}_0)\right)^2+\left(\vec{e}_3^{\,0}\cdot(\vec{r}-\vec{r}_0)\right)^2\right)\right),\end{aligned} \tag{15.21}$$

where

$$C(A_0,\vec{e}_1^{\,0},\vec{r},\vec{r}_0)=\frac{a(\vec{e}_1^{\,0}\cdot(\vec{r}-\vec{r}_0))}{\pi}\vec{A}_0\exp\left(i\,k_0\,\vec{e}_1^{\,0}\cdot(\vec{r}-\vec{r}_0)\right)\cdot\vec{e}_1^{\,0}\,h^2 \tag{15.22}$$

and

$$a(\vec{e}_1^{\,0}\cdot(\vec{r}-\vec{r}_0))=\frac{ak_0}{k_0+2ia\vec{e}_1^{\,0}\cdot(\vec{r}-\vec{r}_0)}. \tag{15.23}$$

Term (15.21) follows from propagation of the Gaussian beam (15.17) along the vector $\vec{e}_1^{\,0}$ between the plane P_0 and plane parallel to it, which contains point $\vec{r} = (x_0, y, z)$. The term $\exp(i\,k_0\vec{e}_1^{\,0}\cdot(\vec{r}-\vec{r}_0))$ is consequence of electromagnetic wave propagation along the vector $\overrightarrow{e_1}^0$ from plane P_0 up to the point $\vec{r}$ on the plane $x = x_0$.

When we take superposition of Gaussian beams (15.21) we get approximation of $A(\vec{r})$ in the form:

$$A(\vec{r}) \approx \sum_{j,k} G(A_{j,k}, \overrightarrow{e_1}^{j,k}, \overrightarrow{e_2}^{j,k}, \overrightarrow{e_3}^{j,k}, \vec{r}, \vec{r_{j,k}}), \tag{15.24}$$

where $A_{j,k} = A(x_0, y_j, z_k)$ and $\vec{r_{j,k}} = (x_0, y_j, z_k)$.

Approximation of $E(\vec{r})$ for any x has similar form to the approximation (15.24):

$$E(\vec{r}) \approx \sum_{j,k} C_{j,k}\, G(A_{j,k}, \overrightarrow{e_1}^{j,k}, \overrightarrow{e_2}^{j,k}, \overrightarrow{e_3}^{j,k}, \vec{r}, \vec{r_{j,k}}), \tag{15.25}$$

where $\vec{r} = (x, y, z)$. If every Gaussian beam propagates along axis OX, $\overrightarrow{e_1}^{j,k} = \vec{e_x}$, then the approximation (15.25) is built as superposition of the Gaussian beams. In order to shorten description, Gaussian beams, which propagate along wave propagation direction further in the paper will be called the *oriented Gaussian beams*. The method using oriented Gaussian beams in order to obtain an approximate solution of electric field we will call the *oriented Gaussian beam method* (OGBM).

It is worth mentioning the recent paper [19], see also references therein, where the Gaussian wave beam and wave packet types of solutions are constructed with help of Maslov's complex germ theory (short-wave or semi-classical asymptotics with complex phases). The term "semi-classical" asymptotics is understood in a broad sense: asymptotic solutions of evolutionary and stationary partial differential equations from wave or quantum mechanics. The results are related to the linearized equations of cold plasma in a toroidal domain (tokamak), but other applications are quite visible.

15.5 Oriented Gaussian beams method application to x-rays propagation through optical elements

In this section we consider the problem of x-ray propagation through lenses, as x-rays optics elements. Materials used for x-ray focusing have so small parameters β and δ, that the equation (15.3) for propagation inside the material can be rearranged to the following form:

$$\frac{\partial A}{\partial x} - i\, k_0(-\delta + i\,\beta)\, A = \frac{i}{2\, k_0}\left(\frac{\partial^2 A}{\partial y^2} + \frac{\partial^2 A}{\partial z^2}\right). \tag{15.26}$$

By substitution:

$$A(x, y, z) = B(x, y, z) \exp(i\, k_0(-\delta + i\,\beta)\, x), \tag{15.27}$$

equation (15.26) can be reduced to the paraxial equation for vacuum:

$$\frac{\partial B}{\partial x} = \frac{i}{2\, k_0}\left(\frac{\partial^2 B}{\partial y^2} + \frac{\partial^2 B}{\partial z^2}\right). \tag{15.28}$$

The method (OGBM) for obtaining the solution of the equation (15.28) is presented in the previous section. This means that to calculate the electric wave field after the lens we can use the approximation (15.25) to obtain the solution of the equation (15.28) describing propagation of waves in vacuum and then obtain the form of the function E' describing the electric wave field after a lens using the formula:

$$E'(\vec{r}) = E(\vec{r})\exp(ik_0(-\delta + i\,\beta)|\vec{r_A} - \vec{r_B}|), \tag{15.29}$$

where $|\vec{r_A} - \vec{r_B}|$ denotes the local thickness of the lens by which x-rays waves propagate along the local direction of x-ray propagation $\vec{e_1}$.

In x-ray optics problems the direction of wavefront propagation differs slightly from the direction along the optical axis, hence the distances $|\vec{r_A} - \vec{r_B}|$ and the local width of a thin lens are approximately equal. Therefore, the change in the electromagnetic wave field due to the passage of the wave through the thin lens can be described as follows

$$E'(x_C, y, z) = E(x_C, y, z)\exp(i\,O_P\,F(y, z)), \tag{15.30}$$

where $E(x_C, y, z)$ is the electric wave field in front of the optical element and $E'(x_C, y, z)$ is the electric wave field after the lens, $O_P = k_0(-\delta + i\beta)$ is determined by the properties of the lens material, and $F(y, z)$ describes the local thickness of the lens, depending on y and z. The point $x = x_C$ on the OX-axis corresponds to the geometric center of the lens. The factor $O_PF(y, z)$ in the exponent takes into account the local phase and amplitude changes due to the passage of the wave through the lens.

The direction of propagation of electromagnetic waves is consistent with the direction indicated by the vector $\vec{e_1}$. The direction of electromagnetic waves propagation at any point is perpendicular to some phase surface $\phi(\vec{r}) = D$, where D is a constant. For such phase surface, the wave electric field E fulfills the following relationship:

$$\frac{E}{|E|} = \exp(i\,\phi) = D_1, \tag{15.31}$$

where D_1 is a constant. The gradient of the phase function ϕ at a point is perpendicular to the phase surface $\phi(\vec{r}) = D$ at that point, so the phase function gradient indicates the direction of propagation of electromagnetic waves. This means that in order to determine the wave propagation direction, the gradient of ϕ should be calculated. For the electric wave field the following equality is satisfied:

$$\frac{\vec{\nabla} E}{E} = \frac{\vec{\nabla}|E|}{|E|} + i\vec{\nabla}\phi, \tag{15.32}$$

therefore, the propagation direction of the wave electric field E can be determined by the imaginary part of (15.32):

$$\vec{k_0} = \vec{\nabla}\phi = \Im\left[\frac{\vec{\nabla} E}{E}\right]. \tag{15.33}$$

When propagation of x-ray in the vacuum is considered, the propagation direction $\vec{e_1}$ can be determined straight from the electric wave field E using the following formulas:

$$\begin{aligned} e_{1,y} &= \frac{1}{k_0}\Im\left(\frac{E_y}{E}\right), \\ e_{1,z} &= \frac{1}{k_0}\Im\left(\frac{E_z}{E}\right), \\ e_{1,x} &= 1 - \frac{e_{1,y}^2 + e_{1,z}^2}{2}, \end{aligned} \tag{15.34}$$

where $E(\vec{r})$ can be approximated by equation (15.25).

When the x-ray wave propagates through the lens, the electric wave field E changes, and with it the surfaces in which the phase ϕ is constant and also the local wave propagation directions change. The change in the components of the vector $\vec{e_1}$ caused by refraction of x-rays on the surfaces of the thin lens can be determined by applying the formula (15.34) to the equation (15.30). After the passage of the x-ray through a thin lens, a new direction of x-ray propagation $\vec{e_1}'$ can be determined based on the following formula:

$$\begin{aligned} e_{1,y}' &= e_{1,y} - \delta\,\frac{\partial F(y, z)}{\partial y}, \\ e_{1,z}' &= e_{1,z} - \delta\,\frac{\partial F(y, z)}{\partial z}, \\ e_{1,x}' &= 1 - \frac{e_{1,y}'^2 + e_{1,z}'^2}{2}. \end{aligned} \tag{15.35}$$

In calculations, if lenses and their shape are digitized on a mesh, the derivatives $\frac{\partial F(y,z)}{\partial y}$ and $\frac{\partial F(y,z)}{\partial z}$ have to be approximated by finite differences on the mesh.

15.6 Study of accuracy and efficiency of Gaussian beam methods

15.6.1 Estimation of convergence rate of solution obtained with superposition of oriented Gaussian beams to electric field described by boundary condition, which is a fast oscillating function

X-ray propagation can be described in terms of the absolute value of electric field $|E(\vec{r})|$ and the phase function ϕ, where:

$$E(\vec{r}) = |E(\vec{r})|\exp(i\phi(\vec{r})), \tag{15.36}$$

ϕ is a real-valued function. Gaussian beams are exact solutions of the paraxial equation. Since the solution for $x \geqslant x_0$ is obtained by superposition of Gaussian beams propagating along the direction of electromagnetic wave propagation, the estimated error of approximation at distance x will be the same as the error for the boundary condition for x_0. Thus, the accuracy of the solutions given for $B(\vec{r})$

depends directly on the accuracy with which the boundary condition $x = x_0$ is approximated. This means that in order to estimate error of calculation of focusing x-rays we need only to estimate error with which the boundary condition is approximated by the formula (15.24) on the plane $x = x_0$ after the x-rays have propagated through the system of lenses.

It is well known that after the x-ray had propagated through a lens, the phase function ϕ becomes the quadratic function inside the aperture of a lens [1]. The parameters of this function are related to the form of the lens and the material from which it is made of [3]. The form of the parabolic lens can be described with its total width W and the minimal one W_{sm}, as well as the curvature of the lens R. When we consider x-rays propagation, we also have to take into account its frequency ω_0, which determines the value of the refraction δ and the absorption β parameters. Lenses outside the aperture are flat, and the function describing the lens form is a continuous function. This function, however, is not differentiable in points where the parabolic form changes to plane. For the sake of simplicity, let's assume that the lens has a parabolic shape also outside of the aperture. If function $\phi(\vec{r})$ is smooth enough, it can be described approximately using the square function. So its value at any point is given by the expression:

$$\phi(\vec{r}) = \phi_{j,k} + \nabla\phi_{j,k} \cdot (\vec{r} - \vec{r_{j,k}}) + \frac{1}{2}(\vec{r} - \vec{r_{j,k}})^T \mathbf{H}_{j,k}(\vec{r} - \vec{r_{j,k}}), \tag{15.37}$$

where $\mathbf{H}_{j,k}$ is the Hessian matrix at $\vec{r_{j,k}} = (x_0, y_j, z_k)$, $\phi_{j,k} = \phi(\vec{r_{j,k}})$. When we use oriented Gaussian beams we use approximation for the phase in the form:

$$\phi(\vec{r}) \approx \phi_{j,k} + \vec{\nabla}\phi_{j,k} \cdot (\vec{r} - \vec{r_{j,k}}), \tag{15.38}$$

so the error of approximation can be determined by the estimation of the value

$$\text{error}_{j,k}(\vec{r}) = \frac{1}{2}(\vec{r} - \vec{r_{j,k}})^T \mathbf{H}_{j,k}(\vec{r} - \vec{r_{j,k}}). \tag{15.39}$$

In order to obtain convergence $B(\vec{r})$ to the boundary condition $A(x_0, y, z)$ oscillations generated by $\text{error}_{j,k}(\vec{r})$ should have a negligible effect on the solution. For this purpose the following condition should be fulfilled:

$$|\text{error}_{j,k}(\vec{r})| \ll |a(\vec{e}_1^{\,j,k} \cdot \vec{r})((\vec{e}_2^{\,j,k} \cdot (\vec{r} - \vec{r_{j,k}}))^2 + (\vec{e}_3^{\,j,k} \cdot (\vec{r} - \vec{r_{j,k}}))^2)|, \tag{15.40}$$

where $\vec{r} = (x_0, y, z)$. More detailed information about convergence $B(\vec{r})$ to the boundary condition $A(x_0, y, z)$ can be found in [4].

Let's find estimation of $\text{error}_{j,k}(\vec{r})$ after x-ray propagates though the system of lenses. The width of lens through which the x-ray passes is approximately equal:

$$d(y_j, z_k) \approx R((y_j - y_0)^2 + (z_k - z_0)^2) - W_{sm}, \tag{15.41}$$

where W_{sm} is the smallest distance between the parabolic surfaces of the lens and R is the curvature radius of the lens surfaces. Therefore after the x-ray propagates though the single lens

$$\text{error}^{s}_{j,k}(\vec{r}) \approx k_0\, \delta\, R((y - y_j)^2 + (z - z_k)^2), \tag{15.42}$$

and after N lenses:

$$\text{error}^{N}_{j,k}(\vec{r}) \approx k_0 \delta R N((y - y_j)^2 + (z - z_k)^2). \tag{15.43}$$

Taking into account the above considerations we can estimate the value of the spatial step, which should be chosen when determining the electrical field propagating by N lenses. The space step h should satisfy the condition:

$$N k_0 \delta R h^2 / b \ll 1, \tag{15.44}$$

where $b = ah^2$ ($b = O(1)$). The smaller value of b the smaller space step h should be taken. For $a = \frac{1}{2h^2}$, to obtain a good result, it is sufficient that the condition $2N k_0 \delta R h^2 \ll 1$ will be satisfied. Thus, the approximate value of step h can be derived from the relation

$$h \approx \sqrt{\frac{0.05}{N k_0 \delta R}}. \tag{15.45}$$

This is the maximum space step h, which should be used to obtain good approximation of electric field with phase approximation (15.38).

Using approximation (15.25), it is possible to get precise results without reducing the spatial step h with the increasing rate of function oscillation [18]. It is true even if the change of function phase at neighboring points is much greater than π.

15.6.2 Propagation of x-rays through a lens and its aperture boundaries

Let's consider a function that describes the electric field after the x-rays have passed through the lens, that is used in experiments. Such function stands for a superposition of one, associated with the x-rays that propagate through the aperture of the lens, and one, with the x-rays that pass through its flat part. When we consider the parts of lens that are located away from the aperture edges on the flattening of the lens, we get results with very high accuracy because the flat part has a curvature equal 0, so the approximation error connected with change of phase in this parts of lens is also equal 0. The smaller the error, the higher is the accuracy. When we consider the focusing part of lens (far from the aperture edge), the approximation accuracy is associated with the error described by the equation (15.42). X-rays propagating through the edges of the lens aperture generates spherical waves, therefore the use of a particular propagation direction $\vec{e_1} = \vec{\nabla}\phi / k_0$ may reduce the accuracy of the calculations of the electric wave field E near the aperture's edges of the lens. The smaller the width of Gaussian beams used in calculations, the more precisely the direction of the wave propagation $\vec{e_1}$ can be determined, and the higher the accuracy of calculations (because the smaller is the vicinity of the aperture's edges, for which the wave propagation direction can not be unambiguously specified). Additionally, if the absolute value of E in the vicinity of the aperture edges is much smaller than the function E amplitude on the lens, then the influence of the aperture boundaries has a negligible effect on the electric wave field.

15.7 Numerical calculations scheme

15.7.1 Implicit Runge–Kutta scheme

In this section we consider the simplified model of the x-ray propagation through a lens, following [5] for the wave field, that varies progressively along the x-axis only. The basic Maxwell equations for isotropic conducting medium are derived in chapter 6, section 6.5. The resulting equation from this section, equation (6.109) for the non-magnetic medium, inside the lenses takes the form:

$$\boxed{\nabla^2\vec{E} - \varepsilon_r\mu_0\frac{\partial^2\vec{E}}{\partial t^2} = \sigma\mu_0\frac{\partial\vec{E}}{\partial t}.} \tag{15.46}$$

After substitution the single frequency ω_0 wave

$$\vec{E}(\vec{r}, t) = \vec{A}_l(\vec{r}, t)\exp(i(k_0x - \omega_0t)) + c.c., \tag{15.47}$$

for the field inside the lenses, for the relative permittivity related to the index of refraction n with the relation is, by the definition is equal to $n = \sqrt{\varepsilon_r}$, having:

$$\begin{aligned}\frac{\partial\vec{A}_l}{\partial x} + \frac{i\omega_0}{2c_0}\left(1 - \varepsilon_r\mu_0c_0^2 - \frac{i\mu_0\sigma c_0^2}{\omega_0}\right)\vec{A}_l &= \frac{\partial\vec{A}_l}{\partial x} + \frac{i\omega_0}{2c_0}(1 - n^2)\vec{A}_l\\ &= \frac{ic_0}{2\omega_0}\left(\frac{\partial^2\vec{A}_l}{\partial y^2} + \frac{\partial^2\vec{A}_l}{\partial z^2}\right).\end{aligned} \tag{15.48}$$

Using the notation of x-ray propagation theory (see previous section),

$$n^2 = 1 - 2(\delta - i\beta) + (\delta - i\beta)^2,$$

we can neglect the term $(\delta - i\beta)^2$ since δ and β are very small quantities. Now equation (15.48) takes the final form:

$$\boxed{\frac{\partial\vec{A}_l}{\partial x} + B\vec{A}_l = \frac{ic_0}{2\omega_0}\left(\frac{\partial^2\vec{A}_l}{\partial y^2} + \frac{\partial^2\vec{A}_l}{\partial z^2}\right),} \tag{15.49}$$

where $B = \frac{\omega_0}{c_0}(\beta + i\delta)$. We intend to solve equation (15.49) numerically using the finite difference method, and the implicit Runge–Kutta scheme of the second order with iterative procedure implementation [8].

Let's start with equation (15.49) for the propagation inside the lens. Its Runge–Kutta discretization yields

$$\frac{A_{j,k}^{n+1} - A_{j,k}^{n}}{h_l} + BA_{j,k}^{n+\frac{1}{2}} = \frac{ic_0}{2\omega_0 h_r^2}\left(A_{j+1,k}^{n+\frac{1}{2}} + A_{j-1,k}^{n+\frac{1}{2}} + A_{j,k+1}^{n+\frac{1}{2}} + A_{j,k-1}^{n+\frac{1}{2}} - 4A_{j,k}^{n+\frac{1}{2}}\right), \quad (15.50)$$

where h_l is the space step in x-axis inside the lens. In order to calculate intermediate quantities $A_{j,k}^{n+\frac{1}{2}}$ we apply the implicit scheme

$$A_{j,k}^{n+\frac{1}{2}}\left(1 + \frac{BD}{2} + \frac{ic_0 h_l}{\omega_0 h_r^2}\right) = A_{j,k}^{n} + \frac{ic_0 h_l}{4\omega_0 h_r^2} \times \left(A_{j+1,k}^{n+\frac{1}{2}} + A_{j-1,k}^{n+\frac{1}{2}} + A_{j,k-1}^{n+\frac{1}{2}} + A_{j,k+1}^{n+\frac{1}{2}}\right), \quad (15.51)$$

where D is the distance between two points inside the lens, that needs to be defined according to the shape of our parabolic lens.

Finally we arrive at a simple formula

$$A_{j,k}^{n+1} = 2A_{j,k}^{n+\frac{1}{2}} - A_{j,k}^{n}. \quad (15.52)$$

The system of equations (15.51) and (15.52) gives us the implicit Runge–Kutta method of the second order [17].

For the propagation in free space, $B = 0$, the equation (15.51) takes the form:

$$2\frac{A_{j,k}^{n+\frac{1}{2}} - A_{j,k}^{n}}{h_x} = \frac{ic_0}{2\omega_0 h_r^2}\left(A_{j+1,k}^{n+\frac{1}{2}} + A_{j-1,k}^{n+\frac{1}{2}} + A_{j,k+1}^{n+\frac{1}{2}} + A_{j,k-1}^{n+\frac{1}{2}} - 4A_{j,k}^{n+\frac{1}{2}}\right). \quad (15.53)$$

15.7.2 Algorithm: parameters of integration choice

Inside the lenses. From the dispersion relation for equation (15.49) the following conditions must be satisfied for the propagation inside the lenses

$$h_l \ll \frac{\omega_0 h_r^2}{c_0},$$

and

$$h_l <<< \frac{1}{|B|}.$$

We take $h_x = h_l$ as the space step for x-axis inside the lens and outside of it.

The algorithm of calculation consists of three steps: first step

$$A_{j,k}^{n+\frac{1}{2}(0)}\left(1 + \frac{BD}{2} + \frac{ic_0 h_l}{\omega_0 h_r^2}\right) = A_{j,k}^{n} + \frac{ic_0 h_l}{4\omega_0 h_r^2} \times \left(A_{j+1,k}^{n} + A_{j-1,k}^{n} + A_{j,k+1}^{n} + A_{j,k-1}^{n}\right), \quad (15.54)$$

second step (we need to run it for several times)

$$A_{j,k}^{n+\frac{1}{2}(m+1)} = \frac{A_{j,k}^{n} + \frac{ic_0h_l}{4\omega_0h_r^2}(A_{j+1,k}^{n(m)} + A_{j-1,k}^{n(m)} + A_{j,k+1}^{n(m)} + A_{j,k-1}^{n(m)})}{\left(1 + \frac{BD}{2} + \frac{ic_0h_l}{\omega_0h_r^2}\right)}, \tag{15.55}$$

the last step

$$A_{j,k}^{n+1} = 2A_{j,k}^{n+\frac{1}{2}(m)} - A_{j,k}^{n}.$$

Outside the lenses. If we put $B = 0$ in equations (15.54) and (15.55), we will obtain the implicit scheme in vacuum outside the lenses. In calculations it's enough to take $m = 4$ [8].

15.8 The numerical simulations

15.8.1 General description

To provide numerical simulations of the problem by the algorithm of section 15.7.2 we choose the boundary conditions at $x = 0$ as in [6] in order to have a possibility to compare results, following [5]. Hence, the form of the conditions are chosen as Gaussian beams with 'full width at half maximum', $W = 700$ μm in the horizontal direction, and $W = 35$ μm in the vertical direction. We consider here two cases: the first case using the data from [6], and the second case with few numbers of lenses to perform the 2D simulation. For the first case the results showed that we need to consider more than 50 000 points in each direction that will be difficult for the personal computer so, we perform 1D simulation. For the second case we use a smaller number of lenses up to 15 lenses in order to perform the 2D simulation.

15.8.2 The first case, 33 aluminium lenses with 15 keV x-ray

The conventional characteristics of x-rays range is expressed in eV, linked to frequency by Planck formula [2]. For example the energy 12.4 keV corresponds to the frequency about 3 EHz, or the wavelength 100 pm. We present the tests of the program for different number of points in each direction. The results should not depend on the number of points that we use in y, z directions (in 1D case just for y). In order to do that we need to choose h_x, as the space steps in x, and h_r for y, and z directions respectively correctly and try to find the suitable number of points for our mesh. The dispersion relation governs the relation between the space steps in each direction $h_x \ll \frac{\omega_0 h_r{}^2}{c_0}$. In order to choose the number of points in y, z direction we need to calculate $0.001/h_r$, this relation will give us the number of points in the horizontal and vertical direction, h_x here is used as the space step in x-axis to identify the place of the detector. For instance, if the focal plane at 1.298 m from the lenses, it means that we need $1.298/h_x$ points in x-axis. While, h_l is used as the space step inside the lens, and it should satisfy the conditions in the dispersion relation for equation (15.49).

- Number of Al lenses = 33.
- Radius of curvature = 0.2 mm.
- Energy = 15 keV.
- $\delta = 2.414 \times 10^{-6}$.
- $\beta = 1.299 \times 10^{-8}$.
- The distance from the source to the lenses is 63 m.

15.8.3 The choice of space steps and errors

The choice of the space step h_x

Let us choose the best value for the constant C in the relation

$$h_x = C\frac{\omega_0}{c_0}h_r^2,$$

that will give us the best choice for the space step h_x along x. We fix the number of points to be 40 000, which means that we have fixed the space step h_r. We tried with some values for the constant h_r, the results showed that the best choice for the space step is 0.032 911 422 which gives us small error in the W and in the focal distance, see table 15.1. We can go further by choosing a smaller value for the constant C that means smaller value for h_x which will give us smaller error. But, in this case we need more points for the propagation after the lenses. For example, if we choose the constant C to be 0.016 455 711 that gives space step $h_x = 0.000\,000\,781$, and we need 1 661 440 points in x-axis to reach the detector at 1.298 m.

The error for choosing h_x

In order to compute the error we will use the Runge rule [17]. We can assume that $Z_h = 3.55$ and $Z_{2h} = 3.65$ for the spot size. The error can be written as

$$R = \frac{Z_h - Z_{2h}}{3} = 0.033\,333\,333,$$

which is small if we compare it to the $W = 3.65$. We can do the same for the focal distance f if $Z_h = 1.253\,909\,3$, $Z_{2h} = 1.252\,881\,8$, then

$$R = \frac{Z_h - Z_{2h}}{3} = 0.000\,342\,5,$$

Table 15.1. The choice of h_x.

Constant	h_x	Focal distance	W
0.131 645 688	0.000 006 25	1.258 031 2 m	3.25 μm
0.065 822 844	0.000 003 125	1.253 909 3 m	3.55 μm
0.032 911 422	0.000 001 562	1.252 881 8 m	3.65 μm

which is very small if you compare it to 1.252 881 8. From the numerical investigations we fix the best choice for the constant in the relation $C = 0.032\,911\,422$ that corresponds to $h_x = 0.000\,001\,562$.

The choice of $\mathrm{h_l}$

Next, we need to choose the space step inside the lens h_l that will give us small enough error according to the Runge rule for the spot size, and for the focal distance. We fix here the number of points to be 40 000 which means that $h_r = 0.000\,000\,025$. We have seen from the dispersion relation for the propagation inside the lens that the following conditions have to be satisfied

$$h_l \ll \frac{\omega_0 h_r^2}{c_0}.$$

It means that

$$h_l <<< \frac{1}{|B|},$$

for $\frac{1}{|B|} = 5.4533 \times 10^{-6}$. Table 15.2 gives us different values for h_l with the corresponding spot size, and focal distance. As you can see that the best choice for h_l is 0.000 000 125 that will give us small error according to the Runge rule for the spot size, and for the focal distance. The best choice for h_l corresponds to 8000 steps for each lens.

The choice of $\mathrm{h_r}$

Now, we will try to find the best choice for the space step in y direction h_r I mean to find the suitable number of points in the y, and z direction we do that by changing the number of points and try to prove that the results don't depend on the number of points. We fixed here the space step inside the lens $h_l = 0.000\,000\,125$, we use the best choice for the constant to obtain h_x from the relation

$$h_x = C\frac{\omega_0}{c_0}h_r^2$$

which is equivalent to 0.000 001 562 for 40 000 points, and 0.000 001 for 50 000 (table 15.3). Due to the large number of lenses, we need to consider more points we

Table 15.2. The choice of the step within the lens h_l.

h_l	Focal distance	W
0.000 001	1.252 881 8 m	3.65 μm
0.000 000 5	1.249 220 4 m	3.90 μm
0.000 000 25	1.248 300 4 m	3.95 μm
0.000 000 125	1.248 070 7 m	3.95 μm

Table 15.3. The choice of h_r.

Number of points	h_r	Focal distance	W
40 000	0.000 000 025	1.248 070 7 m	3.95 μm
50 000	0.000 000 02	1.247 034 0 m	3.92 μm

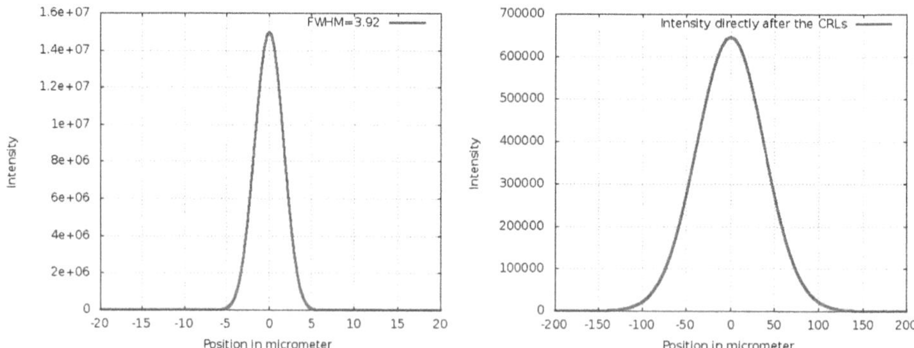

Figure 15.1. Results for focusing hard x-rays with 33 perfect Al lens directly after the lenses, and at 1.298 m from the lenses with W = 3.92 μm, and focal distance = 1.247 034 0 m, courtesy of M Elsawy [5].

started with 40 000 points, and checked with 50 000 points. But, the results for 40 000, and 50 000 points are very close for the focal distance, and for the W. We could take more points which means smaller h_r, and h_x. But, we use here the implicit Runge–Kutta method of the second order it means that we need to take 80 000 points as the next step after 40 000 not 50 000. But, because of the limitations in the computer we will consider only 50 000 points. We can say that as a starting point, it's enough to consider 50 000 points. In the future one can try with more points using supercomputers, and also for the 2D case. In other words, the suitable choice of h_r will be 0.000 000 02.

15.9 Results for ideal lenses and the bulk defects influence

15.9.1 Space steps choice and plots

We have obtained the best choice for the space steps h_x, h_l and h_r which corresponds to use of 50 000 points. Our results give focal distance f = 1.247 034 0 m, and W = 3.92 μm at 1.298 m from the lens, see the figure 15.1. We considered ideal lenses without any defects that's the main reason for having smaller W than the experimental data. We considered only the 1-dim case for the horizontal direction while in the experimental data you will see the horizontal, and vertical results at the figure 15.2. Table 15.4 gives us the comparison between our results, and the experimental data. The results show that we have a good agreement with the experimental data for the focal distance, and for the intensity at the focal plane, while in the case of the Gaussian distribution we have a smaller W value. We believe

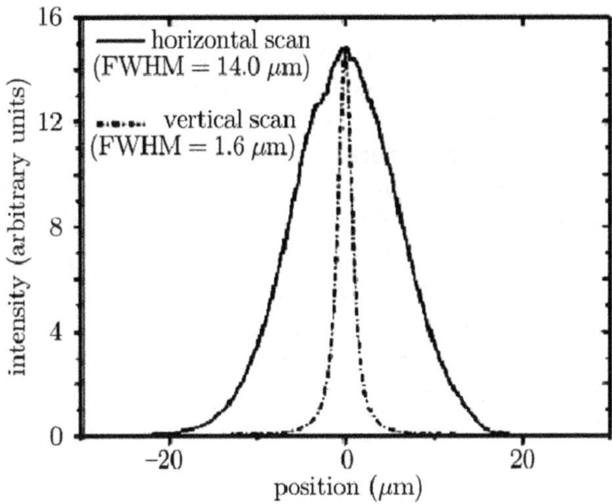

Figure 15.2. Experimental data for the horizontal and vertical direction [5, 6], reproduced from M Elsawy.

Table 15.4. Comparing the results.

Item	Focal distance	W
Our results (Ideal CRLs)	1.247 034 0 m	3.92 μm
Experimental data	1.255 303 6 m	14.0 μm

that it is perhaps because of the ideal lens that we use in simulations, and the fixed frequency of the electromagnetic wave in the modeling.

Remark. For two-dimensional simulation we address the reader to [5].

15.9.2 Final remarks

The FORTRAN program may be used in simulations to compute the focal distance, spot size and cross structure of the x-ray beam. Two cases with the same shape of the lenses as in [6] were considered, having good agreement with the experimental data for the focal distance, and for the intensity at the focal plane while, however, for the W we have smaller results. We believe that the smaller W value for the Gaussian beam appears because we use ideal lenses without any defects. For the presented method, some tools for monitoring accuracy and reliability of calculations are developed, which make possible to obtain results with a given accuracy. The calculation accuracy, that appear due to discretization, is controlled by applying the Runge rule [7] to estimate calculation errors for several values characterizing the focusing.

The computation complexity (number of operations) of the presented method using Gaussian beams is estimated as $M_I \cdot M_F$. Here M_I is the number of points digitizing the electric field after the lenses and M_F is the number of points in which we calculate the electric field at some distance after the lenses. The numbers M_I, M_F may be different; for example, when we calculate the wave picture at a distance equal to a focal length, M_F can be taken much smaller than M_I. After the passage of x-rays through a system of several tens of lenses, the wave field becomes rapidly oscillating, and a rather dense numerical grid is needed for its digitizing. There are no fundamental restrictions on the choice of M_F, only the desired resolution influences the choice of M_F.

Defects in lenses (inclusions, caverns) weakly change locally a refractive index, but the refractive index changes abruptly on inclusions, leading to diffraction effects. As a result of an abrupt change in the refractive index, the corresponding Fourier series may converge slowly; therefore, taking into account the influences of defects in lenses can be a difficult problem for Fourier methods. The method developed in [4, 7, 8], reviewed in this chapter might be useful for solving such difficult research problems.

References

[1] Aristov V V and Shabelnikov L G 2008 Recent advances in x-ray refractive optics *Physics: Uspekhi* **51** 57–77

[2] https://en.wikipedia.org/wiki/Electronvolt

[3] Schurig D and Smith D R 2005 *Negative-refraction Metamaterials: Fundamental Principles and Applications* ed G I Eleftheriades and K G Balmain (New York: Wiley)

[4] Wojda P and Kshevetskii S 2019 Oriented Gaussian beams for high-accuracy computation with accuracy control of x-ray propagation through a multi-lens system *J. Synchrotron Radiat.* **26** 363–72

[5] Elsawy M M R and Leble S 2014 Finite-difference solution of parabolic equation and numerical simulation for x-ray focusing *Task Q.* **18** 101–11

[6] Roth T, Alianelli L, Lengeler D, Snigirev A and Seiboth F 2017 Materials for x-ray refractive lenses minimizing wavefront distortions *MRS Bull.* **42** 430–6

[7] Kshevetskii S, Wojda P and Maximov V 2016 A high-accuracy complex-phase method of simulating x-ray propagation through a multi-lens system *J. Synchrotron Radiat.* **23** 1305–14

[8] Kshevetskii S and Wojda P 2015 Efficient quadrature for the fast oscillating integral of paraxial optics *Math. Appl.* **43** 253–67

[9] Goodman J W 1996 *Introduction to Fourier Optics* 2nd edn (New York: McGraw-Hill)

[10] Kogelnik H 1965 On the propagation of Gaussian beams of light through lenslike media including those with a loss or gain variation *Appl. Opt.* **4** 1562–9

[11] Deschamps G 1972 Ray technics in electromagnetics *Proc. IEEE* **60** 1022–35

[12] Chabory A, Sokoloff J, Bolioli S and Combes P F 2005 Computation of electromagnetic scattering by multilayer dielectric objects using Gaussian beam based techniques *C. R. Phys.* **6** 654–62

[13] Ghannoum I, Letrou C and Beauquet G 2009 A Gaussian beam shooting algorithm for radar propagation simulations *RADAR 2009 Int. Radar Conf. Surveillance for a safer world (Oct 2009) (Bordeaux, France)*

[14] Babich V M and Buldyrev V S 1991 Asimptoticheskie metody v zadachah difrakcii korotkih voln *Short-Wavelength Diffraction Theory. Asymptotic Methods* ed V M Babich and V S Buldyrev (Berlin: Springer)

[15] Levi M 2000 *Parabolic equation Methods for Electromagnetic Wave Propagation* (London: The Institution of Electrical Engineers) p 333

[16] Leontovich M A 1944 A method for solving problems of electromagnetic wave propagation along the surface of the Earth *Izv. AN SSSR, Ser. Phys.* **8** 16

[17] Butcher J C 2008 *Numerical Methods for Ordinary Differential equations* 2nd edn ch 3

[18] Jackson D 1930 *AMS Coll. Publications* vol XI (New York: American Mathematical Society)

[19] Anikin A Yu, Dobrokhotov S Yu, Klevin A I and Tirozzi B 2019 Short-wave asymptotics for Gaussian beams and packets and calarization of equations in plasma physics *Physics* **1** 301–20

IOP Publishing

Practical Electrodynamics with Advanced Applications

Sergey Leble

Chapter 16

Magnetic field dynamics, novel aspects of a theory based on Landau–Lifshitz–Gilbert equations

The electromagnetism has a two limit form, the electrostatics and magnetostatics, which, however, has dynamical generalization that allows to account magnetic field evolution without inclusion of the electric component, if processes under consideration are slow enough. These are such very important problems as domain walls (DW) dynamics or spin waves. The DW motion and its form peculiarities under impact of a matter conditions are studied here. This chapter is written on the basis of the chapter 'Microwaveguides. Magnetic moment transport' from the book [1], that is widened on the base of the new articles [2, 3]. So, the significant part of study will be devoted to DW in microwires [4, 40, 41], as a good didactic platform.

16.1 An exchange interaction concept

In the pioneer papers of Heisenberg [5] it was established that the Weiss model of ferro-magnetism is explained by interaction between electrons, that may be understood on the basis of *exchange interaction* appearing in quantum theory. The first such notion of exchange interaction as a part of mean energy of hydrogen molecule was introduced in the Heitler–London paper [6]. Formally it is considered as the exchange of electrons that move in the unperturbed system with equal energies at different points. The most exact one-particle approximation of such interaction naturally enters the Fock theory [7] and its self-consistent one-particle generalizations that provides better exchange integral evaluation. For the spectrum properties in the framework of the Hartree–Fock approximation, see [8].

The Heisenberg theory, by its own classification [5], takes as the first approximation such a case, that the lattice separations as being very large and assumes that every electron thus belongs to its own atom. In the next approximation, one considers the mentioned exchange of electrons, but ignore its movement along

doi:10.1088/978-0-7503-2576-9ch16

matter. There are a lot of models that either start from free electrons or account its motion combined with ones attached to atoms [9, 10].

The electrons degrees of freedom in the Heisenberg model naturally include spin variables S_i, moreover the space coordinates do not play an important role because of the tight attachment of the particles to atoms. The main idea may be expressed in terms of a 1D chain of atoms, whose electron system interacts by exchange forces, so J_{ik} marks the exchange integral between atoms i, k. Its Hamiltonian is written as

$$\hat{H} = \sum_{i,k} J_{ik}\vec{S}_i\vec{S}_k. \tag{16.1}$$

The minimal version of such a Heisenberg chain account only the interaction between closest neighbors, so only $k = i \pm 1$ terms survive in the sum.

The two-, or three-dimensional Heisenberg network, is indeed an implementation of space geometry and symmetry for atoms position choice. We do not intend in this textbook to study these discrete models [11], one can try to solve the problem, chapter 20, section 20.10, exercise 3. Its abundant applications for a modeling of magnetic phenomena are based on transition to continuum limit that is called the Landau–Lifshits–Gilbert equation. In an approach of energy balance base similar equations may be derived as shown in [13, 14]. It also naturally includes the general thermodynamic consideration, see chapter 18, section 18.1, that apart from pure magnetic variables and the magnetic susceptibilities, may include electric components, that introduces the constants describing the magneto-electrical effect.

The DW dynamics, being a phase transition by nature, in some specific conditions resembles a solitary wave propagation. As it was pointed out, the theory of the domain walls dynamics is based on the Heisenberg model, that is one of the versions of a matter atomic system quantum description.Within some restrictions frame, the particular case of Heisenberg chain are integrable it is widely investigated namely as a true soliton model with many typical features of this intriguing phenomenon [11, 14]. In the one-particle approximation after space and spin variables division and fixed atoms position, one arrives to the so called Heisenberg network model. Its main term in nonrelativistic Hamiltonian contains a sum of closest neighbors spin products. Because of the fundamental link between spin and corresponding magnetic moment, the action of the external magnetic field is described by the Zeeman energy term. The magnetic interaction between electrons may also be taken into account as a spin–spin one.

In the case of magnetic moment vector perturbation with a scale significantly large compared to interatomic distance, a transition to continuous equations may be delivered. We introduce the magnetic moment space density and approximate the terms of the discrete equation obtained by the Taylor expansion. The resulting vector equation figured in literature under the name Landau–Lifschitz (LL) equations [13, 17]. An extra important term is responsible for relaxation. It originates from thermodynamic exposition that first appeared in the Gilbert paper [15], see also [16]. Finally the LL equation completed by such relaxation term, figure out in literature under the name Landau–Lifschitz–Gilbert (LLG) [17]. It is nonlinear vector equation, which versions are intensely studied. There are transformations that change the form of nonlinear terms [16], that allow to specify the DW form more effective.

16.2 Heisenberg network dynamics

16.2.1 Heisenberg model: anisotropy

Consider the simplest Heisenberg model of a matter with account of anisotropy and Zeeman term with external magnetic field $\vec{B}||x$. Its *Hamiltonian* $\mathcal{H}$ for $\overrightarrow{S_n}$ as unitary vectors representing electron spins is written as the sum by nearest neighbors

$$\mathcal{H} = -\sum_n J_n \overrightarrow{S_n}\overrightarrow{S}_{n+1} + \sum_n K_n (S_n^r)^2 - \mu_B B \sum_n g_n S_n^x, \tag{16.2}$$

The exchange integrals J_n represent quantum Coulomb interaction between closest neighbors, approximately described in Heitler–London model or by more advanced Hartree–Fock theory. The second term accounts easy-axis (r) anisotropy. The last term includes. the coefficients g_n, that are gyromagnetic ratios, μ_B Bohr magneton.

The quantum **Heisenberg dynamics** $i\hbar \dot{S}_n^i = [S_n^i, \mathcal{H}]$, via calculation of commutators of the spin operators components 'i = (x,y,z)' with Hamiltonian (16.4), is described by the Cartesian system of ODE as follows. Simplifying the model, we put $g = g_n$, reading the equation components:

$$\begin{aligned}
i\hbar \dot{S}_n^x &= -J_n\left(S_n^y(S_{n+1}^z + S_{n-1}^z) - S_n^z(S_{n+1}^y + S_{n-1}^y)\right) + 2K_n S_n^y S_n^z \\
i\hbar \dot{S}_n^y &= -J_n(S_n^z(S_{n+1}^x + S_{n-1}^x) - S_n^x(S_{n+1}^z + S_{n-1}^z)) - 2K_n S_n^x S_n^z - g\mu_B B S_n^z \\
i\hbar \dot{S}_n^z &= -J_n\left(S_n^x(S_{n+1}^y + S_{n-1}^y) - S_n^y(S_{n+1}^x + S_{n-1}^x)\right) + g\mu_B B S_n^y.
\end{aligned} \tag{16.3}$$

The index of J_n is retaining to the atom 'n', as exchange energy of interaction between 'n' and '$n + 1$' electrons.

In the case of homogeneous isotropic medium it is often used the more simple model $J_n = J$. having in mind the future plans to study amorphous cylindrical microwires we left the only easy-axis anisotropy x-term, proportional to K_n, of stress distribution origin, that would determine a dynamics of DW in a microwire. It is known from basic fundamentals of quantum mechanics, that for any quantum operator $\hat{L}$, its quantum mean value satisfy the relation

$$\frac{d}{dt}\langle \hat{L} \rangle = \frac{1}{i\hbar}\langle [\hat{L}, \hat{H}] \rangle, \tag{16.4}$$

where $\hat{H}$ is evolution operator (Hamiltonian). The rhs in classical limit goes to the Poisson bracket, that algebraically acts as the commutator [17]. Therefore the form of rhs in terms of mean values coincides with one of (16.3). In the next section we go to the mean values of the spin operators $\vec{S}_n$ and the Hamiltonian $\mathcal{H}$. For the sake of compactness we save notations for the mean values as $\vec{S}_n$. We do understand that in problems of such phenomena as DW dynamics the number of electrons involved is very large, hence the average values of such multi-spin operators are the adequate variables.

16.2.2 General continuum LLG equations

For a **transition to continuum**, let us suppose that scales of the DW structure are large compared to lattice constants, micro-, and nano- peculiarities, that allows to go from individual atomic spins to the density of spin vector: $\vec{S}_n \to \vec{S}(\vec{r})$. In such approach, the phenomenon of the DW propagation is modeled via the modified three dimensional continuous version of the Heisenberg net equation (16.3) that is referred to as the already mentioned LLG equation. All rest parameters we also will consider as space fields. Having in mind applications to amorphous materials, we will use a model in which the principal exchange integral field $J_n \to J(\vec{r})$ is supposed to be isotropic, keeping the only nonzero component of the anisotropy coefficient $K_n \to K(\vec{r})$.

To write the 3D version of (16.3) we should go to summation to multi-index $n \to n, \alpha, \alpha = x, y, z, \alpha$ numerate the Cartesian coordinates of the closest neighbors, see figure 16.1. For an isotropic material we take the distance of a closest neighbor in three orthogonal directions as one equal to a. The approximation of a difference $F_{n+1,x} + F_{n-1,x}$ is written as

$$F(x + a, y, z) + F(x - a, y, z) = 2F(x, y, z) + a^2 F_{xx}(x, y, z), \tag{16.5}$$

that is derived by addition of the two-terms Taylor expansions

$$\begin{aligned} F(x + a, y, z) &= F(x, y, z) + aF_x(x, y, z) + \frac{a^2}{2} F_{xx}(x, y, z), \\ F(x - a, y, z) &= F(x, y, z) - aF_x(x, y, z) + \frac{a^2}{2} F_{xx}(x, y, z). \end{aligned} \tag{16.6}$$

Similar expressions are obtained for a shift along y and z. Such expressions from equation (16.3) as $S_{n+1}^z + S_{n-1}^z$ should be replaced by

$$\sum_{\alpha=1}^{3} \left(S_{n+1,\alpha}^z + S_{n-1,\alpha}^z \right), \tag{16.7}$$

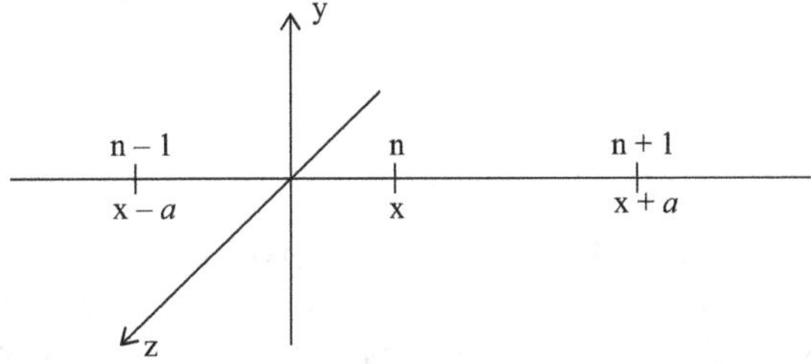

Figure 16.1. On the graph the closest neighbors coordinates are shown.

that goes to

$$\sum_{\alpha=1}^{3}(2S^z(x, y, z) + a^2 S^z_{\alpha\alpha}(x, y, z)) = 6S^z(x, y, z) + a^2\Delta S^z(x, y, z). \tag{16.8}$$

The term $(S_n^y(S_{n+1}^z + S_{n-1}^z) - S_n^z(S_{n+1}^y + S_{n-1}^y))$ yields the x-components of a vector product:

$$a^2(S^y(\vec{r})\Delta S^z(\vec{r}) - S^z(\vec{r})\Delta S^y(\vec{r})). \tag{16.9}$$

We arrive at the system that represents the continuous version of equation (16.3)

$$\begin{aligned} \dot{S}^x &= -a^2 J(\vec{r})(S^y(\vec{r})\Delta S^z(\vec{r}) - S^z(\vec{r})(\Delta S^y(\vec{r})) + 2K(\vec{r})S^y(\vec{r}))S^z(\vec{r})),\\ \dot{S}^y &= -a^2 J(\vec{r})(S^z(\vec{r})\Delta S^x(\vec{r}) - S^x(\vec{r})(\Delta S^z(\vec{r})) - 2K(\vec{r})S^x(\vec{r}))S^z(\vec{r}) - g\mu_B BS^z(\vec{r}),\\ \dot{S}^z &= -a^2 J(\vec{r})(S^x(\vec{r})\Delta S^y(\vec{r}) - S^y(\vec{r})\Delta S^x(\vec{r})) + g\mu_B BS^y(\vec{r}). \end{aligned} \tag{16.10}$$

At the lhs we skip the argument for brevity.

A compact vector form is visible in equation (16.10), having

$$\dot{\vec{S}} = -\frac{a^2}{2}J\vec{S} \times \Delta\vec{S} + 2KS^z\vec{j} - g\mu_B BS^x\vec{i}, \tag{16.11}$$

$\vec{i}, \vec{j}, \vec{k}$ are standard Cartesian unit vectors, under the choice $\vec{B}||\vec{i}$. For brevity the arguments of $\vec{S}$, J, K are omitted. A transition from spin density field $\vec{S}$ to the normalized magnetization vector field $\vec{m}$ assuming that $\vec{S} \to \vec{M} \to \vec{m} = \vec{M}/|\vec{M}_0|$, $B \to H$, a dimensionless form contains the renormalized parameters, e.g. $x' = x/a$, primes below are also omitted. M_0 is the saturation magnetization.

An account of *Gilbert phenomenological damping* [15] adds the term

$$-\alpha\vec{m} \times \frac{\partial\vec{m}}{\partial t}, \tag{16.12}$$

that results in the LLG equation. One can include other types of interactions like spin–spin interactions, spin–phonon coupling. Also equation (16.11) can be generalized to the case of 2D square and cubic 3D lattices (see exercise 3, chapter 20, section 20.10).

16.3 Walker theory

16.3.1 Walker solution of 1D LLG

A simplest version of the LLG equation that consists of equations (16.11) and (16.12) has the class of exact solutions found first by Walker [18], with two anisotropy terms $K_{||}$ and $K_{\perp}$. This theory is presented in a recent paper [25] with standard notations: external magnetic field $H < H_c = 27\alpha M_0$- critical field value, with M_0 as the saturation magnetization, α—Gilbert loss parameter, J—exchange integral.

If we denote the magnetization vector $\vec{M}$ under the angle θ; the angle between z, $\vec{M}$ plane and x-axis is marked as ϕ.

Then, the solution of 1D LLG equation may be found under the assumption $\phi = \phi_w$, independent of x and t. It is also supposed, that the angle θ to be a function only of the characteristic variable $x - vt$, velocity v is constant. The coordinate x is measured in units of $(J/K)^{1/2}$ and v has the dimension t^{-1}.

A special numerical investigation of a Cauchy problem shows that the DW evolution tends smoothly to this explicit solution.

The Walker solution described in [25] in implicit form is given by:

$$\sin(2\phi_w) = H/H_c,$$
$$\tan\theta/2 = \exp\frac{z - vt}{\Delta}, \qquad (16.13)$$

where $H_c = \alpha K_{\perp}/2$ is the Walker *breakdown field*, $H/H_c \leqslant 1$ and $v = H\Delta/\alpha$ is the Walker rigid-body DW speed. The parameter $\Delta = (\frac{K_{||} + K_{\perp}\cos^2(\phi_w)}{J})^{-1/2}$ is the DW width, we, in the next section will use it as length unit. The angle θ as function of $\frac{x - vt}{\Delta}$ is shown in figure 16.2. Note, that as show computer simulations, for the field H that exceeds the critical value H_c the magnetization begins to precess about the field direction and a periodic component appears in the DW motion.

16.3.2 Walker solution instability

An important phenomena in the DW dynamics theory and experiment related to a solutions instability. Such breakdown of a solution physically means the appearance of waves emission. In magnetism these are spin waves [26]. General theory of stability is rather many variants [27], we present here the version that starts from

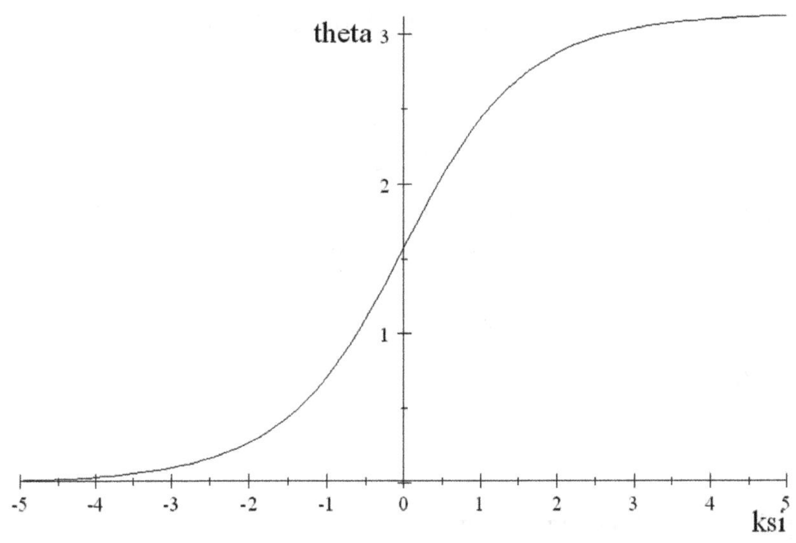

Figure 16.2. Walker solution, by (16.13), $\theta = 2\arctan\xi$, $\xi = \frac{x - vt}{\Delta}$.

linearization of basic equation (LLG in our case) on a solution background. The next step is investigation of obtained linear equation, its evolution operator eigenproblem. The behavior of the deviation from the given solution depends on the eigenvalues position at the complex plane. Along this line, the *analytic study of stability* is performed, by definition, on the base of LLG [24], As prescribed, we linearize the 1D LLG over Walker solution, given by the pair $\theta_W(x, t)$, $\phi_W(x, t)$. We add to these angle variables small variables θ, ϕ, so as $\theta_W + \theta$, $\phi_W + \phi$ and put it into the basic system, keeping the first order terms with respect to the pair θ, ϕ.

We rewrite the LLG equation in matrix form, writing the unknown variables as the vector $\Lambda = (\theta, \phi)^T$ is written as:

$$\Lambda_t = L(\theta, \phi, \partial, \partial^2)\Lambda, \tag{16.14}$$

where $\partial = \frac{\partial}{\partial \xi}$, $\xi = x - vt$. It is the system of partial derivative equation with ξ-dependent coefficients.

We search solutions, going to the new vector Λ_1 by

$$\Lambda = \Lambda_1 \exp[\lambda t], \tag{16.15}$$

for the coefficients of (16.14) does not depend on t.

The vector Λ_1 satisfies the ordinary differential equation

$$L(\theta, \phi, \partial, \partial^2)\Lambda_1 = \lambda\Lambda_1. \tag{16.16}$$

Apart from θ, ϕ we introduce variables $\partial\theta$, $\partial\phi$ in (16.16) it leads to 4×4 first order ODE system

$$\Lambda'_\xi = \Gamma(\lambda)\Lambda' \tag{16.17}$$

for

$$\Lambda' = (\theta, \phi, \partial\theta, \partial\phi)^T \tag{16.18}$$

and

$$\Gamma(\lambda) = \begin{pmatrix} 0 & I \\ B & C \end{pmatrix} \tag{16.19}$$

here I denotes 2×2 identity matrix, B and C are 2×2 matrices, which have matrix elements, expressed in terms of material constants and variable ξ. For example

$$B_{11} = \frac{\alpha\lambda}{2J} + \frac{1}{2J}(K_{\|} - 2K_{\perp} - K_{\|}\sqrt{1 - \rho^2})\cos[2G(\xi)],$$

where $\rho = H/H_c$

$$G(\xi) = \int_0^\xi \frac{dt}{\cosh(t)}, \tag{16.20}$$

is the Gudermanian function, see figure 16.3.

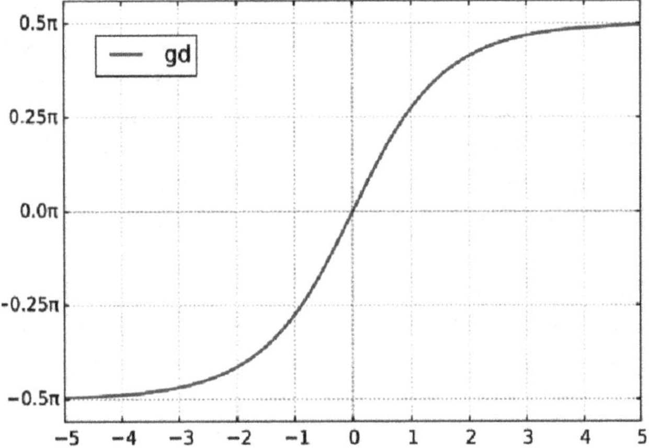

Figure 16.3. Gudermanian function, https://wikimedia.org/api/rest_v1/media/math/render/svg/20d9588835e5d0c4211ce099d3be72fa6538e063.

Other matrix elements also have stable asymptotic behavior, and tend to constant values, see [24], that is the direct corollary of the Walker solution behavior at infinities. Therefore, limits at infinities of the 4×4 matrix Γ are $\Gamma^{\pm}$ that are constant 4×4 matrices, as is visible from example (16.3.2) and the definition (16.20). As it follows from general theory of ODE systems with constant coefficients, the asymptotic solutions of (16.19) are linear combinations of $\Lambda_0 \exp[\kappa\xi]$ with κ being complex numbers. The plane wave solutions with ($\kappa^{\pm} = ik$, $\Im k = 0$) exist only when the pair λ, κ is a solution of

$$det[\Gamma^{\pm}(\lambda) + \kappa] = 0.$$

Each of the two equations has two branches of allowed $\lambda^{\pm}_{1,2}$, known as the Fredholm borders [28]. According to the general theory, the essential spectrum of an operator is restricted by the Fredholm borders.

The paper [24] presents numerical results for the typical example of yttrium iron garnet (YIG), with values of parameters $J = 3.8 \times 10^{-12}$ J m^{-1}, $K_{\|} = 2 \times 10^{3}$ J m^{-3}, $\gamma = 35.1$ kHz (A/m)$^{-1}$, $M_c = 1.94 \times 10^{5}$ A m^{-1}. $\alpha = 0.001$, the parameter $K_{\perp}$ is varying. It may be summarized as

- H = 0, the two branches of the spectrum of $\lambda^{\pm}_{1,2}$ are the same.
- The spectrum encroaches the right half plane, unstable plane waves shall exist and spin-wave emission is expected. A similar conclusion is issued in [20], but for $H > H_c$.
- If $\kappa^{\pm}$ with positive (negative) real part [27], then the essential spectrum of an operator is observed in the regimes with $n^{+}_{-} + n^{-}_{+} \neq 4$, where $n^{\pm}_{+}(n^{\pm}_{-})$ being the number of $\kappa^{\pm}$ with positive (negative) real part.

For λ on boundaries $\lambda^{\pm}_{1,2}$, the associated eigenmode is plane (spin) wave, while for eigenmodes that are not on the boundaries, it represents spin-wave packets.

The group velocity conventionally is determined by $\partial\lambda/\partial k$; thus, for its *negative* values these are stern modes. The *positive* group velocity indicates bow modes.

Thus, Walker propagating DW will always emit stern waves in a low field, and both stern and bow waves in a higher field. The true propagating DW is always dressed with spin waves. The emitted spin waves shall be damped away during their propagation, making them hard to detect in realistic wires. In [18] the phenomenon is studied by numeric simulations for infinite 1D medium, whose magnetization at any point is M, where $|M| = M_c$, is constant. Results are shown on the plot in [18], the curves are plotted for various values of the material parameter $2\pi M_0^2/K$.

Note, yet, the corresponding *Walker DW* propagating speed will deviate from its predicted value, agreeing very well with recent simulations [24], that would be subject to experimental verification.

16.3.3 Nanowires as Heisenberg chain

There are investigations on spin-wave emission delivered directly on the level of discrete model: Heisenberg chain for mesoscopic problems. For example, it is done in the work on field-driven domain wall dynamics [26]. In it, the 1D case is simulated by Heisenberg chain, supplied by Gilbert damping in discretized form. Applied to that a DW in a nanowire can propagate under a longitudinal magnetic field by emitting spin waves (SWs). The authors of [26] numerically investigated the properties of SWs emitted by the DW motion, such as frequency and wave number, and their relation with the DW motion. A case of zero damping $\alpha = 0$ is of some special interest.

In a similar way, domain wall dynamics in magnetic nanotubes of *tubular geometry* is studied numerically in [29]. Suppression of Walker breakdown and Cherenkov-like spin-wave emission is simulated. It is found that DWs in the tubular geometry are much more robust than ones in flat strips studied in [25]. This is explained by topological considerations. Simulations show that a breakdown of the DW can be suppressed over the Walker critical value of magnetic field. The DW velocities above 1000m s^{-1} are achieved by fields under H_c. A different velocity barrier of the DW propagation is observed if it reaches the phase velocity of spin waves. This effect occurs thereby triggering a Cherenkov-like emission of spin waves.

16.4 Propagation of domain wall in cylindrical amorphous ferromagnetic microwire

16.4.1 Introductory remarks

There are many important applications of amorphous ferromagnetic microwires [22, 30, 32] based on phenomenon of magnetization reversal via fast DW propagation along them [21, 23, 31].

To achieve the competitiveness of microwire-based magnetic devices it is necessary to use microwires possessing a high-speed and controllable magnetization reversal process. The magnetization reversal depends on the space micromagnetic structure of microwire's metallic nucleus induced during the manufacturing process [33, 35] by tension or compression [36]. Experimental study of stress distribution is done through indirect methods [42] because the nuclei of microwires are made from amorphous material that does not allow one to recover internal stress distribution, for example, by x-rays scattering. Experimentally the micromagnetic structure is confirmed by results of magnetic properties study: Fe-based glass-covered microwires exhibit magnetic bistablility—perfect rectangular hysteresis loops [42]. According to [35], the magnetic anisotropy coefficient exhibits an essential radial dependence. The authors of the papers [47, 49] assume an elongated cylindrical symmetric 2D-DW, for 3D space distribution of DW magnetization vector [3, 48] based on LLG.

As for an analytical approach the LLG equation is used in this case [50], we focus on the properties of linearized LLG and weakly nonlinear [2] as well as its particular solutions for DW model of full LLG [48]. Having explicit formula for DW velocity that contain the matter parameters, one, for example, can estimate the anisotropy coefficient having the velocity value by experimental data [39, 43].

16.4.2 The LLG equation for cylindric microwires

As it was announced, we study applications of the LLG equation mainly to DW dynamics in cylindric amorphous Fe-rich microwires. It is a good example of the practical electrodynamics because of the isolated DW may be excited and detected by few-turn coils with current [21]. Hence, such *magnetic wave guide* may transfer the DWs along a wire, which dynamics and form may be investigated. Such a wire conventionally has a diameter of micrometer order, the magnetic domain can fill it as a whole, restricted by DWs. The DW may be used as an information unit, i.e. the transmission along the wire to be used in micromagnetic devices [22].

The vector equation (16.11) after account the Gilbert term and described in section 16.2.2 rescaling, reads

$$-\frac{\partial \vec{m}}{\partial t} = J\vec{m} \times \Delta\vec{m} + K(\vec{m} \cdot \vec{k})\vec{m} \times \vec{k} + \gamma\vec{m} \times \vec{H} - \alpha\vec{m} \times \frac{\partial \vec{m}}{\partial t}, \tag{16.21}$$

where $\vec{k}$ is the unit vector along z, $\gamma = \mu g \mu_B$ is the gyromagnetic ratio, proportional to the gyromagnetic factor g, the Bohr magneton μ_B and the magnetic permeability μ. The Gilbert damping parameter $\alpha \ll 1$ is dimensionless, $J > 0$ is the isotropic exchange integral that stands for energy coupling constant, K is the strength of an effective easy-axis anisotropy. The vector $\vec{H} = (0, 0, H)$ represents a longitudinal magnetic field.

16.4.3 LLG transforms: statement of problem

Recall, that the LLG equation (16.21) is nonlinear. A search of optimal form for it means to do transformations that change nonlinearity character, look the overview [17]. Indirect estimations from experiments show that for Fe-based microwires the distribution of magnetization in the metallic nucleus has two areas: an inner core, where the magnetization is axially directed and an outer shell, where magnetization is directed radially. The last observation corresponds the general boundary condition of electrodynamics [10], see also chapter 9. The process of magnetization by means of DW propagation takes place in the inner core and corresponding changes in outer shell should match it, it should be taken into account as boundary conditions.

Consider the inner core of the wire, following [47] and [48] on the basis of the 3D LLG equation (16.21). As in preceding sections we write it in a dimensionless form for the normalized magnetization vector $\vec{m}(\vec{r}, t)$ dependent on the dimensionless position vector $\vec{r}$ in terms of mean space distance between atoms a and dimensionless time t, assuming that the wire axis coincides with the z-axis of the coordinate system, for the parameters description see section 16.4.2.

Let use introduce a complex variable $\omega(\vec{r}, t)$ such that

$$\omega = \frac{m^x + im^y}{1 + m^z}, \qquad \text{or inversely, } m^x + im^y = \frac{2\omega}{1 + \omega\omega^*}, \qquad m^z = \frac{1 - \omega\omega^*}{1 + \omega\omega^*}, \tag{16.22}$$

[17], where the superscript '*' stands for complex conjugation.

As a next steps, introduce a new variable Ω such that

$$\omega = \exp(\Omega). \tag{16.23}$$

Adding the conjugate relation with account of equation (16.22) yields

$$\omega\omega^* = \exp[2\Re\Omega] = \frac{1 - m^z}{1 + m^z}, \tag{16.24}$$

and, subtracting,

$$\exp[2\Im\Omega] = \frac{m^x + im^y}{m^x - im^y}. \tag{16.25}$$

It divides the components of the vector $\vec{m}$ for complex $\Omega = \Re\Omega + i\Im\Omega$.

Then equation (16.21) takes the partially linearized form with respect to Ω

$$(i + \alpha)\frac{\partial\Omega}{\partial z} + J\nabla^2\Omega + (K - J(\nabla\Omega)^2)\tanh\left(\frac{\Omega + \Omega^*}{2}\right) = \gamma H, \tag{16.26}$$

that opens a way to built particular solutions.

The equation (16.26) have inhomogeneity term γH, that we have moved to the rhs. The cylindrical symmetry implies the boundary conditions at $r = 0$ for the magnetization vector $\vec{m}$, that can be transferred to the complex variable Ω, having in mind (16.22) and (16.23). In cylindric coordinates, the components read

$$\begin{aligned} m_z &= m_z, \\ m_\rho &= m_x \cos\phi + m_y \sin\phi, \\ m_\phi &= m_x \sin\phi + m_y \cos\phi. \end{aligned} \tag{16.27}$$

In outer shell in the boundary vicinity the vector $\vec{m}$ should be orthogonal to the surface, hence

$$\begin{aligned} m_z &= 0, \\ m_\phi &= 0. \end{aligned} \tag{16.28}$$

A statement of problem for (16.26) should include either an initial condition (Cauchy problem), or a boundary regime at an end of the wire.

16.4.4 Basic equation in quadratic-linear approximation and its general solution

The basic equation (16.26) for the complex variable Ω is nonlinear. Let us rewrite it in quadratic approximation for nonlinear terms, keeping the only term in the $\tanh(\Re\Omega)$ Taylor expansion:

$$\tanh(\Re\Omega) = \Re\Omega - \frac{(2\Re\Omega)^3}{12} + \cdots$$

Then the equation (16.26) takes linear form

$$(\alpha + i)\Omega_t + J\nabla^2\Omega + K\Re\Omega = L\Omega = \gamma H. \tag{16.29}$$

A validity of finite number of terms approximation restricts values of $\Re\Omega$ by a condition $\Re\Omega \leqslant 1$, that restricts $m_z \geqslant 0$, see the relation (16.24). The form of expression of magnetic moment components m_i via Ω is such that even the bigger values of it do not change much the solution form, because of behavior of $\tanh(\Re\Omega)$ and $\frac{1}{\cosh(\Re\Omega)}$ functions at large arguments range. So, we have the complete linearization (16.29), that is the inhomogeneous equation with the magnetic field as external source.

Introducing the notations for the real and imaginary parts of the variable Ω as

$$\Omega = \Re\Omega + i\Im\Omega, \tag{16.30}$$

yields the system of dependent equations:

$$\alpha\Re\Omega_t + J\nabla^2\Re\Omega + K\Re\Omega = \gamma H + \Im\Omega_t, \tag{16.31}$$

$$\Re\Omega_t + \alpha\Im\Omega_t + J\nabla^2\Im\Omega = 0. \tag{16.32}$$

Both equations are of parabolic type, as heat (diffusion) equation. There is a general theory of Cauchy problem solution of such equations [51] with obvious particular solution of inhomogeneous equation (16.31).

$$\Re\Omega_p = \frac{\gamma}{K}H. \tag{16.33}$$

The general solution is the sum

$$\Re\Omega = \Re\Omega_h + \Re\Omega_p = \Re\Omega_h + \frac{\gamma}{K}H. \tag{16.34}$$

with

$$\alpha\Re\Omega_{ht} + J\nabla^2\Re\Omega_h + K\Re\Omega_h = \Im\Omega_{ht}. \tag{16.35}$$

Boundary conditions in terms of $\Re\Omega$ are found from equations (16.28) and (16.24), that gives at the wire surface $\rho = \rho_0$

$$\Re\Omega = 0. \tag{16.36}$$

Similarly one derives conditions for $\Im\Omega$, see exercise 5, in chapter 20, section 20.10.

The components of the magnetization vector are evaluated from

$$\begin{aligned} m^x &= \frac{\exp(\Omega) + \exp(\Omega^*).}{1 + \exp[2\Re\Omega]} = \frac{\cos[\Im\Omega]}{\cosh[\Re\Omega]}, \\ m^y &= -i\frac{\exp(\Omega) - \exp(\Omega^*).}{1 + \exp[2\Re\Omega]} = -\frac{\sin[\Im\Omega]}{\cosh[\Re\Omega]}, \\ m^z &= \frac{1 - \exp[2\Re\Omega]}{1 + \exp[2\Re\Omega]} = \tanh[\Re\Omega], \end{aligned} \tag{16.37}$$

that define the space structure of a DW. Try exercise 5, in chapter 20, section 20.10.

The system of equations (16.35) and (16.32) gives an approximate general solution, while the system of equations from section 16.6 gives an exact, but particular one.

16.5 Average magnetization fields and DW dynamics

16.5.1 The third order nonlinearity account

We left now both terms in the tanh Taylor expansion, having:

$$(\alpha + i)\Omega_t + J\nabla^2\Omega + (K - J(\nabla\Omega)^2)\left(\Re\Omega - \frac{(\Re\Omega)^3}{3}\right) = \gamma H. \tag{16.38}$$

To explain the main idea of the DW propagation treatment we take the simplified version of the theory similar to [3, 48]. The structure of the equation (16.29) is such that a linear in z term

$$\Omega^z = a + bz \tag{16.39}$$

does not contribute in $(\alpha + i)\Omega_t + J\nabla^2\Omega$. This term, however, satisfies the equation

$$K - J(\nabla\Omega^z)^2 = 0, \tag{16.40}$$

if the coefficient b is chosen as

$$K - Jb^2 = 0, \tag{16.41}$$

or, in complex notations

$$b = b_1 + ib_2, \tag{16.42}$$

next, plugging into equation (16.40) yields

$$K - J(b_1^2 + 2ib_1b_2 - b_2^2) = 0, \tag{16.43}$$

that gives the system

$$
\begin{aligned}
-2Jb_1b_2 &= 0, \\
K - J(b_1^2 - b_2^2) &= 0,
\end{aligned}
\tag{16.44}
$$

which solution yields

$$
\begin{aligned}
b_2 &= 0, \\
K - Jb_1^2 &= 0.
\end{aligned}
\tag{16.45}
$$

The quadratic equation (16.45) solution yields

$$
b_1 = \pm\sqrt{\frac{K}{J}}. \tag{16.46}
$$

In the last part of the section we omit the term Ω^z defined by equation (16.39) in Ω. We, simplifying, also the use of the linear approximation, having in mind that basic nonlinearities are taken into account by transformations (16.22) and (16.23). In [2] we have shown that the contribution to expression for acceleration for the inhomogeneity distribution along z vanishes if only the second order nonlinearity (whose presence is due to the Dzialoshinski–Morija term account) is taken into account.

16.5.2 Stationary background introduction

The process of magnetic field switching that forces the evolution of magnetic moment field proceeds inhomogeneously along z, that is visible in the figure of [39]. It is explained by the presence of inhomogeneities of real matter, e.g. defects, the coefficients of the basic equations depend on coordinates as well as the exchange and other matter parameters [35, 40, 41]. A mathematical description of this transition as non-stationary process is extremely complicated in 3D, therefore we model the DW propagation as a stationary process, but account the named inhomogeneity ones for coefficients of the LLG equation. Let the inhomogeneity be a stationary field that appears as a result of the magnetic nucleation on defects, described by a term Ω_0, hence the resulting field would be a superposition of both fields, stationary and one for a DW.

Split now $\Omega = \Omega_0 + \Omega'$, separating a stationary part that we will take as a background for a DW propagation. Such background magnetization may appear if a wire contains a significant number of defects [2, 44, 45].

Then, for the stationary Ω_0 part we write

$$
\mathcal{L}_0(\Omega_0) = J\nabla^2\Omega_0 + K\Omega_0 - K\frac{\Omega_0^3}{3} - J(\nabla\Omega_0)^2\Omega_0 = 0. \tag{16.47}
$$

16.5.3 A linearization on a nonzero background

Then, for non-stationary part, we left terms linear in Ω'

$$\begin{aligned}\mathcal{L}(\Omega') = (\alpha + i)\Omega_t' + J\nabla^2\Omega' + K\mathfrak{R}\Omega' \\ - K\Omega_0{}^2\mathfrak{R}\Omega' - J\mathfrak{R}\Omega_0\nabla\Omega_0 \cdot \nabla\Omega' - J(\nabla\Omega^0)^2\mathfrak{R}\Omega' = \gamma H.\end{aligned} \tag{16.48}$$

This equation, after averaging we shall use for a DW mobility study.

16.5.4 Averaging procedure and DW mobility

For the effective inhomogeneities account, we would apply the averaging procedure considering the DW as a large-scale phenomenon compared to the ones attained to inhomogeneities. The averaging acts the variable Ω_0 as

$$\Omega_0 \rightarrow \bar{\Omega}^0 = \frac{1}{l}\int_0^l \Omega^0 dz, \tag{16.49}$$

where l is typical scale of Ω^0 oscillations spectrum along z. For the further analysis, after averaging and the subdivision $\Omega' = R + iS$ splits the equation (16.48) in two

$$\alpha R_t + [K(1 - \overline{\Omega_0{}^2}) - J\overline{(\nabla\Omega^0)^2}]R - J\overline{\Omega^0\nabla\Omega^0} \cdot \nabla R - J(\nabla\Omega^0)^2 R = \gamma H + S_t, \tag{16.50}$$

$$R_t + \alpha S_t + J\nabla^2 S - J\overline{\Omega^0\nabla\Omega^0} \cdot \nabla S = 0. \tag{16.51}$$

To built a general solution of the system (16.50) and (16.51) we first solve equation (16.51), then plug a solution into the rhs of (16.50).

Generally, such a system is coupled, the variable R enters in equation (16.51). There is an important case of averaged-homogeneous Ω^0, more exactly

$$\overline{\Omega^0\nabla\Omega^0} = 0, \tag{16.52}$$

that yields the following simplifications

$$\alpha R_t + [(1 - \overline{\Omega_0{}^2})K - J\overline{(\nabla\Omega^0)^2}]R + J\nabla^2 R = \gamma H + S_t, \tag{16.53}$$

and

$$R_t + \alpha S_t + J\nabla^2 S = 0. \tag{16.54}$$

The equation (16.53) is inhomogeneous, so by a general theorem its solution is a sum of particular solution of it R_0 and a general solution of the homogeneous equation r. Suppose that for the particular solution the function S_0 depends only on t and linear on z, we have

$$S_{0t} = -R_{0t}/\alpha, \tag{16.55}$$

then

$$(\alpha^2 + 1)R_{0t} + \alpha\left[\left(1 - \frac{1}{2}\overline{\Omega_0^2}\right)K - J\overline{(\nabla\Omega^0)^2}\right]R_0 + \alpha J\nabla^2 R_0 = \alpha\gamma H. \tag{16.56}$$

Next, looking for a particular solution, having in mind a DW propagation, choose R_0 as a linear function of z. The term $\frac{\partial^2 R_0}{\partial z^2}$ gives zero. The form of Laplace operator in cylindrical coordinates admits the following choice of equation (16.56) for R_0

$$R_{0t} + \frac{\alpha}{(\alpha^2+1)}\left[\left(1 - \frac{1}{2}\overline{\Omega_0^2}\right)K - J\,\overline{(\nabla\Omega^0)^2}\right]R_0 = \frac{\alpha\gamma}{(\alpha^2+1)}H. \tag{16.57}$$

The equation has the following solution. Denote

$$\frac{\alpha}{(\alpha^2+1)}\left[\left(1 - \frac{1}{2}\overline{\Omega_0^2}\right)K - J\,\overline{(\nabla\Omega^0)^2}\right] = \kappa. \tag{16.58}$$

After simple transformations for the zero initial condition choice leads to

$$R_0 = H\frac{\alpha\gamma}{\kappa(\alpha^2+1)}(1 - e^{-\kappa t}) + Wz/2, \tag{16.59}$$

where W is defined by the initial condition. It is equal the width of the DW. For a small parameter κ the two first terms of expansion give

$$R_0 = H\frac{\alpha\gamma}{(\alpha^2+1)}(-t + \kappa t^2 e^{-\kappa t}) + Wz. \tag{16.60}$$

We skip some details of calculations (see [2]). The addition of Ω^z from (16.39) gives a different solution with the parameter κ, that may not depend on K.

16.5.5 Velocity and acceleration

Let us rewrite the expression of equation (16.60) as

$$R_0 = W\left(z - \frac{\alpha\gamma H}{W(\alpha^2+1)}(t - \kappa t^2/2)\right), \tag{16.61}$$

the velocity of DW propagation is then

$$v_{dw} = \frac{\alpha\gamma}{W(\alpha^2+1)}H, \tag{16.62}$$

that exhibits the linear dependence on the field H, observed in experiments with homogeneous wires [21, 37–39]. If we look at the quadratic term in t, the acceleration is extracted as

$$a_{dw} = \frac{\alpha\gamma H}{2W(\alpha^2+1)}\kappa = v_{dw}\kappa. \tag{16.63}$$

An estimation of the velocity and acceleration is suggested as exercise 7 in chapter 20, section 20.10.

16.6 Exact particular solutions of LLG equation

The transformed LLG equation (16.26) consists of two parts—linear and nonlinear one. It opens a possibility to impose a condition such that let both parts of the equation vanish simultaneously. On the way of an exact particular solution search it is possible to put more strong conditions [48]:

$$
\begin{aligned}
(i + \alpha)\frac{\partial \Omega}{\partial t} &= \gamma H, \\
\nabla^2 \Omega &= 0, \\
K(\vec{r}) - J(\nabla \Omega)^2 &= 0.
\end{aligned}
\tag{16.64}
$$

It is important to note, that the equations of the system (16.64) are compatible. We also admit a dependence of anisotropy coefficient on coordinates.

We continue to work with the simplest case of constant field H. The general solution for such a case of equation (16.64) has the form

$$
\Omega(\vec{r}, t) = \Sigma(\vec{r}) - \frac{i\gamma}{1 - i\alpha} Ht. \tag{16.65}
$$

This substitution leads to stationary equations for $\Sigma(\vec{r})$ of the same form:

$$
\begin{cases}
\nabla^2 \Sigma = 0, \\
K(\vec{r}) - J(\nabla \Sigma)^2 = 0.
\end{cases}
\tag{16.66}
$$

The second term of the relation (16.65) originated from the rhs of the first of equation (16.64) fix the linear time dependence of the solution. Note, that Ω depends linearly on the applied magnetic field H. Let us note that the magnetization vector $\vec{m}$ components, as a function of Σ can be expressed by means of relations (16.37).

16.6.1 Anisotropy coefficient coordinate dependence impact

As the next step, let us analyze the system (16.66), within an assumption about anisotropy coefficient coordinate dependence. The results of a wire cooling problem solution [34] allow to suppose that K linearly depends only on the radial variable ρ;

$$
K = K_0 + K_1\rho, \tag{16.67}
$$

where K_0 and K_1 are constants.

The problem, taken from equation (16.66), we solve next, is the Laplace equation

$$
\nabla^2 \Sigma = 0, \tag{16.68}
$$

that in a view of the whole cylindric wire problem, is written in cylindrical coordinates ρ, φ, z. As it is seen from the method of variables separation, its general solution is a superposition of terms with arbitrary coefficients. These coefficients choice let to satisfy the last equation of the system (16.66):

$$
K_0 + K_1\rho - J(\nabla \Sigma)^2 = 0. \tag{16.69}
$$

The compatible solution of equations (16.68) and (16.69) contains two terms of the Laplace equation solution expansion, the first is $\varphi-$ independent Ω^z, see equation (16.39), in new notations [48] it reads as:

$$\Sigma = s + \beta z + \rho^\nu[A\cos(\nu\varphi) + B\sin(\nu\varphi)]. \tag{16.70}$$

where s is the phase of the solution, it is chosen as zero, the rest are complex: $\beta = \beta' + i\beta''$, $A = A' + iA''$, $B = B' + iB''$. The primed ones are real parts. The general case $K_1 \neq 0$, gives the condition $\nu = 3/2$, while $K_1 = 0$ corresponds to $\nu = 1$ that determines the planar DW solution, found in [47], and one, for $\nu = 3/2$ determines flexural planar DW from the Vereshchagin paper [48].

16.6.2 The illustrations of DW form in 3D

As was already mentioned, the magnetization vector $\vec{m}$ may be restored by (16.37), plugging in it (16.65) and (16.70). The DW form may be determined by the condition $m_z = 0$, that may be read as the equation of a surface that divides areas with opposite directions of magnetic moments. Constants, β, A and B are not arbitrary. The part of the relations between them is obtained by substitution of equation (16.70) into equation (16.66).

The authors of [47, 48] in a sense mimic boundary conditions, because to pose the correct ones (16.28) and (16.36) is impossible for the particular solutions under consideration. Following [47] we assume the magnetization is radially directed at few points $\varphi = 0, \pi/2, \pi, -\pi/2$, having in mind that transition from inner to outer core would correct the conditions. This condition fix the values of the parameters: $A'' = 0$, $B'' = \pi/(2R^\nu)$, $B' = 0$, where R is the radius of inner core of the wire $d = 2R$. In addition for the flexural planar DW it is required that the surface $m^z = -\tanh(\Re\Omega) = 0$, which defines the shape of DW is a connected one. This requirement guarantee the uniqueness of the DW form which is similar to the planar DW case. This choice of parameters simplifies the equation $\Re\Omega = 0$, as in equation (16.36), so that from equation (16.70) we write

$$z + c\rho^\nu\cos(\nu\varphi) = 0, \tag{16.71}$$

where $c = A'/\beta'$.

This leads to the following relation between the solution coefficients and material parameters for the planar DW

$$(A')^2 + (\beta')^2 = \frac{K}{J} + \frac{\pi^2}{4R^2}, \quad \beta'' = 0. \tag{16.72}$$

and for the flexural planar DW

$$K_1 = \frac{9}{4}J\left((A')^2 - \frac{\pi^2}{4R^3}\right), \quad (\beta')^2 = \frac{K_0}{J}, \quad \beta'' = 0. \tag{16.73}$$

Explicitly the surface $\Sigma = 0$ (16.71) may be pictured as the 3D plot of the function $z(\rho, \varphi)$

$$\frac{z}{c} = \rho^\nu\cos(\nu\varphi). \tag{16.74}$$

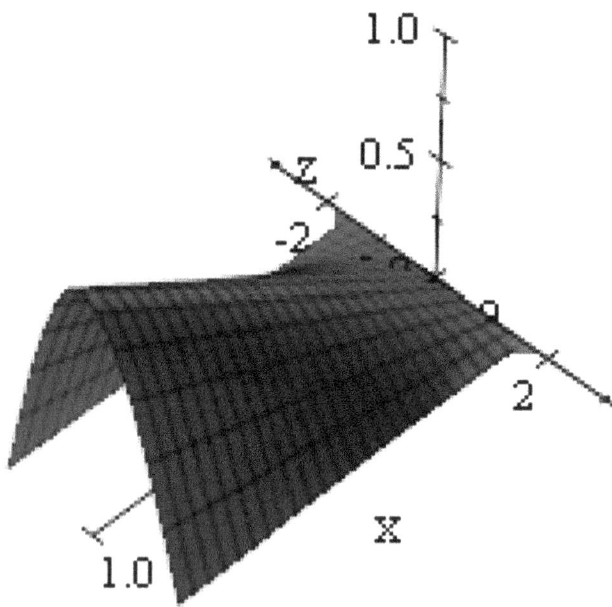

Figure 16.4. $\nu = 1$. The planar DW, z-axis of the wire is vertical.

Those surfaces are presented in figure 16.4 for $\nu = 1$ and for $\nu = 3/2$, in figure 16.5, respectively.

The 3D plots (figures 16.4 and 16.5) represent the surface where the magnetization vector is directed perpendicular to the axis of the wire. It shows the shape of DWs. The surfaces $z_\nu(\rho, \varphi)/c$ are shown in such geometry, that z-axis of the wire is vertical, the choice of units is such that $\frac{z}{c} \in [-1, 1]$, while $\rho \in [0, 1]$ directed left and $\varphi \in [0, 2\pi]$ directed right.

A simple analysis shows that in the first ($\nu = 1$) case the surface representing DW form is planar, it is ellipse with semi-minor axis 1 and semi-major axis $\sqrt{2}$ in dimensional (wire radius) units. Let us add, that for $\nu = 1, \varphi = 0$, $\frac{dz}{d\rho} = -c$, $\frac{\beta'}{A'} = \tan\theta = \frac{d}{L}$, then, the length of the DW is $L = \frac{A'}{\beta'}d$. This parameter, marked in figure 16.6 may be evaluated from measurements of a Faraday induction pulse length, as in [46]. In the second ($\nu = 3/2$) case the surface is more complicated, that is visible in figure 16.5. The curve at surface $\rho = 1$ returns to its starting value $\frac{z}{c} = 1$ at the angle $\varphi = \frac{4\pi}{3}$. Alternative pictures of the DW form are presented in [3]

16.6.3 Velocity of DW propagation: anisotropy constants determination

The real part of Ω as it follows from equation (16.37) determines the propagation of the DW. Its argument contains a linear combination of variables z and t, hence the velocity of propagation follows from explicit expression (16.65) after isolation the term $z - Vt$ we can write

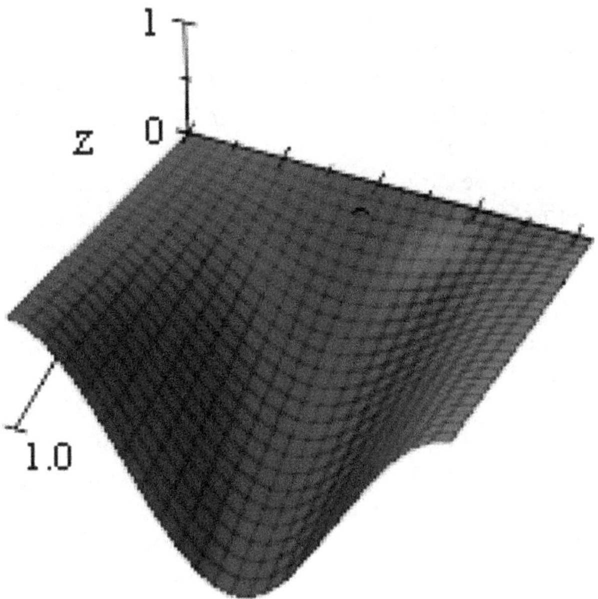

Figure 16.5. $\nu = 3/2$. The flexural DW, z-axis of the wire is vertical.

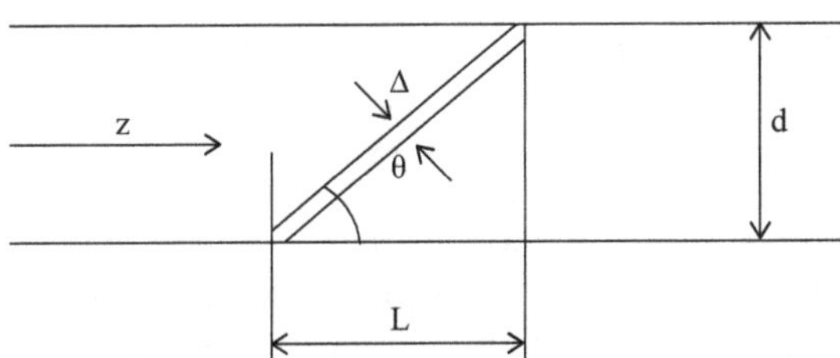

Figure 16.6. The geometry of the DW position in the wire, z-axis of the wire is horizontal, d-diameter of the wire ferromagnetic core, θ-angle between the DW and the wire axis.

$$V = SH, \quad S = -\frac{\gamma\alpha}{1+\alpha^2}\frac{1}{\beta'},$$
$$= \begin{cases} -\dfrac{\gamma\alpha}{1+\alpha^2} \Big/ \sqrt{\dfrac{K}{J} + \dfrac{\pi^2}{4R^2} - (A')^2}, & \text{if } K = K_0, \\ -\dfrac{\gamma\alpha}{1+\alpha^2} \Big/ \sqrt{\dfrac{K_0}{J}}, & \text{if } K = K_0 + \rho K_1, \end{cases} \tag{16.75}$$

where $S = dV/dH^z$ is named the mobility of DW. Note that these expressions are very similar to celebrated Walker's formula, however, an essential difference is hidden in β' parameter. According to equation (16.73) for flexural planar case the relation (16.75) becomes exactly the Walker's formula, see section 16.3.1. For the planar case to transform the expression (16.75) into the Walker's formula in addition one should put the parameter A' to be exactly $\pi/2R$. In the general case equation (16.75) modifies Walker's formula. Let us also point out a more significant difference in the statement of problem: Walker's formula has been deduced for the 1D case with strong condition for the vector $\vec{m}$ direction, whereas we deal with the three dimensional problem [19, 24].

Going to experiment, we suppose that a real microwire has the radial dependence of anisotropy coefficient as (16.67), i.e. the DW is of the case, $K_1 \neq 0$. Having experimental data, see figure 16.6 against the dependence shown in equation (16.75) and the relations (16.73) we would extract the parameter $\beta'/A' = c$ from it, having A' directly from the first equality from equation (16.73) and β' directly from the second one. In both cases we consider the parameters J, D, $K_{0,\,1}$ as prescribed by the DW form and the wire matter.

After the procedure of restoration through equations (16.37), (16.65) and (16.70) m^z is a tanh function with linear argument with respect to z, the DW width Δ then is an inverse multiplier of z, i.e. $\Delta = 1/\beta'$. This parameter can be also extracted from experiment likely one in [49], where the width is found via duration of a pulse excited in the measuring pick-up coil. As for the influence on inclination of the DW principle plane, according to equation (16.73) it depends on both K_0 and K_1 and also on J. The larger K_1 is, the more upright the DW becomes. In contrast, the larger K_0 is, the more inclined the DW becomes.

Note yet, that experimental curves, as shown, e.g. in [21] start from a values of H which correlate with the demagnetization fields. In more advanced theory it should be account in the expressions of the theoretical formula (16.75).

In experiments made to verify the theoretical results there have been analyzed two versions of the same wires, namely with glass coating and when the glass is removed [43]. From a theoretical point of view the removal of the glass essentially decreases stresses in the metallic nucleus. In turn, the anisotropy coefficient is proportional to applied stress and, hence, decreases as well. Then according to equation (16.73) β' also decreases and from equation (16.75) it follows that the DW propagation velocity becomes larger. This behavior is confirmed by experimental measurement of the DW velocity [3].

16.6.4 The field strength and induced magnetization by a coil

A good exercise of magnetostatic theory, chapter 9, section 9.1, application relates directly to the experiments with DW in microwires, for a few-turn coils and solenoids are used to induce the magnetic field inside the wires core. The textbook formula for the field inside the isolated one-turn coil plane [52] of radius R and the current I is given by

$$H_z = \frac{\mu I}{2\pi} \frac{1}{R + r}\left(K + \frac{R + r}{R - r}E\right), \tag{16.76}$$

for $r > 0$, where $K(k)$, $E(k)$ are elliptic integrals of first and second kind, $k^2 = \frac{4Rr \sin\theta}{(R + r)^2 + 2Rr \sin\theta}$, the plane xy is at $\theta = \frac{\pi}{2}$.

In the vicinity of $r = R$ the following (more general) expression [52]

$$H_z = \frac{\mu I}{2\pi} \frac{1}{\sqrt{(R + r)^2 + z^2}} \left(K + \frac{R^2 - r^2 - z^2}{(R - r)^2 + z^2} E \right), \tag{16.77}$$

has finite limit

$$H_z(R) = \frac{\mu I}{2\pi} \frac{1}{2R} (K - E). \tag{16.78}$$

The field at $r = 0$, $z = 0$ tends to the value:

$$\begin{aligned} H_z(0) &= \lim_{r=0, z=0} \frac{\mu I}{2\pi} \frac{1}{\sqrt{(R + r)^2 + z^2}} \left(K + \frac{R^2 - r^2 - z^2}{(R - r)^2 + z^2} E \right) \\ &= \frac{\mu I}{2\pi R} (K + E). \end{aligned} \tag{16.79}$$

The ratio of the fields for the elliptic integral module $k = \sqrt{\frac{4RR}{(R+R)^2 + 2RR}} = \sqrt{\frac{2}{3}}$ at $r = R$ is evaluated as

$$\frac{H_z(0)}{H_z(R)} = \frac{2(K(0) + E(0))}{K(0.816) - E(0.816)}, \tag{16.80}$$

where

$K(0) = \frac{\pi}{2}$, $E(0) = \frac{\pi}{2}$,

$K(\frac{2}{3}) = 2.03$, $E\left(\frac{2}{3}\right) = 1.26$,

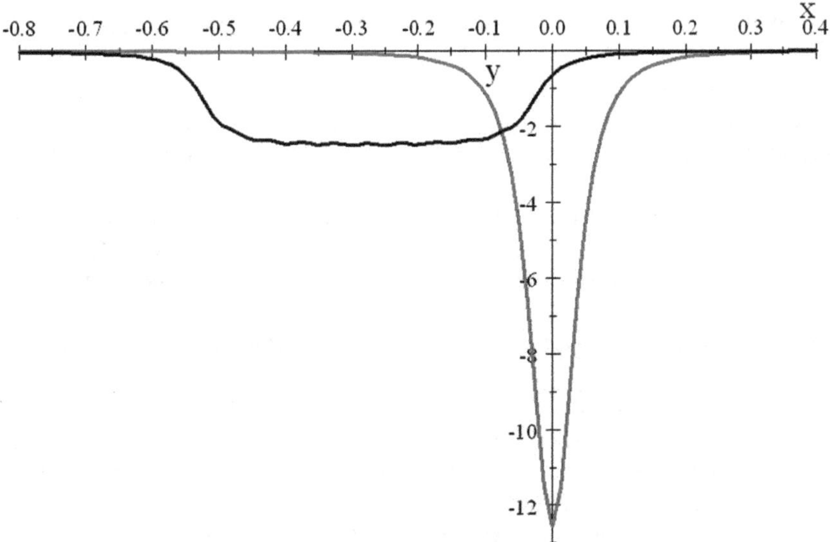

Figure 16.7. On the graph, in red, is the magnetic field component H_z in the center of a single turn with a current of 0.1. The black one—a field of 10 turns with current of 0.01 on the axis of the wire.

that gives $\frac{2\pi}{2.03-1.26} = 8.16$. See, for example, keisan.casio.com https://keisan.casio.com/exec/system/1180573454. A coil of N turns has inhomogeneous field as it follows from equation (16.79).

$$H_z^N(0) = -\sum_{n=0}^{n=N} \frac{\mu I}{8} \frac{\pi}{\sqrt{R^2 + (z+nd)^2}} \left(1 + \frac{R^2 - (z+nd)^2}{R^2 + (z+nd)^2}\right). \tag{16.81}$$

Hence, doing calculations on base of experiment, some correction factor should be taken into account, see the figure 16.7. The graph is built by the formula (16.81) for $N = 1, 10$; $R = 0.05$; $d = 0.05$. It gives the scaling coefficient about 2.

References

[1] Leble S 2019 *Waveguide Propagation of Nonlinear Waves Impact of Inhomogeneity and Accompanying Effects* (Berlin: Springer)

[2] Leble S B and Rodionova V V 2020 Dynamics of domain walls in a cylindrical amorphous ferromagnetic microwire with magnetic inhomogeneities *Theor. Math. Phys.* **202** 252–64

[3] Vereshchagin M, Baraban I, Leble S and Rodionova V 2020 Structure of head-to-head domain wall in cylindrical amorphous ferromagnetic microwire and a method of anisotropy coefficient estimation *J. Magn. Magn. Mater.* **504** 166646

[4] Rodionova V, Ilyn M, Ipatov M, Zhukova V, Perov N and Zhukov A 2012 Spectral properties of electromotive force induced by periodic magnetization reversal of arrays of coupled magnetic glass-covered microwires *J. Appl. Phys.* **111** 07E735

[5] Heisenberg W 1928 Zur Theorie des Ferromagnetismus *Z. Phys.* **49** 619–36
Heisenberg W 1928 *Zur Quantentheorie des Ferromagnetismus. Probleme der Modernen Physik, A. Sommerfeld Festschrift* (Leipzig: Hirzel) pp 114–22
Heisenberg W 1931 Zur Theorie des Magnetpstriktin und der Magnetisierungkurve *Z. Phys.* **69** 287–97

[6] Heitler W and London F 1927 *Z. Phys.* **44** 455–72

[7] Fock V 1930 An approximate method for solving the quantum many-body problem (Reported at the session of the Russian Phys.-Chem. Soc. on 17 December 1929) *Z. Phys.* **61** 126 (TOI 5, N 51, 1, 1931, UFN 93, N 2, 342, 1967)

[8] Popov I Y and Melikhov I F 2014 The discrete spectrum of the multiparticle Hamiltonian in the framework of the Hartree–Fock approximation *J. Phys.: Conf. Ser.* **541** 012099-1–4

[9] Hubbard J 1963 Electron correlations in narrow energy bands *Proc. R. Soc. Lond.* **276** 238–57

[10] Jiles D C 1998 *Introduction to Magnetism and Magnetic Materials* 2nd edn (London: Taylor and Francis)

[11] Porsezian K 1993 On the discrete and continuum integrable Heisenberg spin chain models *Future Directions of Nonlinear Dynamics in Physical and Biological Systems. NATO ASI Series (Series B: Physics)* vol 312 ed P L Christiansen, J C Eilbeck and R D Parmentier (Boston, MA: Springer) pp 243–8

[12] Heremans J, Thrush C M, Lin Y-M, Cronin S, Zhang Z and Dresselhaus M S

[13] Landau L D and Lifshitz E 1935 On the theory of the dispersion of magnetic permeability in ferromagnetic bodies *Phys. Z. Sowjet.* **8** 153–69

[14] Guo B 2008 Landau–Lifshitz equations *Frontiers of Research with the Chinese Academy of Sciences: Volume 1* (Singapore: World Scientific)
[15] Gilbert T L 2004 A phenomenological theory of damping in ferromagnetic materials *IEEE Trans. Magn.* **40** 3443–9
[16] Lakshmanan M and Nakamura K 1984 Landau–Lifshitz equation of ferromag- netism: Exact treatment of the Gilbert damping *Phys. Rev. Lett.* **53** 2497–9
[17] Lakshmanan M 2011 The fascinating world of the Landau–Lifshitz–Gilbert equation: an overview *Philos. Trans. R. Soc.* A **369** 1280–300
[18] Walker I R 1956 *Bell Telephone Laboratories Memorandum* unpublished. An account of this work is to be found in Dillon J F Jr Magnetism
[19] Schryer N L and Walker L R 1974 The motion of 180° domain walls in uniform dc magnetic fields *J. Appl. Phys.* **45** 5406
[20] Bouzidi D and Suhl H 1990 *Phys. Rev. Lett.* **65** 2587
[21] Zhukov A and Zhukova V 2009 *Magnetic Properties and Applications of Ferromagnetic Microwires with Amorphous and Nanocrystalline Structure. Nanotechnology Science and Technology Series* (New York: Nova Science)
[22] Vázquez M 2015 *Magnetic Nano- and Microwires: Design, Synthesis, Properties and Applications. Woodhead Publishing Series in Electronic and Optical Materials* (Amsterdam: Elsevier)
[23] Alam J, Bran C, Chiriac H, Lupuc N, Óvári T A, Panina L V, Rodionova V, Varga R, Vazquez M and Zhukov A 2020 Cylindrical micro and nanowires: fabrication, properties and applications *J. Magn. Magn. Mater.* **513** 167074
[24] Wang X S *et al* 2012 *Phys. Rev. Lett.* **109** 167209
[25] Hu B and Wang X R 2013 Instability of Walker propagating domain wall in magnetic nanowires *Phys. Rev. Lett.* **111** 027205
[26] Wang X S and Wang X R 2014 *Phys. Rev.* B **90** 184415
[27] Kato T 1980 *Perturbation Theory for Linear Operators* (Berlin: Springer)
[28] Henry D 1981 *Geometric Theory of Semilinear Parabolic equations* (Berlin: Springer)
[29] Yan M *et al* 2011 *Appl. Phys. Lett.* **99** 122505
[30] Allwood D A and Cowburn R P 2010 *Magnetic Domain Wall Logic* (New York: Wiley)
[31] Varga R, Zhukov A, Zhukova V, Blanco J M and Gonzalez J 2007 Supersonic domain wall in magnetic microwires *Phys. Rev.* B **76** 132406
[32] Omelyanchik A *et al* 2020 Ferromagnetic glass-coated microwires for cell manipulation *J. Magn. Magn. Mater.* **512** 166991
[33] Vazquez M and Chen D X 1995 The magnetization reversal process in amorphous wires *IEEE Trans. Magn.* **31** 1229–38
[34] Chiriac H, Óvári T-A and Zhukov A 2003 Magnetoelastic anisotropy of amophous microwires *J. Magn. Magn. Mater.* **254–255** 469–71
[35] Antonov A S, Borisov V T, Borisov O V, Pozdnyakov V A, Prokoshin A F and Usov N A 1999 Residual quenching stresses in amorphous ferromagnetic wires produced by an in-rotating-water spinning process *J. Phys. D: Appl. Phys.* **32** 1788
[36] Severino A M, Gómez-Polo C, Marín P and Vźquez M 1992 Influence of the sample length on the switching process of magnetostrictive amorphous wire *J. Magn. Magn. Mater.* **103** 117–25
[37] Zhukova V, Blanco J M, Rodionova V, Ipatov M and Zhukov A 2014 Fast magnetization switching in Fe-rich amorphous microwires: Effect of magnetoelastic anisotropy and role of defects *J. Alloys Compd.* **586** S287–90

[38] Zhukov A, Blanco J M, Ipatov M, Rodionova V and Zhukova V 2012 Magnetoelastic effects and distribution of defects in micrometric amorphous wires *IEEE Trans. Magn.* **48** 1324–6

[39] Rodionova V *et al* 2012 The defects influence on domain wall propagation in bistable glass-coated microwires *Physica* B **407** 1446–9

[40] Zhukova V, Blanco J M, Rodionova V, Ipatov M and Zhukov A 2012 Domain wall propagation in micrometric wires: Limits of single domain wall regime *J. Appl. Phys.* **111** 07E311

[41] Zhukov A, Blanco J M, Chizhik A, Ipatov M, Rodionova V and Zhukova V 2013 Manipulation of domain wall dynamics in amorphous microwires through domain wall collision *J. Appl. Phys.* **114** 043910

[42] Rodionova V, Baraban I, Chichay K, Litvinova A and Perov N 2017 The stress components effect on the Fe-based microwires magnetostatic and magnetostrictive properties *J. Magn. Magn. Mater.* **422** 216–20

[43] Baraban I, Leble S, Panina L and Rodionova V Control of magneto-static and -dynamic properties by stress tuning in Fe-Si-B amorphous microwires with fixed dimentions *J. Magn. Magn. Mater.* **477** 415–19

[44] Zhukova V, Blanco J M, Rodionova V, Ipatov M and Zhukov A 2014 Fast magnetization switching in Fe-rich amorphous microwires: Effect of magnetoelastic anisotropy and role of defects *J. Alloys Compd.* **586** S287–90

[45] Zhukov A, Blanco J M, Ipatov M, Rodionova V and Zhukova V 2012 Magnetoelastic effects and distribution of defects in micrometric amorphous wires *IEEE Trans. Magn.* **48** 1324–26

[46] Jiménez A, del Real R P and Vázquez M 2013 Controlling depinning and propagation of single domain-walls in magnetic microwires *Eur. Phys. J.* B **86** 113

[47] Janutka A and Gawroński P 2015 Structure of magnetic domain wall in cylindrical microwire *IEEE Trans. Magn.* **51** 1–6

[48] Vereshchagin M 2018 Structure of domain wall in cylindrical amorphous microwire *Physica B: Condens. Matter* **549** 91–3

[49] Panina L V, Ipatov M, Zhukova V and Zhukov A 2012 Domain wall propagation in Fe-rich amorphous microwires *Physica B: Condens. Matter* **407** 1442–5

[50] Klein P, Richter K, Varga R and Vázquez M 2013 Frequency and temperature dependencies of the switching field in glass-coated microwire *J. Alloys Compd.* **569** 9–12

[51] Mathews J and Walker R L 1970 *Mathematical Methods of Physics* 2nd edn (New York: Benjamin)

[52] Stratton J A 1941 *Electromagnetic Theory* (New York: McGraw-Hill)

Sergey Leble

Chapter 17

Condensed matter electrodynamics: equations of state by partition function

A necessity of the material relations in electrodynamics of a medium was explained in chapter 5, section 5.1.7, see also chapter 18, section 18.1. There are two origins for such relations, one—empiric with the use of thermodynamic principles, second—based on fundamental microscopic principle of statistical physics. The first was presented in many problems, considered in this book, the second we will outline in this chapter.

It was also mentioned that the derivation *ab initio* of an equation of state is based on statistical physics. Often, such course study goes after the electrodynamics. We, however, in the advanced approach would sketch the principles of both classical and quantum statistical mechanics of equilibrium that yields the key formulas of the link between conjugate (external–internal pairs) thermodynamic parameters. In electrodynamics, such pairs that enter the work specific form (see energy balance equation of section 5.2.5) are

(1) the electric induction $\vec{D}$ as the external parameter and electric field $\vec{E}$ as internal one;
(2) the magnetic induction $\vec{B}$ as external and the magnetic field $\vec{H}$ as internal parameters [1].

In problems, that unify electrodynamics and, say, hydrodynamics, we should add pairs like pressure/density, and others.

17.1 On derivation *ab initio* of an equation of state

17.1.1 The first law of thermodynamics forms by classical statistical physics

In equilibrium, the distribution function of a closed N-particle system is described by Gibbs formula

$$f(q, p, a) = \frac{1}{Z}\exp\left[-\frac{H(q, p, a)}{kT}\right], \tag{17.1}$$

where H is the Hamilton function, Z is a normalization constant, named as statistical integral (partition function), generalized coordinates $q = \{q_1, \ldots q_{3N}\}$ and momentum $p = \{p_1, \ldots p_{3N}\}$ sets q, p as a point at phase space, $a = \{a_1, \ldots a_s\}$ is a set of external parameters, T is the Kelvin temperature, k the Boltzmann constant. The normalization condition

$$\frac{1}{Z}\int \exp\left[-\frac{H(q, p, a)}{kT}\right]dqdp = 1, \tag{17.2}$$

guarantees the probability interpretation of the distribution function [2]. Such property allows to define the mean value for arbitrary function $A(q, p)$ as

$$\bar{A} = \int A(q, p)f(q, p, a)dqdp = \frac{1}{Z}\int A(q, p)\exp\left[-\frac{H(q, p, a)}{kT}\right]dqdp. \tag{17.3}$$

Changing the parameter Z to F, as

$$F = kT\ln\frac{1}{Z}, \tag{17.4}$$

we rewrite the normalization condition in terms of the new function F:

$$e^{\frac{F}{kT}}\int \exp\left[-\frac{H(q, p, a)}{kT}\right]dqdp = 1. \tag{17.5}$$

Let us calculate variation of the identity (17.5), taking a tiny change of T and all the variables that enter the condition:

$$e^{\frac{F}{kT}}\delta\left(\frac{F}{kT}\right)\int e^{-\frac{H}{kT}}dqdp - e^{\frac{F}{kT}}\int e^{-\frac{H}{kT}}\left[\frac{1}{kT}\sum_k \frac{\partial H}{\partial a_k}\delta a_k + H\delta\left(\frac{1}{kT}\right)\right]dqdp = 0 \tag{17.6}$$

we obtain using the main property of this operation, that acts as differentiation. Applying the definition of the mean value (17.3), now for energy,

$$U = \int H\exp\left[\frac{F - H}{kT}\right]dqdp = \bar{H} \tag{17.7}$$

we write

$$\delta\frac{F}{kT} - \frac{1}{kT}\sum_k \frac{\partial \bar{H}}{\partial a_k}\delta a_k - \bar{H}\delta\frac{1}{kT} = 0, \tag{17.8}$$

where the mean energy by the canonical Gibbs distribution with probability density (17.1) is equivalent to the internal energy of thermodynamics.

Introduce the internal parameters as

$$\frac{\partial \bar{H}}{\partial a_k} = -B_k, \tag{17.9}$$

hence the elementary work is

$$\delta A = \sum_k B_k \delta a_k. \tag{17.10}$$

The variation of functions F, T coincides with differentials [5]. The equality (17.8) then reads:

$$kTd\frac{F}{kT} - kTUd\frac{1}{kT} = dF + dT\frac{U - F}{T} = -\delta A, \tag{17.11}$$

that point out an interpretation of $F = U - TS$ as free energy, where S is entropy [1]. We arrive at the conventional form of the equilibrium first law of the thermodynamics for a closed system

$$dU + \delta A = TdS. \tag{17.12}$$

Further, the parameter F depends on a, T as follows from (17.5)

$$dF(a, T) = \sum_k \frac{\partial F}{\partial a_k} da_k + \frac{\partial F}{\partial T} dT. \tag{17.13}$$

Differentiating in equation (17.11)

$$d\frac{F}{kT} == \frac{1}{kT} \sum_k \frac{\partial F}{\partial a_k} da_k + \frac{1}{kT}\frac{\partial F}{\partial T} dT + Fd\frac{1}{kT}, \tag{17.14}$$

yields

$$\sum_k \frac{\partial F}{\partial a_k} da_k + \frac{\partial F}{\partial T} dT + SdT = -\sum_k B_k \delta a_k. \tag{17.15}$$

From this identity, apart from thermodynamic relation

$$S = -\frac{\partial F}{\partial T}, \tag{17.16}$$

it follows the important link

$$B_k = -\frac{\partial F}{\partial a_k}. \tag{17.17}$$

As a bridge between statistical physics and thermodynamics, on the basis of equation (17.4), we take

$$B_k = -kT\frac{\partial\left(\ln \frac{1}{Z}\right)}{\partial a_k}, \tag{17.18}$$

that formally defines the equations of state.

For the simplest example of only a magnetic field action, we ignore here other external parameters action, getting for isotropic medium, $B_z = B$, $H_z = H$, $B_x = B_y = 0$, put $a_1 = B$, taking the expression for the work (5.56) from the energy balance equation for matter (5.52) (section 5.2.5), we arrive at the first thermodynamic law having the form:

$$\frac{\partial F}{\partial B}dB + \frac{\partial F}{\partial T}dT + SdT = -\frac{1}{4\pi}HdB, \tag{17.19}$$

that gives

$$\frac{\partial F}{\partial B} = -\frac{1}{4\pi}H, \tag{17.20}$$

then, the equation of state is written as

$$H = 4\pi kT\frac{\partial lnZ}{\partial B}. \tag{17.21}$$

The alternative thermodynamic potential, convenient for a direct magnetization evaluation, is built by Legendre transformation as

$$\tilde{F} = F - \frac{1}{4\pi}\vec{B}\vec{H} + \frac{1}{8\pi}H^2, \tag{17.22}$$

used in the theory of magnetics. For such case

$$\begin{aligned} d\tilde{F} &= dF - \frac{1}{4\pi}\vec{B}\,d\vec{H} - \frac{1}{4\pi}\vec{H}\,d\vec{B} + \frac{1}{4\pi}\vec{H}\,d\vec{H} \\ &= dF - \frac{1}{4\pi}(\vec{B} - \vec{H})d\vec{H} - \frac{1}{4\pi}\vec{H}\,d\vec{B}, \end{aligned} \tag{17.23}$$

where $\vec{B} - \vec{H} = 4\pi\vec{M}$. The 1st law of thermodynamics in terms of F, S

$$dF + dT\frac{U - F}{T} = \delta A = \frac{1}{4\pi}\vec{H}\,d\vec{B}, \tag{17.24}$$

is rewritten as either

$$d\tilde{F} + \vec{M}d\vec{H} + \frac{1}{4\pi}\vec{H}\,d\vec{B} + \frac{U - F}{T}dT = \delta A = \frac{1}{4\pi}\vec{H}\,d\vec{B}, \tag{17.25}$$

or as

$$\sum_k \frac{\partial \tilde{F}}{\partial H_k}dH_k + \frac{\partial \tilde{F}}{\partial T}dT + SdT = -\vec{M}d\vec{H}, \tag{17.26}$$

that gives

$$\frac{\partial \tilde{F}}{\partial H_k} = -M_k,\ S = -\frac{\partial \tilde{F}}{\partial T}, \tag{17.27}$$

see, e.g. [1].

17.1.2 The first law of thermodynamics by quantum statistical physics

This is a normalization condition (for classical statistics) by Gibbs distribution, which is equivalent to the quantum statistics if to replace the integration with respect to the coordinates and the impulses of a system with the summation by quantum states of the quantum distribution operator (density matrix).

17.1.3 The scheme for two subsystems

Let a Hamiltonian $\hat{H}$ be a function of external thermodynamic parameter a, which is conjugate to the internal parameter B, so that the elementary work is equal to Bda. Within the framework of the equilibrium quantum statistical physics, for given Boltzmann constant k and temperature T, the Gibbs operator

$$\hat{f} = \frac{1}{Z}\exp\left[-\frac{\hat{H}}{kT}\right],$$

contains the normalization constant,

$$Z = \mathrm{Tr}\left[\exp\left[\frac{-\hat{H}}{kT}\right]\right], \tag{17.28}$$

called a statistical sum (partition function). Then, similar to the algorithm of the preceded section,

$$B_i = -kT\frac{\partial \ln Z}{\partial a_i}. \tag{17.29}$$

In this text we shall consider two pairs of the parameters: a, $B\rightarrow B$, H; E, P, these are the component of magnetic field versus component of the field strength or magnetization vector; the component of electric field versus electric displacement or of polarization vector. For the sake of clarity we act in a projection of all fields to one direction, so the Hamiltonian is a function of the magnetic and electric fields components along the marked direction $\hat{H} = \hat{H}(B, E)$. If the Hamiltonian is divided in few terms as $\hat{H} = \hat{H}_1 + \hat{H}_2$, the partition function is factorized as $Z = Z_1 Z_2$, hence the thermodynamic variable

$$-\frac{B}{kT} = \frac{\partial \ln Z_1 Z_2}{\partial a} = \frac{\partial \ln Z_1}{\partial a} + \frac{\partial \ln Z_2}{\partial a}, \tag{17.30}$$

correspondingly splits.

17.2 Spin system and equations of state

17.2.1 Classical Langevin theory

Historically, the theory, based on classical statistical physics, appeared in Langevin's paper [12]. It is based on a classical distribution function

$$Z(H) = \left[\int_0^{2\pi} d\phi \int_0^{\pi} \exp\left[\frac{\mu H \cos\vartheta}{kT}\right] \sin\vartheta d\vartheta\right]^N = \left[\frac{4\pi kT}{\mu H}\sinh\left(\frac{\mu H}{kT}\right)\right]^N, \quad (17.31)$$

where N is a full number of electrons of a body, the formula for partition function in classical theory. The part of expression, which depends on kinetic energy doesn't make an impact in magnetization, as it is independent from magnetic field. For thermodynamic potential—the free energy F, depending on the field is:

$$F(H, T) = -NkT \ln\left(\frac{4\pi kT}{\mu H} \sinh\frac{\mu H}{kT}\right). \quad (17.32)$$

Correspondingly for magnetization M is:

$$M = N\mu\left(\coth\frac{\mu H}{kT} - \frac{kT}{\mu H}\right) = N\mu L\left(\frac{\mu H}{kT}\right), \quad (17.33)$$

$$L(x) = \coth x - \frac{1}{x}; \; L(0) = 0; \; L(\infty) = 1, \quad (17.34)$$

where L is a Langevin function with its border conditions.

Energy of a system relatively to the field is $E = -M \cdot H$, so for the entropy depending from magnetic field by the Langevin theory we have:

$$S(H) = \frac{N}{T}\left[H\mu\left(\coth\frac{\mu H}{kT} - \frac{kT}{\mu H}\right) + kT \ln\left(\frac{4\pi kT}{\mu H} \sinh\frac{\mu H}{kT}\right)\right]. \quad (17.35)$$

The formula (17.35) obtained by classical theory contradicts with the third beginning of thermodynamics, which proclaims that entropy of a system does not change in absolute zero temperature:

$$S_{T=0} = 0, \quad (17.36)$$

otherwise in equation (17.35):

$$S_{\frac{\mu H}{kT}\to\infty} \to -\infty, \quad (17.37)$$

in external magnetic field spins get oriented along the field direction. However, even in the pre-quantum theory the orientations should be 'quantized', the direction of spins cannot be arbitrary. The account of such 'space quantization' was made in the work of Brillouin [13].

17.2.2 Brillouin theory: space quantization

To take spatial quantization into account, it suffices to assume that in equation (17.31) the quantity $\cos\vartheta$, does not change continuously, but takes a discrete series of possible values. Suppose for simplicity that each atom of a paramagnet has one valence electron participating in paramagnetism, and that there are only two energy levels with spin projection quantum numbers 1/2 and −1/2, respectively. As for the magnetic moment projection to the direction of the external magnetic field for the upper level will be $-g\mu_B/2$, and for the bottom $g\mu_B/2$.

Then at equilibrium, with temperature T, relative populations of these levels will be

$$N_- = \frac{\exp\left[\frac{g\mu_B H}{2kT}\right]}{\exp\left[\frac{g\mu_B H}{2kT}\right] + \exp\left[-\frac{g\mu_B H}{2kT}\right]} \tag{17.38}$$

and

$$N_+ = \frac{\exp\left[-\frac{g\mu_B H}{2kT}\right]}{\exp\left[\frac{g\mu_B H}{2kT}\right] + \exp\left[-\frac{g\mu_B H}{2kT}\right]}, \tag{17.39}$$

while

$$N_- + N_+ = 1. \tag{17.40}$$

Thus for magnetization M according to Brillouin we obtain:

$$M = \frac{Ng\mu_B}{2}\frac{\exp\left[\frac{g\mu_B H}{2kT}\right] - \exp\left[-\frac{g\mu_B H}{2kT}\right]}{\exp\left[\frac{g\mu_B H}{2kT}\right] + \exp\left[-\frac{g\mu_B H}{2kT}\right]} = \frac{Ng\mu_B}{2}\tanh\frac{g\mu_B H}{2kT}. \tag{17.41}$$

For $\frac{g\mu_B H}{2kT} \ll 1$ with the increase of temperature T, $\tanh\frac{[g\mu_B H]}{2kT} \approx \frac{[g\mu_B H]}{2kT}$, so

$$M = \frac{Ng^2\mu_B^2 H}{4kT}. \tag{17.42}$$

The account of orbital moment, needs spin and orbital numbers s and l, its vectorial addition gives full momentum quantum number j, which space quantization account. The energy levels of the electron system in magnetic field are characterized by

$$\cos(\vec{j}, \vec{H}) = \frac{m_j}{\sqrt{j(j+1)}}, \tag{17.43}$$

m_j—resultant full momentum projection quantum number.

Magnetization M is, after some algebra [14], yields

$$M = Ng\mu_B jB_j\left(\frac{\varepsilon_z}{kT}\right), \tag{17.44}$$

where

$$\varepsilon_z = g\mu_B jH, \tag{17.45}$$

and the Brillouin function $B_j(\frac{\varepsilon_z}{kT})$, that generalizes the Langevin function (the limit $j \to \infty$), is determined as

$$B_j\left(\frac{\varepsilon_z}{kT}\right) = \frac{2j+1}{2j}\coth\left(\frac{2j+1}{2j}\frac{\varepsilon_z}{kT}\right) - \frac{1}{2j}\coth\left(\frac{1}{2j}\frac{\varepsilon_z}{kT}\right), \tag{17.46}$$

g—g-factor (which equals 2 for fermions from gyro-magnetic relation), μ_B—Bohr's magneton, B—magnetic induction. Let us mention also the already quantum theory of Van Vleck [15], we however would go to more advanced Heisenberg theory.

17.3 Heisenberg theory

The problem of theoretical description of magnetization takes the significant place in modern studies, beginning from seminal results of Weiss [10]. In the paper of Heisenberg [3] on ferromagnetism it was established that the Weiss electric forces are originated from exchange effect of quantum mechanics primarily introduced in Heitler–London results. This paper [3] contains also very deep results of general significance. Multi-electron terms theory was built using the very common symmetry: in respect to group of electrons permutations. It has its extension based on joint symmetry group of permutations and space symmetry group [4], where the exchange integral notion is 'lifted' up the Hartree–Fock equations level [6]. The irreducible representation theory of permutation group allows to express energy via its characters [11] and, generalizing this result, statistical distribution function is constructed. As was shown in section 17.1.2, its derivative gives the magnetization M value conjugate to magnetic field H, that results in equation of state $M(H)$. These results present a very good exercise of practical construction as magnetization curves both for paramagnetic and ferromagnetic matter. It opens the way to built and draw hysteresis loops in a more complicated case when solutions of the Heisenberg equations are not unique.

We, in such a practice, having the transcendent equations of Heisenberg, search a graphical solutions to be applied for the paramagnetism and ferromagnetism cases. It admits a unique solution in the paramagnetism range, that gives magnetization curve. In the ferromagnetism domain few branches of solutions take place, that yields the hysteresis loops.

Practically, notching the interaction points (IP) of functions into the so called matrix of values, which depends on the changes of external magnetic field strength and characteristics of the material. Connecting the extreme values of the matrix in the case of multiple intersection between the same graphs, we managed either to build a hysteresis loop for a bulk ferromagnetic, or to identify some special cases, such as double loops with the central symmetry [18, 19]. In this section we partially follow the exposition of [4, 5].

17.3.1 Partition function

The Hamiltonian spectrum parameters evaluation, as the key ingredient of the equilibrium (Gibbs) distribution is expressed in terms of irreducible representations (numbered by ν) of symmetry group and exchange integrals. As the author of [3] notes 'generally somewhat arbitrary', assume that the distribution of energy values about the mean one E^ν is given by the Gaussian function.

$$\frac{n_\nu}{\sqrt{2\pi(\Delta\bar{E}^\nu)^2}} \exp\left\{-\frac{\Delta E^2}{2(\Delta\bar{E}^\nu)^2}\right\}, \tag{17.47}$$

as the number of terms, that belongs to a spin s, that is equal to n_ν, within the interval $E^\nu + \Delta E$, $E^\nu + \Delta E + d\Delta E$. The basic eigenvalue equation written for multi-electron system for a given symmetry group $g \in G$ with matrices of irreducible representations D^ν_{ik} reads [3, 8]

$$\det[\sum_{g \in G} D^\nu_{ik}(g) J_g - \delta_{ik} E] = 0, \tag{17.48}$$

where J_g are exchange integrals, related to the atoms at $\vec{r_0}$ and $g\vec{r_0}$. It defines the system of terms Γ_ν numerated by the irreducible representations of the symmetry group G. The Pauli principle yields: each value of the system spin corresponds one system of terms,

$$2n = 2 + \cdots + 2 + 1 + \cdots + 1. \tag{17.49}$$

The mean energy is expressed as the sum of the roots $\sum E_i$ of the equation (17.48) as a coefficient by $E^{n_\nu - 1}$

$$E^\nu = \frac{1}{n_\nu} \sum_g \chi^\nu(g) J_g, \tag{17.50}$$

here $\chi^\nu(g) = D^\nu_{ii}(g)$ is the character of the group. Below, the symmetry group is taken as equivalent to the group of all permutations of the electrons with the corresponding characters [3, 8].

In the external magnetic field the additional energy of the system is

$$E' = -\frac{e\hbar}{\mu} Hm, \tag{17.51}$$

where the spin projection quantum number lies in the limits $s \geqslant m \geqslant -s$,

e—an electron charge;
$\hbar$—Plank constant;
μ—an electron mass.

The partition function is built via a product of the Gaussian distribution (17.47) and the exponential term from Gibbs distribution with the Hamiltonian in energy representation, proportional to $E' + E^\nu$

$$\sum_{s=0}^{n}\sum_{m=-s}^{s}\int_{-\infty}^{\infty} d\Delta E \frac{n_\nu}{\sqrt{2\pi(\Delta\bar{E}^\nu)^2}} \exp\left\{\alpha m + \beta\frac{s^2}{2n} - \frac{\Delta E}{kT} - \frac{\Delta E^2}{2(\Delta\bar{E})^2}\right\}$$
$$= \sum_{s=0}^{n}\sum_{m=-s}^{s} n_\nu \exp\left\{\alpha m + \beta\frac{s^2}{2n} + \frac{\Delta(\bar{E}^\nu)^2}{2k^2T^2}\right\}. \tag{17.52}$$

where the dimensionless parameters

$$\alpha = \frac{eh}{2\pi\mu kT}H,$$
$$\beta = \frac{zJ}{kT}, \tag{17.53}$$

z—a number of closest neighbors;
k—Boltzmann constant;
T—temperature;
J—the exchange integral;

are introduced. Plugging the leading term of the expression for $\Delta(\bar{E}^\nu)^2$, also constructed by means of equation (17.48), the next term of the determinant expansion, see again [3], gives

$$Z = \sum_{s=0}^{n}\sum_{m=-s}^{s} n_\nu \exp\left\{\alpha m + \beta\frac{s^2}{2n} - \beta^2\frac{s^2(4n^2 - s^2)}{8n^3 z}\right\}. \tag{17.54}$$

Denoting

$$g(s) = \exp\left\{\frac{\beta s^2}{2n} - \beta^2\frac{s^2(4n^2 - s^2)}{8n^3 z}\right\} \tag{17.55}$$

and taking into account

$$n_\nu = \binom{2n}{n+s} - \binom{2n}{n+s+1}, \tag{17.56}$$

after reordering summation, it yields

$$Z = \sum_{m=-n}^{n}\sum_{s=|m|}^{n} \exp\{\alpha m\}g(m)\left[\binom{2n}{n+s} - \binom{2n}{n+s+1}\right]. \tag{17.57}$$

After transformation similar to integrating by parts we have

$$Z = \Phi\sum_{m=-n}^{n} \exp\{\alpha m\}g(m)\binom{2n}{n+m}. \tag{17.58}$$

The factor Φ has the order of α and β.

One may show that the expression $\exp\{\alpha m\}g(m)\binom{2n}{n+m}$ has the sharp maximum at a point m_0, hence the Taylor expansion in the vicinity of this point may be a good approximation in the exponent of $g(m)$. Approximately

$$g(m) = \exp\left\{\frac{\beta m_0^2}{2n} + \beta\frac{m_0}{n}(m - m_0) - \frac{\beta^2 m_0}{8n^3 z}\left[4m_0 n^2 - m_0^3 + 8n^2(m - m_0) - 4m_0^2(m - m_0)\right]\right\}, \tag{17.59}$$

taking the leading term with m_0^3, then Z takes the form:

$$Z = \Phi \sum_{m=-n}^{n} \exp\left\{\left(\alpha + \beta\frac{m_0}{n} - \beta^2\frac{m_0^3}{2n^3 z}\right)m\right\}g(m)\binom{2n}{n+m} = \Phi\left[2\cosh\left(\frac{\alpha + \beta\frac{m_0}{n} - \beta^2\frac{m_0}{nz} + \beta^2\frac{m_0^3}{2n^3 z}}{2}\right)\right]^{2n}. \tag{17.60}$$

The most probable value is determined as

$$m_0 = \frac{\partial \ln Z}{\partial \alpha} = n\tanh\left(\frac{\alpha + \beta\frac{m_0}{n} - \beta^2\frac{m_0}{nz} + \beta^2\frac{m_0^3}{2n^3 z}}{2}\right). \tag{17.61}$$

The neglected term with $\ln \Phi$ is of order '1'.

17.3.2 On Heisenberg equation solution

Graphical presentation and parameters

The Heisenberg equation (17.61), derived in [3] is equivalent to the couple of explicit functions, depending on M [4].

$$y_1 = M, \tag{17.62}$$

$$y_2 = \tanh\left(\frac{\alpha + \beta M - \beta^2\frac{M}{z} + \beta^2\frac{M^3}{2z}}{2}\right), \tag{17.63}$$

solutions of (17.61) is obtained as cross points M-projection of plots for $y_{1,2}$. The introduction of different variables introduction is also possible, see section 17.5.

The values of the parameters e, m, h, are universal physics constants, while the z, J, T ones choice needs some discussion. So, Heisenberg [3] estimates the parameter J by the note, that the exchange energy should be about kT, giving the value $J \approx 10^{-13}$ erg. For such the value the parameter β is of order 1 for the room temperature. We here use an estimation of the parameter via Curie temperature (section 17.5.1). A definition of the exchange integral within Hartree–Fock theory is given in [4].

Discussion

The expression (17.61) differs from the Weiss formula [10] by $\cot x - 1/x \to \tanh x$ because of the two spin orientation and cubic term. The Heisenberg theory may be applied to such types of metals of feeble conductivity. It takes as first approximation, the lattice separations as being very large and assumes that every electron belongs to its own atom. In the next approximation, one considers the quantum exchange of electrons that move in the unperturbed system with equal energies at different points.

Alternative methods (started for Pauli, Sommerfeld), applied to metals with very good conductivity, take as in the first approximation that electrons can be assumed to be completely free. In the second approximation, one might add in the interactions with the lattice points.

An intermediate approach starts from the Bloch states of electrons in periodic lattice, next it takes into account, as perturbation interaction with lattice: it is either defects of the lattice, or lattice oscillations (phonons).

Both last models contribute in electrodynamics first of all via conductivity, that enter the Maxwell system as an expression for current density.

17.4 Para-, and ferro-magnetic matter

17.4.1 The magnetization curve for a paramagnetic

The paramagnetic case domain in $z\beta$ plane is fixed by the condition:

$$\beta\left(1 - \frac{\beta}{z}\right) < 2, \tag{17.64}$$

see the figure 17.1, it is the domain outside the curve.

To build the magnetization curve the values of M for each α are required. We take them by means of solving the system of equations (17.62) and (17.63) graphically (figure 17.2). There is the only intersection point for each straight line and the curve for paramagnetic materials, for the choice of z, β, such that equation (17.64) holds.

Next, the magnetization curve by the mentioned algorithm is built in figure 17.3.

17.5 Problem of ferromagnetic state

In the case of ferromagnetic state the domain by equation (17.64) is above the curve $z(\frac{J}{kT})$ in figure 17.1, whence the solution of the equation (17.63) is not unique. In order to find out the values of joint points of equations (17.62) and (17.63) graphs, we should follow the bends of hysteresis loop curves more accurately, using a more

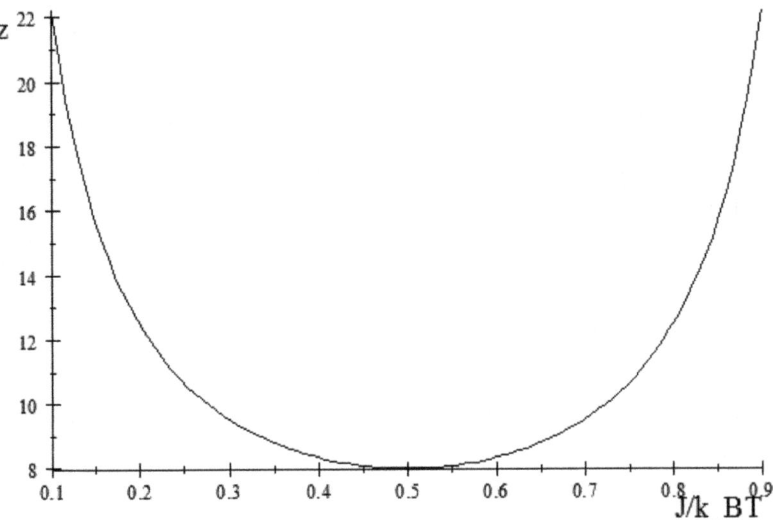

Figure 17.1. The plot of the function $z = \frac{2}{(\frac{J}{kT} - \frac{1}{2}(\frac{J}{kT})^2)}$.

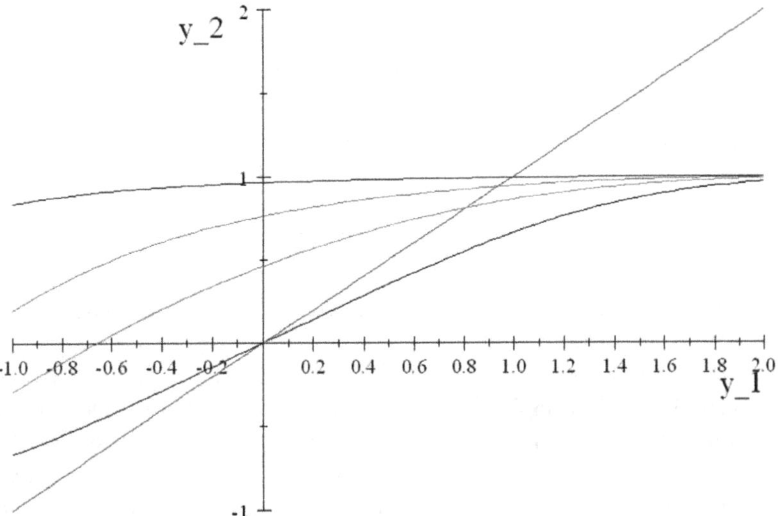

Figure 17.2. The parameters choice: $\beta = 1$, $z = 10$. The intersection points of the red line with curves (black, green, brown) are the solution of the system of equations (17.62) and (17.63). The curves correspond to α, that is proportional to the magnetic field H, substitution from 0 value (the lowest curve) to 3 (the highest curve). The value of α between the neighbor curves differs in 0.1. Reproduced with permission from [5].

convenient version of graphical solution of the following equations equivalent to equation (17.62):

$$\tilde{y}_1 = \text{arctanh}\, M, \tag{17.65}$$

with inverse to $M = \tanh \frac{x}{2}$

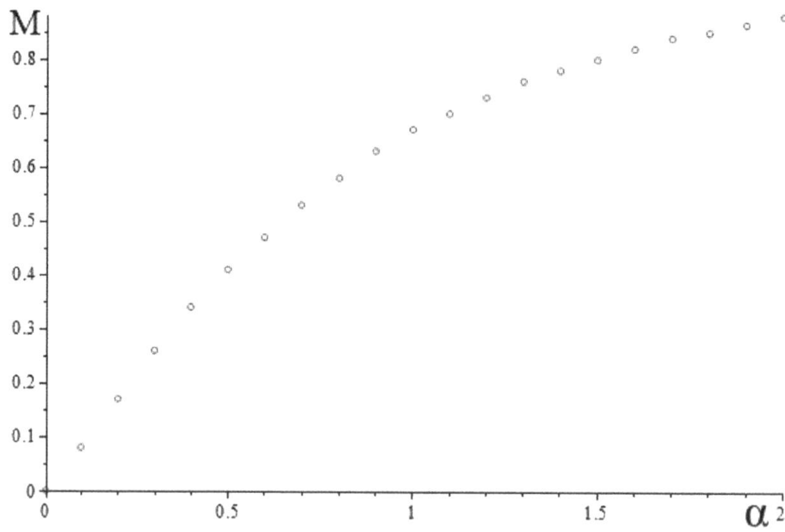

Figure 17.3. Magnetization curve for para-magnets looks typical [16]. The dependence of M on α in range $\alpha = 0\ldots2$ is shown for same parameters values $\beta = 1$, $z = 10$ as in figure 17.2. Created with A Chychkalo.

$$x = \ln\frac{1 + M}{1 - M},\ M \in (-1,\ 1), \tag{17.66}$$

which double intersections with the curve

$$\tilde{y}_2 = \frac{\alpha + \beta M - \beta^2\frac{M}{z} + \beta^2\frac{M^3}{2z}}{2}. \tag{17.67}$$

As it is observed in figure 17.4—the resultant curves (red and blue), built by the means of equations (17.65) and (17.67) system, are rather convex and curved, comparing with equations (17.62) and (17.63) system. The results in two different cases are presented in figures 17.4–17.6. The Heisenberg theory does not take into account domain walls presence in explicit form. It uses the Gauss distribution of states per energy level with parameters expressed in terms of characters of the universal symmetry group, the group of permutations. Few interesting properties however are exhibited in some range of the parameters z and T, see again [18, 19].

17.5.1 Towards the Curie law

To add, the ferromagnetism existence is demonstrated via the set of features, including the Curie law [16]. Its analog is derived in Heisenberg's paper [3] as well. According to it, there exists the critical temperature for switching from para-magnetic into ferromagnetic properties appearance and vice versa, by the boundary of the domain (17.64). It is marked as θ in the original text:

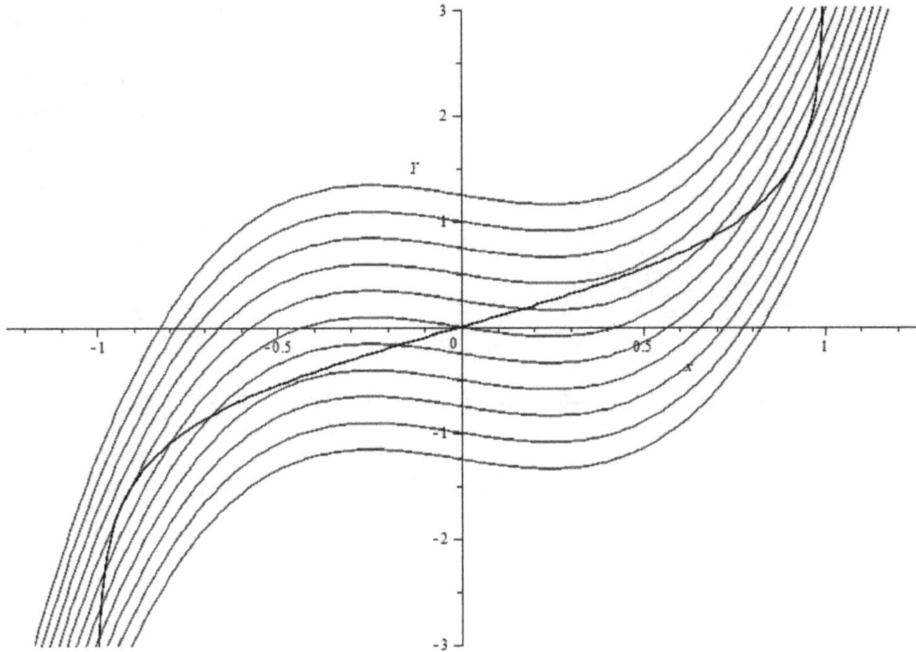

Figure 17.4. The intersection points of equations (17.65) and (17.67) curves are carried out for ($\alpha = -0.75...0.75$, $\beta = 10$, $z = 8$). Red curves show the change of α value in 0.25 one by one. There are cases of five intersections of a single red curve with the black—this is a critically needed condition to obtain the loop, illustrated at figure 17.5. Created with A Chychkalo.

$$\theta = \frac{2J_0}{k\left(1 - \sqrt{1 - \frac{8}{z}}\right)}. \tag{17.68}$$

The solution of the quadratic equation $(1 - \frac{\beta}{z})\beta = 2$ gives the roots:

$$\beta_{1,2} = \frac{z}{2} \pm \sqrt{\frac{z^2}{4} - 2z}, \tag{17.69}$$

the minimal temperature yields for the case

$$\beta_{\max} = \frac{zJ}{kT_{\min}},\ T_{\min} = \frac{zJ}{k\beta_{\max}}, \tag{17.70}$$

and the maximum is equal to

$$\beta_{\min} = \frac{zJ}{kT_{\max}},\ T_{\max} = \frac{zJ}{k\beta_{\min}}. \tag{17.71}$$

For the minimal case of $z = 8$, (iron) a direct application of the theory is impossible, because of the ferromagnetic range is restricted by one point, look e.g. [18, 21].

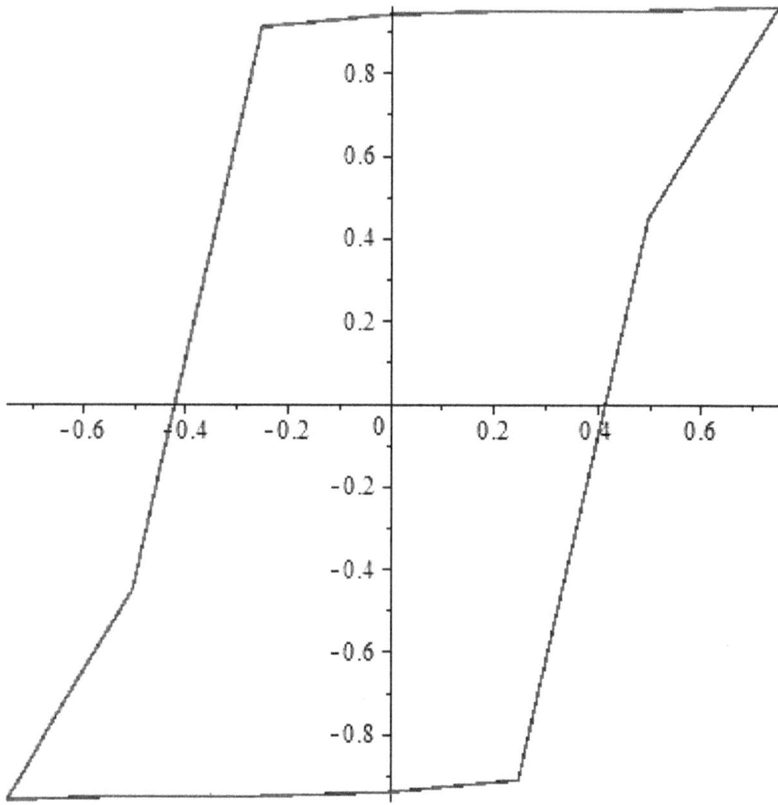

Figure 17.5. Hysteresis loop for common ferromagnetic materials. The values for its compilation are obtained from figure 17.4. The presence of 5 intersection points (IP) described at 17.4, gives the expansion of a loop at the approaching area to abscissa axis. The further away from each other are the extreme points from the set of 5 IP (for a single red line), the wider is the entire loop. The medium IP are not pictured and are not taken into account here. Though, medium IP are all in the loop, bounded by red and blue lines. Reprinted from [42].

Outside the range the magnetization curve, marked at the figure 17.1, $z = 12$, we obtain $\beta = 6 \pm 2\sqrt{3}$, hence the minimal temperature yields for this case

$$T_{min} = \frac{12J_{12}}{k(6 + 2\sqrt{3})}, \tag{17.72}$$

and the maximum

$$T_{\max} = \frac{12J_{12}}{k(6 - 2\sqrt{3})}. \tag{17.73}$$

We would estimate the exchange integral from (17.73) for Ni, having

$$J_{\rm Ni} = \frac{T_{\rm maxNi}k(6 - 2\sqrt{3})}{12}. \tag{17.74}$$

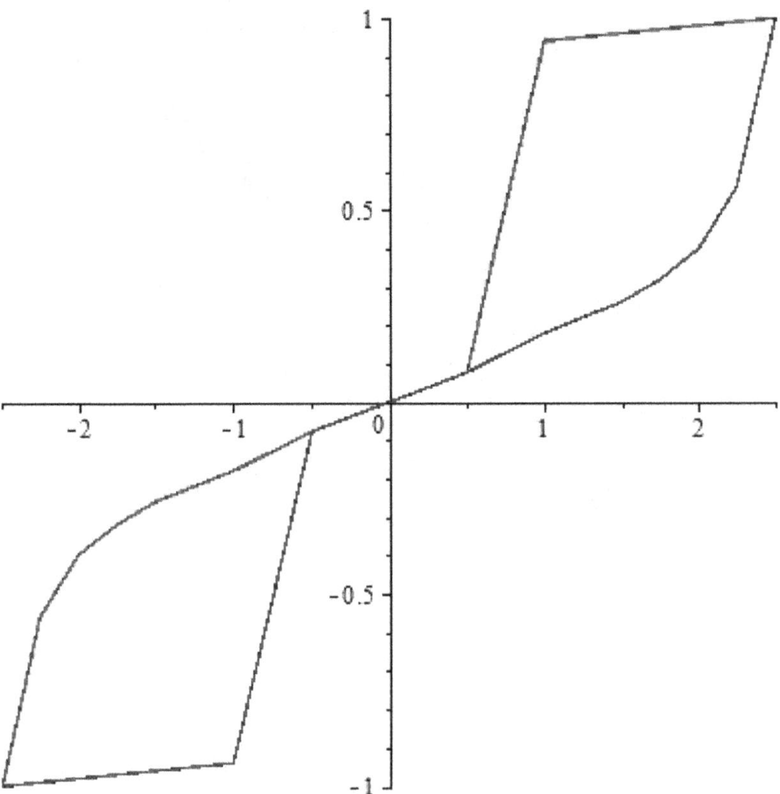

Figure 17.6. Double loop with the center symmetry of sub-loops. The represented graph is the Hysteresis loop in case of maximum three intersections of equation (17.62) with the single curve (17.63) ($\alpha = -2.50...2.50$, $\beta = 11$, $z = 8$). There is no tangible difference if we take either equations (17.65) and (17.67) or (17.62) and (17.63) for compilation. The vivid peculiarity is in 3 IP at once for a single curve (equations (17.63) and (17.67)) with the auxiliary functions from appropriate for each system of equations. The critical condition of double loop compilation: all three IP must be found in the same quarter area. The curves have a central symmetry for positive and negative α values, so the sub-loop at the third quarter ($-\alpha$) always repeats the sub-loop at the first ($\alpha > 0$) if α number is the same, having the difference only in its sign. Reprinted from [42].

Plugging the result into equation (17.72) we compute the critical temperature

$$T_{\min} = \frac{12J_{\mathrm{Ni}}}{k(6 + 2\sqrt{3})}, \tag{17.75}$$

This case, valid for Ni, Co, is described in [20], where the curve of critical magnetization as function of temperature is shown and close to experiments of [21, 22].

As well, the author emphasizes the similarity of the Curie Law with modification to the Weiss theory. Thus, the critical point is taken from the following conclusion, where m_0, T, z are mathematically connected:

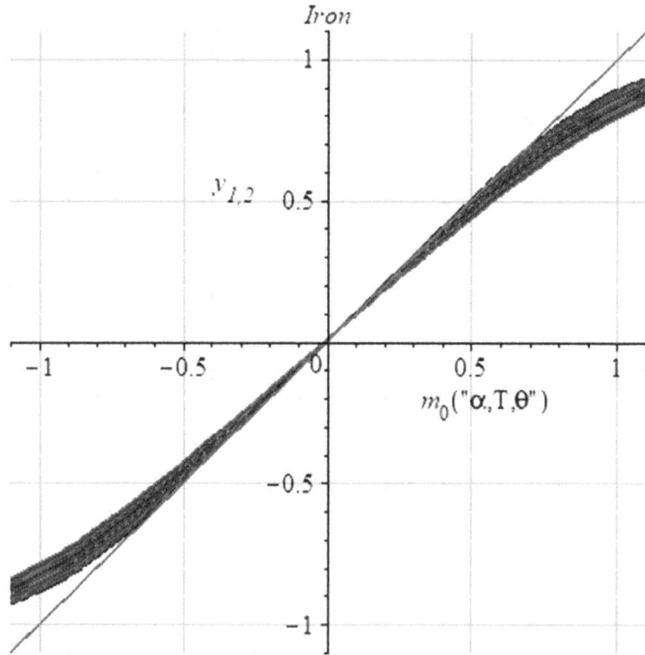

Figure 17.7. The IP-printing for building figure 17.8 (magenta curve) via our approximation analog formula (5.45). The x axis is for the designation of magnetization recession (Fe: $z = 8$, $\alpha = 0.1$, $\theta = 1043$ K, $T = 1163...2043$ K). Created with A Chychkalo.

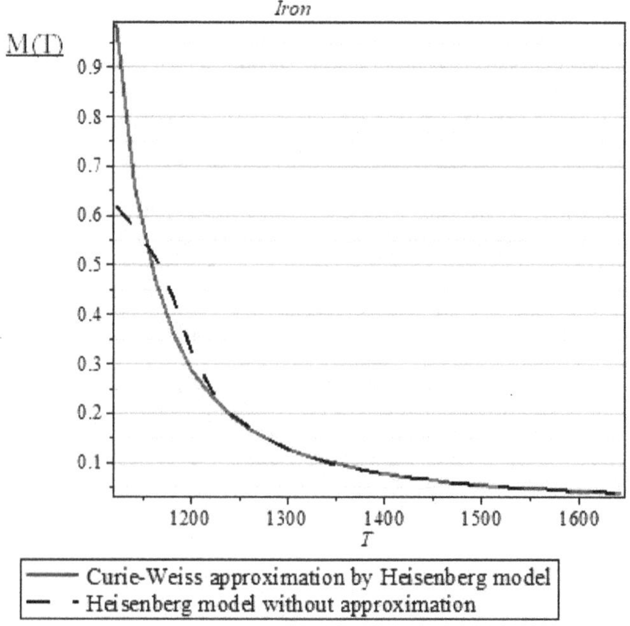

Figure 17.8. The dependence of magnetization on temperature (for Fe), for $T > \theta$. The comparison of two methods is graphed ($z = 8$, $\alpha = 0.1$, $\theta = 1043$ K, $T = 1163...2043$ K). Violet—the solutions of the equation system (17.62) and (17.63) after the substitution of β, which already includes the exact values of z, T and θ. Magenta—the approximation result of figure 17.7. Created with A Chychkalo.

$$m_0 \sim \frac{1}{T-\theta}\frac{T}{T\left(1+\sqrt{\frac{8}{z}}\right)-\theta\left(1-\sqrt{\frac{8}{z}}\right)}. \tag{17.76}$$

As a result of following the Heisenberg's algorithm of transformations, we have obtained the resembling formula (17.76).We use it for graphical analysis and further substitutions of contingent constants, proceeding practical situations imitations with real materials. Iron (Fe) was chosen as the primary sample of the corresponding section of this article out of its lattice structure. Thus the approximation formula got vastly simplified throughout the graphing procedure, because of its number of closest neighbors—8. The processed variant is represented here:

$$M = \frac{z\alpha T^2}{4(\theta - T)\left(z\left(1-\frac{4}{z}+\sqrt{1-\frac{8}{z}}\right)\theta - T\right)}. \tag{17.77}$$

17.6 Multiferroics

In the previous sections we considered the action of magnetic field that exhibits two kinds of phenomena, both explained by spin of electron presence: paramagnetism and ferromagnetism. The action of electric field can demonstrate similar effects, called polarization and ferroelectricity, both coined to constant dipole momentum of a matter constituents as, for example, molecules. A joint action of both fields study have long history, started from Curie's paper of 1895 [23]. This field of interest, called multiferroics, includes study of materials that show at least two simultaneous ferroic (electric, magnetic and elastic, see chapter 18, section 18.4.4) orders. It touches mainly materials showing a coexistence of dielectric and magnetic ordering, i.e. magneto-electric ones potentially have attracting applications in *spintronics* [24].

17.6.1 Electric field action: Stark effect

A perturbation theory by the external electric field E, applied along x, for the perturbed Fockian H_F [4, 6]

$$H\phi = (H_F + V)\phi, \tag{17.78}$$

where $V = eE(x + d_x)$, $\vec{p} = e\vec{d}$, to be a dipole momentum of a matter unit, ϕ reads as the known expansion by small parameter $\varepsilon = eE$

$$\phi = \phi_0 + \varepsilon\phi^{(1)} + \cdots. \tag{17.79}$$

In the first order it gives the following expression for an eigen function of the energy level E_n perturbation

$$\phi_n^{(1)} = \sum_m{}' \frac{(m|x + d_x|n)}{E_m - E_n}\phi_{0m}, \tag{17.80}$$

where $(m|x|n)$ are matrix elements of the Cartesian coordinate x and, similar for dipole momentum $(m|d_x|n)$ in nonperturbed states, the eigenfunctions ϕ_{0m} of the Fockian H_F. Account of constant dipole momenta open a way to ferroelectricity effects like magnetic dipole momentum of spin origin—to ferromagnetism of section 17.5.

17.6.2 Exchange integrals

The key element of Heisenberg theory is exchange integral, that enter the Hamiltonian (17.78) and, therefore to partition function. Plugging equation (17.79) into exchange integrals J_{ik}, defined for the electrons coupled to the centers i, k, by the expression from [6]

$$J_{ik} = \frac{e^2}{2}\int\int\frac{\rho_{ik}(\vec{r}, \vec{r}\,')}{|\vec{r} - \vec{r}\,'|}d\vec{r}\,d\vec{r}\,', \tag{17.81}$$

where, in the spirit of the Fock paper, it may include all interacting electrons of the closest neighbors, as

$$\begin{aligned}\rho(\vec{r}, \vec{r}\,') &= \sum_p \phi_p^*(\vec{r})\phi_p(\vec{r}\,') \\ &= \sum_p (\phi_{0p}^*(\vec{r}) + \varepsilon\phi_p^{(1)*}(\vec{r}))(\phi_{0p}(\vec{r}\,') + \varepsilon\phi_p^{(1)}(\vec{r}\,')) \\ &= \sum_p \phi_{0p}^*(\vec{r})\phi_{0p}(\vec{r}\,') + \varepsilon\sum_p[\phi_p^{(1)*}(\vec{r})\phi_{0p}(\vec{r}\,') \\ &\quad + \phi_{0p}^*(\vec{r})\phi_p^{(1)}(\vec{r}\,')] \\ &\quad + o(\varepsilon^2) = \rho_0(\vec{r}, \vec{r}\,') + \varepsilon\rho_1(\vec{r}, \vec{r}\,') + \cdots.\end{aligned} \tag{17.82}$$

The electric field perturbs the exchange integral as it is prescribed by (17.82) with $J_1 = \frac{e^2}{2}\int\int\frac{\rho_1(\vec{r}, \vec{r}\,')}{|\vec{r} - \vec{r}\,'|}d\vec{r}\,d\vec{r}\,'$, whence

$$\beta = \frac{z(J_0 + \varepsilon J_1)}{kT} = \beta_0 + \varepsilon\beta_1, \tag{17.83}$$

where J_1 is proportional to matrix elements as in (17.80) see [5].

17.6.3 Magneto-electric effect: material equation of state

Here we rely upon the analog of equation of state via equation (17.76) or (17.29). Plugging equation (17.83) into (17.60) yields

$$\frac{Z}{F} = \left[2\cosh\left(\frac{\alpha + \beta_0 M - \beta_0^2\frac{M}{z} + \beta_0^2\frac{M^3}{2z} + \varepsilon\beta_1 M(1 - \frac{2}{z} + \frac{M^2}{2z}) + \cdots}{2}\right)\right]^{2n}, \tag{17.84}$$

the quadratic terms in ε are shown further.

The mean (most probable) value of the magnetization per electron is defined by the Heisenberg relation (17.85)

$$M(\alpha,\,\varepsilon) = \tanh \frac{\alpha + \left(1 - \frac{\beta(\varepsilon)}{z}\right)\beta(\varepsilon)M + \beta(\varepsilon)^2\frac{M^3}{2z}}{2}, \tag{17.85}$$

that now includes not only magnetic field H as the parameter, but also the electric field E via ε. It constitutes the magnetic material relation in implicit form.

Let us illustrate the solution of the transcendent equation (17.85) by the figure 17.9. It shows that there is a minimal value of ε for a solution existence. The derivative of $\ln Z$ by ε gives the polarization on electron as:

$$\begin{gathered} P = \frac{p_0}{n} = e\frac{\partial \ln Z}{n\partial\varepsilon} = e\frac{\partial\omega}{\partial\varepsilon}\tanh\frac{\omega}{2}, \\ \frac{\partial\omega}{\partial\varepsilon} = \beta_1 M - 2\beta\frac{\beta_1}{z}M + 2\beta\beta_1\frac{M^3}{2z} + \beta\left[\left(1 - \frac{\beta}{z}\right) + \frac{3M^2}{2z}\beta\right]\frac{\partial M}{\partial\varepsilon}, \end{gathered} \tag{17.86}$$

evaluation of $\frac{\partial M}{\partial\varepsilon}$ is delivered on the basis of equation (17.85), it is suggested as an exercise in chapter 20, section 20.11.

If we write the material relations in approximation of Taylor expansion of both P, M in $\alpha\varepsilon$ plane up to the first order:

$$\begin{aligned} P &= P_0 + \varepsilon P_1 + \alpha P^1, \\ M &= M_0 + \alpha M_1 + \varepsilon M^1, \end{aligned} \tag{17.87}$$

where

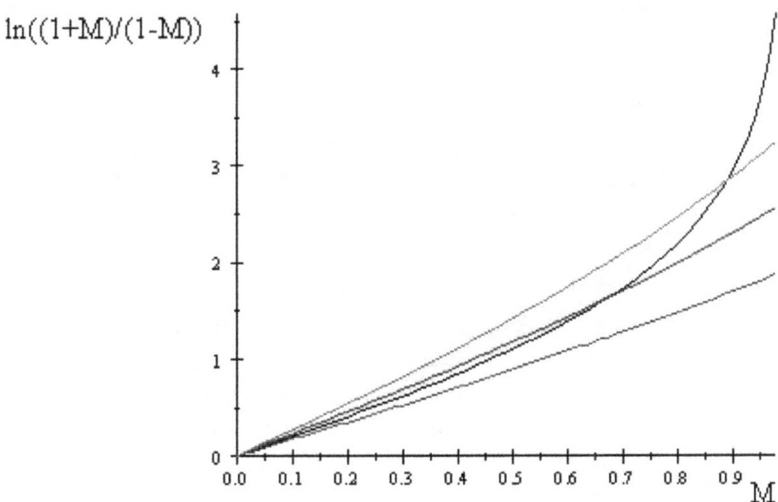

Figure 17.9. The points of intersections of the inverse tanh function (black line) and $\omega(M)$, green line—1 for $\varepsilon = 0.1$, the red one—2 for $\varepsilon = 1$ and the sienna—3 for $\varepsilon = 2$ are shown for $\alpha = 0$. The case of $\varepsilon = 0.1$ is out of the solution range. Reprinted with permission from [5].

$$
\begin{aligned}
P_0 &= e\frac{\partial\omega}{\partial\varepsilon}\tanh\frac{\omega}{2}|_{\varepsilon,\alpha=0} \\
&= e(\beta_1 M - 2\beta_0\frac{\beta_1}{z}M + 2\beta_0\beta_1\frac{M^3}{2z} \\
&\quad + \beta_0\left[(1-\frac{\beta_0}{z}) + \frac{3M^2}{2z}\beta_0\right]\frac{\partial M}{\partial\varepsilon}|_{\varepsilon=0})\mathcal{T}, \\
P_1 &= \frac{\partial P}{\partial\varepsilon}|_{\alpha,\varepsilon=0} = e\left[\frac{\partial^2\omega}{\partial\varepsilon^2}\tanh\frac{\omega}{2} + \left(\frac{\partial\omega}{\partial\varepsilon}\right)^2\frac{1}{2}(1-\mathcal{T}^2)\right]|_{\varepsilon=0}, \\
P^1 &= \frac{\partial P}{\partial\alpha}|_{\alpha,\varepsilon=0} = \frac{e}{2}\frac{\partial\omega}{\partial\varepsilon}(1-\mathcal{T}^2), \\
M_0 &= \mathcal{T}, \\
M_1 &= \frac{1}{2}(1-\mathcal{T}^2), \\
M^1 &= \frac{\partial M}{\partial\varepsilon}|_{\alpha,\varepsilon=0},
\end{aligned}
\tag{17.88}
$$

with $\mathcal{T} = \tanh\frac{\beta_0 M}{4z}(M^2\beta_0 - 2\beta_0 + 2z)$, that demonstrates magneto-electric effects.

17.6.4 On ferroelectricity

As we see, the argument of the hyperbolic tangency function in the rhs of equation (20.16), contains the quadratic polynomial in parameter ε, proportional to the electric field. The roots of the polynomial are $\varepsilon_1 = -\frac{\beta_0}{\beta_1}$ and $\varepsilon_2 = \frac{1}{2\beta_1 - M^2\beta_1}(2z - 2\beta_0 + M^2\beta_0)$. The polarization vector component as a function of the electric field (20.16) is plotted in figure 17.11 together with the parabola figure 17.10. We observe the nonlinear dependence of $P(\varepsilon)$.

17.7 Fine particles case

An important application of the Heisenberg theory and its modifications [4, 20] is based on a possibility to transfer the algorithm to the investigation of tiny particles, see equation (17.30), which is an important area of modern physics.

17.7.1 Energy distribution

To explain the modification of the statistical sum with the surface-bulk subdivision we return to its origin. The partition function by Heisenberg includes the terms with an energy and its Gauss distribution parameter. The result of Heitler for the energy is expressed as the sum by n-particle permutation group $g \in P$ [8], the exchange integrals marked by group elements, see the relation (17.50). Mean square deviation of energy, as $\sum \Delta E_n = 0$, yields

$$
\sum_{n=1}(\Delta E_n)^2 = -2\sum_{n>m}(\Delta E_n)(\Delta E_m). \tag{17.89}
$$

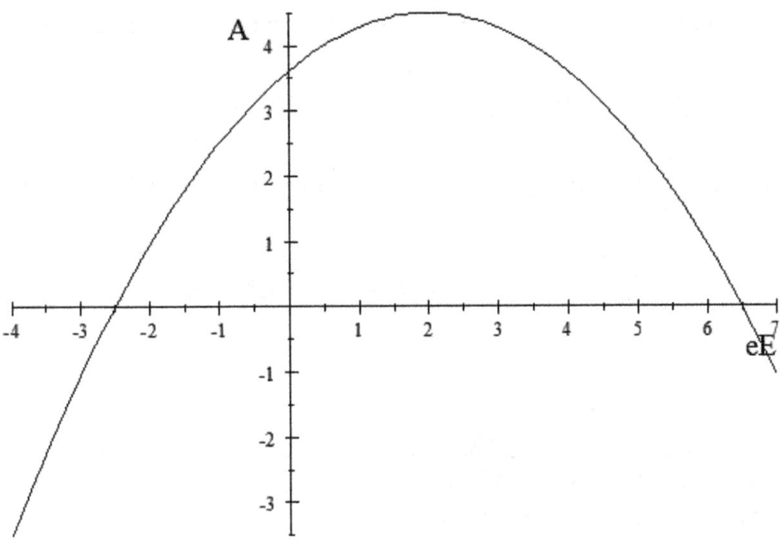

Figure 17.10. The dependence of the polynomial A on ε for the following values of the parameters; $\beta_0 = 5$, $\beta_1 = 2$, $M = 1$.

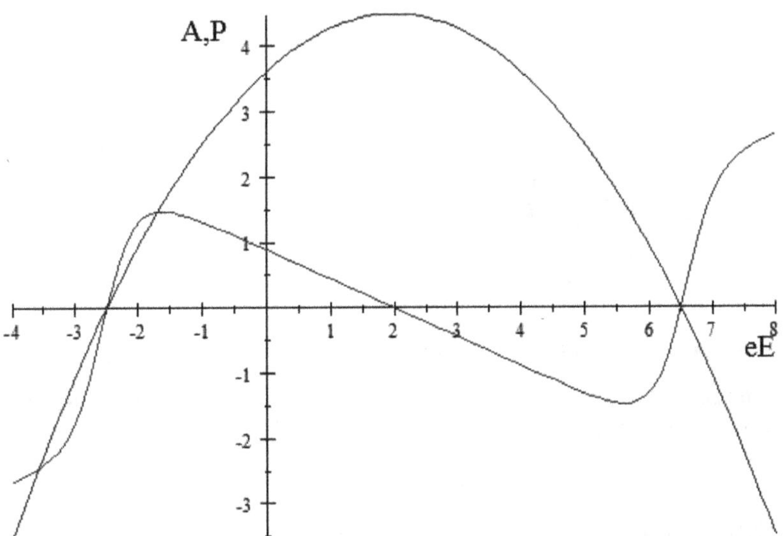

Figure 17.11. The parabola (red) represents the quadratic form A with the same parameters as in figure 17.10. The green curve shows the polarization vector component as function of electric field for zero magnetic field.

From equation (17.89) and $\chi^\nu(e) = n_\nu$, again in terms of the characters, we derive

$$\sum_{n=1}^{n_\nu}(\Delta E_n)^2 = \frac{-1}{n_\nu}\sum_{g,g'}(\chi^\nu(g)\chi^\nu(g') - \chi^\nu(gg'))J_gJ_{g'}. \tag{17.90}$$

Finally the mean square deviation from the mean value E^ν (see [3] for more details) reads

$$(\Delta\bar{E}_n)^2 = \frac{1}{n_\nu^2}\sum_{g,g'}(\chi^\nu(gg') - \chi^\nu(g)\chi^\nu(g'))J_gJ_{g'}. \tag{17.91}$$

17.7.2 Back to statistical sum

The partition function (statistical sum of states) is built as in section 17.3.1 with an extra explanation of equation (17.91). After the principal model simplification for the 'bulk' case $J_g = J$ and some algebra we arrive at equation (17.60).

Now, let us take into account the difference between exchange integrals for surface and bulk particles. Such division is stressed in [17]. The energy (17.50) and the Gauss distribution parameter (17.91), that is also expressed in terms of the joint symmetry group characters [4]. For the case with absence of space symmetry one have only the mentioned surface-bulk division. For simplification we left only two different exchange integrals, marked by indices a, b, a for bulk and b for surface. Then the mean energy reads

$$E^\nu = \frac{MN}{n_\nu}\left[\sum_P \chi^\nu(P)J_b + \sum_{P'}\chi^{\nu_1}(P')J_a\right]. \tag{17.92}$$

The mean square deviation from the mean value E^ν, $\nu = \{0, \nu_1\}$ also splits as follows

$$(\Delta\bar{E}_n)^2 = \frac{1}{n_\nu^2}\sum_{g,g'}(\chi^\nu(gg') - \chi^\nu(g)\chi^\nu(g'))J_b^2 + \cdots, \tag{17.93}$$

in the first term the pair g, g' runs a surface set of atoms.

17.7.3 Partition functions for a tiny particle

Let the spherical nanoparticle radius be Nd, with N—number of layers and d is the atom diameter, then the number of 'surface' particles is estimated as:

$$n_a = 4\pi(1 + 3N^2 - 3N), \tag{17.94}$$

and the bulk ones as

$$n_b = \frac{4}{3}\pi(N-1)^3. \tag{17.95}$$

The energy (17.92) and the distribution parameter (17.91) enter the partition function exponent (17.54) linearly, hence the exponential property admits that the partition function may be factorized as in equation (17.30). The modification of the whole construction could be similarly done for Z^a and Z^b partition functions:

$$M^{a+b} = \frac{\partial \ln(Z_a Z_b)}{\partial \alpha} = \frac{\partial \ln(Z_a)}{\partial \alpha} + \frac{\partial \ln(Z_b)}{\partial \alpha} \tag{17.96}$$

$$= \frac{\partial \ln\left(\left[2\cosh\frac{\omega_a}{2}\right]^{2n_a}\right)}{\partial \alpha} + \frac{\partial \ln\left(\left[2\cosh\frac{\omega_b}{2}\right]^{2n_b}\right)}{\partial \alpha} \tag{17.97}$$

$$= \frac{n_a}{n}\tanh\frac{\omega_a}{2} + \frac{n_b}{n}\tanh\frac{\omega_b}{2}, \tag{17.98}$$

where n_a is given by equation (17.94) and n_b by equation (17.95) correspondingly, so that

$$\frac{n_a}{n} = \frac{4\pi(3N^2 - 3N + 1)}{4\pi N^3} = \frac{3N^2 - 3N + 1}{N^3}, \tag{17.99}$$

similarly the relative number of 'bulk' particles is equal to (see also exercise 4, chapter 20, section 20.11):

$$\frac{n_b}{n} = \frac{(N-1)^3}{N^3}. \tag{17.100}$$

$M^{a+b} = M_a + M_b$ is a most probable value of magnetic quantum number projection per electron for a joint bulk and a surface parts. The functions $\omega_{a,b}$ are described by

$$\omega_a = \alpha + \beta_a M_a - \beta_a^2 \frac{M_a}{z_a}\left(1 - \frac{M_a^2}{2}\right), \tag{17.101}$$

similarly for the b case. The parameters $\beta_{a,b} = \frac{z_{a,b} J_{a,b}}{kT}$ depend on exchange integrals. The bulk one may be defined via Curie point temperature, the surface one are evaluated with correction factor, see the next section and [4].

17.7.4 Numerical estimations and plot

For the numeric estimations let us choose the number of layers $N = 5$ for the Ni nanoparticle. For such layers number we have an equal number of surface and bulk atoms. The parameters we have chosen are the following: the bulk exchange integral, evaluated via Curie temperature value ($\theta_{\mathrm{Ni}} = 627$ K) $J_{\mathrm{Ni}} = 1.8 \times 10^{-21}$ J. The parameter $\beta_b = \frac{12 J_{\mathrm{Ni}}}{k_B T} = 2$ for $T = 795$ K outside the ferromagnetic range. For the surface condition we take $z = 8$, $\beta_a = \frac{8 \times 1.15 J_{\mathrm{Ni}}}{k_B T} = 1.53$; the exchange integral corrected up 15%, for the minor distance between the surface and bulk atoms.

The magnetization of the surface layer is presented in figure 17.12.

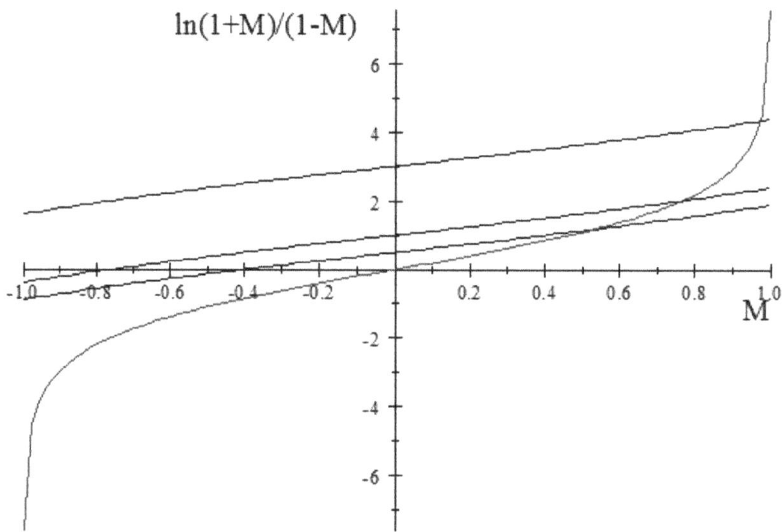

Figure 17.12. The surface layer contribution to magnetization of a nanoparticle model. The three points of intersections of the inverse tanh function (red line) and ω_a black lines are shown for $\alpha = 0.5$, 1.0, 3.0 growth up. The last point $\alpha = 3$ stands close to magnetization saturation. Reprinted with permission from [5].

For the joint magnetization we can use the function (20.11), (see exercise 5, chapter 20, section 20.11).

References

[1] Guggenheim E A 1967 *Thermodynamics An Advanced Treatment for Chemists and Physicists* (Amsterdam: Elsevier)

[2] Reichl L E 1998 *A Modern Course in Statistical Physics* 2nd edn (New York: Wiley)

[3] Heisenberg W 1928 Zur Theorie des Ferromagnetismus *Z. Phys.* **49** 619–36
Heisenberg W 1928 Zur Theorie des Ferromagnetismus *Probleme der Modernen Physik, A. Sommerfeld Festschrift* (Leipzig: Hirzel) pp 114–22
Heisenberg W 1931 Zur Theorie des Magnetostriktion und der Magnetisierungkurve *Z. Phys.* **69** 287–97

[4] Leble S 2019 Heisenberg chain equations in terms of Fockian covariance with electric field account and multiferroics in nanoscale *Nanosystems: Phys. Chem. Math.* **10** 1–13

[5] Leble S 2020 Magnetoelectric effects theory by Heisenberg method based on permutation group symmetry of nanoparticles *Nanosystems: Phys. Chem. Math.* **11** 50–64

[6] Fock V 1930 An approximate method for solving the quantum many-body problem (Reported at the Session of the Russian Phys.-Chem. Soc. on 17 December 1929) *Z. Phys.* **61** 126 (TOI 5, N 51, 1, 1931,UFN 93, N 2, 342, 1967)

[7] Fock V 1933 *Z. Phys.* **81** 195

[8] Heitler W 1928 Zur Gruppentheorie der homopolaren chemischen Bindung Z *Z. Phys.* **47** 835

[9] Martinez-Garzia J C, Rivas M and Garcia J A 2015 Induced ferro-ferromagnetic exchange bias in nanocrystalline systems *J. Magn. Magn. Mater.* **377** 424–9

[10] Weiss P 1907 L'hypothèse du champ moléculaire et la propriété ferromagnétique *J. Phys.* **6** 661
Weiss P 1908 *Phys. Z.* **9** 358
[11] Heitler W 1927 Stäorungsenergie und Austausch beim Mehrkäorperproblem Z *Z. Phys.* **46** 47
[12] Langevin P 1905 Sur la théorie du magnétisme *J. Phys.* **4** 678
Langevin P 1905 *Ann. Chim. Phys.* **5** 70
[13] Brillouin L 1927 *J. Phys. Radium* **8** 74
[14] Vonsovskij S V 1971 *Magnetism, Monograph* (Moscow: Nauka) pp 16–20
[15] Van Vleck J H 1926 The dielectric constant and diamagnetism of hydrogen and helium in the new quantum mechanics *Proc. Natl. Acad. Am.* **12** 662–70
Van Vleck J H 1927 On Dielectric Constants and Magnetic susceptibilities in the new quantum mechanics. Part I. A general proof of the Langevin-Debye formula *Phys. Rev.* **29** 727
Van Vleck J H 1927 On dielectric constants and magnetic susceptibilities in the new quantum mechanics. Part II—Application to dielectric constants *Phys. Rev.* **30** 31
Van Vleck J H 1928 On dielectric constants and magnetic susceptibilities in the new quantum mechanics. Part III—Application to dia- and paramagnetism *Phys. Rev.* **31** 587
[16] Kittel C 2019 *Introduction to Solid State Physics* 8th edn (New York: Wiley)
[17] Leble S 2019 Heisenberg chain equations in terms of Fockian covariance with electric field account and multiferroics in nanoscale *Nanosystems: Phys. Chem. Math.* **10** 1–13
[18] Hirsch A A 1959 Double hysteresis loops in ferromagnetic crystals *J. Phys. Radium* **20** 262–3
[19] Brandão J *et al* 2019 Observation of magnetic skyrmions in unpatterned symmetric multilayers at room temperature and zero magnetic field *Sci. Rep.* **9** 4144
[20] Inglis D R 1932 The Heisenberg theory of ferromagnetism *Phys. Rev.* **42** 442
[21] Bitter F 1932 On the interpretation of some ferromagnetic phenomena *Phys. Rev.* **39** 337–45
[22] Tyler F 1931 The magnetization-temperature curves of iron, cobalt and nickel *Philos. Mag.* **11** 596
[23] Curie P 1984 Sur la symétrie dans les phénomènes physiques, symétrie daun champ électrique et daun champ magnétique *J. Phys. Colloq.* **3** 393
[24] Spaldin N A and Fiebig M 2005 The renaissance of magnetoelectric multiferroics *Science* **309** 391–2
[25] Heremans J, Thrush C M, Lin Y-M, Cronin S, Zhang Z and Dresselhaus M S 2000 Bismuth nanowire arrays: synthesis and galvanomagnetic properties *Phys. Rev. B* **61** 2921
[26] Landau L D and Lifshitz E 1935 On the theory of the dispersion of magnetic permeability in ferromagnetic bodies *Phys. Z. Sowjet.* **8** 153–69
[27] Porsezian K 1993 *Future Directions of Nonlinear Dynamics in Physical and Biological Systems. NATO ASI Series (Series B: Physics)* vol 312 ed P L Christiansen, J C Eilbeck and R D Parmentier (Boston, MA: Springer) pp 243–8
[28] Guo B 2008 Landau–Lifshitz equations *Frontiers of Research with the Chinese Academy of Sciences: Volume 1* (Singapore: World Scientific)
[29] Gilbert T L 2004 A phenomenological theory of damping in ferromagnetic materials *IEEE Trans. Magn.* **40** 3443–9
[30] Lakshmanan M and Nakamura K 1984 Landau-Lifshitz equation of ferromagnetism: Exact treatment of the Gilbert damping *Phys. Rev. Lett.* **53** 2497–9

[31] Jiles D C 1998 *Introduction to Magnetism and Magnetic Materials* 2nd edn (London: Taylor and Francis)

[32] Muehlbauer S, Binz B, Jonietz F, Pfleiderer C, Rosch A, Neubauer A, Georgii R and Boeni P 2009 *Science* **323** 915

[33] Cheong S-W and Mostovoy M 2007 *Nat. Mater.* **6** 13

[34] Kim S *et al* 2018 Correlation of the Dzyaloshinskii—Moriya interaction with Heisenberg exchange and orbital asphericity *Nat. Commun.* **9** 1648

[35] Bouzidi D and Suhl H 1990 *Phys. Rev. Lett.* **65** 2587

[36] Zhukov A and Zhukova V 2009 *Magnetic Properties and Applications of Ferromagnetic Microwires with Amorphous and Nanocrystalline Structure. Nanotechnology Science and Technology Series* (New York: Nova Science)

[37] Vázquez M 2015 *Magnetic Nano- and Microwires: Design, Synthesis, Properties and Applications. Woodhead Publishing Series in Electronic and Optical Materials* (Amsterdam: Elsevier)

[38] Allwood D A and Cowburn R P 2010 *Magnetic Domain Wall Logic* (New York: Wiley)

[39] Olivera J, Gonzalez M, Fuente J, Varga R, Zhukov A and Anaya J J 2014 *An embedded stress sensor for concrete shm based on amorphous ferromagnetic microwires* **14** 19963–78

[40] Zhukov A 2001 Domain wall propagation in a fe-rich glass-coated amorphous microwire *Appl. Phys. Lett.* **78** 3106–8

[41] Varga R, Zhukov A, Zhukova V, Blanco J M and Gonzalez J 2007 Supersonic domain wall in magnetic microwires *Phys. Rev.* **B 76** 132406

[42] Leble S B and Chychkalo A 2020 Hysteresis loops for a bulk ferromagnetic by Heisenberg model Waves *in Inhomogeneous Media and Integrable Systems: Proc. IX Int. Workshop.* IKBFU

IOP Publishing

Practical Electrodynamics with Advanced Applications

Sergey Leble

Chapter 18

More general material relations

18.1 A concept

18.1.1 E–D–B–H relations

The equations of electrodynamics for a matter are not closed; too many basic fields, twelve,—the variables for eight non-independent equations. A closure in its classical form is realized via material equations. Material equations as it is written in chapter 5, section 5.1.7, see also chapter 18, sections 18.1 and 17.1 link the components of the fields as follows

$$D_i = D_i(E_1, \dots B_3), \quad H_i = B_i(E_1, \dots B_3). \tag{18.1}$$

Its linear approximation for an anisotropic matter reads as

$$D_i = \varepsilon_{ik}E_k + \varepsilon'_{ik}B_k, \quad H_i = \tilde{\mu}_{ik}B_k + \tilde{\mu}'_{ik}E_k, \tag{18.2}$$

with summation by repeated indices implied. Links to mechanics may be written similarly

$$j_i = j_i(E_1, \dots B_3), \quad \rho = \rho(E_1, \dots B_3) \tag{18.3}$$

with its natural linearization

$$j_i = \sigma^e_{ik}E_k + \sigma'_{ik}B_k, \rho = a_kE_k + b_kB_k. \tag{18.4}$$

A nonlinear generalization is conventionally investigated via the Taylor expansion of the rhs functions of equations (18.1) and (18.3), see chapter (6), section 6.5.3.

18.1.2 Hydrodynamics–electrodynamics material relations

The classical hydrodynamics of momentum–mass–energy conservation equations for *a* fluid, written for velocity $\vec{v}$, pressure p, temperature T, densities of mass ρ and energy e also need equations of state [1] for a closure, a classical example of such equations of state gives relations $p(\rho, T)$, $e(T)$ for ideal gas mole:

$$p = \rho RT, \quad e = C_v T. \tag{18.5}$$

If a fluid changes its thermodynamic state under action of electric and/or magnetic field, the equations of state should contain the fields $\vec{E}$, $\vec{B}$ components. So, we generalize the equations as

$$p = p(E_1, \dots B_3, \rho, T), \quad e(E_1, \dots B_3, \rho, T). \tag{18.6}$$

More general types of hydrodynamics contain tensors of pressure and energy, hence a closure of such systems of equation need inclusion of a dependence of the tensors on electromagnetic field components. Such relations could be also tensorial [13].

18.1.3 Continuum medium-electrodynamics material relations

The next class of equations for a continuum represents a link between stress and strain [14]. A particular case is the theory of plasticity [15]. The following problems are formulated as:

- Choice of optimal form of basic equation with a symmetry account. General investigation.
- Micro-, nano-peculiarities, surface influence account.
- Links between mechanics and electrodynamics.
- Including the heat transfer and heating.

Equation of motion

A continuum medium operates with tensor fields relates to a point. We need both algebraic and geometric aspects, see e.g. [13]. Let $\vec{t}_i$—be vectors of stress (is a surface traction per unit area) and $\vec{e}_j$—unit coordinate vectors. Let the stress tensor components $\sigma_{ik} = \sigma_{ki}$ are defined as coefficients of the expansion.

$$\vec{t}_i = \sigma_{ij}\vec{e}_j; \tag{18.7}$$

known as the Cauchy stress law. For normal and shearing stresses definition illustration, see figure 18.1; whence the resultant surface force is $\vec{F}_s = \int_S \vec{t}_i \, n_i dS$. By Newton law, the time rate of change of linear momentum $\int_V \rho\vec{v}dV$ equals the resultant force $\vec{F}_s + \rho\vec{b}$. External force, or a body force is denoted as $\vec{b}$ per unit mass, such as electric or magnetic ones. The integrals $\int_S \vec{t}_i\vec{n}_i dS$ may be transformed to the volume one by Gauss theorem

$$\int_S t_{ij}n_j dS = \int_V \frac{\partial \sigma_{ik}}{\partial x_k} dV.$$

It leads to the local form of the Newton law equation

$$\rho\frac{\partial v_i}{\partial t} = \frac{\partial \sigma_{ik}}{\partial x_k} + \rho b_i. \tag{18.8}$$

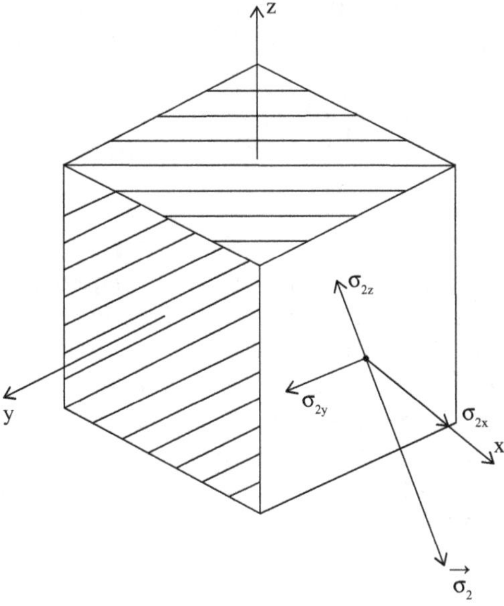

Figure 18.1. The stress components.

Hooke's law

Let $u_i = u_i(y_1; y_2; y_3)$; $i = 1; 2; 3$ denote a displacement field which describes the displacement of each point within the material. The stress–deformation relation:

$$e_{jk} = \frac{1}{2}\left(\frac{\partial u_j}{\partial x_k} + \frac{\partial u_k}{\partial x_j}\right), \tag{18.9}$$

named the Lagrangian strain tensor. If the continuum material is a linear elastic material, we introduce the *generalized Hooke's law* in Cartesian coordinates as

$$\sigma_{jk} = c_{jkil} e_{il}, \tag{18.10}$$

$i; j; k; l = 1; 2; 3$: the Hooke's law is a statement that the stress is proportional to the gradient of the deformation field occurring in the material (linearity).

The fourth-rank tensor with components c_{jkil} characterizes an arbitrary solid. If one substitutes the general Hooke law into the equation of motion (18.8), one arrives at the general elastic continuum equation

$$\rho\frac{\partial v_i}{\partial t} = c_{ikjl}\frac{\partial e_{jl}}{\partial x_k} + \rho b_i. \tag{18.11}$$

In electrodynamics the force is the Lorentz

$$b_i = eE_i + \frac{e}{c}\varepsilon_{ikl} v_k B_l. \tag{18.12}$$

A classification of phenomena related to the interplay of mechanical properties and electric action is based on the stationary version of the equation (18.11), i.e.

$$c_{ikjl}\frac{\partial e_{jl}}{\partial x_k} + \rho\left(eE_i + \frac{e}{c}\varepsilon_{ikl}v_k B_l\right) = 0. \tag{18.13}$$

18.1.4 Energy balance

Changes in temperatures cause thermal effects on materials. These thermal effects include thermal stress, strain, and deformations. Thermal deformation simply means that as the thermal energy (and temperature) of a material increases, so does the vibration of its atoms/molecules; and this increased vibration results in what can be considered a stretching of molecular bonds—that causes the material to expand. Of course, if the thermal energy (and temperature) of a material decreases, the material will shrink. Combinations of such phenomena with external electromagnetic field action (see equation (18.12)) lead to many interesting fields for investigation.

The speed of energy $E = K + U$ change is described by the equation of energy balance

$$\frac{dE}{dt} = W + \sum_k \frac{\delta Q_k}{dt}, \tag{18.14}$$

where $W = \frac{\delta A}{dt} = \int_V wdV$ is the power, kinetic energy is $\frac{1}{2}\int_V \rho\vec{v}\vec{v}dV$, internal energy is $U = \int_V \rho udV$. The power constitutes two parts $W = W_1 + W_2$, those are the surface part $W_1 = \int_S \sigma_k^n \frac{du_k}{dt} dS$, while the bulk part is $W_2 = \int \rho b_k v_k dV$. A quasistationary processes include the work of electromagnetic forces

$$\delta A = \frac{1}{4\pi}(\vec{E}\cdot\delta\vec{D} + \vec{H}\cdot\delta\vec{B}), \tag{18.15}$$

as it follows from energy balance at section 5.2.5. If the material relations are linear, the energy density is expressed by (5.54), that we reproduce here

$$W_m = \frac{1}{8\pi}(\varepsilon_{ik}E_iE_k + \mu_{ik}H_iH_k). \tag{18.16}$$

Now the heat production $\frac{\delta Q_1}{dt} = \int_V \rho s dV$ and the heat transfer $\frac{\delta Q_2}{dt} = -\int_S q_i n_i dS$, where s, q are corresponding densities. Within the formulation of the energy balance (18.14), see also section 5.2.5 for electromagnetic energy, or via momentum balance we can go to such phenomena as flexoelectricity or magnetoelasticity (section 18.5 and 18.4.4).

18.2 Symmetry and groups

18.2.1 Crystallographic symmetry and groups

A geometry of atom positions determine the symmetry of a matter. Such symmetry we define by means of a system of equivalent points notion, for example atoms at the same state may be considered as such points. Transformations, that leave the set of such points (figure) invariant (link equivalent points) we include the symmetry group with the natural product of transformations, as a sequence of such transforms. The space symmetry discrete subgroup (named also crystallographic) is the fundamental base of crystal structure and properties [5]. The related phenomena, such as magnetic are also covered by the same or extended symmetry group (metacrystallographic) or cambiant groups [6].

18.2.2 Tensorial symmetry with respect to indexes transpositions

Consider the symmetry impact for the tensor c_{ikjl}.

The general number of tensor c_{ijkl} components is $3^4 = 81$. Pose the question: how many different components does this tensor have? The answer: it depends on symmetry of the atomic system.

1. Symmetry of the tensor σ_{ij}, as a rotational group object, yields reduction to $6 \times 6 = 36$.
2. Symmetry with respect of the indexes transposition $\sigma_{ij} \Rightarrow \sigma_{ji}$ restricts to minimal components number 21.
3. For a cubic crystal—only three independent ones.
4. Isotropic material (glass, generally amorphous matter), rests only two. Liquid (ideal): the tensor reduces to the only one component: pressure.

Let us give an example of the tensor c for this last case:

$$c_{jkil} = a\delta_{jk}\delta_{il} + b(\delta_{ij}\delta_{kl} + \delta_{ik}\delta_{jl}). \tag{18.17}$$

More examples can be found in [13, 14].

18.2.3 Pauli symmetry with respect to electrons permutations

The most general statement of this kind, related to electrons is formulated as the *Pauli principle*: *The N-electron wave function is antisymmetric with respect to arbitrary permutation of the electron's coordinates, including spin variable.* Its direct application to the multielectron's spin system is made in Heisenberg theory of the ferromagnetism, see section 17.3. Its two-electron form leads to the key notion of magnetism, the exchange integral, for its Fockian implementation see [7].

18.3 Euclidean and Lorentz symmetry

18.3.1 Euclidean group covariance

In the basis of physics lies the general *covariance principle* with respect to the basic coordinate space group of motion [2]. We would name the group the *space symmetry group*. In 3D position space, if a material relation represents a link between vectors, a conditions for the related tensor is based on the Euclidean group symmetry, that formally is a condition of covariance. The group definition is given in chapter 1, section 20.1. If it is applied to equation (18.2), the covariance means that the coefficients ε_{ik}, ε'_{ik}, μ_{ik}, μ'_{ik} are tensors of the second rank with respect to rotations of the group $A \in O^{+}(3)$. Exemplary rotation is shown by the formula (1.4). It, in turn, means, that there is a reference frame for each of the mentioned tensors to be diagonal. However simultaneous diagonalization is possible, if matrices of some tensors commute, go to exercise 1, chapter 20, section 20.12. This way we can widen the group symmetry, including, for example reflections, to investigate the corresponding conditions for the material relations.

18.3.2 Lorentz group covariance

The electrodynamics, being in its origin the Lorentz-covariant, have special relativistic tensor form, that proves that the electromagnetic fields enter the tensor in pseudo-Euclidean space. The derivation of the Lorentz transformation and necessary properties are posed in chapter 4. Such field tensor of the second rank is antisymmetric, it may be conveniently presented as the matrix:

$$F = \begin{pmatrix} 0 & -E_1 & -E_2 & -E_3 \\ E_1 & 0 & B_3 & -B_2, \\ E_2 & -B_3 & 0 & B_1, \\ E_3 & B_2 & -B_1 & 0, \end{pmatrix}. \tag{18.18}$$

It is important that the charge density and current form the four-vector

$$j_0, \vec{j}\,. \tag{18.19}$$

The explicit formula for Lorentz transformation matrix is given in chapter 4, the formula (4.30). Let us reproduce it for the field tensor (the matrix (18.18))

$$F'_{\mu\nu} = \sum_{\mu',\nu'=0}^{3} L_{\mu\mu'} L_{\nu\nu'} F_{\mu'\nu'}. \tag{18.20}$$

The general covariance principle claims: a relation of mathematical physics should have tensor form, i.e. both sides of the relation should transform identically with respect to the space symmetry group action [2]. The material relations (18.2) are written in vectorial form. Its tensorial analog looks as;

$$G_{\mu\nu} = E_{\mu\nu,\mu'\nu'} F_{\mu'\nu'}, \tag{18.21}$$

the matrix of the 4-tensor G differs from the 4-tensor F by the change $E_i \to D_i$, $B_i \to H_i$ with obvious antisymmetry property. This property for both tensors

$$F_{\mu'\nu'} = -F_{\nu'\mu'},\ G_{\mu\nu} = -G_{\nu\mu}$$

gives the following condition

$$G_{\mu\nu} = -G_{\nu\mu} = -E_{\nu\mu,\mu'\nu'} F_{\mu'\nu'} = E_{\nu\mu,\mu'\nu'} F_{\nu'\mu'} = E_{\mu\nu,\mu'\nu'} F_{\mu'\nu'}. \tag{18.22}$$

From it we derive

$$E_{\nu\mu,\mu'\nu'} = -E_{\mu\nu,\mu'\nu'}. \tag{18.23}$$

Changing the indexes of summation as $\mu' \to \nu'$ and vice versa in the middle equality of (18.22) yields antisymmetry by the right pair of indexes

$$E_{\nu\mu,\mu'\nu'} = -E_{\nu\mu,\nu'\mu'}. \tag{18.24}$$

18.3.3 General tensors relations

There is a link between 4-tensor $E_{\mu\nu,\mu'\nu'}$ in space time and the space 2-tensors ε_{ik}, ε'_{ik}, μ_{ik}, μ'_{ik} of relations (18.2). The particular case of equation (18.22)

$$G_{0k} = E_{0k,0k'}F_{0k'} + E_{0k,i'0}F_{i'0} + \frac{1}{2}(E_{0k,i'k'}F_{i'k'} + E_{0k,k'i'})F_{k'i'}, \tag{18.25}$$

from equation (18.18) it is visible, that $F_{0k'} = -E_{k'}$, $F_{i'0} = E_{i'}$, then

$$\begin{aligned} G_{0k} &= -E_{0k,0k'}E_{k'} + E_{0k,i'0}E_{i'} + \frac{1}{2}(E_{0k,i'k'} - E_{0k,k'i'})F_{i'k'} \\ &= E^{a}_{0k,\ k'0}E_{k'} + 2\sum_{i'>k'} E_{0k,k'i'}F_{k'i'}, \end{aligned} \tag{18.26}$$

where $E^{a}_{0k,\ k'0} = \frac{1}{2}(E_{0k,k'0} - E_{0k,0k'})$. Taking equation (18.23) into account gives $E^{a}_{0k,\ k'0} = \frac{1}{2}(E_{0k,k'0} - E_{0k,0k'}) = E_{0k,k'0}$, $E_{0k,k'i'}F_{k'i'} = E_{0k,i'k'}F_{i'k'}$, while

$$G_{0k} = D_k = E^{a}_{0k,\ k'0}E_{k'} + 2(E_{0k,12}B_3 - E_{0k,13}B_2 + E_{0k,23}B_1). \tag{18.27}$$

Comparing with equation (18.2) gives expressions for components of the space tensors ε, ε' in terms of the components of the space–time tensor E, for example

$$\varepsilon_{ki} = E_{0k,i0},\ \varepsilon'_{k1} = 2E_{0k,23},\ \varepsilon'_{k2} = -2E_{0k,13},\ \varepsilon'_{k3} = 2E_{0k,12}. \tag{18.28}$$

Similar,

$$G_{k0} = E_{k0,\ 0k'}F_{0k'} + E_{k0,i'0}F_{i'0} + E_{k0,k'i'}F_{k'i'}, \tag{18.29}$$

see exercise 2 in section 20.12.

Writing equation (18.4) as a relation for current four-vector and the field 2-tensor by means of the third rank tensor,

$$j_\alpha = \Sigma_{\alpha\beta\gamma}F_{\beta\gamma}, \tag{18.30}$$

we can derive the condition, that follow from F antisymmetry:

$$j_\alpha = \frac{1}{2}(\Sigma_{\alpha\beta\gamma}F_{\beta\gamma} + \Sigma_{\alpha\gamma\beta}F_{\gamma\beta}) = \frac{1}{2}(\Sigma_{\alpha\beta\gamma}F_{\beta\gamma} - \Sigma_{\alpha\gamma\beta}F_{\beta\gamma}) = \frac{1}{2}(\Sigma_{\alpha\beta\gamma} - \Sigma_{\alpha\gamma\beta})F_{\beta\gamma}. \tag{18.31}$$

Particularly, for the space components,

$$j_k = \frac{1}{2}(\Sigma_{k\beta\gamma} - \Sigma_{k\gamma\beta})F_{\beta\gamma}, \tag{18.32}$$

and the temporal one

$$j_0 = c\rho = \frac{1}{2}(\Sigma_{0\beta\gamma} - \Sigma_{0\gamma\beta})F_{\beta\gamma}. \tag{18.33}$$

Introducing the tensor, antisymmetric by the pair of right indexes

$$\Sigma^a_{\alpha\beta\gamma} = \frac{1}{2}(\Sigma_{\alpha\beta\gamma} - \Sigma_{\alpha\gamma\beta}), \tag{18.34}$$

we rewrite equations (18.35) and (18.39) in a more compact form

$$j_\alpha = \frac{1}{2}\Sigma^a_{\alpha\beta\gamma}F_{\beta\gamma}, \tag{18.35}$$

having the *generalized Ohm law*. Taking into account the structure of the field tensor $F_{\beta\gamma}$ and $\Sigma^a_{\alpha\beta\gamma}$, both antisymmetric, we go to

$$j_\alpha = \sum_{\beta>\gamma} \Sigma^a_{\alpha\beta\gamma}F_{\beta\gamma}, \tag{18.36}$$

or

$$j_\alpha = \Sigma^a_{\alpha k0}F_{k0} + \sum_{i>k} \Sigma^a_{\alpha ik}F_{ik}. \tag{18.37}$$

The first sum contains only electric field components and the second one only magnetic. More details may be shown if put for the space part

$$j_j = \Sigma^a_{jk0}E_k + \sum_{i>k} \Sigma^a_{jik}F_{ik}, \tag{18.38}$$

and for temporal, i.e. charge density

$$j_0 = c\rho = \Sigma^a_{0k0}E_k + \sum_{i>k} \Sigma^a_{0ik}F_{ik}. \tag{18.39}$$

It shows the polarization and magnetoelectric effects. Do exercises 3, 4, section 20.12.

Final remark: we can widen the Lorentz group symmetry, including, reflections in space–time and charge conjugation, to investigate the corresponding conditions for the material relations. Some general development of the symmetry approach to the electrodynamics generalizations a reader can find in [3].

Further, in the following sections we give an example of short description of phenomena and materials related to the classification base developed in section 18.1.2, 18.1.3 and 18.3.3.

18.4 Active dielectrics

Controlled dielectrics or active dielectrics, are usually called such dielectrics whose properties substantially depend on external conditions–external fields strength, temperature, pressure, and so on. Such dielectrics can serve as working fluids in a variety of sensors, converters, generators, modulators, and other active elements.

A strict classification of active dielectrics is impossible, since the same material can exhibit signs of different active dielectrics, but application of relations (18.30), especially, in particular, equation (18.39) would be helpful. Active dielectrics include ferroelectrics, piezoelectrics, electrets, superionic conductors, etc. So, ferroelectrics often combine the properties of piezoelectrics. In addition, there is no sharp boundary between active and passive dielectrics. The same material, depending on operating conditions, can either fulfill the functions of a passive insulator or the active functions of a conversion or control element.

18.4.1 Ferroelectrics

A good example of dielectrics with nonlinear material relation like general (18.1) has a non-unique polarization curve similar to ferromagnetic. Hence, a ferroelectric is a material with spontaneous polarization, the direction of which can be changed if an external electric field is applied.

In the absence of an external electric field, ferroelectrics, as a ferromagnetic, conventionally have a domain structure, that is, they are divided into microscopic regions with spontaneous polarization. In principle, ferromagnets also have domains —areas of spontaneous magnetization; therefore, the behavior of ferroelectric in an electric field is similar to the behavior of ferromagnets in a magnetic field. The only difference between a ferroelectric and ferromagnet is that when they are placed in an electric field, the electric polarization vector $\vec{P}$ changes, and for a ferromagnet when placed in a magnetic field, the magnetization $\vec{M}$ changes.

18.4.2 Piezoelectricity

The brothers P Curie and J Curie discovered in 1880 the direct piezoelectric effect—the appearance of electrostatic charges on a plate cut from a quartz crystal under the mechanical stress action [4]. It was established, that the charges are proportional to the stress. Apart of the direct piezoelectric effect, the inverse piezoelectric effect is observed when a mechanical deformation of the crystal occurs under the influence of an electric field, and the magnitude of the mechanical deformation is directly proportional to the electric field strength. Both phenomena may be described by (18.39), the sign of Σ^{a}_{0k0} points to either the direct or inverse phenomenon.

The inverse piezoelectric effect should not be confused with electrostriction—the deformation of dielectrics under the influence of an electric field. Electrostriction is observed both in solid dielectrics and in liquid, while the piezoelectric effect is observed only in solid dielectrics with a certain crystalline structure. In addition, during electrostriction, a quadratic dependence between the field strength and strain is observed, see equation (18.3), and in the piezoelectric effect, the dependence is linear.

The piezoelectric effect is observed only when the crystal lattice is asymmetric. The absence of a center of symmetry of the crystal lattice is a necessary but insufficient condition for the appearance of the piezoelectric effect. Such symmetry with respect to orthogonal group $O \in O(3)$, that contains rotations and reflections (do exercise 5 in section 20.12.1).

Piezoelectricity is the electric charge that accumulates in certain domains of solid materials (such as crystals, certain ceramics, and biological matter such as bone, DNA and various proteins) in response to applied mechanical stress [16]. The word piezoelectricity means electricity resulting from pressure and latent heat. It is derived from the Greek word *πιέζειν*; piezein, which means to squeeze or press, and e, elektron, which means amber, known in ancient times as a source of electric charge.

Piezoelectricity is exploited in a number of useful applications, such as the production and detection of sound, piezoelectric inkjet printing, generation of high voltages, electronic frequency generation, microbalances, to drive an ultrasonic nozzle, and ultrafine focusing of optical assemblies. It forms the basis for a number of scientific instrumental techniques with atomic resolution, the scanning probe microscopies, such as STM, AFM, MTA, and SNOM. It also finds everyday uses as push-start propane barbecues, used as the time reference source in quartz watches, and in amplification pickups for some guitars.

18.4.3 Paraelectricity

Paraelectricity is in its fundamental feature for all materials (see chapter 17, section 17.6.1) the ability of many materials (specifically ceramics) to become polarized under an applied electric field. Unlike ferroelectricity, this can happen even if there is no permanent electric dipole that exists in the material, and removal of the fields results in the polarization in the material returning to zero [8], similar to paramagnetics. The mechanisms that cause paraelectric behavior are the distortion of individual ions (displacement of the electron cloud from the nucleus) and polarization of molecules or combinations of ions or defects.

Paraelectricity can occur in crystal phases where electric dipoles are unaligned and thus have the potential to align in an external electric field and weaken it.

An example of a paraelectric material of high dielectric constant is strontium titanate.

The $LiNbO_3$ crystal is ferroelectric below 1430 K, and above this temperature it transforms into a disordered paraelectric phase. Similarly, other perovskites also exhibit paraelectricity at high temperatures.

Paraelectricity has been explored as a possible refrigeration mechanism; polarizing a paraelectric by applying an electric field under adiabatic process conditions raises the temperature, while removing the field lowers the temperature, likely general Carnot cycle.

18.4.4 Magnetoelasticity

Bending of the plates are presented in many papers, which especially include smart materials which have wide range of applications. They can be found in such area as industry, everyday objects like toys, sport equipments. Plates filled in with ferrofluid, which are classified as smart materials, can be treated as smart materials as well. See the relation (18.13).

In [9] the two-dimensional equations of magnetoelasticity for a magnetoelastic ferromagnetic plate of finite size are derived. The assumption that magnetic susceptibility is very large compared with unity allows to obtain analytical representation for the components of the excited magnetic field in the case of a very thin plate. As for dynamical behavior in [10] there are considered problems of magnetoelastic oscillations of thin electrically conducting plates and shells situated in a stationary magnetic field. The authors formulated a hypothesis relative to the character of the variation of the electromagnetic field and of the elastic displacements along the thickness of the shell which allows to reduce the three-dimensional equations of magnetoelasticity to two-dimensional ones.

A general theory of the ferroelasticity is reviewed in [11].

18.5 Flexoelectricity

Flexoelectricity is a property of a dielectric material whereby it exhibits a spontaneous electrical polarization induced by a strain gradient. Flexoelectricity is closely related to piezoelectricity, but while piezoelectricity refers to polarization due to uniform strain, flexoelectricity refers specifically to polarization due to strain that changes from point to point in the material. This nonuniform strain breaks centrosymmetry, meaning that unlike in piezoelectiricty, flexoelectric effects can occur in centrosymmetric crystal structures [13]. Flexoelectricity is not the same as ferroelasticity.

The electric polarization due to mechanical stress in a dielectric, in terms of equations (18.10) and (18.14) is given by

$$P_i = d_{ijk}\sigma_{jk} + c_{ijkl}\frac{\partial e_{jk}}{\partial x_l},$$

where the first term corresponds to the direct piezoelectric effect and the second term corresponds to the flexoelectric polarization induced by the strain gradient.

Here, the flexoelectric coefficient, c_{ijkl}, is a fourth-rank polar tensor and d_{ijk} is the coefficient corresponding to the direct piezoelectric effect.

18.6 Ferroelasticity

Ferroelasticity is a phenomenon in which a material may exhibit a spontaneous strain. In ferroics, ferroelasticity is the mechanical equivalent of ferroelectricity and ferromagnetism [12]. When stress is applied to a ferroelastic material, a phase change will occur in the material from one phase to an equally stable phase, either of different crystal structure (e.g. cubic to tetragonal), or of different orientation (a 'twin' phase). This stress-induced phase change results in a spontaneous strain in the material.

The shape-memory effect and superelasticity are manifestations of ferroelasticity [17]. Nitinol (nickel titanium), a common ferroelastic alloy, can display either superelasticity or the shape-memory effect at room temperature, depending on the nickel/titanium ratio [11].

References

[1] Leble S 1990 *Nonlinear Waves in Waveguides* (Heidelberg: Springer)
[2] Fock V and Kemmer N 1964 *Theory of Space, Time and Gravitation* 2nd rev (Oxford: Pergamon)
[3] Duplij S, Goldin G A and Shtelen V M 2008 Generalizations of nonlinear and supersymmetric classical electrodynamics *J. Phys. A: Math. Gen.* A **41** 304007
[4] Curie J and Curie P 1880 Développement, par pression, de laélectricité polai re dans les cristaux hémièdres à faces inclinées *C. R. Acad. Sci.* **91** 383
Curie J and Curie P 1881 Contractions et dilatations produites par des tens ions électriques dans les cristaux hémièdres à faces inclinées *C. R. Acad. Sci.* **93** 1137–40
[5] Trifonov E D and Petrashen M I 2017 *Primenenie teorii grupp v kvantovoy mehanike* (Moscow: Knizhnyy dom "LIBROKOM")
Petrashen M I and Trifonov E D 2013 *Applications of Group Theory in Quantum Mechanics* (New York: Dover)
[6] Cracknell A P 2016 *Magnetism in Crystalline Materials: Applications of the Theory of Groups of Cambiant Symmetry* (Amsterdam: Elsevier)
[7] Leble S 2019 Heisenberg chain equations in terms of Fockian covariance with electric field account and multiferroics in nanoscale *Nanosystems: Phys. Chem. Math.* **10** 1
[8] Chiang Y *et al* 1997 *Physical Ceramics* (New York: Wiley)
[9] Hasanyan D J and Piliposyan G T 2001 Modelling and stability of magnetosoft ferromagnetic plates in a magnetic field *Proc. R. Soc.* A **457** 2063–77
[10] Ambartsumian S A, Bagdasarian G E and Belubekian M V 1973 On the magnetoelasticity of thin shells and plates *J. Appl. Math. Mech.* **37** 102–18
[11] Rosensweig R E 1985 *Ferrohydrodynamics* (New York: Cambridge University Press)
[12] Heremans J P, Thrush C M, Lin Y-M, Cronin S, Zhang Z, Dresselhaus M S and Mansfield J 2000 Bismuth nanowire arrays: Synthesis and galvanomagnetic properties *Phys. Rev.* B **61** 2921
[13] Heinbockel J H 2001 *Introduction to Tensor Calculus and Continuum Mechanics* (Bloomington, IN: Trafford)
[14] Chadwick P 1999 *Continuum Mechanics: Concise Theory and Problems* (New York: Dover)
[15] Chakrabarty J 2006 *Theory of Plasticity* (Amsterdam: Elsevier)
[16] Heywang W, Lubitz K and Wersing W (ed) 2008 *Piezoelectricity* (Berlin: Springer)
[17] Jiles D C 1998 *Introduction to Magnetism and Magnetic Materials* 2nd edn (London: Taylor and Francis)

IOP Publishing

Practical Electrodynamics with Advanced Applications

Sergey Leble

Chapter 19

On direct and inverse problems of electrodynamics

19.1 Direct problem of plane wave propagation in a layered medium

To explain the problem we address the results of the reflection and refraction problem solution of chapter 10, in its simplest version of a plane wave propagating orthogonally to a boundary, section 10.1.3. Let us return to the problem of a plane wave propagation along the z-axis in layers of isotropic matter with different and constant ε_i, μ_i and σ_i (see the definitions in chapter 5) divided by plane boundaries $z = z_i$ at Cartesian coordinates x, y, z (figure 19.1) and denote:

$$E_{0x} = u(z), \quad B_{0y} = \frac{ic}{\omega}\frac{\partial u}{\partial z}, \tag{19.1}$$

the relation (5.11b) from section 5.1.3 is taken into account. The function $u(z)$ satisfies the Helmholtz equation (10.5)

$$u_{zz} + \frac{i\omega\mu}{c^2}(i\varepsilon\omega - 4\pi\sigma)u = 0. \tag{19.2}$$

The principle difference with the field representation (10.13) for the internal layers is the account of the reflected wave

$$\begin{aligned} u_i &= P_i e^{ik_i z} + R_i e^{-ik_i z}, \\ u_{i+1} &= P_{i+1} e^{ik_{i+1} z} + R_{i+1} e^{-ik_{i+1} z}, \end{aligned} \tag{19.3}$$

while for the initial and external half-planes one could leave the ansatz

$$\begin{aligned} u_0 &= P_0 e^{ik_0 z} + R_0 e^{-ik_0 z}, \\ u_\infty &= P_\infty e^{ik_\infty z}. \end{aligned} \tag{19.4}$$

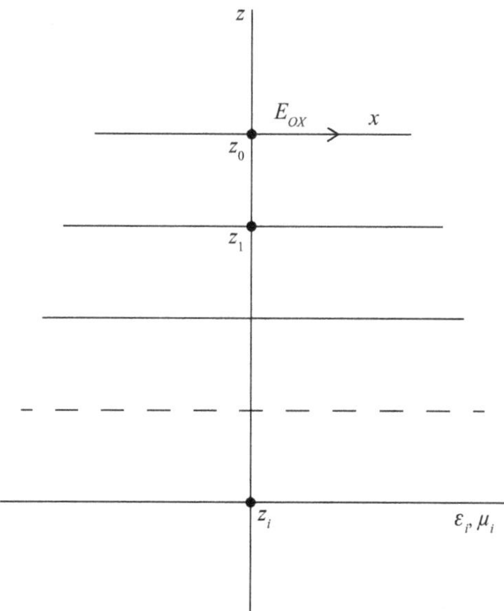

Figure 19.1. Scheme of layers.

The boundary conditions in terms of the function $u_i(z)$ (see equation (19.2)) for the internal points $z = z_i$, $i \neq 0, \infty$, read in discontinuity points

$$
\begin{aligned}
u_i(z_i) &= P_i e^{ik_i z_i} + R_i e^{-ik_i z_i} \\
&= u_{i+1}(z_i) \\
&= P_{i+1} e^{ik_{i+1} z_i} + R_{i+1} e^{-ik_{i+1} z_i}, \\
&\left.\frac{\partial u_i}{\partial z}\right|_{z=z_i} = ik_i P_i e^{ik_i z_i} - ik_i R_i e^{-ik_i z_i} \\
&= \left.\frac{\partial u_{i+1}}{\partial z}\right|_{z=z_{i+1}} = ik_{i+1} P_{i+1} e^{ik_{i+1} z_i} - ik_{i+1} R_{i+1} e^{-ik_{i+1} z_i}.
\end{aligned}
\tag{19.5}
$$

The equations (19.5) form the system that may be solved with respect to P_{i+1}, R_{I_1}. If, for each layer, the vector

$$
\alpha_i = \begin{pmatrix} P_i \\ R_i \end{pmatrix}, \tag{19.6}
$$

is introduced, the transition from one layer to the adjacent one may be expressed in term of the scattering matrix S_i. The linear dependence of the reflection and transmission coefficients is expressed in the matrix form:

$$
\alpha_{i+1} = S_i(k_i, k_{i+1}; \varepsilon_i, \varepsilon_{i+1}, z_i, z_{i+1})\alpha_i. \tag{19.7}
$$

This chain starts from $i = 0$ and finishes with $i = n$ as

$$
\alpha_\infty = S_n \alpha_n, \tag{19.8}
$$

with

$$\alpha_\infty = \begin{pmatrix} P_\infty \\ 0 \end{pmatrix}. \tag{19.9}$$

The total relation between α_0 and α_∞ is expressed by

$$\alpha_\infty = \prod_{i=0}^{i=n} S_i \alpha_0. \tag{19.10}$$

The transmission matrix

$$S = \prod_{i=0}^{i=n} S_i, \tag{19.11}$$

solves the *direct problem* of multi-layer medium propagation in the conditions described above. The matrix S evaluation is not difficult principally, but the result dependence of the parameters k_i, k_{i+1}; ε_i, ε_{i+1}, z_i, z_{i+1} is complicated, that imply numerical computation for large number of layers. Solve exercise 1 of section 20.13 for two layers, as an example.

19.2 On inverse problem

19.2.1 Remarks

An inverse problem relies upon the actual results of some measurements to infer the values of the parameters that characterize the body under observation. Such very important tasks as radar/lidar measurements (see section 19.3.2), x-ray tomography [1], etc, are the problems of such class. While the forward problem has (in deterministic physics) a unique solution, the inverse problem generally does not [2].

They create issues of bad-conditioned ('ill-posed'—English, 'schlecht aufgestelltes Problem'—German, 'non bel posti'—Italian, nekorrektnye—Russian). Often, this class of problems is called reverse ('inverse problems'—ang.) [3]. As for the basic element of such problems solution there are some modifications to these problems—formally close, but well-conditioned (classical issue). It carries out this procedure the name of regularization and allows, on one hand, control potential uncertainties in the formulation, and on the other—on measuring errors in the very context of this formulation.

19.2.2 Some details of problem formulation

Let us give some details in notations of the previous section: The problem (direct or inverse) is based on equation (19.2), expression for E_x, B_y and boundary conditions (19.5).

So, the direct problem is formulated in terms of amplitudes as:

$$A \to R, A_\infty$$

And the inverse problem:

$$A(\omega),\ R(\omega) \to \varepsilon_i;\ z_i,$$

that are determination of points of discontinuity, by which we divide the medium to layers with constant permeabilities and the permeabilities itself.

The initial form of the plane wave in the 'left' half-plane is

$$z < 0$$
$$Ae^{ikz} + Re^{-ikz} = u(z),$$

with the continuity condition of tangential component of electric field (E_x) at $z = 0$:

$$A + R = A_1 + B_1,$$

where B_1—coefficient of reflection at $z = z_1$. For the magnetic field B_y component we write

$$u_z = ik_1A_1e^{ik_1z} - ik_1B_1e^{-ik_1z},$$

with

$$k_1 = \sqrt{\varepsilon_1}\frac{\omega}{c}.$$

Both continuity conditions for E_x, B_y at $z = 0$ yields:

$$A + R = A_1 + B_1,$$
$$k(A - R) = k_1(A_1 - B_1).$$

Solving the system gives

$$A_1 = \frac{1}{2}\left[\left(1 + \frac{k}{k_1}\right)A + \left(1 - \frac{k}{k_1}\right)R\right],$$
$$B_1 = \frac{1}{2}\left[\left(1 - \frac{k}{k_1}\right)A + \left(1 + \frac{k}{k_1}\right)R\right].$$

The result determines the matrix of scattering S in the space of the vectors:

$$\alpha_0 = \begin{pmatrix} A \\ R \end{pmatrix},\ \alpha_1 = \begin{pmatrix} A_1 \\ B_1 \end{pmatrix}$$

The matrix of scattering S links the vectors as

$$\alpha_1 = S\alpha_0,$$

that gives the amplitudes at

$$z \in [0, z_1].$$

Consider now the boundary at $z = z_i$, the conditions looks similar, but more complicated, forming the system,

$$A_i e^{ik_i z_i} + B_i e^{-ikz_i} = A_{i+1} e^{ik_i z_i} + B_{i+1} e^{k_i z_i},$$
$$k_i(A_i e^{ik_i z_i} - B_i e^{-ikz_i}) = k_{i+1}(A_{i+1} e^{ik_i z_i} - B_{i+1} e^{-ik_i z_i}).$$

19.2.3 Equations of the inverse problem

The system of the equations in matrix form reads:

$$\alpha_{i+1} = S((k_i, k_{i+1}), \varepsilon_i, \varepsilon_{i+1}, z_i, z_{i+1})\alpha_i, \quad (19.12)$$

where the wavevectors

$$k_i = \sqrt{\varepsilon_i} k = \sqrt{\varepsilon_i}\frac{\omega}{c},$$

are also expressed in terms of ε_i, ω, that means

$$S_i = S_i(\varepsilon_i, \varepsilon_{i+1}, z_i, z_{i+1}, \omega), \quad (19.13)$$

after iterations we arrive at

$$\alpha_n = \prod\nolimits_{i=1}^{n} S_i \alpha_0 = S\alpha_0,$$

that defines the function

$$\alpha_n = \alpha_n(\varepsilon_1, \dots \varepsilon_n, z_1, \dots , z_n, \omega, A).$$

At the final point

$$z = H,$$

the system simplifies a little:

$$\varepsilon_n(Ae^{ik_n H} + Be^{-ik_1 H}) = A_\infty e^{-ik_\infty H}, \quad (19.14)$$

$$k_n(Ae^{ik_1 H} - B_n e^{-ik_n H}) = k_\infty A_\infty e^{ik_\infty H}, \quad (19.15)$$

complete the system, that gives

$$\begin{pmatrix} A \\ R \end{pmatrix} = \alpha_0 \rightarrow \alpha_1 \dots \rightarrow \alpha_\infty = \begin{pmatrix} A_\infty \\ 0 \end{pmatrix}.$$

Finally we can formulate:
Solution of *direct problem*:

Equations (19.12), (19.14) $\Rightarrow$ solution of direct problem R, A_∞ (of the whole medium).

Solution of *inverse problem*:

Given the reflection coefficient for a necessary set of frequencies $R(\omega_i)$, $i = 1, \dots,$ N, $N > 2n$, we calculate the set of the medium parameters $\varepsilon_1, \dots \varepsilon_n$, $z_1, \dots z_n$; take exercise 2 of chapter 20, section 20.13. A stability of the problem solution depends on a set of data position and its errors.

19.3 Data collection methods: examples

Any experiment in physics has the ill-posed problems features [4]. The same we can claim about signals transmissions [5].

19.3.1 Plasma Langmuir probe

To outline the mathematical properties of a typical inverse problem of such kind, we suggest to study the classical experiment of plasma physics. Technically the metal probe is used, see figure 19.2, the basic idea is the current measurement through the probe metal ball. We will calculate the current as a function of the sphere's potential. A plasma is an ionized gas, for its description see chapter 13.

A gas of ions creates a medium where the characteristic is certain distribution function: $f(\vec{r}, \vec{v})$, while $V(\vec{r})$—potential of electric field, that is determined at each point. Phase space is a set of points

$$\{\vec{r}, \vec{v}\} \in \Gamma,$$

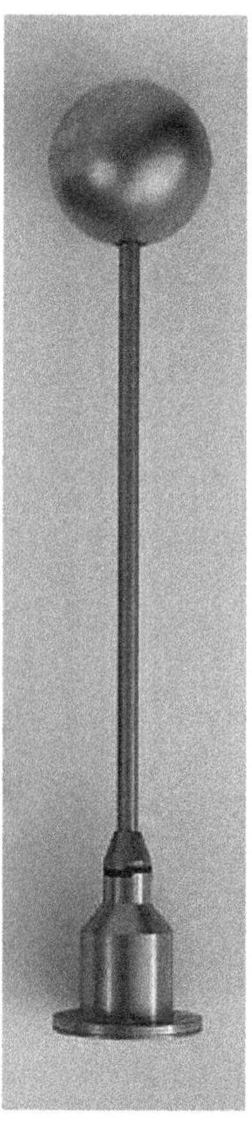

Figure 19.2. Typical Langmuir probe (https://en.wikipedia.org/wiki/Langmuir_probe, author: AndersIE).

on which the distribution function $f(\vec{r}, \vec{v})$ has a sense of density of probability of finding an electron at the point $\{\vec{r}, \vec{v}\}$ of phase space.

For example, $f(\vec{r}, \vec{v}, t)$, in classical kinetics of gases is a solution to Boltzmann's equation

$$\frac{\partial f}{\partial t} + (\vec{v}, \nabla)f = F(f),$$

where $F(f)$—is collision integral. Generally, the equation describes the evolution of f in time for gases. We here consider the stationary case, the time is omitted. By definition, density of the plasma (electron) is obtained by the integration by velocity subspace:

$$n_e(\vec{r}) = \int f(\vec{r}, \vec{v})d\vec{v}. \tag{19.16}$$

Similarly, the density of the electron stream (current) is calculated as

$$\vec{j}(\vec{r}) = e\int \vec{v}f(\vec{r}, \vec{v})\, d\vec{v}. \tag{19.17}$$

Let's consider an example of diluted plasma. If the density of electrons is small, ($n_e \ll 1$) or the track length the free electron λ is less than the probe ball radius $\lambda \ll a$ or (in presence of magnetic field) Langmuir's radius is significantly more than $a \Rightarrow \rho_e \gg a$:

$$f(\vec{r}, \vec{v}) = n_0(\vec{r})f_0(E), \quad E = \frac{mv_0^2}{2} = \frac{mv^2}{2} + eV(\vec{r}).$$

Enter spherical coordinates with origin '0' in the middle of the spherical part of the probe, see figure 19.3:

$$f(\vec{v}, \vec{r})|_{r=a} = n_0 f_0\left(\frac{mv^2}{2} + eV(a)\right), \quad \theta < \frac{\pi}{2},$$

$$f(\vec{v}, \vec{r})|_{r=a} = 0, \quad \theta > \frac{\pi}{2},$$

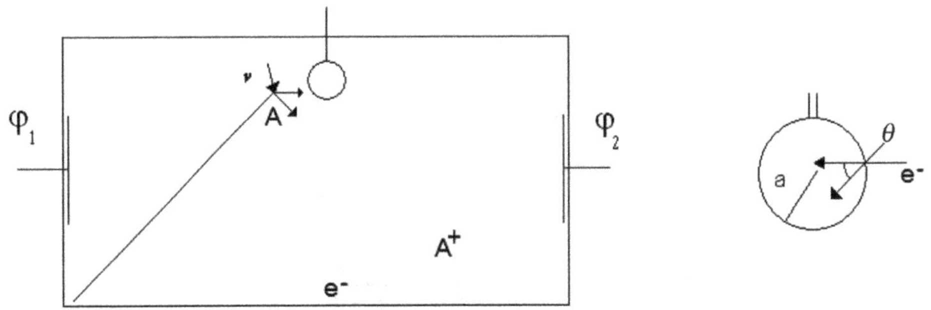

Figure 19.3. The scheme of the measuring chamber with Langmuir probe (marked as a circumference). The angle θ is shown in the right picture.

where θ is the angle between the speed vector $\vec{v}$ and a vector normal to the surface of the ball $\vec{n}$. Then, the stream of electrons through the surface of the ball J is evaluated as surface integral $\int_S \vec{j}\,\vec{n}dS$. We substitute equation (19.17)

$$J = 2\pi e \int_0^{\frac{\pi}{2}} \sin\theta d\theta \int_0^{\infty} v^2(v\cos\theta)f(a, \vec{v})dv = \pi e n_0 \int_0^{\infty} v^2 f_0\left(\frac{mv^2}{2} + eV\right)vdv,$$

$$J = \frac{2\pi e n_o}{m^2}\int_{eV}^{\infty}(E - eV)f_0(E)dE. \tag{19.18}$$

The equation (19.18) is the integral equation of Volterra of first kind in respect to the function $f_0(E)$. We obtain the distribution $f_0(E)$ by measuring the current as function of voltage, i.e. Volt–Ampère characteristics (figure 19.4). A function in a certain area between vertical lines is a function that we can trust

$$V \in [A, B], \tag{19.19}$$

Denote

$$eV = x, \quad E = s,$$

$$f_0(E) = \varphi(s), \quad J(V) = f(x), \ \frac{x}{e} \in [A, B],$$

then

$$f(x) = \int_x^{\infty} K(x, s)\varphi(s)ds, \tag{19.20}$$

where:

$$K(x, s) = \frac{2\pi e n_0}{m^2}(s - x).$$

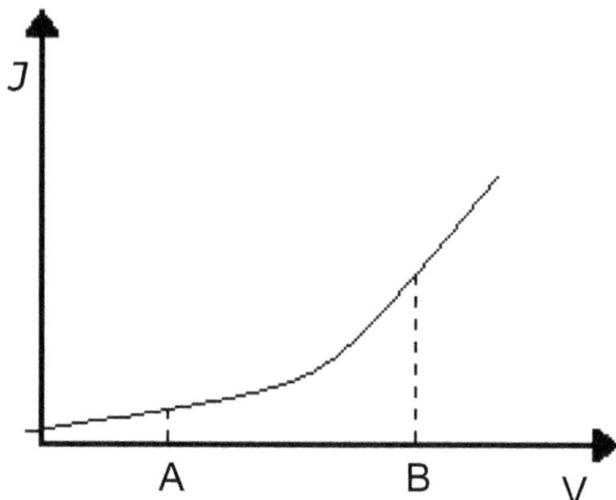

Figure 19.4. Volt–Ampère characteristics.

Such notations give us the conventional form of the first kind Volterra equation. Remind that

$$f(x) = \int_x^\infty K(x, s)\varphi(s)ds + \varphi(x)\text{—Volterra equation of the second kind.}$$

Formulation of the inverse problem
We are looking for $\varphi(s) = ?$ as a solution to equation (19.20) with the condition (19.19) and the given function $f(x)$.

Comment. Volterra equations with a continuous first-order kernel are unstable (see next section, section 19.4).

19.3.2 Radar

There are a few similar devices:

- Radar (RAdio Detection And Ranging)—uses radio waves (radio frequency).
- Sonar (SOund Navigation And Ranging)—uses sound.
- Lidar (LIght Detection And Ranging)—uses light (LASER + teleskop) (figure 19.5).

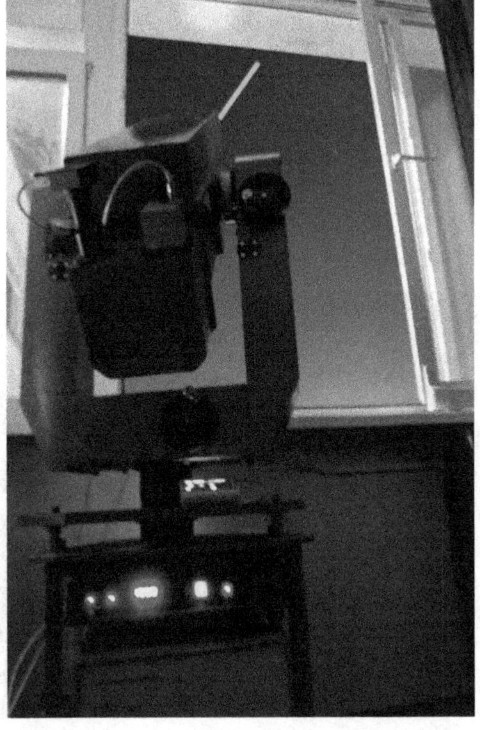

Figure 19.5. LIDAR at Immanuel Kant Baltic Federal University, Kaliningrad.

Lidar monitoring: European Aerosol Research Lidar Network to Establish an Aerosol Climatology: EARLINET http://www.earlinet.org (figures 19.6 and 19.7). We start with Helmholtz equation, see chapter 15, section 15.2:

$$\Delta p + k^2 p = 0, \tag{19.21}$$

where:

$$k = \frac{2\pi}{\lambda}.$$

19.3.3 Huygens' and Kirchhoff' formulas

Let ϕ, ψ be solutions to the Helmoltz equation (19.21):

$$\Delta\phi = -k^2\phi, \quad \Delta\psi = -k^2\psi.$$

There is a Green identity that combines values within the area with surface values (Green's theorem [6]):

$$\int_V (\phi\Delta\psi - \psi\Delta\phi)dV = \int_S \left(\phi\frac{\partial\psi}{\partial n} - \psi\frac{\partial\phi}{\partial n}\right)dS. \tag{19.22}$$

Hence, for solving the Helmholtz equation, we get:

$$\int_S \left(\phi\frac{\partial\psi}{\partial n} - \psi\frac{\partial\phi}{\partial n}\right)dS = 0. \tag{19.23}$$

$\vec{r}$—location of any point, unit vector, i.e. $\vec{n} = \frac{\vec{r}}{r}$ (figure 19.8).

The integral over the whole S is equal to the total of S_1 and S_2 (see also figure 19.8). You can check that $\psi = \frac{e^{ikr}}{r}$, $r = |\vec{r} - \vec{r}'|$ is the solution of equation (19.21), the total is then

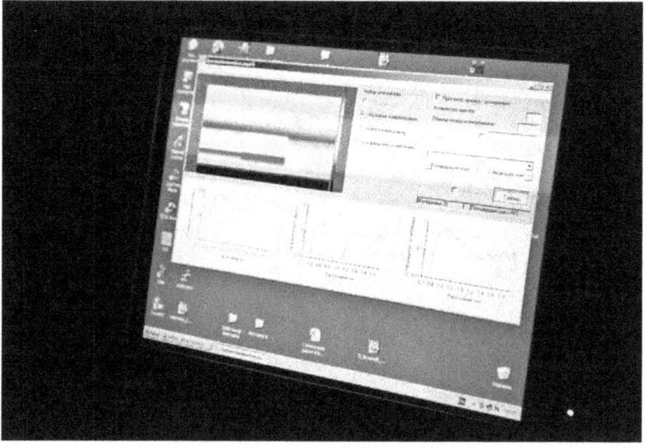

Figure 19.6. Lidar. Data on the monitor. Aerosol in air testing.

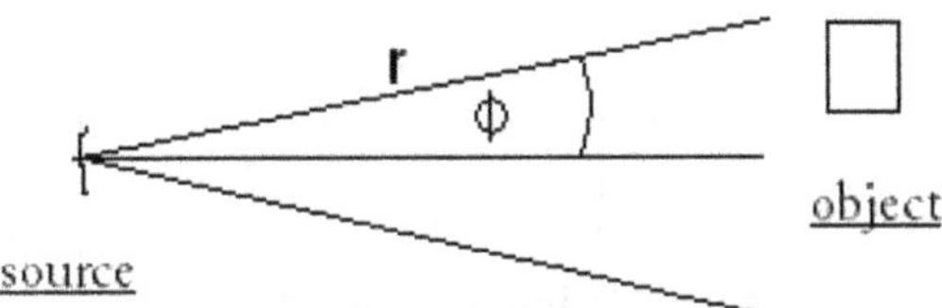

Figure 19.7. Electromagnetic wave generation.

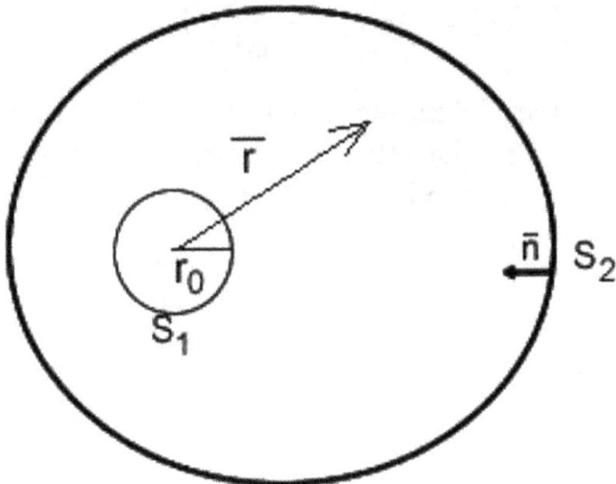

Figure 19.8. Integration areas: S_1—small sphere of radius r_0, n_1—normal unit vector directed into the big sphere S_2.

$$\int_{S_1}\left[\phi\frac{\partial\frac{e^{ikr}}{r}}{\partial n}-\frac{e^{ikr}}{r}\frac{\partial\phi}{\partial n}\right]dS, \tag{19.24}$$

integration goes along $\vec{r}'$, $dS = d\vec{r}'$. Use the identities:

$$\frac{\partial}{\partial n}=-\frac{\partial}{\partial r},\qquad \frac{\partial}{\partial r}\left(\frac{e^{ikr}}{r}\right)=\left(\frac{ik}{r}-\frac{1}{r^2}\right)e^{ikr},\qquad dS=r_0{}^2d\Omega,$$

deriving from equation (19.24):

$$\begin{aligned}&-\int_{S_1}\phi\left(\frac{ik}{r_0}-\frac{1}{r_0{}^2}\right)e^{ikr_0}\,dS-\int_{S_1}\frac{e^{ikr_0}}{r_0}\frac{\partial\phi}{\partial n}\,dS\\&=-\frac{ikr_0-1}{r_0{}^2}e^{ik_0r_0}r_0{}^2\int_{S_1}\phi\,d\Omega-\frac{e^{ikr_0}}{r_0}r_0{}^2\int_{S_1}\frac{\partial\phi}{\partial n}d\Omega.\end{aligned} \tag{19.25}$$

Considering the limit at $r_0 \to 0$, we get:

$$\phi(0)\int d\Omega = 4\pi\phi(0).$$

In general for $\vec{r}$, we substitute for (19.23) an equation (19.25), remembering that $\psi = \frac{e^{ikr}}{r}$ (Kirchhoff formula):

$$\phi(\vec{r}) = -\frac{1}{4\pi}\int_{S_2}\left[\phi\frac{\partial\frac{e^{ikr}}{r}}{\partial n} - \frac{e^{ikr}}{r}\frac{\partial\phi}{\partial n}\right]dS, \tag{19.26}$$

the environment is determined by individual points.

We can take the field $\phi|_{S_1} = 0$ (Huygens formula):

$$\phi(\vec{r}) = \frac{1}{4\pi}\int_S \frac{e^{ikr}}{r}\frac{\partial\phi}{\partial n}\, dS. \tag{19.27}$$

ϕ, ψ—as solutions to the Helmholtz equation (19.21), using the Green formula (19.22), we get Kirchoff formulas (19.26) and Huygens (19.27).

For electric field $\vec{E}$, you can enter the potential (ϕ):

$$\vec{E} = -\nabla\phi. \tag{19.28}$$

Its normal component is:

$$(\vec{n},\, \vec{E}) = E_n = -\frac{\partial\phi}{\partial n}. \tag{19.29}$$

19.3.4 Direct and inverse problems for a radar/lidar

Using the expression for normal component of the electric field (19.29), write the Huygens formula (19.27) as:

$$\phi(\vec{r}) = \frac{1}{4\pi}\int_S E_n(\vec{r}')\,\frac{e^{ikr}}{r}\, dS. \tag{19.30}$$

Next, insert equation (19.30) to the relation (19.28):

$$\vec{E}'(\vec{r}) = -\frac{1}{4\pi}\nabla\int_S E_n(\vec{r}')\,\frac{e^{ikr}}{r}\, dS. \tag{19.31}$$

This means that we have obtained a formula for the electric field at any point observations, only normal components to the antenna surface are needed to obtain the field.

1. *Direct problem.*

   Based on the formula (19.31). Getting $\vec{E}'(\vec{r})$ (directional diagram) from information with the transmitter S surface and its points, as well as field distributions $E_n(\vec{r}')$ at the surface. Typically, enter the spherical coordinates: $\vec{r} \Rightarrow r,\ \theta,\ \varphi$.
2. *Inverse problem.*

   Also based on the same formula (19.31), that we should consider as integral equation of the first kind. Search for S and E_n based on measurements of $\vec{E}'(\vec{r})$ at a set of points. Particularly, a flat area may be the radiating antenna as a transmitter (figure 19.9).

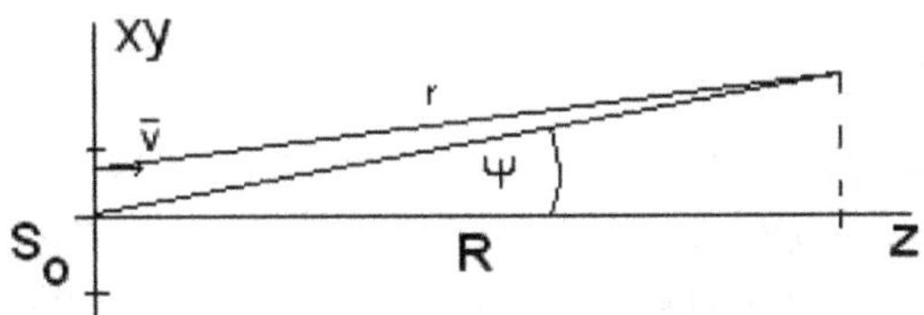

Figure 19.9. The propagation of the wave coming from the antenna, its surface marked S_0.

19.4 Inverse problems as ill-posed one

Materials common with ill-posed inverse problems, including information on the principles of the theory of radars (lidars) data processing and tomography [3, 7].

19.4.1 The Tikhonov regularization

Wiki: 'Tikhonov regularization, named for Andrey Tikhonov, is the most commonly used method of regularization of ill-posed problems. In statistics, the method is also known as ridge regression'.

Consider the Fredholm equation of the first kind for $x \in [c, d]$,

$$\int_a^b K(x, s)\phi(s)ds = f(x), \tag{19.32}$$

or in the operator's form

$$\hat{K}\phi = f.$$

It is well-known that such an equation solution is generally unstable [1, 3].

It can be said that the Tikhonov method combines the Gauss and Moore–Penrose approach [3]. In such context it is looking for a minimum sum of the square of distance and norm with weight α

$$T^{\alpha}[\phi] = \alpha\|\phi\|^2 + \|\hat{K}\phi - f\|^2. \tag{19.33}$$

Definition 1. Tikhonov Functional in L_2. According to the definition of a norm by scalar product in Hilbert space $\phi \in L_2[a, b]$,

$$\|\phi\|^2 = (\phi, \phi) = \int_a^b \phi^2(s)ds,$$

then, putting

$$\hat{K}\phi, f \in L_2[c, d],$$

we rewrite equation (19.33) as

$$T^{\alpha}[\phi] = \int_c^d \left[\int_a^b K(x, s)\phi(s)ds - f(x)\right]^2 dx + \alpha\int_a^b \phi^2(s)ds. \tag{19.34}$$

Discretization of the Tikhonov functional

Discretization of the Tikhonov functional (19.34) at the grid $x = c + hi$, $s = a + \tau j$, $h = \frac{d-c}{m}$, $\tau = \frac{b-a}{n}$; $\phi(a + \tau j) = \phi_j$, $f(c + hi) = f_i$, $K(c + hi, a + \tau j) = K_{ij}$, $i = 0, 1, m$; $j = 0, 1, \ldots, n$ gives functions (functional) of variables ϕ_j that represent search functions

$$T_d^\alpha = \sum_{i=1}^{m} h\left[\tau \sum_{j=1}^{n} K_{ij}\phi_j - f_i\right]^2 + \alpha \sum_{j=1}^{n} \tau\phi_j^2, \qquad (19.35)$$

where α is called a regularization parameter (it may be a matrix factor).

A prerequisite for a minimum of discretized functional (function (19.35)) is the Euler system which arises after differentiating with respect to ϕ_j and equalize to 0.

$$\frac{\partial T_d^\alpha}{\partial \phi_j} = 0. \qquad (19.36)$$

The result looks as:

$$\sum_{i=1}^{m}\sum_{j=1}^{n} h\tau K_{ij}K_{ik}\phi_j + \alpha\phi_k - \sum_{i=1}^{m} hK_{ik}f_i = 0. \qquad (19.37)$$

Matrix form of solution writes

$$\phi = (K^TKh\tau + \alpha I)^{-1}K^Thf. \qquad (19.38)$$

It is clearly seen that the Euler equation (19.37) and its solution goes to equivalent Gaussian theory, when $\alpha \to 0$.

Wiki: 'The pseudoinverse matrix can be computed via a limiting process:

$$A^{MP} = lim_{\alpha\to 0}[A^TA + \alpha]^{-1} \qquad (19.39)$$

if the limit exists. (see Tikhonov regularization). These limits can exist even if $(AA^*)^{-1}$ and $(A^*A)^{-1}$ do not exist.'

A choice of α, different selection methods

- Estimation using test examples. Scheme: we choose the ϕ_0 function which can easily be scaled. We calculate appropriate

$$f_0(x) = \int_c^d K(x, s)\phi_0(s)ds. \qquad (19.40)$$

By f_0 evaluate $\phi(\alpha) = (K^TKh\tau + \alpha hI)^{-1}K^Thf_0$. The functional derivative

$$\delta(\alpha) = \left\|\phi_0 - \phi(\alpha)\right\|^2, \qquad (19.41)$$

is minimal, if

$$\frac{d\delta(\alpha)}{d\alpha} = 0,$$

that is an equation for α. Based on this statement we can deliver:

- Conducting numerical experiments, develop a technology of such experiments, simulations.
- Iterative method, subsequent approximations $z_\alpha \to z$, test z as significant physical solution.
- Estimate using mathematical inequalities (see, e.g. [1, 7]).

See examples of the regularization of the ill-posed problems solutions in [8].

References

[1] Engl H W, Hanke M and Neubauer A 1996 *Regularization of Inverse Problems* (Dordrecht: Kluwer)

[2] Sabatier P C 1996 Nonuniqueness in inverse problems (English. English summary) *J. Inverse Ill-Posed Probl.* **4** 307–16

[3] Tikhonov A N and Goncharskij A V (ed) 1987 *Ill-posed Problems in Natural Sciences* (Moscow: Mir)

[4] Petrov Yu P and Sizikov V S 2005 *Well-posed, Ill-posed, and Intermediate Problems with Applications (Inverse and Ill-Posed Problems)* (The Netherlands: V.S.P. Intl Science)

[5] Kotelnikov V A 1933 On the carrying capacity of the ether and wire in telecommunications *Material for the First All-Union Conference on Questions of Communication* (Moscow: Izd. Red. Upr. Svyazi RKKA) (Russian) (English translation, PDF–Wiki)
Kotelnikov V A 1956 Teoria potencialnoj pomechoustojchivosti (Moskva: Gosenergoizdat)

[6] Tikhonov A N and Samarskii A A 2011 *Equations of Mathematical Physics, Dover Books on Physics* (New York: Dover)

[7] Vogel C R 2002 *Computational Methods for Inverse Problems* vol 10 (Philadelphia, PA: SIAM)

[8] Lapinski T and Leble S 2014 A Mathematica Program for heat source function of 1D heat equation reconstruction by three types of data arXiv: 1410.7066v1 [math.NA]

Chapter 20

Advanced exercises

It would be important to follow the book [1] for plenty basic definitions, identities and advanced exercises on electrodynamics.

20.1 Short list of useful vector and tensor relations

Vectors

1. bac–cab

$$\vec{a} \times [\vec{b} \times \vec{c}] = \vec{b}(\vec{a}, \vec{c}) - \vec{c}(\vec{a}, \vec{b}),$$

2. mixed product

$$(\vec{a}[\vec{b} \times \vec{c}]) = (\vec{c}[\vec{a} \times \vec{b}]) = -(\vec{b}[\vec{a} \times \vec{c}]).$$

For definitions and identities see also [2].

Tensors

The unit tensor of the second rank is δ-symbol of Kronecker. Its components are

$$\delta_{ik} = \begin{cases} 1, \ i = k, \\ 0, \ i \neq k. \end{cases}$$

The antisymmetric unit tensor of the third rank ε_{isl}, $\varepsilon_{123} = 1$ (symbol of Levy–Civita) is a useful tool in vector calculus. For example: tensor notations link Cartesian components of the vectors and its direct product as follows

$$\varepsilon_{isl} x_s p_l = [\vec{r} \times \vec{p}]_i. \tag{20.1}$$

Exercise 1. Prove that the Kronecker symbol satisfies the definition of the tensor of the second rank.

Exercise 2. Prove that the Levy–Civita satisfies the definition of the tensor of the third rank.
Exercise 3. Prove the identity of bac–cab by the relation (20.1).

20.2 A few definitions: curves, surfaces, integrals, etc

The equations of electrodynamics are systems of integral or differential equations. Formulations of problems in electrodynamics in practice need to use definitions of volume, surface and linear integrals, that in turn needs definitions of geometrical notions such as curves and surfaces.

20.2.1 Curves

It is convenient to set a curve in Cartesian coordinates by three functions

$$x_1 = x(t),\ x_2 = y(t),\ x_3 = z(t), \tag{20.2}$$

where t is a parameter, that 'follows' the curve. If the curve $x_i(t)$ is finite, $t \in [0,\ 1]$ is a possible choice. Let t be the arc length parameter. Then the vector

$$T_i = \frac{dx_i}{dt}, \tag{20.3}$$

represents the unit tangent vector to the curve at a point $x_i(t)$ on the curve. Further,

$$N_i = \frac{1}{K}\frac{dT_i}{dt}, \tag{20.4}$$

is the unit vector, normal to the curve, if K is a scale factor, called the curvature. The important example of a curve is a trajectory of a material point movement. In such a case the parameter t has the sense of time. The tangent vector then is velocity, proportional to $\vec{T}$, and the second derivative is acceleration, that is proportional to $\vec{N}$. Both derivatives are important in electrodynamics, see section 3.7.

Exercise 1. Prove that the vector T_i is unit.
Exercise 2. Derive an expression for K, supposing that N_i is unit. Prove that T_i and N_i are orthogonal.
Exercise 3. Evaluate the acceleration vector for the position vector of the curve of figure 20.1 and the Pointing vector by equation (10.81). Plot the dependence of its components on angle θ.

20.2.2 Surfaces

Similarly one defines a surface by the triple of coordinates as functions of two parameters.

$$x(u,\ v),\ y(u,\ v),\ z(u,\ v), \tag{20.5}$$

or as $x_i(u,\ v)$, or as $x_i(u_1,\ u_2)$, introducing the pair u_μ, $\vec{r} = x_i\vec{e_i}$.

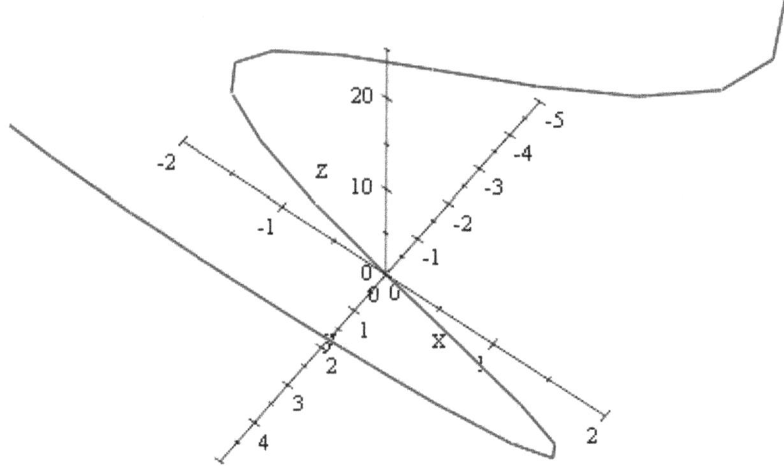

Figure 20.1. An exemplary curve, parameterized by $x = t$, $y = 2\sin t$, $z = t^2$.

The tangent vector to the coordinate curve at the surface is given by

$$\vec{T_\mu} = \frac{\partial x_i}{\partial u_\mu}\vec{e_i}, \tag{20.6}$$

the normal vector is defined as

$$\vec{N} = \frac{\vec{r_u} \times \vec{r_v}}{|\vec{r_u} \times \vec{r_v}|}. \tag{20.7}$$

For pictures see [3].

Exercise 4. It is instructive to plot a few exemplary curves and surfaces by means of a program for symbolic computations. For example, the equations

$$x = a\sin u\cos v;\ y = a\sin u\sin v;\ z = a\cos v, \tag{20.8}$$

determine a sphere of radius a.

20.2.3 Integrals

On integrability

The Pfaff form in 2D for pair $A(x, y)$, $B(x, y)$ is defined as

$$p = Adx + Bdy, \tag{20.9}$$

it is exact, if

$$\frac{\partial A}{\partial y} = \frac{\partial B}{\partial x}, \tag{20.10}$$

an integral of such form does not depend on integration path.

Integrals by a curve in a plane and space
The parameterization of a curve is useful for evaluation of contour integrals in integral form of Maxwell equations (section 2.2). For its definition we need the formulas for $\vec{d}\,l = \vec{t}\,dl$, where $\vec{t}$ is unit vector, tangent to the curve, see equation (20.3).

Exercise 5. Evaluate integral $\int_L \vec{E}\vec{dl}$ by the line from figure 20.1 for $t \in [0, 1]$ for a constant vector field $\vec{E}$.

Integrals by a surface
The parameterization of a surface is useful for evaluation of surface integrals in integral form of Maxwell's equations (section 2.2). For its definition we need the formulas for $\vec{d}\,S = \vec{n}\,dS$, where $\vec{n}$ is the unit vector orthogonal to the surface, see equation (20.7).

Exercise 6. Evaluate integral $\int_S \vec{E}\vec{dS}$ by the upper hemisphere from figure 20.2 for the constant vector field $\vec{E}$.

20.2.4 Dirac delta-function

There are lot of misunderstandings in textbooks on the use and interpretation of this nice trick, that is often written as

$$\int_{-\infty}^{\infty} \delta(x - y)f(y)dy = f(x),\ x,\ y \in (-\infty,\ \infty). \tag{20.11}$$

It is important to note at once in the relation (20.11) the term '$\int$' is not integration at all as well as $\delta(x - y)$ is not a function!

The map $f(y) \to f(x)$ is a linear continuous functional [2, 4].

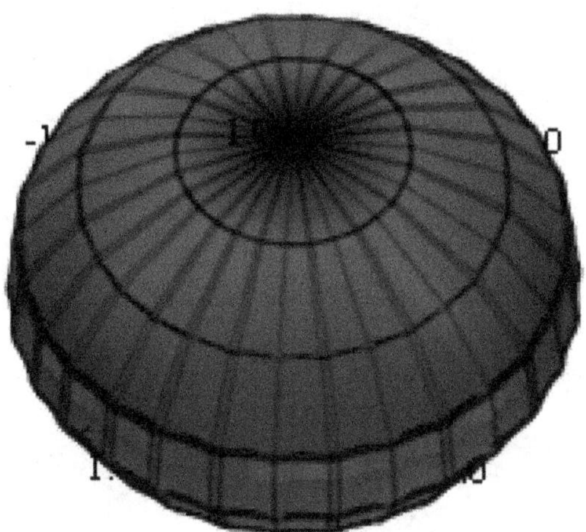

Figure 20.2. The sphere of unit radius $a = 1$.

Generally, a linear continuous functional defines 'generalized function' or 'distribution'.

An alternative definition, more visual for physics is given by the limiting procedure on a class of 'good' functions as Gaussian $G(x, w_n)$, $x \in (-\infty, \infty)$. Such a way, having a sequence of G_n, such that $\int_{-\infty}^{\infty} G_n dx = 1$ with the width w_n tending to zero with $n \to \infty$, plotted in figure 20.3. We write

$$\lim_{n\to\infty} \int_{-\infty}^{\infty} G_n(x-y)f(y)dy = f(x),\ x, y \in (-\infty, \infty), \tag{20.12}$$

that defines the delta-function action. The 'normal' Riemann integrals at the lhs tend to the rhs, but the limit $\lim_{n\to\infty} G_n(x-y)$ does not exists.

There are lot of modifications of such distributions, that allows to solve mathematical problems in compact algebraic form [4]. For example, a general relation

$$\int_{-\infty}^{\infty} \delta(\phi(y))f(y)dy = \sum_s \int_{-\infty}^{\infty} f(y)\frac{\delta(y-y_s)}{|\phi_y(y_s)|}dy,\ y \in (-\infty, \infty), \tag{20.13}$$

that is derived by Taylor expansion of the 'good' function ϕ in vicinities of the points (roots) $y = y_s$, $\phi(y) = 0$. The function has only the simple roots and the Taylor expansions exist.

It allows to give natural and rigorous definitions of such important but 'misty' notions as point particle or point sources.

Exercise 1. Consider a few simple examples of the function ϕ at equation (20.13), such as linear $\phi = ay + b$, and quadratic $\phi = ay^2 + by + c$.

Exercise 2. Integrating by parts in equation (20.12), derive the relation

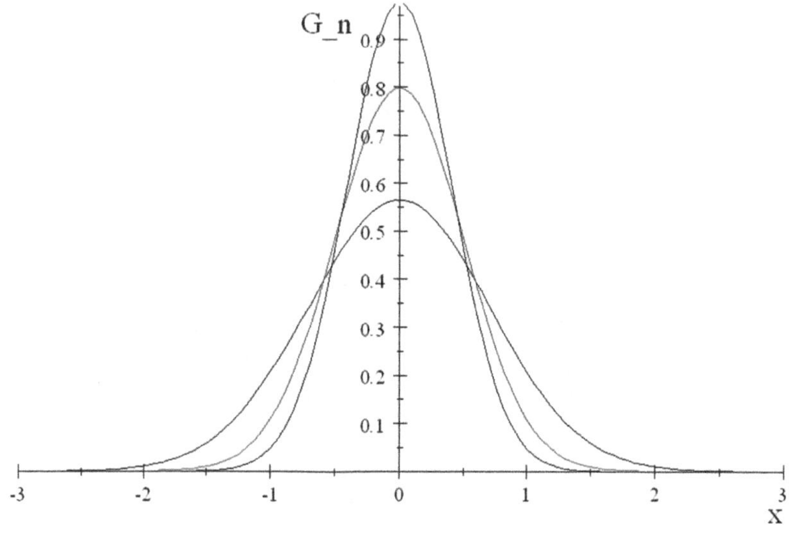

Figure 20.3. The Gauss functions $\frac{\sqrt{n}}{\sqrt{\pi}}e^{-nx^2}$ for n = 1-black, n = 2-red, n = 3-green.

$$\int_{-\infty}^{\infty} \delta'(y)f(y)dy = -f'(0).$$

It defines the point dipole.

Exercise 3. Generalize the result of exercise 2 for arbitrary order of the derivative, sequentially integrating by parts in equation (20.12).

20.3 Projecting operators

Starting from a general exposition of the method described in [5, 6], let us consider a 2×2 matrix with operator-valued elements;

$$P = \begin{pmatrix} p & \pi \\ \xi & \eta \end{pmatrix}, \tag{20.14}$$

with basic determining idempotent condition

$$P^2 = P.$$

It immediately yields

$$P = \begin{pmatrix} p & \pi \\ \pi^{-1}(p - p^2) & 1 - \pi^{-1}p\pi \end{pmatrix}. \tag{20.15}$$

There are possibilities of operators p, π choice, that fix the projection subspaces

$$\begin{pmatrix} p & \pi \\ \pi^{-1}(p - p^2) & 1 - \pi^{-1}p\pi \end{pmatrix}\begin{pmatrix} u \\ v \end{pmatrix} = \begin{pmatrix} \tilde{u} \\ \tilde{v} \end{pmatrix}. \tag{20.16}$$

Using this equality and the condition of completeness $P_+ + P_- = I$, we obtain the explicit form of the projection operators, that corresponds to two copies of linear independent vectors.

Exercise 1. Take the operators p, π as numbers and build $P_\pm$.

Exercise 2. Prove that the projectors are orthogonal in the sense, that

$$P_+P_- = 0.$$

Exercise 3. Build the operators $P_\pm$ for arbitrary non-commuting p, π.

20.4 Dressing method

Exercise 1. Prove that the Korteveg–de Vries (KdV) equation

$$u_t + uu_x + u_{xxx} = 0, \tag{20.17}$$

is the compatibility condition of the Lax pair:

$$\begin{aligned} -\psi_{xx} + u\psi &= \lambda\psi, \\ \psi_t + a\psi_{xxx} + bu\psi_x + c\psi &= 0, \end{aligned} \tag{20.18}$$

for the appropriate choice of a, b, c.

Remark: Compatibility condition writes as $\psi_{tx} = \psi_{xt}$. For general information see [7].

Exercise 2. Check that the system (20.18) is covariant with respect to the Darboux transformation

$$\begin{aligned} \psi[1] &= \psi_x + \sigma\psi, \\ u[1] &= u + \sigma_x, \end{aligned} \tag{20.19}$$

where $\sigma = -\phi_x/\phi$ is built with a solution ϕ of the system (20.18) for some $\lambda = \mu$.
Exercise 3. Taking a solution of equation (20.18) with $u = 0$ to built $u[1]$ and check that it is a solution of the KdV equation (20.17).

20.5 Dielectric waveguides

New fiber optic techniques, including—possible applications in future computer techniques with the light use—non-linear effects in dispersive media.

20.5.1 Dielectric slab

Exercise 4. Build and plot the dielectric slab transversal guide eigenfunctions. The functions $Y(y) = Y_0 \sin(\alpha y)$, $y \in [-h, h]$ satisfy the equation

$$Y_{yy} = -\alpha^2 Y, \tag{20.20}$$

generally

$$Y(y) = \begin{cases} Y_0 \sin(\alpha y), \ y \in [-h, h] \\ Y_0 \exp[-py], \ y \in [h, \infty) \\ -Y_0 \exp[py], \ y \in (-\infty, -h]. \end{cases} \tag{20.21}$$

Solving the system from chapter 11, section 11.1.2

$$\begin{cases} p = \dfrac{\alpha}{\varepsilon} \tan(\alpha h), \\ \alpha^2 + p^2 = \omega^2(\varepsilon - 1)/c^2 \end{cases}, \tag{20.22}$$

for given $\omega = 10^{11}\ \mathrm{s}^{-1}$, $h = 10^{-3}$ m, $\varepsilon = 2$. Calculate eigenvalues, using a graphic solution of equation (20.22) similar to figure 11.2.
Exercise 5. Prove orthogonality of the basic functions (20.21) and normalize them.

20.5.2 Dielectric cylinder—optical fibers

Exercise 1. Do the transition of the Maxwell system from Cartesian to the cylindrical coordinates ρ, φ, z [2].

20.5.3 Rectangular waveguide

Exercise 1. Consider the waveguide of rectangular cross-section. Build waveguide modes as basic functions, considering equations (12.11) and (12.12).

Exercise 2. Plugging the expansion of a electric field component by the obtained basis into equations (12.4) and (12.5), derive the equations of a mode propagation similar to ones of section 12.3.

Exercise 3. Having the 1 + 1 linear D'Alambert equation, write a solution of the boundary regime propagation problem for given polarization of the electric field component by the scheme of section 12.3.1.

20.6 Electromagnetic waves in metamaterials

Exclusive dispersive media such as metamaterials with important modification of energy relations in conditions of negative dielectric permittivity and magnetic permeability.

20.6.1 Electromagnetic waves in metal rectangular waveguide, system derivation

In this section we suggest as an exercise to follow the details of the evolution system derivation. Start with equation (12.26) with constant ε, μ.

Transforming equation (12.31) as

$$\frac{\partial \varphi}{\partial x} = \frac{q}{p}\chi - \frac{1}{p}\frac{\partial \alpha}{\partial t}, \tag{20.23}$$

$$\frac{\partial \chi}{\partial x} = \frac{1}{pr}\frac{\partial^2 \varphi}{\partial x^2} - \frac{r}{p}\varphi - \frac{1}{p}\frac{\partial \beta}{\partial t}, \tag{20.24}$$

$$\frac{\partial \alpha}{\partial x} = -\frac{1}{p}\hat{a}^2\frac{\partial \varphi}{\partial t} - \frac{q}{p}\beta, \tag{20.25}$$

$$\frac{\partial \beta}{\partial x} = -\frac{r}{p}\alpha - \frac{1}{rp}\frac{\partial^2 \alpha}{\partial x^2} + \frac{1}{p}\hat{a}^2\frac{\partial \chi}{\partial t}, \tag{20.26}$$

we see the x-derivatives at the rhs. Let us use the shorthands as

$$\partial_x \equiv \frac{\partial}{\partial x}, \ \partial_t \equiv \frac{\partial}{\partial t}, \ \partial_{tt} \equiv \frac{\partial^2}{\partial t^2}, \ \partial_{xx} \equiv \frac{\partial^2}{\partial x^2},$$

and, also go to notations of the partial derivatives by indices as for example $U_x = \frac{\partial U}{\partial x}$. We show some details below.

To express α_{xx} we differentiate equation (20.25) by x

$$\alpha_{xx} = -\frac{1}{p}\hat{a}^2\varphi_{xt} - \frac{q}{p}\beta_x,$$

and plug equation (20.23) in the result

$$\alpha_{xx} = -\frac{1}{p}\hat{a}^2\left(\frac{q}{p}\chi_t - \frac{1}{p}\alpha_{tt}\right) - \frac{q}{p}\beta_x,$$

rewriting equation (20.26) as

$$\begin{aligned}\beta_x &= -\frac{r}{p}\alpha - \frac{1}{rp}\left(-\frac{1}{p}\hat{a}^2\left(\frac{q}{p}\chi_t - \frac{1}{p}\alpha_{tt}\right) - \frac{q}{p}\beta_x\right) + \frac{1}{p}\hat{a}^2\chi_t \\ &= -\frac{r}{p}\alpha - \frac{1}{rp}\left(-\frac{q}{p^2}\hat{a}^2\chi_t + \frac{1}{p^2}\hat{a}^2\alpha_{tt} - \frac{q}{p}\beta_x\right) + \frac{1}{p}\hat{a}^2\chi_t;\end{aligned}$$

rearranging yields

$$\beta_x = -\frac{r}{p}\alpha + \frac{q}{rp^3}\hat{a}^2\chi_t - \frac{1}{rp^3}\hat{a}^2\alpha_{tt} + \frac{q}{rp^2}\beta_x + \frac{1}{p}\hat{a}^2\chi_t,$$

and, next, obtain

$$\left(1 - \frac{q}{rp^2}\right)\beta_x = -\frac{r}{p}\alpha + \frac{q}{rp^3}\hat{a}^2\chi_t - \frac{1}{rp^3}\hat{a}^2\alpha_{tt} + \frac{1}{p}\hat{a}^2\chi_t.$$

So, as a final relation we have

$$(rp^3 - qp)\beta_x = -r^2p^2\alpha - \hat{a}^2\alpha_{tt} + (q + rp^2)\hat{a}^2\chi_t.$$

Similar

$$\varphi_{xx} = \frac{q}{p}\chi_x - \frac{1}{p}\alpha_{xt},$$

sequentially

$$\varphi_{xx} = \frac{q}{p}\left(\frac{1}{pr}\varphi_{xx} - \frac{r}{p}\varphi - \frac{1}{p}\beta_t\right) - \frac{1}{p}\left(-\frac{1}{p}\hat{a}^2\varphi_{tt} - \frac{q}{p}\beta_t\right),$$

continue,

$$\varphi_{xx} = \frac{q}{p^2r}\varphi_{xx} - \frac{qr}{p^2}\varphi + \frac{1}{p^2}\hat{a}^2\varphi_{tt},$$

multiplying by the factor, we write

$$\left(1 - \frac{q}{rp^2}\right)\varphi_{xx} = -\frac{qr}{p^2}\varphi + \frac{1}{p^2}\hat{a}^2\varphi_{tt},$$

then

$$\left(p^2 - \frac{q}{r}\right)\varphi_{xx} = \hat{a}^2\varphi_{tt} - qr\varphi,$$

and next

$$\chi_x = \frac{1}{rp^3 - pq}(\hat{a}^2\varphi_{tt} - qr\varphi) - \frac{r}{p}\varphi - \frac{1}{p}\beta_t.$$

The notation

$$\frac{1}{rp^3 - pq} = \nu,$$

leads to

$$\chi_x = \nu\hat{a}^2\varphi_{tt} - \nu qr\varphi - \frac{r}{p}\varphi - \frac{1}{p}\beta_t.$$

Finally,

$$\beta_x = -\nu r^2p^2\alpha - \nu\hat{a}^2\alpha_{tt} + \nu(q + rp^2)\hat{a}^2\chi_t.$$

Exercise 1. Plot the real and imaginary parts of the function $\tilde{a}(\omega)$, defined by equation (12.92), for the values of the parameters, given in [6].

Exercise 2. Evaluate $(S^{mn}, S^{n'm'}S^{n''m''}S^{m'''n'''})$, plugging the eigenmode functions (12.84) by the scalar product definition (12.81). Write the particular case for one-mode waveguide via S, with $n = m = 1$.

Exercise 3. Evaluate the nonlinear constants in equation (12.99).

Exercise 4. Derive the nonlinear Shrödinger equation for a wavepacket on the basis of equation (12.100) by the technique of chapter 11, section 11.1.2.

20.7 Plasma confinement

Exercise 1. Read the introduction, explanation and a discussion.

Exercise 2. Check the commutation relation (13.52).

Exercise 3. Check the statement that L defined by equation (13.54) is the symmetry operator for equation (13.51).

Exercise 4. Do the transition from original x, v_x, v_y to the invariant variables ϕ, ψ, x in equation (13.51).

Exercise 5. Prove the relation (13.61).

Exercise 6. Evaluate the function $P_\perp$, defined by equation (13.62) for the Maxwellian distribution (13.64).

20.8 Wave propagation at plasma

Some aspects of wave propagation at plasma, comprehensive description of helicoidal and other plasma wave phenomena. Accompanying effects. Plasma heating.

Exercise 1. Consider linear waves in 1D plasma. Deliver a complete classification of electromagnetic waves in it.

Exercise 2. Build the projecting operators to subspaces of the classified solutions.

Exercise 3. Take into account weak nonlinearity in the problem of exercise 1. Apply the projecting operators of exercise 2 to the obtained system.

Exercise 4. Plugging the Maxwell distribution in equation (13.26) derive the dispersion relations (13.27) and (13.32) for Langmuir waves.

Exercise 5. Transform the expression (13.73) in the spirit of equation (13.49) and derive the dispersion relation as by equation (13.22).

Exercise 6. Study the nonlinear resonance conditions for wave packets, propagating in different directions, for equation (14.22).

Exercise 7. Simplify the expression (14.51) for the correction to the distribution function (ion acoustic-Langmuir waves interaction), integrating in:

$$\Phi^{a}_{int} d\vec{p}\,, \tag{20.27}$$

that pave the way to derive the system (14.52)–(14.53), that also may be a task for the exercise.

20.9 Refraction in presence of conductivity

X-rays manipulation and focusing.

Exercise 1. Check that Kogelnik solution (15.7) satisfies the equation (15.5).

Exercise 2. Plot the Kogelnik solution (15.7) with different values of the parameter σ.

Exercise 3. Implement the Kogelnik solution (15.7) into the set of Maxwell's equations (6.1), (6.2), (8.4), and (6.4) solutions. Write the initial and boundary functions for this solutions set.

Exercise 4. Repeat calculations of the x-ray beam parameters for different choice of

(a) lens number, e.g. 30,

(b) radius of lens curvature, e.g. 0.3 mm.

20.10 Magnetism, a novel aspect

A theory based on Heisenberg's model and the Landau–Lifshitz–Gilbert equations. Ferromagnetic domains dynamics in nanoscale and microwires.

Exercise 1. Derive Bio–Savart* formula from Maxwell's equation.

The algorithm:

1. Start with the Ampère–Maxwell equation in differential form and Gauss units for stationary field.
2. Apply the operator 'rot' to the Ampère equation, having the Poisson equation.
3. Write the Poisson equation solution in integral form**.
4. Define the curve as $\vec{r_0}(t)$, the parameter t is convenient to choose as the length along the curve.
5. Represent the isolated charge q density as $q\delta(\vec{r} - \vec{r_0}(t))$.

6. The corresponding charge density current reads $\vec{j} = q\frac{\vec{r_0}(t)}{dt}\delta(\vec{r} - \vec{r_0}(t))$.
7. Technically it is convenient to use the ∇ operator*** $\mathrm{rot}\,\vec{j} = q\nabla \times \frac{\vec{r_0}(t)}{dt}$ $\delta(\vec{r} - \vec{r_0}(t)) = q\nabla\delta(\vec{r} - \vec{r_0}(t)) \times \frac{\vec{r_0}(t)}{dt}$.
8. The magnetic field then is hence expressed as $\vec{H}(t) = -\frac{4\pi}{c}\int\frac{\mathrm{rot}'\,\vec{j}\,d\vec{r}\,'}{|\vec{r} - \vec{r}\,'|} = -\frac{4\pi q}{c}\int\frac{\nabla'\delta(\vec{r}\,' - \vec{r_0}(t))d\vec{r}\,'}{|\vec{r} - \vec{r}\,'|} \times \frac{\vec{r_0}(t)}{dt}$.
9. Integrating by parts yields: $\vec{H}(t) = \frac{4\pi q}{c}\int\delta(\vec{r}\,' - \vec{r_0}(t))\nabla'\frac{1}{|\vec{r} - \vec{r}\,'|}d\vec{r}\,' \times \frac{\vec{r_0}(t)}{dt}$.
10. Using the delta-function property: $\vec{H}(t) = \frac{4\pi q}{c}\nabla\frac{1}{|\vec{r} - \vec{r_0}(t)|} \times \frac{\vec{r_0}(t)}{dt}$.
11. Next, differentiating results in $\vec{H}(t) = -\frac{4\pi q}{c}\frac{(\vec{r} - \vec{r_0}) \times \frac{\vec{r_0}(t)}{dt}}{|\vec{r} - \vec{r_0}(t)|^3}$.
12. The final result, that will give the field of all moving charges with linear current density $\nu(t)$, $\vec{H_a}(\vec{r}) = \int\vec{H}(t)dt = -\frac{4\pi}{c}\int\nu(t)\frac{(\vec{r} - \vec{r_0}) \times \frac{\vec{r_0}(t)}{dt}}{|\vec{r} - \vec{r_0}(t)|^3}dt$.
13. The case of constant current density $\nu(t) = I$ leads to $\vec{H_a}(\vec{r}) = \int\vec{H}(t)$ $dt = -\frac{4\pi I}{c}\int\frac{(\vec{r} - \vec{r_0}) \times \frac{\vec{r_0}(t)}{dt}}{|\vec{r} - \vec{r_0}(t)|^3}dt$.

Exercise 2. For a given curve, representing thin current I, derive the expression for magnetic field at an arbitrary point. Apply the model to the round circle and solenoid examples.

* The Biot–Savart law is used for computing the resultant magnetic field $\vec{H}$ at position $\vec{r}$ in 3D-space generated by a flexible current I (for example due to a wire)—wiki.

** The Poisson equation $\Delta\vec{A} = \vec{b}$ has solution $\vec{A} = \int\frac{\vec{b}(\vec{r}\,')d\vec{r}\,'}{|\vec{r} - \vec{r}\,'|}$.

*** $\nabla = \vec{i}\frac{\partial}{\partial x} + \cdots$, its algebra you find in [1]. In a few words, You operate with ∇ as with a vector, but always remember that it acts on $\vec{r}$-dependent entries standing right. For example, applying b(ac)–c(ab) relation to vector fields $\vec{a}(\vec{r})$, $\vec{b}(\vec{r})$, as in section 20.1, we obtain

$$\nabla \times [\vec{a} \times \vec{b}] = \vec{a}(\nabla\vec{b}) - \vec{b}(\nabla a).$$

Exercise 3. A real configuration of solid atoms positions should be taken into account when large scale continuous approximation is applied. Its anisotropy may be taken into account by a difference in scales for different directions. It may be understood from the following example for a plane, taking the set of points

$i, j + 1; i + 1, j + 1,$

$i, j; i + 1, j,$

with the following closest neighbors terms

$u(i+1,j)u(i,j)J_{i+1,j} + u(i,j+1)u(i,j)J_{i,j+1} + u(i,j)u(i+1,j+1)J_{i+1,j+1}$.
Expanding the field u in Taylor series, yields
$u(i+1,j) = u(i,j) + \frac{\partial u}{\partial x}|_i a + \frac{\partial^2 u}{\partial x^2} a^2/2$.

Exercise 4. Taking the plane Heisenberg network, derive its continuous version, tending to obtain the anisotropy terms.

Exercise 5. From equations (16.25) and (16.28) derive the boundary condition for $\Im\Omega$.

Exercise 6. Build and draw the DW structure by equation (16.37), and a particular solution of equations (16.35) and (16.32).

Exercise 7. Estimate the velocity and acceleration of DW in conditions of the experiment [8], by the approximation

$$\Omega_0 = A \sin\left(\frac{2\pi z}{l}\right), \tag{20.28}$$

where A is amplitude and l-period of oscillations along z, under condition for $\overline{\Omega}_0$ (16.47), in 1D form:

$$J\Omega_{0zz} + K\Omega_0 - K\frac{\Omega_0^3}{6} - J(\Omega_{0z})^2\Omega^0 = 0. \tag{20.29}$$

20.11 Condensed matter electrodynamics: equations of state by partition function

20.11.1 Paramagnetics and ferromagnetics

Exercise 1. Evaluate the exchange integral by equation (17.73). Build the magnetization curve for Cobalt for $T > \theta_{\max\mathrm{Co}} = 1121$ K, $z = 12$ outside the ferromagnetic range.

Exercise 2. Build the hysteresis loop for Cobalt for $T < \theta_{\max\mathrm{Co}}$ inside the ferromagnetic range.

20.11.2 Multiferroics

Exercise 3. Evaluate $\frac{\partial M}{\partial \varepsilon}$ by means of (17.85).

Exercise 4. Plot the number of surface and bulk particles comparison function on base of equation (17.99).

Exercise 5. Plot the joint surface+bulk magnetization curve by the formula (20.13).

20.12 General material relations

Exercise 1. Derive conditions for ε' for given μ' to be simultaneously diagonalized by rotations.

Exercise 2. Derive the link between components of the relativistic tensor of fourth rank $E_{\alpha\beta\gamma\delta}$ and space tensor ε_{ik} on the base of equation (18.26).

Exercise 3. Derive the link between components of the relativistic tensor of third rank Σ and space tensor σ_{ik} of conductivity on the base of equation (18.38).

Exercise 4. Write the relation (18.39) in more details as a function of the magnetic field and link it with the polarization phenomena and magneto-electric effects.

20.12.1 Piezoelectricity

Exercise 5. Study the reflection and inverse transformations in 3D as symmetry for the material relations (18.39). Transformation of inversion is

$$\vec{r}\,' = -\vec{r},$$

for reflection of x, it is

$$x' = -x,\ y' = y,\ z' = z.$$

20.13 Inverse problems of electrodynamics

Materials common with ill-posed inverse problems, including information on the principles of the theory of radars (lidars) data processing and tomography. The Tikhonov regularization is outlined.

Exercise 1. Build and investigate the S-matrix for orthogonal plane wave reflection from two-layer medium as the function of frequency for a given thickness of the layers, equal to wavelength and dielectric constants $\varepsilon_1 = 2$, $\varepsilon_2 = 3$.

Exercise 2. Write and analyze an explicit solution of inverse problem of orthogonal plane wave reflection from two-layer medium.

References

[1] Batygin V and Toptygin I 1964 *Problems in Electrodynamics* (London: Academic)

[2] Korn G A and Korn T M 1968 *Mathematical Handbook for Scientists and Engineers* (New York: McGraw-Hill)

[3] Heinbockel J H 2001 *Introduction to Tensor Calculus and Continuum Mechanics* (Bloomington, IN: Trafford)

[4] Demidov A S 2001 Generalized functions in mathematical Physics: main ideas and concepts (Huntington: Nova Science). With an addition by Egorov Yu V.

[5] Leble S 2016 General remarks on dynamic projection method *Task Q.* **20** 113–30

[6] Leble S and Perelomova A 2018 *Dynamical Projectors Method in Hydro- and Electrodynamics* (Boca Raton, FL: CRC Press)

[7] Doktorov E and Leble S B 2007 *Dressing Method in Mathematical Physics* (Berlin: Springer)

[8] Rodioniova V *et al* 2012 The defects influence on domain wall propagation in bistable glass-coated microwires *Physica B* **407** 1446–9

CPSIA information can be obtained
at www.ICGtesting.com
Printed in the USA
BVHW051857050221
599342BV00003B/7

9 780750 325745